Lecture Notes on Multidisciplinary Industrial Engineering

Series Editor

J. Paulo Davim, Department of Mechanical Engineering, University of Aveiro, Aveiro, Portugal

“Lecture Notes on Multidisciplinary Industrial Engineering” publishes special volumes of conferences, workshops and symposia in interdisciplinary topics of interest. Disciplines such as materials science, nanosciences, sustainability science, management sciences, computational sciences, mechanical engineering, industrial engineering, manufacturing, mechatronics, electrical engineering, environmental and civil engineering, chemical engineering, systems engineering and biomedical engineering are covered. Selected and peer-reviewed papers from events in these fields can be considered for publication in this series.

Jhareswar Maiti · Mamata Jenamani ·
Jitendra Kumar Jha · Akhilesh Kumar ·
Sayak Roychowdhury · Goutam Sen
Editors

Operations Research and Data Analytics: Current Trends and Future Perspectives

Selected Papers from International Conference on Industrial Engineering and Analytics (ICONIEA) 2024

Springer

Editors
Jhareswar Maiti
Department of Industrial and Systems Engineering
Indian Institute of Technology Kharagpur
Kharagpur, West Bengal, India

Mamata Jenamani
Department of Industrial and Systems Engineering
Indian Institute of Technology Kharagpur
Kharagpur, West Bengal, India

Jitendra Kumar Jha
Department of Industrial and Systems Engineering
Indian Institute of Technology Kharagpur
Kharagpur, West Bengal, India

Akhilesh Kumar
Department of Industrial and Systems Engineering
Indian Institute of Technology Kharagpur
Kharagpur, West Bengal, India

Sayak Roychowdhury
Department of Industrial and Systems Engineering
Indian Institute of Technology Kharagpur
Kharagpur, West Bengal, India

Goutam Sen
Department of Industrial and Systems Engineering
Indian Institute of Technology Kharagpur
Kharagpur, West Bengal, India

ISSN 2522-5022 ISSN 2522-5030 (electronic)
Lecture Notes on Multidisciplinary Industrial Engineering
ISBN 978-981-96-8326-0 ISBN 978-981-96-8327-7 (eBook)
https://doi.org/10.1007/978-981-96-8327-7

This Springer imprint is published by the registered company Springer Nature Singapore Pte Ltd.
The registered company address is: 152 Beach Road, #21-01/04 Gateway East, Singapore 189721, Singapore

Preface

In today's rapidly evolving technological and industrial landscape, Operations Research and Data Analytics (ORDA) have emerged as vital tools for solving complex real-world challenges. As fields that draw on mathematical, statistical, and computational approaches, ORDA offers a robust decision-making and optimization framework, allowing organizations to operate more efficiently and effectively. This book, titled *Operations Research and Data Analytics: Current Trends and Future Perspectives*, is a collective endeavor by the Department of Industrial and Systems Engineering (ISE) at IIT Kharagpur, which seeks to showcase the recent advancements and applied research in these fields. It is one of the two volumes, published from a curated selection of peer-reviewed articles presented at the International Conference on Industrial Engineering and Analytics, 2024 (ICONIEA, 2024). The other volume, titled *Recent Advances in Industrial and Systems Engineering*, presents a collection of research articles in areas such as Supply Chain Management, Manufacturing, Human Factors and Ergonomics, Safety, Quality, and Reliability.

This compilation of ORDA articles aims to provide insights into how OR and DA methodologies are being applied in innovative ways to solve real-world problems in business, transportation, health care, agriculture, climate science, safety, security, and surveillance.

The book is organized into five thematic parts that represent the significant ways ORDA are reshaping industries and scientific research. Part I focuses on importance of decision sciences and business analytics in industry. This part explores the synergy between data analytics and decision sciences in business settings, covering diverse topics such as data-centric prognosis in Industry 4.0, customer segmentation analysis for homebuyers, customer experience quantification in retail banking, data generation techniques for time series, and sales forecasting for restaurants. It delves into the applications of data-driven decision-making processes, highlighting advancements that support businesses in gaining competitive advantage through analytics, forecasting, and strategic planning.

Part II covers the implementation of operations research (OR) in transportation and logistics sector. With the growing complexity of global supply chains and logistics, OR techniques are essential for optimizing resource allocation, route planning, and transportation management. Articles in this section address real-world problems like optimization of air-cargo for drones. The articles discuss both foundational and cutting-edge methodologies that improve efficiency in transportation and logistics.

Part III explores application of analytics in health care. This part covers health data analysis and predictive modeling in a range of applications, such as sentiment analysis for impact assessment of COVID-19 on mother and children, Alzheimer's disease detection, Parkinson's disease detection, image classification for colorectal cancer histopathology, scheduling of in-home health care, explainable AI applications for mental health, PCOS diagnosis, and early detection of HAVs. These articles show how machine learning techniques have revolutionized healthcare systems, helping the practitioners with more accurate diagnosis leading to effective treatment plans and improved outcomes.

Part IV explains the role of ORDA techniques in agricultural and climate sciences. Articles here focus on enhancing agricultural productivity and climate resilience through data analytics. This part explores important research in the areas rainfall prediction, paddy crop image classification, rice grain type detection, and crop disease detection. Usage of advanced machine learning models in agriculture will enable the practitioners to make more informed decisions, leading to improved productivity, reduced losses, and adoption of sustainable practices.

Part V examines application of ORDA in the area of safety, security, and surveillance. As safety and security concerns grow in both public and private sectors, data analytics is increasingly important in surveillance, threat detection, and risk assessment. This part examines how advanced analytics are applied to identify and mitigate security concerns in different areas, such as anomaly detection in power consumption, object counting for CCTV and drone surveillance, development of perception alert system, data-driven cybersecurity risk assessment, and game theory models for cybersecurity resource allocation for autonomous vehicles.

This volume, *Operations Research and Data Analytics: Current Trends and Future Perspectives*, demonstrates the applicability of ORDA methodologies in solving complex problems, optimizing systems, and advancing research across multiple domains. As ORDA continue to evolve, their contributions will become even more indispensable for driving efficiency, fostering sustainability, and enhancing safety and security in today's digital and data-centric world.

We would like to thank the contributing authors for submitting their research articles for this compilation. We are also thankful to the reviewers for sharing their time and expertise in bringing this collection together. Finally, we express our sincere gratitude to Springer for this opportunity and for their professional support.

Kharagpur, India

Jhareswar Maiti
Mamata Jenamani
Jitendra Kumar Jha
Akhilesh Kumar
Sayak Roychowdhury
Goutam Sen

Contents

About the Editors

Dr. Jhareswar Maiti, Suresh and Vidya Nair Chair Professor, is the former head and professor in the Department of Industrial and Systems Engineering at IIT Kharagpur. He is the founder chairman of the Center of Excellence on Safety Engineering and Analytics (CoE-SEA) at IIT Kharagpur since July 2020 and has developed its academics with Ph.D., M.S., and 2-year M.Tech. programs. He has also developed the Safety Analytics and Virtual Reality Lab at IIT Kharagpur. His research interests include safety engineering, management and analytics, human factors and ergonomics, risk assessment and uncertainty modeling, quality management, and AI/ML/VR applications in safety. He authored the text book on *Multivariate Statistical Modeling in Engineering and Management*. With more than 200 publications in the international journals and conferences of repute, Dr. Maiti is the recipient of several national and international awards such as P. C. Mahalanobis distinguished educator award and IEOM outstanding professor of occupational safety and health award.

Mamata Jenamani is a professor in the Department of Industrial and Systems Engineering, at IIT Kharagpur specializing in E-Business. With a research focus spanning auctions, recommender systems, social network analysis, consumer review analysis, and digital supply chain management, she has authored over 100 journal and conference publications in these areas. She has also led numerous sponsored and consulting projects, driving innovation and practical applications. As the head of an E-Business Center of Excellence, she has played a key role promoting research, industry collaboration, and technological advancements in the field.

Dr. Jitendra Kumar Jha is an associate professor and the current Head of the Department of Industrial and Systems Engineering at IIT Kharagpur. He obtained his Ph.D. from IIT Kanpur where he worked in the area of supply chain management. His main areas of teaching and research include operations research, statistical decision modeling, facility planning, supply chain optimization, and logistics planning. He published/presented more than seventy papers in international journals and conferences, and his publications appeared in the leading journals like *Journal of the*

Operational Research Society, Computers and Operations Research, Applied Mathematical Modeling, Computers and Industrial Engineering, International Journal of Production Research, Journal of Manufacturing Systems, etc. He is serving as an editorial board member of *International Journal of Industrial Engineering: Theory, Applications, and Practice.*

Dr. Akhilesh Kumar is currently working as an associate professor in the Department of Industrial and Systems Engineering at IIT Kharagpur. Previously, he worked as a solution architect in the consulting team at JDA Software, Bengaluru. He received his B.Tech. degree in Manufacturing Engineering from NIFFT (India) and Ph.D. degree in Industrial Engineering from Wayne State University (USA) in 2005 and 2011, respectively. He has authored several technical papers in the area of analytics and optimization. His publications appeared in such journals as the *International Journal of Production Economics, European Journal of Operational Research, Expert System with Applications*, and *IEEE*. He has carried out various research and consultancy projects with Shell, Tata Sons, Tata Metaliks, Tata Hitachi, TM International Logistics Ltd., DHI Government of India, European Union, UKIERI, SPARC, Ford Motor Company, and Delphi Automotive LLP.

Dr. Sayak Roychowdhury is an assistant professor with the Department of Industrial and Systems Engineering, IIT Kharagpur. He received M.S. and Ph.D. degrees from the Department of Integrated Systems Engineering, The Ohio State University, Columbus, OH, USA, in 2014 and 2017, respectively. His areas of interest include statistical learning, operations research, reinforcement learning, quality engineering, and stochastic optimization.

Dr. Goutam Sen is currently working as an assistant professor in the Department of Industrial and Systems Engineering at IIT Kharagpur. He received his Ph.D. in network optimization jointly from Industrial Engineering and Operations Research, IIT Bombay, India, and from the School of Information Technology, Monash University, Melbourne, Australia. His research area includes application of integer programming to solve transportation and network design problems in supply chain and logistics. He has published in top tier Industrial Engineering journals like *Computers and Operations Research, Interfaces (INFORMS)*, and *Networks*. He closely interacts with organizations of national importance such as Indian Railway, ISRO, DRDO, and Police on a variety of discrete optimization problems and secured multiple project awards in the recent past.

Part I
Decision Sciences and Business Analytics

Chapter 1
Data-Centric Prognostics: Challenges and Opportunities in the Era of Industry 4.0

Amit Kumar Jain, Minghao Zhong, Don McGlinchey, Pankaj Kumar, Jay Prakash Srivastava, and Sandeep Kumar

Abstract Data-centric prognostics use historical and current data statistically and probabilistically to predict an asset's remaining useful life. In the era of Industry 4.0, technological advances in new sensors and rapid advances in computing offer an obvious motivation to develop and apply data-centric prognostics. The present paper endeavours to thoroughly examine the challenges at hand. The exclusive challenges addressed encompass explainability, prescriptive analytics, and affordability in data-centric prognostics. For instance, the lack of transparency in data-centric prognostics undermines trust and accountability, hindering informed decision-making and potentially leading to misguided actions. Likewise, developing robust methodologies that predict outcomes and provide actionable insights and recommendations for optimal decision-making is critical. Lastly, demand for affordable solutions can enhance SMEs' operational efficiency and competitiveness. Moreover, common challenges discussed include the availability of sufficient data, uncertainty, time-variant life cycle loading, etc. Another main contribution of this paper is that it offers insightful guidance on future directions. Such as the prospect of federated learning to offset the drawbacks of a data-centric approach is deliberated. Two other promising directions

A. K. Jain (✉)
Department of Mechanical Engineering, Birla Institute of Technology and Science, Pilani, Rajasthan 333031, India
e-mail: amitkumar.jain@pilani.bits-pilani.ac.in

M. Zhong · D. McGlinchey
Department of Mechanical Engineering, School of Computing, Engineering and Built Environment, Glasgow Caledonian University, Glasgow, UK

P. Kumar
Department of Mechanical Engineering, SR University, Warangal, Telangana, India

J. P. Srivastava
Department of Mechanical Engineering, Gandhi Institute of Technology and Management, Hyderabad, Telangana, India

S. Kumar
School of Management, IILM University, Greater Noida, Uttar Pradesh, India

J. Maiti et al. (eds.), *Operations Research and Data Analytics: Current Trends and Future Perspectives*, Lecture Notes on Multidisciplinary Industrial Engineering,
https://doi.org/10.1007/978-981-96-8327-7_1

for future research include prototype-based models and Turing's type-B random machines. The highlighted challenges and opportunities lead to a better understanding for the researcher on further developing and successfully implementing data-centric prognostics in Industry 4.0 systems.

Keywords Prognostics · Industry 4.0 · Maintenance · Explainability · Prescriptive analytics · Affordability

1.1 Introduction

Reducing capital-intensive equipment's life cycle cost (LCC) is a primary concern for Industry 4.0 systems. Prognostics analytics can reduce LCC by enhancing asset performance, avoiding unplanned failures, and significantly dipping maintenance costs [1]. The prognostics is science based on investigating modes of failure, faulty circumstances, and early signs of wear and tear. Investigated signs are correlated with a damage propagation model. Prognostics approaches predict the failures of the component in terms of remaining useful life (RUL), which plays a vital role in maintenance decision-making. Prognostics approaches are mainly classified into physics of failure-based approach and data-centric approach. These approaches have been developed for mechanical, electrical, and electronic systems. The physics of the failure-based approach is also known as the model-based approach, in which damage models are used to determine fault generation and propagation [2]. The major drawback of this method is that model development necessitates a thorough understanding of the system. At the same time, the data-centric approaches are based on machine learning techniques and statistical pattern recognition [1]. This approach can be implemented at the system, subsystem, or component level [3–5]. The versatility of data-centric approaches explains why these are becoming increasingly popular and regarded by some researchers as the most suitable approach for modern Industry 4.0 systems.

In the era of Industry 4.0, sensor technology keeps advancing, and the current trend is for autonomous decision-making to improve asset performance. Thus, applying data-centric prognostics has become vital for today's Industry 4.0 systems. Some of the most recent and novel contributions in data-centric prognostics are given in [6–9]. Moreover, the data-centric approach can be applied in other engineering domains (see [10–13]). However, various challenges should be considered in implementing data-centric prognostics efficiently. Consequently, this paper aims to shed comprehensive light on the challenges associated with implementing data-centric prognostics in the context of Industry 4.0. To do so, the paper reviews data-centric prognostics within Industry 4.0, addressing exclusive and common challenges.

One of the exclusive challenges is the issue of explainability, as the lack of transparency in data-centric prognostics raises concerns about trust and accountability. Without a clear understanding of the underlying processes, stakeholders may find it challenging to make informed decisions, potentially leading to misguided actions. Additionally, the paper highlights the need for robust methodologies that predict outcomes and provide actionable insights and recommendations to enable optimal decision-making. Affordability is another critical challenge in adopting data-centric prognostics, particularly for small and medium-sized enterprises (SMEs). As the demand for cost-effective solutions increases, addressing affordability becomes imperative to enhance operational efficiency and competitiveness among SMEs. The common challenges discussed include the availability of sufficient data, uncertainty, and the time-variant life cycle loading of assets.

Additionally, the paper provides insightful guidance and future directions for researchers. The exploration of federated learning as a potential solution for the drawbacks of a purely data-centric approach opens new avenues for investigation. Furthermore, the paper introduces promising directions for future research, including prototype-based models and Turing's type-B random machines, offering a roadmap for researchers to navigate the evolving landscape of data-centric prognostics.

The contributions of this paper can be summarised as follows:

- We have first identified three key challenges faced exclusively in data-centric prognostics. The first is explainability, an important topic in recent artificial intelligence (AI) research. The second is prescriptive analytics, which advances previous analytics further and features automating optimal decision-making. The last one is affordability, which we have identified as a significant challenge in applying AI in an industry context in the era of Industry 4.0.
- Identification of other common challenges in a broader prognostic context is also provided to equip researchers with a comprehensive view and understanding of the challenges in researching the prognostic field. The challenges serve as a foundation and always require attention regardless of the type of prognostics that are being investigated.
- Lastly and most importantly, we have discussed potentially applying federated learning as a framework to solve some of the exclusive challenges mentioned before, namely prescriptive analytics and affordability. We have justified why federated learning is a good solution by investigating its characteristics and explaining why those characteristics could be a good fit to address SMEs' pain points in building effective AI applications in the Industry 4.0 era. To the best of our knowledge, this paper is the first to propose addressing the specific challenges with federated learning, and therefore, it potentially leads the way for researchers to investigate the application of federated learning in similar scenarios.

This paper contributes to a deeper understanding of data-centric prognostics and its successful integration into Industry 4.0 systems. Through this paper, we aspire to pave the way for advancements that will overcome current challenges and propel the field towards new horizons of innovation and effectiveness.

The paper is organised as follows. Section 1.2 highlights the challenges of implementing a data-centric prognostics approach in Industry 4.0 systems. In Sect. 1.3, opportunities to overcome those challenges are discussed. The conclusion is provided in Sect. 1.4.

1.2 Challenges in Data-Centric Prognostics

The paper employs a meticulous approach to review and categorise challenges. The categorisation of challenges into two distinct groups—exclusive and common—serves a crucial purpose in understanding the nuanced landscape of data-centric prognostics in the context of Industry 4.0. This classification framework helps highlight the exclusive challenges that demand specific attention and those challenges that are universal across various facets of the prognostic domain.

1.2.1 Exclusive Challenges

1.2.1.1 Explainability

Data-centric prognostics need a large amount of data to train models and be effective. The data are essential and serve as ingredients for the models. Recent years have witnessed an exploding increase in the amount and complexity of data available, which is why data-centric prognostics are becoming increasingly popular. Some outstanding data-centric prognostics in artificial intelligence (AI) have achieved higher performance levels. Though impressive, some models have prohibitively large numbers of parameters, especially deep learning ones. Such models are often called 'black box' models as they lack understandable reasons for producing the results. As with applying deep learning in real-world scenarios, there is a need for transparency and explainability. People want models that allow decision-makers to be held responsible for their decisions. Existing investigations on explainable AI (XAI) are typically restricted to sensitivity analysis, local pseudo explanations by LIME, game-theoretic Shapley additive explanations, and Grad-CAM models [14–17]. Thus, there is a pressing need for effective explainability methods that transparently explain the decision-making processes employed in various data-centric prognostic methodologies. Two promising directions for future research include prototype-based models [18] and Turing's type-B random machines [19].

1.2.1.2 Prescriptive Analytics

Prognostics results are the cornerstone of prescriptive analytics. Given the predicted time to failure, proper maintenance action is needed to prevent significant downtime from occurring. It takes more than just human knowledge to know the most appropriate maintenance action; this is where prescriptive analytics comes in. Prescriptive analytics use data to regulate the best decision. It relies on predictions from machine learning models and goes a step further than prediction to prescribe the optimal action and form decision-making. It learns a machine-dependent policy using reinforcement learning [20, 21] and other approaches, such as causal machine learning [22].

Given the increasing demand for prescriptive analytics, developing robust methodologies that predict outcomes and provide actionable insights and recommendations for optimal decision-making is critical.

1.2.1.3 Affordability

Small and medium-sized enterprises (SMEs) are imperative for economic development [23]. Unlike tech giants like Google and Microsoft, they have small budgets and lack analytics skills. With digitalisation being increasingly adopted and so abundant data becoming available, it is therefore critical for SMEs to utilise these data to help bring up productivity first. The successful adoption of AI in SMEs will help them whether the economic downturn the world is currently facing after the pandemic to warrant economic growth [24]. In [5], several organisational challenges are identified: the challenge of human resources, referring to the lack of AI skills; the challenge of financial resources, referring to the cost of digital expertise; the challenge of material resources, referring to the lack of a large amount of high-quality data and the needed infrastructure to acquire them. In light of the growing necessity for cost-effective solutions tailored to small and medium enterprises (SMEs) capabilities, there is a clear demand for accessible and affordable technologies to enhance operational efficiency and competitiveness.

1.2.2 Common Challenges

1.2.2.1 Availability of Sufficient Asset Condition Data

The efficiency of the data-centric approach depends on the quantity and quality of the system's monitored data. It is often impossible to obtain data on failure progressions because it cannot run until failure due to its criticality and failure modes. Data collection instruments like sensors and transmitters are costly [3, 4]. Thus, it is not possible to monitor every component. The need is to monitor only the critical components that cause high downtime. Based on the current literature, the available sample size of the dataset required ranges from 10 to 40. In Industry 4.0 systems, insufficient historical or operational data is available, subject to environmental stress conditions. In various electromechanical systems, failures occur very slowly with a degradation path. Failure can take some time to happen, even years. Monitoring data in such a system is complicated. Accelerated ageing and unnatural failure progressions are required in that case, which has strengths and weaknesses. Maintenance activities (repair, overhaul, etc.) often do not resume the unit's original condition or may degrade at a different rate after maintenance activities. Sometimes, the fault is detected and immediately repaired before it fails. Thus, the point at which the unit will fail is not recorded; only the repair time is known, and it causes a suspension in data.

On the other hand, with tones of the system's monitored data, it becomes challenging to train the algorithm. In that case, samples from the data are taken and analysed to judge the characteristics of the population of data, making the process more difficult and computationally extensive.

Lastly, the system's monitored data has discrepancies from the original values due to various disturbances [4]. Some essential sources are manufacturing and assembly variations, process noise, measurement noise, environment, operator, operational, etc. Some of the disturbances are unavoidable and produce noise in the data. These noise sources make learning the clean and transparent relationships between input and output difficult. We need appropriate signal processing techniques to remove the noise from the data before it is applied to prognostic models. Without denoising scheme learning, meaningful relationships may be hard to acquire.

1.2.2.2 Life Cycle Load

A product experiences various environmental and operational loads, from manufacturing/assembly to the end of its life [5]. Sometimes, the severity of loads before the product is operated is poorly documented, affecting the product. This lack of information about the effects of life cycle load produces noise in the data and leads to inaccuracy in prediction.

1.2.2.3 Signature Identification

A machine or component gives various failure signatures (like vibration, sound, etc.). Separation of the appropriate failure signature, which plays a substantial role in the degradation of the component, is critical [6]. A healthy baseline is an approach to represent that the data till this set line is in good condition, and the data beyond this line is considered as in bad condition. A healthy baseline facilitates verification and validation activities towards certification of the data-centric prognostics approach. However, the problem is setting this healthy baseline due to the lack of benchmark datasets or performance metrics useful for a data-driven prognostic approach. A predefined value called the threshold needs to be set for failure criteria assessment. If the data grows beyond this predefined value, it is a failure. In many cases, the threshold value is not known specifically due to various types of uncertainty in the data.

1.2.2.4 Prognostics Approaches

The data-centric approach can be classified as statistical techniques and artificial intelligence; selecting the appropriate technique is vital as these techniques have their capabilities. Parameters to remember while selecting the appropriate technique are initial wear in the system's accuracy and robustness [7]. Most of the systems come with initial wear, which occurs due to manufacturing inefficiencies. It is not considered abnormal, but it makes a pretty good difference in using an asset. The accuracy of the predicted results is a measure of the closeness to the actual values of failure of the system. The selected technique sometimes lacks the characteristic of being strong enough to withstand intellectual challenge due to a lack of robustness. This can cause a significant loss of resources when constructing the discarded algorithms.

1.2.2.5 Uncertainty

Uncertainty is inherent in the system. At every processing step, the uncertainty gets involved and, if not managed, can grow. Uncertainty can lead to inaccurate prediction, which will delay the maintenance activities; thus, it should be handled with utmost care [8]. Uncertainty is classified as aleatory uncertainty and epistemic uncertainty. The prediction may not be effective without incorporating the measures of uncertainty. The width of the uncertainty bounds is kept in mind for prognostics. It is crucial to have a method capable of handling uncertainties associated with the system.

1.2.2.6 Use of Prognostics Results

The prognostics results support the maintenance decision-making. There are two main cases: case one without applying prognostics and case two with prognostics [9]. If the machine detects the failure, the operation stops, and maintenance preparation is done. After that, the actual maintenance activities are performed to rectify the fault. This whole maintenance period is unscheduled. Thus, it causes high downtime because of the unavailability of machines and inventory. In case two, where the continuous remaining useful life estimation is done, and the potential failure is estimated, maintenance preparation is started without stopping the machine operation because the failure time is known early enough. Then, the actual maintenance is performed. This is the only significant portion of downtime, significantly less than the case one. This activity is scheduled maintenance. Thus, significantly less downtime, enough time for maintenance preparation, and arrangement of resources done on time result in low inventory cost and the life cycle cost of the machine becoming very high because it is used until the end of its life.

1.3 Opportunities in Data-Centric Prognostics

1.3.1 Federated Learning as a Solution

In the era of Industry 4.0, many SMEs face the problem of limited amounts of data they possess and the unwillingness to share and exchange data due to competition or privacy issues, which are also referred to as data silos. With all the challenges and difficulties mentioned previously, particularly affordability and prescriptive analytics, federated learning (FL) emerges as a promising privacy-preserving distributed machine learning framework for a solution. The idea of FL is to utilise fragmented data sources without necessarily having to collect and combine them in a data centre. In FL, data owners collaboratively train a shared model with each participant's local data and only exchange important information, such as model updates. Data stays where they are. With data privacy and security issues in mind, some privacy preservation techniques are needed to allow safe and secure data transmission; the commonly used ones are secure multi-party computation (MPC), homomorphic encryption (HE), and differential privacy (DP). The communication between data parties is encrypted and secured, thus enabling model training on distributed datasets, and only important information, such as model updates, is exchanged. This avoids the massive overhead of transferring raw data and provides data security and privacy that is seriously valued nowadays. Recent research suggests that models trained by FL can achieve comparable performance [26] to those centrally trained in a conventional way and superior to those trained on individual data sources. Depending on the application, FL can be designed in a client–server or peer-to-peer architecture (see Figs. 1.1 and 1.2). Every data owner has full autonomy to operate its data and decide when to join and leave. McMahan et al. practised FL in an edge-server architecture in the context of updating language models on mobile phones [27].

By partitioning the dataset, federated learning can be categorised into horizontal federated learning (HFL), vertical federated learning (VFL), and federated transfer learning (FTL). HFL refers to where datasets overlap many features, so joining multiple datasets can increase the total number of data samples. VFL refers to where

Fig. 1.1 Client–server architecture

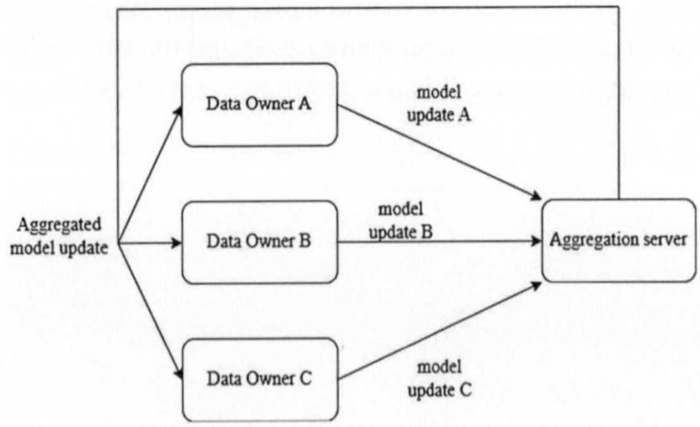

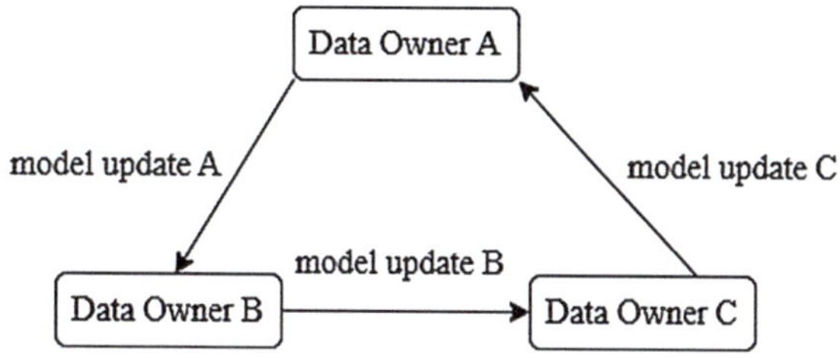

Fig. 1.2 Peer-to-peer architecture (cyclic)

data samples overlap in different datasets, so the model can be trained with more features than a single dataset, thus making more accurate models. FTL is applicable where neither data samples nor data features overlap significantly. Figures 1.1, 1.2 and 1.3 show the three FL categories in a two-party scenario (Figs. 1.4 and 1.5).

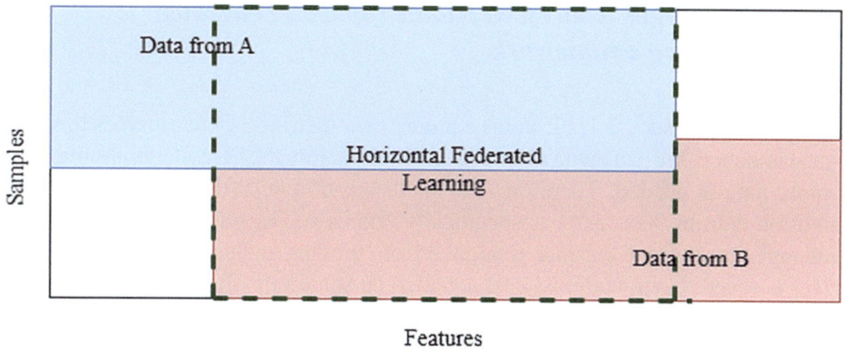

Fig. 1.3 Illustration of the HFL use case includes scenarios where data features overlap largely between data sources, such as client data from different banks

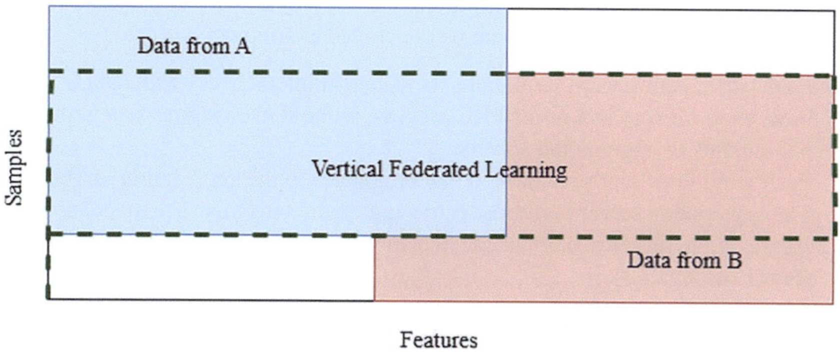

Fig. 1.4 Illustration of VFL use case includes scenarios where individual data entry is contained in multiple data sources but possesses different data features, such as client data from a bank and an e-commerce Website

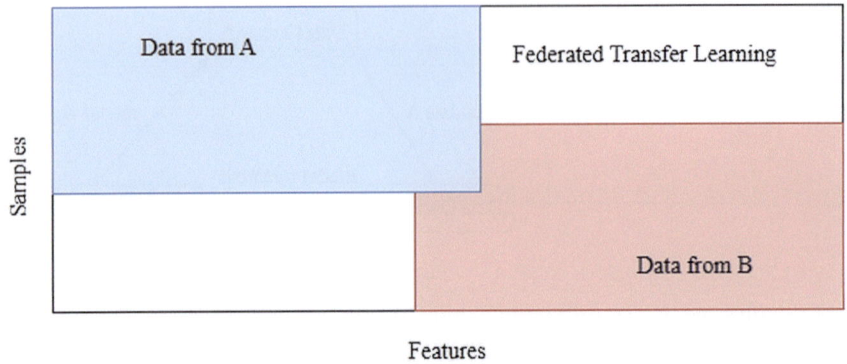

Fig. 1.5 Illustration of FTL

1.3.2 Challenges Addressed from Proposed Federated Learning Framework

As introduced in Sect. 3.1, FL trains a model over distributed data sources to secure data abundance. We can invite more organisations to join the FL collaboration scheme if more data is needed. Therefore, the challenge of availability of sufficient asset condition data in Sect. 2.2.1 is specifically addressed, as well as the affordability challenge, because the expense generated from moving and storing data in a central server is saved. Computational cost can also be shared by all participants, whether equally or according to the relative size of each participant's data. The concept of federated learning can also be applied to reinforcement learning (RL), a mainstream methodology in prescriptive analytics, to form federated reinforcement learning (FRL). Similarly, FRL can be categorised into horizontal federated reinforcement learning (HFRL) and vertical federated reinforcement learning (VFRL). Here, we give a simple example of wind farms along the coast of the UK to illustrate HFRL. The basic steps in a FRL process can be concluded as follows:

1. Each participating wind farm trains its RL agent interactively with data from the local wind farm as in a normal RL process, without exchanging data with other participants or aggregation servers.
2. Each wind farm sends its masked model updates to the aggregation server.
3. The aggregation server performs secure aggregation over masked model updates.
4. Aggregation server sends an updated model to each participating wind farm (Fig. 1.6).

By leveraging RL in a federated manner, we can get the following specific benefits besides those stated earlier:

- Avoid non-i.i.d samples. In a single-agent setting, RL agents commonly encounter non-i.i.d samples during the learning process, as the previous experience can strongly affect the experience gained afterwards, which would break the i.i.d

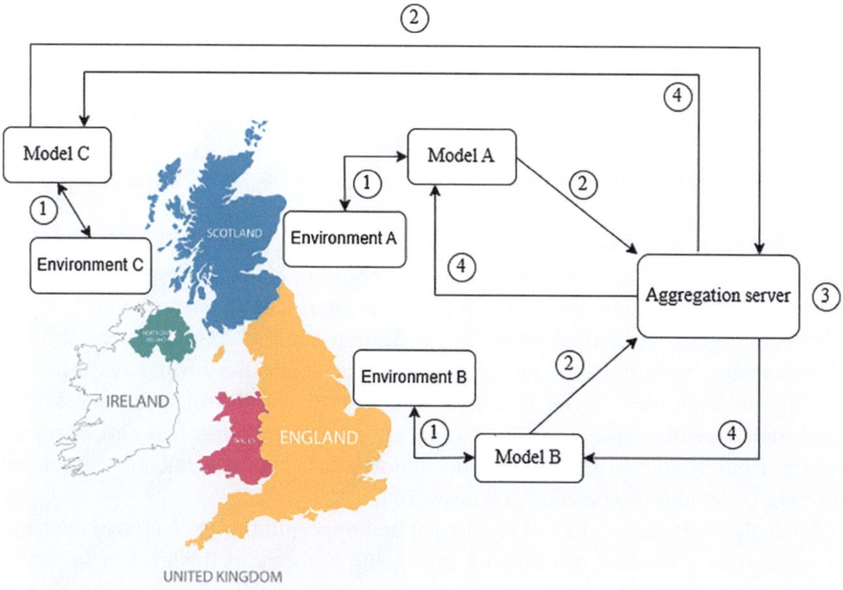

Fig. 1.6 FRL example of wind farms in the UK

assumption in machine learning settings. HRFL can build more robust and accurate models that prevent agents from following the hidden track by absorbing knowledge from diverse environments.

- Boost sample efficiency. In the conventional single-agent setting of RL, each sample is only used by one single agent, so training can take a very long time and, therefore, hard to converge. With HFRL, knowledge is extracted from multiple agents, aggregated in the central server, and handed out to each agent to share one piece of knowledge with all the agents, increasing sample efficiency.
- Accelerate the learning process and overall better performance. With better sample efficiency and more diverse samples to learn from, the HFRL agent has faster learning and more robust performance.

1.3.3 Limitations of the Proposed Framework

In this section, the limitations of HFL and VFL are discussed, respectively.

For HFL, there are three main challenges. First, since encryption provides data security, we cannot inspect data for each participating party, which brings difficulties in setting the hyperparameters such as learning rate, batch size, and structure of neural networks. Second, applying FL needs the effort to bring companies and organisations together to share data and information they possess, which can be a big concern as big companies lack the motivation to collaborate and share the data, which is usually larger than those smaller participants. This will lead to larger companies' unwillingness to join the FL scheme. Third, we assume all participants are honest, but this might not be the case in real-life situations. Participants can be dishonest and submit fake model updates so that they can still get rewards from the jointly trained model without, in effect, contributing anything.

VFL suffers from its vulnerability to communication failures and reliable and effective communication mechanisms by its design: geographically, VFL requires closer and more direct interaction between participating parties. When data transfer is slow and unstable, utilisation of computing resources is inefficient as the training of the model cannot start until it receives all the intermediate information.

In FRL, agents are trained in the same, distinct, similar, or completely different environments. Most existing research works investigate the former two kinds of learning environments. In our previous examples of wind farms, the question of what environments wind farms at different sites are in remains. It is almost certain that those environments are not the same, but whether they are similar or completely different is yet to be researched and answered.

The holistic understanding of challenges and opportunities presented in the paper positions it as a valuable resource for advancing the state of the art of data-centric prognostics.

1.4 Conclusion

This paper employs a meticulous approach to review the challenges in the data-centric prognostics approach. The categorisation of challenges into exclusive and common provides a structured approach to understanding and addressing the intricacies of data-centric prognostics. This classification facilitates targeted problem-solving, efficient resource allocation, a holistic understanding of the field, and guidance for future research directions.

By addressing exclusive challenges such as explainability, prescriptive analytics, and affordability, the paper contributes to the existing body of knowledge, offering an in-depth understanding of the field. Moreover, incorporating insights on data availability, uncertainty, and time-variant life cycle loading demonstrates a comprehensive consideration of relevant common challenges.

State-of-the-art prescriptive analytics is introduced for intelligent maintenance. Additionally, the paper provides insightful guidance for researchers and identifies innovative avenues like federated learning, to boost the successful integration of data-centric prognostics in Industry 4.0 systems. The exploration of federated learning as a potential solution indicates a forward-looking perspective, showcasing the paper's effort to integrate cutting-edge technologies into data-centric prognostics.

The identified challenges and opportunities are expected to help researchers and industries practice data-centric prognostics in Industry 4.0 systems more effectively and efficiently.

References

1. Kumar, S., Torres, M., Chan, Y.C., Pecht, M.: A hybrid prognostics methodology for electronic products. In: 2008 IEEE International Joint Conference on Neural Networks (IEEE World Congress on Computational Intelligence), pp. 3479–3485. IEEE (2008)
2. Pecht, M., Gu, J.: Physics-of-failure-based prognostics for electronic products. Trans. Inst. Meas. Control. **31**(3–4), 309–322 (2009)
3. Gu. J., Vichare, N., Tracy, T., Pecht, M.: Prognostics implementation methods for electronics. In: 2007 Annual Reliability and Maintainability Symposium, pp. 101–106. IEEE (2007)
4. Pecht, M.: Prognostics and health management of electronics. In: Encyclopedia of Structural Health Monitoring (2009)
5. Dragomir, O.E., Gouriveau, R., Dragomir, F., Minca, E., Zerhouni, N.: Review of prognostic problem in condition-based maintenance. In: 2009 European Control Conference (ECC), pp. 1587–1592. IEEE (2009)
6. Jain, A.K., Lad, B.K.: Data driven models for prognostics of high speed milling cutters. Int. J. Performability Eng. **12**(1), 3 (2016)
7. Jain, A.K., Lad, B.K.: Dynamic optimization of process quality control and maintenance planning. IEEE Trans. Reliab. **66**(2), 502–517 (2017)
8. Jain, A.K., Lad, B.K.: Prognosticating RULs while exploiting the future characteristics of operating profiles. Reliab. Eng. Syst. Saf. **1**(202), 107031 (2020)
9. Jain, A.K., Lad, B.K.: Predicting remaining useful life of high speed milling cutters based on artificial neural network. In: 2015 International Conference on Robotics, Automation, Control and Embedded Systems (RACE), pp. 1–5. IEEE (2015)
10. Saikiran, M., Ravali, G.: Comparative study of vegetable based and conventional cutting fluids in machining of copper alloys. Mater. Today: Proc. **1**(19), 611–614 (2019)
11. Hussain, M., Gupta, P., Kumar, P., Das, A.K.: Selective laser melting of single track on Ti–6Al–4V powder: experimentation and finite element analysis. Arab. J. Sci. Eng. **45**, 1173–1180 (2020)
12. Kumar, P., Ashok, G.: Design and fabrication of smart seed sowing robot. Mater. Today: Proc. **1**(39), 354–358 (2021)
13. Kumar, P., Ravi, S.: Investigation on effects of vegetable-based cutting fluids in turning operation of "EN 24 Steel." Mater. Today: Proc. **1**(39), 95–99 (2021)
14. Arrieta, A.B., Díaz-Rodríguez, N., Del Ser, J., Bennetot, A., Tabik, S., Barbado, A., García, S., Gil-López, S., Molina, D., Benjamins, R., Chatila, R.: Explainable artificial intelligence (XAI): concepts, taxonomies, opportunities and challenges toward responsible AI. Inf. Fusion **1**(58), 82–115 (2020)
15. Selvaraju, R.R., Cogswell, M., Das, A., Vedantam, R., Parikh, D., Batra, D.: Grad-cam: visual explanations from deep networks via gradient-based localization. In: Proceedings of the IEEE International Conference on Computer Vision, pp. 618–626 (2017)
16. Dieber, J., Kirrane, S.: Why model why? Assessing the strengths and limitations of LIME. arXiv preprint arXiv:2012.00093 (2020)
17. Chen, H., Lundberg, S., Lee, S.I.: Explaining models by propagating Shapley values of local components. In: Explainable AI in Healthcare and Medicine: Building a Culture of Transparency and Accountability, pp. 261–270 (2021)
18. Angelov, P.P., Gu, X.: Toward anthropomorphic machine learning. Computer **51**(9), 18–27 (2018)
19. Jiang, R., Crookes, D.: Shallow unorganized neural networks using smart neuron model for visual perception. IEEE Access **9**(7), 152701–152714 (2019)
20. Goby, N., Brandt, T., Neumann, D.: Deep reinforcement learning with combinatorial actions spaces: an application to prescriptive maintenance. Comput. Ind. Eng. **1**(179), 109165 (2023)
21. Lepenioti, K., Pertselakis, M., Bousdekis, A., Louca, A., Lampathaki, F., Apostolou, D., Mentzas, G., Anastasiou, S.: Machine learning for predictive and prescriptive analytics of operational data in smart manufacturing. In: Advanced Information Systems Engineering Workshops: CAiSE 2020 International Workshops, Grenoble, France, 8–12 June 2020, Proceedings 32 2020, pp. 5–16. Springer International Publishing

22. Vanderschueren, T., Boute, R., Verdonck, T., Baesens, B., Verbeke, W.: Failure Prediction Versus Maintenance Prescription: Optimizing Maintenance Interventions by Learning Individual Treatment Effects. Pre-Print Friday (2022)
23. Rakshit, S., Mondal, S., Islam, N., Jasimuddin, S., Zhang, Z.: Social media and the new product development during COVID-19: an integrated model for SMEs. Technol. Forecast. Soc. Chang. **1**(170), 120869 (2021)
24. Bettoni, A., Matteri, D., Montini, E., Gładysz, B., Carpanzano, E.: An AI adoption model for SMEs: a conceptual framework. IFAC-PapersOnLine **54**(1), 702–708 (2021)
25. Lu, X., Wijayaratna, K., Huang, Y., Qiu, A.: AI-enabled opportunities and transformation challenges for SMEs in the post-pandemic era: a review and research agenda. Front. Public Health **29**(10), 885067 (2022)
26. Li, W., et al.: Privacy-preserving federated brain tumour segmentation. In: International Workshop on Machine Learning in Medical Imaging, pp. 133–141 Springer (2019)
27. Konečný, J., McMahan, H.B., Yu, F.X., Richtárik, P., Suresh, A.T., Bacon, D.: Federated learning: strategies for improving communication efficiency. arXiv preprint arXiv:1610.05492 (2016)

Chapter 2
Customer Segmentation for Marketing: An Extensive Case Study of New Homebuying Customers

Dishant Banga

Abstract Customer segmentation is a vital marketing strategy that groups target markets based on shared characteristics for tailored marketing, enhancing customer satisfaction, loyalty, and growth. It uses advanced analytics to identify meaningful segments across demographic, psychographic, behavioral, and geographic factors. The process involves data collection, analysis, segment profiling, and evaluating of segment viability, necessitating continuous updates and integration with other marketing strategies. Despite challenges like data quality and dynamic markets, successful real-world applications across industries showcase its efficacy in improving targeting, personalization, and customer experiences.

Keywords Machine learning · K-means clustering · Principal component analysis (PCA) · Unsupervised learning · Return on investment (ROI)

2.1 Introduction

Customer segmentation is a strategic approach in marketing where businesses categorize their diverse customer base into distinct groups based on shared traits and behaviors, enhancing personalized marketing and customer relations. This segmentation facilitates a deeper understanding of customer preferences and needs, allowing for more targeted and effective marketing strategies. It employs various methods such as demographic, psychographic, behavioral, geographic, and firmographic segmentation, each providing unique insights into consumer behavior [5]. Effective customer segmentation leads to improved customer acquisition, enhanced loyalty, better resource allocation, and the identification of new growth opportunities [4]. This targeted approach optimizes marketing efforts and boosts overall business

---

D. Banga (✉)
Bridgetree, USA
e-mail: dishant.u.banga@gmail.com

J. Maiti et al. (eds.), *Operations Research and Data Analytics: Current Trends and Future Perspectives*, Lecture Notes on Multidisciplinary Industrial Engineering,
https://doi.org/10.1007/978-981-96-8327-7_2

performance by fostering stronger customer connections and increasing engagement. By referencing studies such as [1, 3], this research underlines the pivotal role of data-driven customer segmentation in transforming retail marketing. It emphasizes how personalized marketing, product customization, and enhanced customer experiences are key to increasing revenue and building lasting customer loyalty, ultimately reshaping the retail sector. Theoretical Frameworks and Approaches: Numerous theoretical frameworks have been proposed to guide customer segmentation efforts. These frameworks include demographic, psychographic, behavioral, geographic, and firmographic segmentation. Each approach offers unique insights into customer behavior and preferences. For instance, demographic segmentation divides customers based on age, gender, income, and other demographic variables, while psychographic segmentation considers lifestyle, values, and personality traits [2, 6].

2.2 Methodology

Framework for the process of customer segmentation has been used for this case study.

1. Step 1: Define Objectives: Setting goals for segmentation like marketing effectiveness and customer satisfaction.
2. Step 2: Collect Data: From sources like CRM systems, surveys, and analytics
3. Step 3: Identify Segmentation Variables: Using demographic, psychographic, and behavioral criteria.
4. Step 4: Analyze the Data: Through statistical analysis and clustering algorithms.
5. Step 5: Develop Segment Profiles: Create detailed profiles for each segment.
6. Step 6: Tailor marketing strategies, deploy them across channels, and track metrics like customer engagement, conversion rates, and ROI. In the case study, segments were tested with test and control groups in marketing campaigns and evaluated through response analysis, ensuring that ROI and sales parameters were effectively measured.

2.3 Data Collection and Measurement Metrics

Data: The case study utilizes data from 500 k unique homebuyers' customers, predominantly aged 25–35, for a marketing campaign. The consumers were divided into test and control groups. The dataset includes 100 demographic variables (age, gender, and income), one year of sales data, and distance to store information, providing comprehensive insights for analysis. In this study, significant variables included income, home price, wealth rating, and credit range. Jupyter Python Notebook facilitated the analysis process.

Algorithm: K-means clustering, to identify the optimal number of clusters and principal components analysis (PCA), for dimensionality reduction.

Final Segmentation Variables: Home price, current home value, income, wealth rating, credit range, net worth range, and vista type.

Performance Metrics: Response analysis to measure the ROI and other sales metrics.

2.4 Findings and Contribution

In Fig. 2.1, the elbow graph displays the optimal cluster value derived from the provided dataset. It is evident from the graph that the ideal number of clusters is 2.

Continuing with the process's subsequent stages, Fig. 2.2 illustrates segments through lines on a spider plot, and each segment distinguished by distinct colors. The enclosed area within each line signifies each feature's average values of the respective segment. The segment's audience has the following characteristics.

Cluster 0: Customers with low income, net worth, current home value, and price.

Key Characteristics: Home-centered, lower home values, interest in autos, and motorcycles.

Cluster 1: Customers with high income, net worth, current home value and price, and net worth.

Key Characteristics: Higher income, college educated, drive luxury SUVs, and high credit rating.

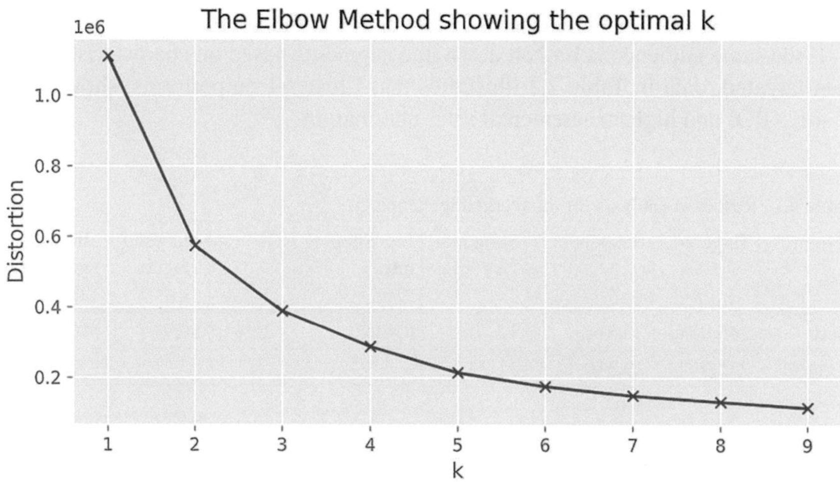

Fig. 2.1 K-means clustering elbow graph

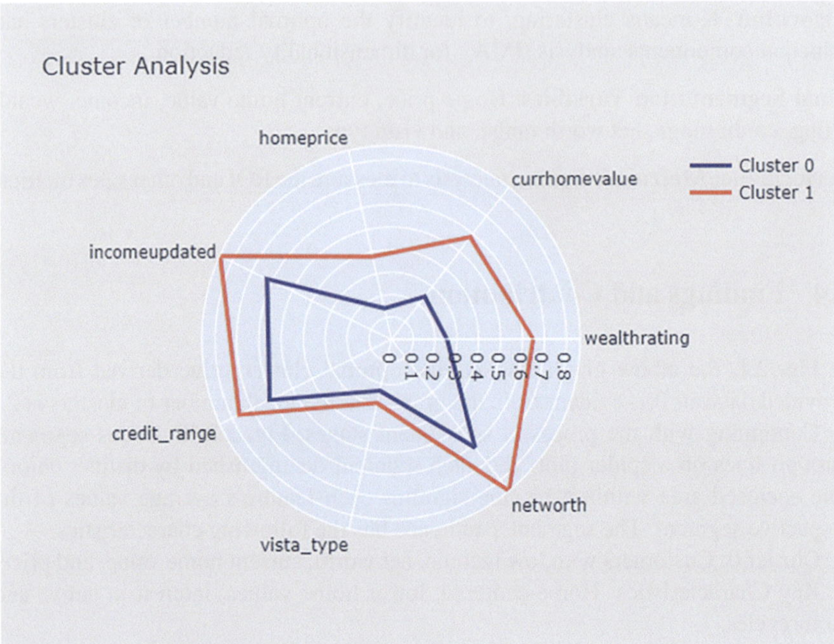

Fig. 2.2 Spider graph representing segments by K-means clustering

Table 2.1 presents the response analysis of the audience targeted in the marketing campaign, indicating their performance. The overall performance of the campaign was negative based on metrics like ROI and incremental and margin.

If the same audience is broken down into segments based on characteristics and then targeted, data in Table 2.2 illustrates that Cluster 1 outperforms, showcasing positive ROI and higher incremental sales and margin.

Table 2.1 Response analysis for the marketing campaign

Group	Targets	Responder	Shop rate (%)	Inc. Shop rate	ROI	Inc. sale/ Target	Inc. margin/ Target
Test	245,984	30,514	12.40	0.49%	– 20%	$0.91	$0.30
Control	245,984	29,310	11.92				

Table 2.2 Response analysis for the marketing campaign broken down by segments

Cluster	Group	Targets	Responder	Shop rate (%)	Inc. shop rate	ROI	Inc. sale/ Target	Inc. margin/ Target
Cluster 0	Test	72,014	8187	11.37	0.54%	–52%	$0.87	$0.29
	Control	72,275	7823	10.82				
Cluster 1	Test	52,726	9975	18.92	0.60%	80%	$1.43	$0.48
	Control	53,066	9721	18.32				

2.5 Limitations and Future Scope

The limitations of customer segmentation include data quality issues, which can skew segment accuracy, and the dynamic nature of consumer behavior that requires continuous updates to segmentation models. There is also a risk of oversimplification, missing subtle but critical nuances within customer groups. Future scope involves enhancing data integration and employing real-time analytics to adapt to changing consumer patterns swiftly, thus improving the responsiveness and effectiveness of segmentation strategies. Additionally, integrating AI and machine learning could refine segmentation precision and operational efficiency. Customer segmentation is a powerful tool that enables targeted marketing efforts, enhanced engagement, and improved conversions, fostering strong customer relationships. This case study illustrates its application by showing how understanding a customer base's unique characteristics and preferences allows for the development of highly effective marketing strategies, thus boosting the return on investment in marketing activities.

References

1. Johnson, S., Lee, M.: Data-driven marketing strategies: a paradigm shift in retail. Int. J. Bus. Res. **15**(2), 78–93 (2020)
2. Kotler, P., Armstrong, G.: Principles of Marketing. Pearson (2017)
3. Smith, A., Johnson, B., Davis, C.: Understanding customer segmentation in retail: a comprehensive study. J. Retail Market. **25**(3), 45–62 (2021)
4. Wedel, M., Kannan, P.K.: Marketing analytics for data-rich environments. J. Mark. **80**(6), 97–121 (2016)
5. Wells, W.D., Prensky, D.: Consumer Behavior. Wiley (1996)
6. Wind, Y., Mahajan, V.: Convergence Marketing: Strategies for Reaching the New Hybrid Consumer. Financial Times Prentice Hall (2001)

Chapter 3
Quantification of Customer Experience in Retail Banking

Sandeep Dey, Indranil Mukherjee, and Prasun Das

Abstract The prevailing challenging nature of directly measuring customer satisfaction in retail banking encourages this research to propose a novel metric that leverages customer behavior and demographics to directly measure customer experience (CX) in banks, replacing traditional surveys. Previous studies acknowledge the importance of customer satisfaction (CSAT) but mention challenges in accurately measuring it. The framework includes two models: the customer satisfaction prediction model, which analyzes transactional data to predict a customer satisfaction score, and the customer churn propensity model, which identifies those customers at risk of leaving the bank. By combining these two models, the framework calculates a net satisfaction score that represents a customer's overall CX. This score eliminates the bias from self-reported surveys and allows banks to evaluate marketing campaigns, proactively analyze customer sentiment, and prioritize interactions based on experience scores. The accuracy of this framework is validated through high-performing XGBoost models. Future research could explore quantifying customers' digital inclination for further enhancements in CX measurement. Overall, this framework enables data-driven strategies for managing customers by quantifying their experiences accurately.

Keywords Customer satisfaction · Retail banking · Churn · Marketing analytics

S. Dey (✉) · I. Mukherjee
School of Management Sciences, Maulana Abul Kalam Azad University of Technology, West Bengal, Kolkata, West Bengal, India
e-mail: sandeepdey1981@gmail.com

I. Mukherjee
e-mail: indranil.mukherjee@makautwb.ac.in

P. Das
SQC and OR Division, Indian Statistical Institute, Kolkata, Baranagar, West Bengal, India
e-mail: prasun@isical.ac.in

J. Maiti et al. (eds.), *Operations Research and Data Analytics: Current Trends and Future Perspectives*, Lecture Notes on Multidisciplinary Industrial Engineering,
https://doi.org/10.1007/978-981-96-8327-7_3

3.1 Introduction

Every service industry is going through major transformations due to emergence of advanced and disruptive technologies. The change in technology landscape has forced all the service industries to foster divergent thinking in order to fulfill the enhanced customer expectations. All service industries such as retail, E-commerce, travel, hospitality, insurance, and telecom have started to adopt and implement the cutting-edge technologies. Banking industry happens to be the most conservative in embracing the new technologies. However, most of the world's leading banks have now started thinking of implementing the new technologies such as advanced data science, analytics, machine learning, and cloud computing across all fields. Today customers have a number of options for monetary transactions other than banks, prominent examples being GPay, Amazon Pay, Airtel Money, PayZapp, Alibaba, Paytm, etc. Today's customers are extremely sensitive toward the services rendered, and hence, meeting and possibly exceeding customer satisfaction become highly challenging for all the banks (including leading banks across the globe).

Banks are spending millions of dollars in their marketing activities and performing all the cross-sell and up-sell campaigns to enhance revenue, increase market penetration, and reduce the customer churn/attrition. Every bank has its marketing team primarily split into two parts, viz. customer value management (CVM) and customer relationship management (CRM). These units jointly work toward single organizational goal to improve customer experience. In order to meet this organizational goal, traditional marketing analytics exercises are performed by most of the banks. However, it is important to redefine customer 360° and improve customer lifecycle management. Customer life cycle can broadly be classified into four parts, viz.

- Customer acquisition,
- New to bank onboarding (normally it covers first 6 months to one year),
- Product and service subscription and generating customer value, and
- Customer attrition or customer churn.

In each of these components, there exists the opportunity to introduce and apply new cutting-edge methodology and standardize the process so that every bank can follow the same operating procedure and reap the benefits of optimal customer satisfaction.

Customer churn in the banking domain describes a lost customer who unsubscribes from a bank service and subscribes to another bank. Customer churn happens for several reasons. For instance, if an account with a bank was created for a specific purpose, the customer was likely to close it after achieving the said purpose. Customer churn also happens if a customer is dissatisfied with a bank's service or is relocating or moving to another location. Policy changes initiated by a bank may also be responsible for customer churn. Banks and other financial institutions keep a regular check on their customers' transactions to detect common warning signs in a customer's behavior before churn occurs. Such behaviors include a reduction in the volume of transactions and dormancy of accounts. To identify these customers and implement targeted retention programs, banks can leverage the new customer experience (CX) index framework which is conceptualized and implemented in this study. This CX index analyzes customer transaction behavior to identify those customers who are truly satisfied but may still be at risk of churning due to a specific policy or reason. By factoring in dissatisfaction with this policy, banks can refine the CX score to reflect a more accurate customer experience.

Churn management has become an integral component of customer relationship management (CRM) because of the serious challenge of customer churn in the banking sector. Churn management emphasizes the need for banks to take steps to prevent or minimize customer churn through several customer retention programs. This also helps to establish long-term relationships with customers and maximize the value of their customer base. Banks are affected when valuable customers leave taking their investment or capital to competitors. By keeping regular checks on customers' transaction status, banks generate a huge amount of data on customers' transaction behavior. However, the real challenge is to analyze this stupendous volume of data and extract meaningful knowledge from it, which is often not possible using traditional statistical methods. This necessitated the development of powerful algorithms that use machine learning techniques to discover hidden patterns in the data and predict the likelihood of a customer unsubscribing from an organization's services. This is particularly important in view of the fact, as noted from several independent studies, that it is more expensive to acquire new customers than to retain the existing ones.

The rest of the paper is organized as follows: Sect. 3.2 delves into a comprehensive literature review, examining past research conducted in this domain. Sections 3.3 and 3.4 establish a clear foundation for the study by defining its scope and outlining the research objectives. Section 3.5 sheds light on the proposed methodology employed in the research. Section 3.6 then dives deep into the data preparation process, detailing aspects such as data source identification, campaign management, the rationale behind data synthesis, underlying assumptions made, data aggregation techniques, data quality improvement methods, and finally, the creation of the workable database used for analysis. Section 3.7 undertakes an exploratory data analysis. Sections 3.8 and 3.9 focus on predictive modeling, discussing the performance of various algorithms used, the importance of different features, model deployment strategies, the resulting findings, and their interpretation. Section 3.10 explores the implications of the study, considering both its managerial and theoretical aspects. Finally, Sect. 3.11 summarizes the key findings and outlines potential avenues for future research in this area.

3.2 Literature Review

Extensive research has delved into the multifaceted aspects of customer satisfaction, exploring critical factors like loyalty and trust. Leninkumar [1] discovered substantial positive correlations between customer trust and loyalty, customer satisfaction and loyalty, and customer satisfaction and trust. This evidence underscores that the pivotal role customer satisfaction plays in influencing loyalty and establishes satisfaction as a precursor to trust. Trust is not merely an isolated aspect of the customer experience; it is deeply interwoven with the overall satisfaction that customers derive from their interactions with a business. Leninkumar's findings highlight the interconnectedness of these elements, suggesting that businesses should prioritize trust-building measures as part of their broader customer satisfaction strategies. Meanwhile, Hossain and Leo [2] evaluated the service quality in retail banking within the Middle East, specifically in Qatar, finding that customers rated tangibles highest and competence lowest. Tangibles include physical facilities, equipment, and appearance of personnel, all of which are immediately perceptible to customers and therefore form a significant part of their service quality evaluation. On the other hand, competence, which involves the skills, knowledge, and ability of service providers, was perceived as lacking. Hossain and Leo emphasized the pressing need to improve service quality to enhance customer satisfaction amidst growing competition. Their study implies that banks in the Middle East need to focus more on training and developing their employees' competencies to meet and exceed customer expectations. Moving on to strategic planning, Mittal and Frennea [3] provided valuable insights for managers, emphasizing the incorporation of customer satisfaction into strategic planning processes. Their work offers guidance on measuring and analyzing satisfaction while crafting customer-focused strategies. They argue that understanding customer satisfaction is not just about tracking metrics but involves a strategic approach that aligns with overall business goals. By integrating customer satisfaction metrics into strategic planning, businesses can create more effective plans that address customer needs and enhance loyalty. Similarly, Zalatar [4] investigated service quality in Philippine commercial banks, revealing significant gender differences in customers' expectations and perceptions of service quality. This research aids in developing strategies to enhance customer perceptions. For example, understanding that male and female customers may prioritize different aspects of service quality can help banks tailor their services and marketing efforts more effectively. Gender-specific strategies might include personalized communication, differentiated service approaches, and tailored product offerings to meet the unique expectations and needs of diverse customer segments. The literature illustrates that while service quality is paramount to satisfaction, other factors such as quality, dependability, and responsiveness are equally crucial. Pasha and Razashah [5] emphasized the importance of maintaining service quality to manage client expectations and foster loyalty, identifying key determinants like tangibility, responsiveness, reliability, and assurance. Tangibility relates to the physical aspects of service, responsiveness to the willingness to help customers promptly, reliability to the ability to perform the promised service dependably, and assurance to the knowledge and courtesy of employees and their ability to convey trust and confidence. Pasha and Razashah findings highlight the multifaceted nature of service quality and the need for a balanced approach to managing all these dimensions effectively. In a related study, White [6] examined how different motivational types from self-determination theory (SDT) influence the antecedents of customer satisfaction, highlighting the critical role of motivation. SDT distinguishes between intrinsic and extrinsic motivation, where intrinsic motivation involves engaging in an activity for its inherent satisfaction, and extrinsic motivation involves performing an activity to achieve some separable outcome. White's study suggests that understanding the different motivational drivers of customers can help businesses design better engagement strategies. For instance, businesses might foster intrinsic motivation by enhancing the enjoyment and satisfaction customers derive from their interactions, while extrinsic motivation can be leveraged through rewards and incentives. Exploring the realm of Internet banking, Raza et al. [7] studied service quality dimensions and their impact on e-customer satisfaction and loyalty. They found that all dimensions positively influence customer satisfaction, which in turn boosts loyalty. The dimensions studied include efficiency, reliability, fulfillment, privacy, responsiveness, compensation, and contact. Their research indicates that ensuring high service quality across these dimensions can significantly enhance the customer experience in online banking. For example, providing a user-friendly interface (efficiency), ensuring system reliability, and protecting customer data (privacy) are critical to fostering e-loyalty. Kaura [8] further investigated service quality, perceived price fairness, and service convenience, identifying key dimensions of service quality and emphasizing their significance in satisfying customers. Service convenience was examined through dimensions like decision convenience, access convenience, transaction convenience, benefit convenience, and post-benefit convenience. Kaura's study highlights that not only the quality of service but also the convenience of service access and perceived fairness of pricing play essential roles in shaping customer satisfaction. This suggests that banks should focus on streamlining processes and ensuring transparent pricing strategies to enhance customer perceptions and satisfaction. In the Turkish banking sector, Dincer and Hacioglu [9] evaluated banks' performance based on customer service and satisfaction, noting that performance varied with satisfaction levels and ownership types. Their study pointed out that state-owned banks, private banks, and foreign banks might have different service quality perceptions among customers, which affects overall satisfaction and performance outcomes. This implies that banks need to consider ownership structures and tailor their service strategies accordingly to meet diverse customer expectations. Meanwhile, Eren [10] focused on customer satisfaction with bank chatbots, finding that customer expectations influence perceived performance but do not directly impact satisfaction. The study suggested that while chatbots are becoming an essential part of customer service, their effectiveness largely depends on meeting customer expectations regarding interaction quality, response accuracy, and usability. Eren emphasized the need for continuous improvements in chatbot technology and training to align with customer expectations and enhance satisfaction levels. Muyeed [11] assessed service quality in Bangladeshi retail banking, revealing that prompt and accurate transactions had the highest impact on customer perception, while modern equipment had the least. This finding underscores the importance of operational efficiency in banking services. Customers highly value swift and precise transaction processing, which directly affects their satisfaction. Conversely, while modern equipment and décor contribute to the overall service ambiance, they do not significantly impact customer satisfaction if the core banking services are not efficient and reliable. Athanassopoulos et al. [12] delved into the behavioral consequences of customer satisfaction, examining its impact on customer behavior. They found that satisfied customers are more likely to exhibit loyalty behaviors, such as repeat purchases and positive word-of-mouth. This study highlights the broader implications of customer satisfaction for business growth and brand reputation. Ensuring high levels of customer satisfaction can lead to increased customer retention, reduced churn rates, and enhanced brand advocacy. In another industry, Lee et al. [13] applied the Taiwan customer satisfaction index model to tourism, proposing two satisfaction levels to enhance satisfaction effectively. They identified overall satisfaction and attribute-specific satisfaction as critical levels to address. This dual approach allows businesses to target specific areas for improvement while maintaining a holistic view of customer satisfaction. The model provides a structured framework for measuring and managing satisfaction in the tourism industry, which can be adapted to other service-oriented sectors. Korkutata [14] analyzed the satisfaction of clients in personal exercise training, finding no significant differences based on demographic factors. This study suggests that personal trainers and fitness centers should focus on personalized service delivery and consistent quality across all customer segments to enhance satisfaction. Regardless of gender, membership period, or frequency of use, customers value personalized attention and high-quality service. On a broader scale, Aimee [15] conducted a literature review to deepen the understanding of customer satisfaction, defining key factors and metrics. The review identified critical dimensions such as service quality, perceived value, trust, and emotional connection as fundamental to customer satisfaction. Razafimanjary's work provides a comprehensive overview of existing research and highlights the need for a multifaceted approach to understanding and improving customer satisfaction. Identifying service satisfaction measures that are applicable across industries, Gilbert and Veloutsou [16] provided a basis for cross-industry benchmarking, aiding practitioners in capturing best practices. Their study identified core characteristics of customer satisfaction that are consistent across different sectors, including banking, retail, government, hospitality, sports, and restaurants. These findings enable businesses to benchmark their performance against industry standards and adopt best practices to enhance customer satisfaction. Mittal and Frennea [3] once again emphasized the strategic importance of customer satisfaction, offering managers guidelines for effective implementation. They highlighted the need for a structured approach to measuring and managing customer satisfaction, including regular surveys, feedback mechanisms, and continuous improvement initiatives. By embedding customer satisfaction into the strategic framework, businesses can ensure that their efforts are aligned with customer needs and expectations. Rahi and Ghani [17] integrated the DeLone and McLean information system success model and self-determination theory to examine the continuance of Internet banking usage, contrasting traditional studies focused on initial acceptance. Their study found that factors like system quality, information quality, and service quality significantly influence continued usage intentions. Additionally, intrinsic and extrinsic motivational factors from SDT play a crucial role in sustaining customer engagement with Internet banking services.

Chen et al. [18] incorporated the FAIRSERV model to understand the role of fairness in service delivery for customer satisfaction, highlighting its equivalence to service quality. Their findings suggest that perceptions of fairness in service interactions significantly impact customer trust and perceived value, which in turn influence overall satisfaction. By ensuring fair treatment and transparent processes, businesses can enhance customer trust and satisfaction. Furthermore, Rahi and Ghani [19] developed an integrated model combining expectation confirmation theory (ECT) and self-determination theory (SDT) to predict customers' continued use of Internet banking, demonstrating that both theories significantly influence continued usage intention. This comprehensive approach provides a robust framework for understanding customer behavior in the context of technology and motivation. ECT focuses on customers' expectations and their confirmation or disconfirmation post-usage, while SDT emphasizes intrinsic and extrinsic motivational factors driving continuous engagement.

Customer satisfaction is crucial for business success, leading researchers to identify key factors, often abstract and intangible, typically measured indirectly through surveys. This study aims to develop a composite metric using transactional and demographic data to address customer dissatisfaction and create an overall customer experience score. By doing so, it informs marketing campaigns and initiatives, ensuring a comprehensive understanding of customer needs and improving satisfaction levels. Understanding customer satisfaction allows businesses to tailor their approaches to better meet customer expectations and enhance overall performance. The integration of various theories and models highlights the complexity of customer satisfaction and the importance of a multifaceted approach. Developing a composite metric for customer experience, considering both transactional and demographic attributes, significantly advances how businesses can measure and improve satisfaction. This metric provides valuable insights, helping businesses identify improvement areas and implement effective strategies to enhance customer experiences. Understanding different motivational factors, service quality dimensions, and the impact of fairness in service delivery can help businesses create more targeted and effective customer satisfaction initiatives. The findings underscore the need for continuous improvement and adaptation to changing customer needs and expectations. Leveraging these insights, businesses can enhance their customer satisfaction efforts, leading to increased loyalty and long-term success. In conclusion, customer satisfaction is a vital aspect of business strategy. Ongoing research provides valuable insights for improving customer experiences. By understanding the various factors influencing satisfaction and developing comprehensive metrics, businesses can better meet customer needs and foster lasting relationships. This research underscores the importance of a customer-centric approach and the need for businesses to continuously evolve and adapt to maintain high levels of satisfaction.

3.3 Scope

The review of literature shows that it is imperative to quantify the customer satisfaction for retail banking so that it may be compared with the benchmark and appropriate measures can be initiated to improve customer satisfaction to the maximum extent possible. There are many factors which contribute to customer satisfaction specific to retail banking. Broadly, all the factors can be classified into two areas, products and services. In traditional marketing analytics of retail banking, various cross-sell and up-sell models are developed, and campaigns are rolled out to address the potential good customers. This traditional methodology operates with the underlying assumption that relatively high profitable customers are more satisfied as compared to the less profitable ones, and hence, it is necessary to always address the former category for product and service offerings. During the customer selection phase, customer satisfaction is generally not considered because the quantified value is not available. Customer satisfaction is an intangible feature, and there exists a great deal of fuzziness in customer-provided feedbacks on satisfaction levels.

During research investigation, the importance of customer dissatisfaction is understood which results in customer churn. Addressing this issue will lead to customer loyalty and retention. Today's retail banking industry is putting a great deal of emphasis on self-service technology which saves significant resources for the banks but can lead to a certain degree of dissatisfaction among the customers. Banks need to specify the quantum of dissatisfaction generated as a result of such strategy so that suitable remedial measures may be put in place.

Thus, the scope of the present work lies in developing an integrated measure for customer satisfaction based on collected customer data, analysis of the data, factoring in customer satisfaction, and /or the lack of it, thereby helping retail banks to plan their promotional and administrative strategies so as to sustain and enhance their organizational growth.

3.4 Objective

The objective of this study is to develop a framework that would be able to quantify the customer experience from their transactional and behavioral information. The framework, thus developed, would be able to quantify customer experience without taking any feedback from the customers. The quantification framework will be developed in such a way that would be able to capture both customer satisfaction and customer dissatisfaction together, thereby providing an integrated metric for customer satisfaction index.

3.5 Proposed Methodology

There are two types of data that will be leveraged to develop the framework: banking transaction data of active customers and churn customers. There is some commonality of the variables pertaining to the banking transaction data of existing customers and churn customers, but there are some variables related to churn customers for which the behavior is significantly different from that of the existing customers. The methodology adopted in this work will quantify the customer satisfaction using all of these variables. Generally, customer satisfaction is directly proportional to the following factors: number of purchases (usage), duration associated (loyalty), and purchase amount (trust). All these three factors will be combined to make a tentative customer satisfaction index. The respective variables will be identified from the existing transaction data as follows:

- One-off purchase (purchases had a high correlation with one-off purchases)
- Purchase frequency
- One-off purchase frequency (frequency of purchases made in one go)
- Purchase trx (number of purchase transactions made)
- Tenure.

These variables are considered as direct/estimating variables as these variables help to classify the customer satisfaction levels (ref. Table 3.1).

From the above table, it is assumed that tenure is a representation of loyalty, number of purchases indicates potential promoter, and purchase amount reflects trust. A suitable transformation and aggregation of the direct variables will be done to quantify the customer satisfaction. The exact cutoff values for high and low satisfaction levels will be identified by analyzing the transactional data.

The churn propensity model will be developed based on the existing and churn customers. This model will compute the churn propensity score for each customer. The churn score indicates the dissatisfaction. A discounting factor will be computed so that the net satisfaction can be computed from the predicted satisfaction.

Table 3.1 Classification of satisfaction levels

Tenure	Number of purchases	Purchase amount	Satisfaction level
H	H	H	8
H	H	L	7
H	L	H	6
H	L	L	5
L	H	H	4
L	H	L	3
L	L	H	2
L	L	L	1

H—high, L—low

The intuition behind the discounting factor formalism is to rationalize the true customer satisfaction. The fundamentals of customer satisfaction are grounded in high customer stickiness and customer loyalty which together results in higher customer lifetime value (CLV). The initial satisfaction score is based on transactional and behavioral information about the customers. However, each customer has a probability to churn owing to dissatisfaction. This dissatisfaction is quantified using the churn propensity score. However, the challenge lies in incorporating the churn propensity score in the customer satisfaction score, the latter being normalized and able to take values in [0, 1]. Moreover, churn probability cannot be directly subtracted from satisfaction score since the fundamental basis for these two predictive models is not same. The concept of discounting factor is, therefore, introduced to factor in the impact of churn propensity in the satisfaction score and computed as {(Satisfaction Score * Churn score)/(Satisfaction score + Churn Score)}.

$$\text{Discounting Factor} = \frac{\text{Initial Satisfaction Score} * \text{Churn Propensity Score}}{\text{Initial Satisfaction Score} + \text{Churn Propensity Score}}$$

Finally, the customer experience will be computed by combining customer satisfaction and customer dissatisfaction and is defined as the net satisfaction score.

$$\text{Net Customer Satisfaction score} = \text{Initial Satisfaction score} - \text{Discounting Factor}$$

3.6 Data Preparation

A. Source Identifiation

Retail banking customer data is an extremely confidential information and is stored under customer relationship management (CRM). Hence, banks do not allow any third party to have access to this information. So, the data collection exercise has been performed through research on various banking Web sites, world bank data, blogs, data science, and machine learning forums such as Kaggle, GitHub, and Medium.

The idea behind this data collection exercise is to develop a synthetic dataset which will be in alignment with the business objectives. The retail banking customer value management data can broadly be classified into seven parts which are stored in separate data bases, viz. demographics, transactional, products and subscriptions, savings and investments, insurances, loans, and credit cards.

B. Campaign Management

It plays a pivotal role in customer value management. It helps in the cross-sell/up-sell of various products and services as well as in improving customer experience by engaging the customers with tailor made products and services/business concepts. Hence, campaign data is equally important to drive customer engagements. There are two types of campaign data that have been considered in this work:

1. Direct marketing
2. Email campaign.

There are couple of datasets distributed across all the dimensions as mentioned above, which are essentially actual data from CRM systems but the data will be masked to maintain customer confidentiality.

C. Basis for Data Synthesis

There are total of 12 datasets that are extracted from Kaggle and distributed across all the dimensions mentioned above, which are essentially actual data from CRM systems.

D. Dataset developed by Liz Petrocelli (https://data.world/lpetrocelli/retail-banking-demo-data)

Datasets are stitched together from real-world data sources (CRM + core banking) by Liz Petrocelli. Built off of that 1999 Czech data.

E. Aggregation: Retail banking datasets

- Source-1: completedacct
- Source-2: completedcard
- Source-3: completedclient
- Source-4: completeddisposition
- Source-5: completeddistrict
- Source-6: completedloan
- Source-7: completedorder
- Source-8: completedtrans
- Source-9: CRM Call Center Logs
- Source-10: CRM Events
- Source-11: CRM Reviews
- Source-12: LuxuryLoanPortfolio

All the individual datasets are aggregated into one dataset as retail bank based on client_id, account_id, and other primary keys.

F. Cleaning for Quality

Post aggregation, the duplicate records are identified and removed. There are few records for which most of the attributes are null. All such records are dropped from the dataset. All the numeric attributes are analyzed and checked for outliers. The six-sigma approach is taken for outlier removal. There are few numeric attributes, similar in nature, which have been standardized using normality check and subjected to log transformation. All the variables are converted into lower case and space replaced with underscore.

G. Workable Database

All the 12 data sources are aggregated in four stages. The objective of this aggregation is to bring all the attributes of retail banking customers on a common platform.

Stage-1: Mapping the district dataset with client data.

Stage-2: Aggregate transaction data by account_id, year, and months.

Stage-3: Aggregate the demographic data with transaction data.

Stage-4: Aggregate the demographic data and transaction data with tenure.

The merged dataset contains all the essential attributes of banking such as demography, purchasing behavior of the customer, bank balance of the customer, employment status, gender, income, location code, and marital status. The final dataset comprises ~4500 customers and 22 attributes.

3.7 Exploratory Data Analysis

After analyzing the existing attributes with respect to the research objective, it is understood that some attributes can be derived which will be helpful to develop the predictive models for customer satisfaction quantification and churn propensity. The new derived variables are tenure, total number of transactions since account opening, average monthly transaction, total transaction amount since account opening, average monthly transaction amount, maximum number of transactions ever done in a month, and number of transactions made per month (transaction frequency) (ref. Table 3.2). Transaction frequency is an indication of customer's transaction pattern and explains on an average how many transactions are performed by the consumer and the frequency distribution (ref. Fig. 3.1). These variables can be termed as the direct variables as they provide direct estimation of customer satisfaction. The dependent variable defined as customer satisfaction is derived from three direct variables, viz. average transaction per month, average monthly transaction amount, and tenure (ref. Table 3.4). It is to be noted that the cutoff value based on which the satisfaction levels are defined is selected based on the summary statistics. Both for average monthly transaction amount and tenure, the median is selected since the standard deviation is high (ref. Table 3.2). However, for average transaction per month, the cutoff is selected as 75th percentile since the mean, and 25th and 50th percentile is same as 3 (ref. Table 3.3).

It can be observed that average transaction per month ranges from 0 transaction to maximum of 8 transactions. However, average transaction is 3 for approx. 35% customers (ref. Fig. 3.1). Also, customer distribution across various satisfaction levels indicates that it is not uniformly distributed (ref. Fig. 3.2). This will lead to class imbalance during predictive model development using this derived attribute. This is dealt with using oversampling technique. The correlation analysis (ref. Fig. 3.3) reveals a strong positive correlation between total transaction, transaction frequency, average balance, and maximum transaction in a month. This high correlation leads to a higher standard deviation and inconsistency in the predictability of the model. Hence,

Table 3.2 Summary statistics: derived variables

Descriptive statistics	Total transactions (#)	Avg balance	Total transaction amount	Transaction frequency	Avg. monthly transaction amount	Tenure (months)
Mean	193.69	5052.61	13,84,508	0.5	28,765	40.17
Std	108.54	3046.91	13,25,214	0.1	23,141	18.14
Min	9	98.35	29,400	0.23	817	10
25th percentile	108	2664.22	4,21,248	0.43	9520	25
50th percentile	172	4445.43	8,85,206	0.48	23,000	35
75th percentile	269	6793.94	19,45,954	0.56	41,238	60
Max	594	20,735.48	76,00,375	1.00	1,27,260	71

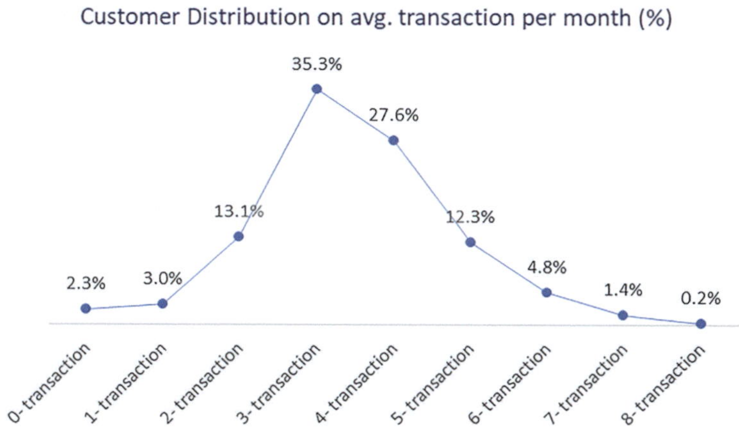

Fig. 3.1 Avg. transaction per month

Table 3.3 Summary statistics of three important attributes for deriving dependent variable

Summary statistics	Average monthly transaction amount	Average transaction per month	Tenure
Count	4500	4500	4500
Std	23,141	1	18
Minimum	817	0	10
Mean	28,765	3	40
25th percentile	9520	3	25
50th percentile	23,000	3	35
75th percentile	41,238	4	60
Max	1,27,260	8	71

Table 3.4 Number and magnitude of average monthly transactions and the corresponding satisfaction levels

Actual attributes			Derived dependent variable
Average transaction per month	Average monthly transaction amount	Tenure	Satisfaction level
<4	<23,000	<35	1. Worse than expected
<4	≥23,000	<35	2. Terrible
≥4	<23,000	<35	3. Dissatisfied
≥4	≥23,000	<35	4. Neither satisfied nor dissatisfied
<4	≥23,000	≥35	5. Satisfied
<4	≥23,000	≥35	6. Very satisfied
≥4	<23,000	≥35	7. Excellent
≥4	≥23,000	≥35	8. Better than expected

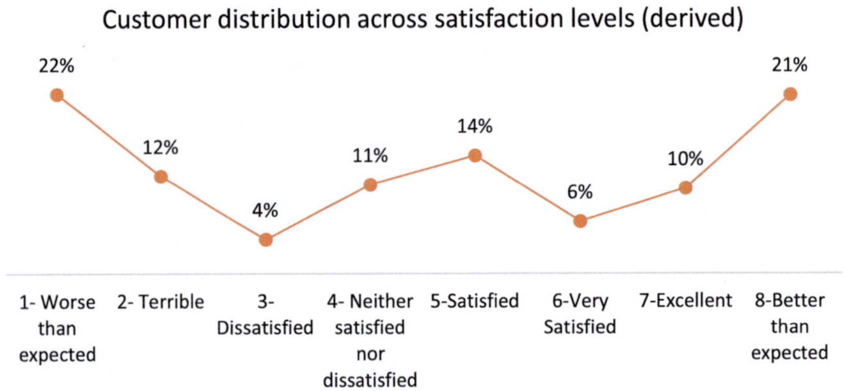

Fig. 3.2 Distribution customers across various satisfaction levels derived from transaction pattern

multicollinearities have to be removed using dimension reduction technique such as principal component analysis (PCA). It is to be noted that the portfolio contains 16% churn customers (ref. Fig. 3.4). Churn propensity model will be developed using all the demographic and transactional attributes (Table 3.4).

3.8 Model Building—Predictive Analytics

It is to be noted that two models would be developed as part of the predictive analytics exercise. The first predictive model will quantify the extent of satisfaction, and the second predictive model will predict the churn propensity. Finally, both scores will

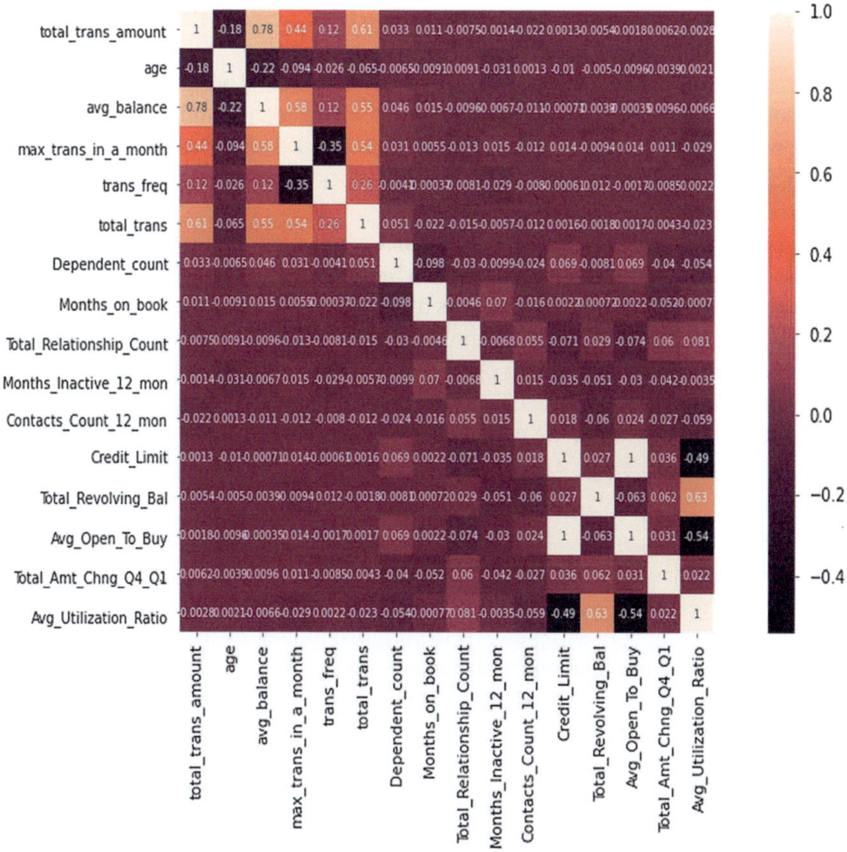

Fig. 3.3 Correlation analysis

Fig. 3.4 Distribution of customer (active versus churn)

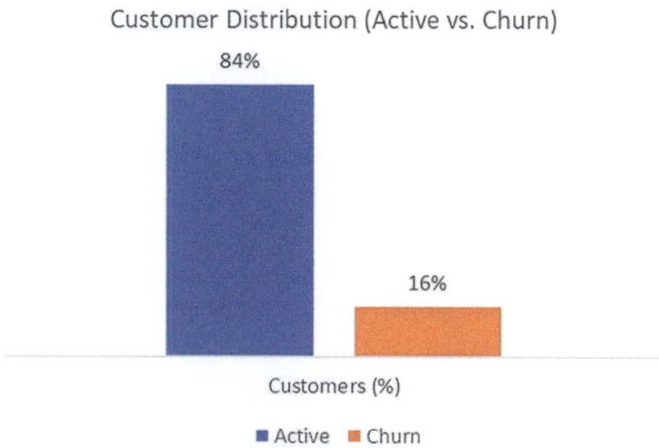

be combined to get the net satisfaction or customer experience. Both the models will be classification models, the former is the multinomial classification model, and the latter is binary classification model. Hence, extreme gradient boosting will be used as the predictive engine for both the models. There are other algorithms such as decision tree, random forest, and gradient boosting machines. However, XGBoost is capable of managing the complex datasets. Hence, XGBoost is chosen as the algorithm for implementation of the predictive models.

XGBoost is a gradient boosting machines which uses gradient boosting trees (gbtree) as the error predictor. It starts off with a simple predictor which predicts an arbitrary number (usually 0.5) regardless of the input. Needless to say, that predictor has a very high error rate. Then, the above idea is applied until the error is brought to a minimum. In XGBoost, training of error prediction model is not done by trivially optimizing the predictor on (feature, error) pairs. In XGBoost, a gbtree is learned such that the overall loss of the new model is minimized while keeping in mind not to overfit the model.

$$\text{Loss}(o) = \min_{o} \sum_{l=1}^{N} \text{loss}(y_i, f(x_i) + o) + \frac{1}{2}\lambda o^2$$

where N is the number of samples, f is the original model, λ is the $L2$ regularization parameter and o is the value which we want to find.

Model 1 (Prediction of customer satisfaction): During customer satisfaction prediction, the explicit dependent variable was not available. Hence, an extensive study of the transactional behavior of the customers was performed to understand and identify the factors directly responsible for customer satisfaction. As an outcome of this exercise, three factors are identified, viz. number of purchases (usage), duration associated (loyalty), and purchase amount (trust). These three factors are combined to set up a derived customer satisfaction index.

The dependent variable is defined as "satisfaction levels" which has 8 levels. There is a total of 4500 customers which is split into two parts for model development (70%) and model validation (30%). The development sample consists of 3150 customers, and validation sample consists of 1350 customers.

The distribution of customers across all the eight satisfaction levels is consistent except level-3 and level-6. Before up-sampling, the distribution for level-3 was 4% and that for level-6 was 6% (ref. Table 3.5 and Fig. 3.5). Since all the 8 levels are not uniformly distributed, hence an up-sampling exercise is performed to reduce the biasness. The distribution improved slightly following the application of up-sampling, as level-3 became 7% and level-6 increased to 9% (ref. Table 3.5 and Fig. 3.6).

Model 2: Prediction of customer churn.

The development sample consists of 3,150 customers, and validation sample consists of 1,350 customers which is identified during customer satisfaction prediction model. The same development sample is considered for churn prediction model development. The dependent variable is "Attrition_Flag" which is the actual variable

Table 3.5 CUstomer distribution across 8 levels of dependent variable

Satisfaction levels	Before up-sampling		After up-sampling	
	Customer (#)	Customer (%)	Customer (#)	Customer (%)
1. Worse than expected	664	21	664	21
2. Terrible	370	12	370	12
3. Dissatisfied	138	4	250	7
4. Neither satisfied nor dissatisfied	350	11	350	11
5. Satisfied	423	13	423	13
6. Very satisfied	202	6	300	9
7. Excellent	318	10	318	10
8. Better than expected	685	22	685	22

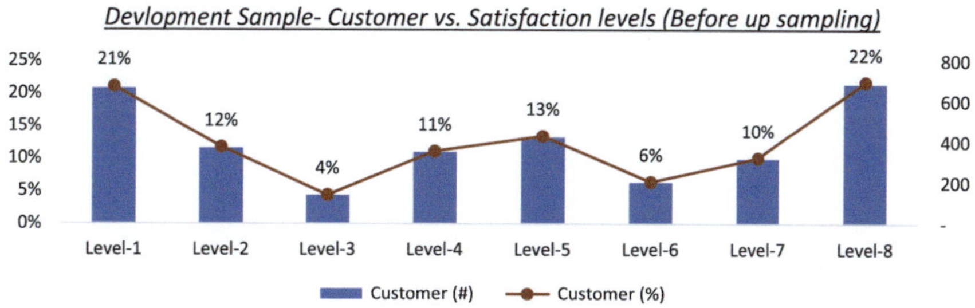

Fig. 3.5 Distribution of satisfaction levels and class imbalance—before up-sampling

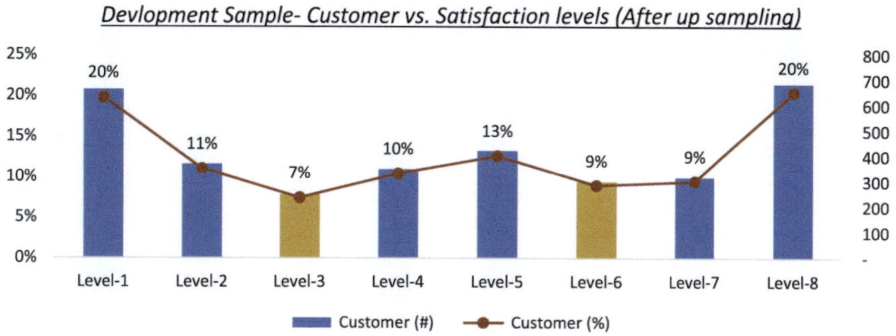

Fig. 3.6 Distribution of satisfaction levels and class imbalance—after up-sampling

available in the customer relationship database of the retail bank. The distribution of the churn customers is skewed which is initially imbalanced. Using oversampling technique, this is balanced (ref. Table 3.6 and Fig. 3.7). After up-sampling, the distribution is significantly improved toward uniformity, and the bias gets reduced (ref. Table 3.7 and Fig. 3.8).

Table 3.6 Customer distribution (active versus CHURN)

Status	Before up-sampling		After up-sampling	
	Customer (#)	Customer (%)	Customer (#)	Customer (%)
Active	2660	84	2660	57
Churn	490	16	2000	43

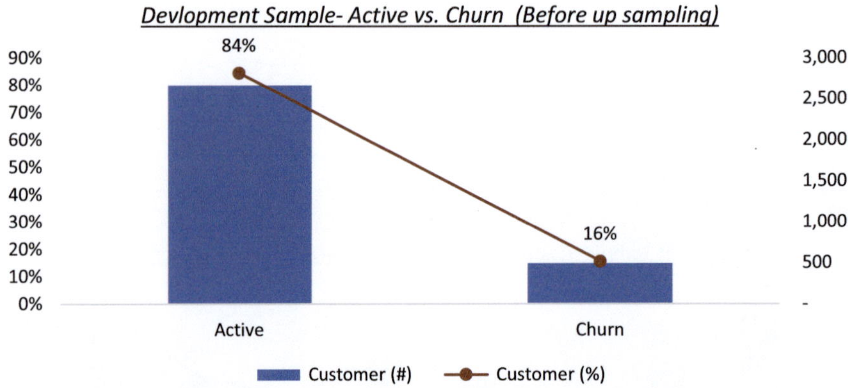

Fig. 3.7 Distribution of churn and class imbalance—before up-sampling

Table 3.7 Classification report on customer satisfaction level after up sampling

Satisfaction level	Precision	Recall	F1-score	Support
1. Worse than expected	0.89	0.97	0.93	306
2. Terrible	0.92	0.97	0.95	157
3. Dissatisfied	0.73	0.65	0.69	46
4. Neutral	0.80	0.75	0.77	130
5. Satisfied	0.91	0.79	0.84	214
6. Very satisfied	0.80	0.72	0.76	78
7. Excellent	0.85	0.89	0.87	141
8. Better than expected	0.88	0.91	0.90	278
Accuracy			0.87	1,350
Macro avg	0.85	0.83	0.84	1,350
Weighted avg	0.87	0.87	0.87	1,350

3.9 Results and Discussions

Three algorithms were used to identify the most suitable model out of which gradient boosted decision tree performed the best. A two-step process has been executed as mentioned in the methodology section. In the first step, a predictive model is

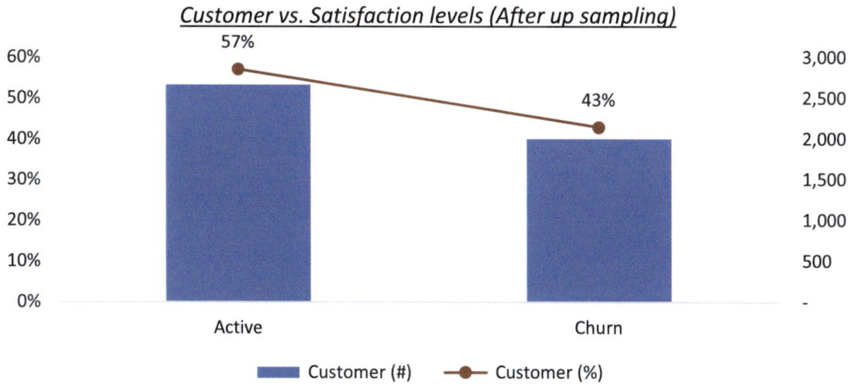

Fig. 3.8 Distribution of churn and class imbalance—after up-sampling

developed with the help of XGBoost algorithms that provides 87.4% accuracy (ref. Fig. 3.9 and Table 3.7). Also, in order to bring out the control mechanism, the factors which are driving the satisfaction levels are also identified. The SHapley Additive exPlanations (SHAP) method is a game theoretic approach that can be used to explain the output of any machine learning model and determine feature importance. In this case, it is applied to identify the features that have the most influence on customer satisfaction. It is observed that age, months_on_book, total transactions, dependent count, transaction frequency, maximum transaction in a month, and avg_balance are the important factors which drive the customer satisfaction (ref. Figs. 3.10, 3.11).

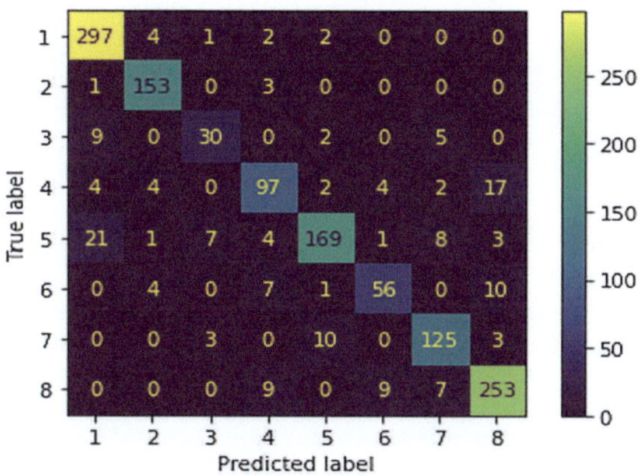

Fig. 3.9 Confusion matrix heatmap showing predicted versus true labels for a classification model. The matrix is color-coded, with darker shades indicating lower values and lighter shades indicating higher values. The diagonal from top left to bottom right shows higher values, indicating correct predictions. The color bar on the right ranges from 0 to 250, with green representing higher values. Key data points include values like 297, 153, and 253 on the diagonal

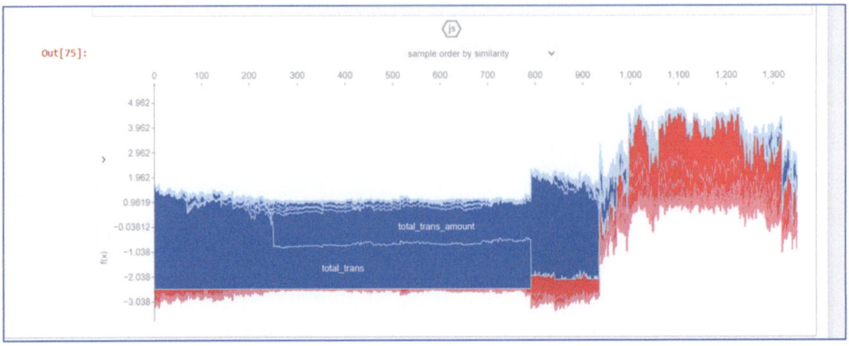

Fig. 3.10 Area chart displaying data over time with two colored sections, blue and pink, representing different data sets. The x-axis shows time intervals, while the y-axis indicates numerical values. The chart highlights fluctuations and trends, with a noticeable increase in values towards the right side. The legend identifies the pink section as "W1"

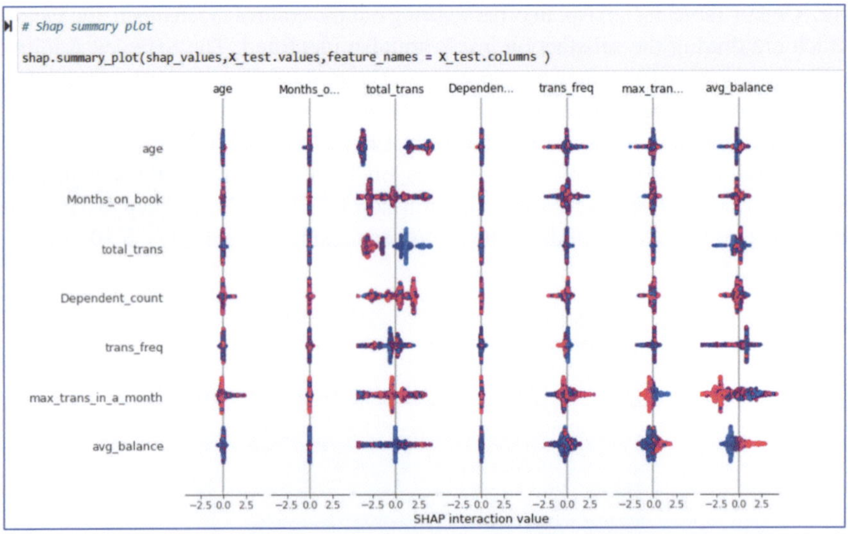

Fig. 3.11 Violin plot displaying the distribution of various variables across different interaction levels. The variables include age, interating_book, total_items, interation1_count, trans_freq, YAC_trans1_U_count, and zero_balance. Each variable is represented by a pair of mirrored plots in blue and pink, indicating different data groups. The x-axis represents interaction levels, while the y-axis shows the variable values. The plot highlights the density and spread of data points for each variable across interaction levels

In the second step, a churn propensity predictive model is developed using XGBoost with an accuracy of 85.48%, and area under the curve (AUC) is 86% (ref. Table 3.8, Figs. 3.12, 3.13). This predictive model provides the churn probability

Table 3.8 Classification report on customer categories showing active and churn customers

Customer category	Precision	Recall	F1-score	Support
Active	0.92	0.91	0.92	1159
Churn	0.49	0.51	0.50	191
Accuracy			0.85	1350
Macro avg	0.70	0.71	0.71	1350
Weighted avg	0.86	0.85	0.86	1350

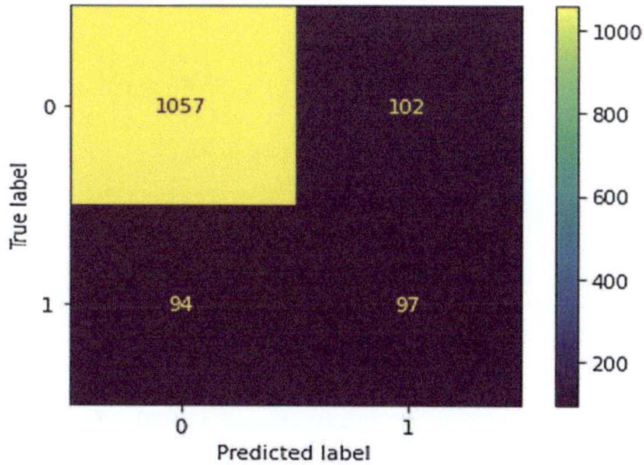

Fig. 3.12 Confusion matrix heatmap showing predicted versus true labels. The matrix is divided into four quadrants: top-left (1057), top-right (102), bottom-left (94), and bottom-right (97). The color gradient ranges from dark purple to yellow, with a scale bar on the right indicating values from 0 to 1000. The x-axis is labeled "Predicted label" and the y-axis is labeled "True label"

which is considered as dissatisfaction score in the methods of customer satisfaction quantification. Since both predictive models have achieved significantly high accuracy and precision, hence the models are recommended for quantification of customer satisfaction.

3.10 Theoretical and Managerial Implications

The benefit of this new intuitive framework extends in multiple directions, offering a range of advantages for businesses. One of the most crucial benefits is its ability to reduce bias by leveraging transactional data. By utilizing objective and accurate information, this framework provides a more reliable basis for evaluating the impact of various marketing campaigns on customer satisfaction. The use of this framework in strategy development for customer management is particularly valuable. It enables businesses to assess the effectiveness of their marketing efforts and make informed decisions on how to optimize their strategies. The ability to estimate customer satisfaction through this framework allows organizations to gain insights into the success or failure of different campaigns, enabling them to refine their approach and allocate resources more effectively. Furthermore, this intuitive framework empowers companies to be proactive in identifying and understanding customer sentiment. By

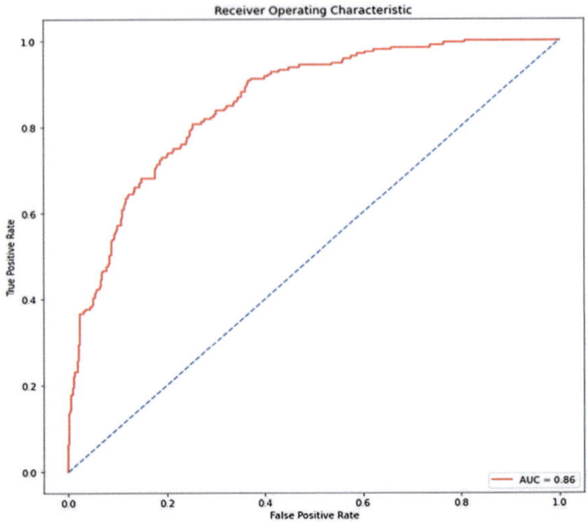

Fig. 3.13 ROC curve with AUC 86%

analyzing transactional data and employing advanced analytics techniques, businesses can gain a deep understanding of how customers feel about their products or services. This not only helps in quantifying sentiment but also aids in predicting future behavior. One significant advantage provided by this framework is its ability to risk rank customers based on their level of satisfaction and engagement. By categorizing customers into different risk segments, organizations can prioritize their efforts toward those who are at higher risk of churn or dissatisfaction. This proactive approach allows companies to address potential issues before they escalate, thus reducing voluntary churn rates significantly. Additionally, by accurately identifying customer sentiment and preferences through this framework, businesses can design personalized value propositions that resonate with individual customers' needs and expectations. This tailored approach enhances overall customer experience while creating a positive impact among consumers who feel understood and valued by the company.

The adoption of this new intuitive framework offers numerous advantages to businesses involved in customer management strategies. By reducing bias through the use of transactional data and estimating customer satisfaction levels accurately, organizations can evaluate marketing campaign impacts effectively during strategy development stages. With proactive identification and quantification of customer sentiment enabled by this framework, companies can confidently prioritize actions based on risk rankings while providing personalized value propositions that prevent voluntary churns and create a lasting positive impact on customers' experiences with wow factors incorporated into each interaction.

3.11 Conclusions and Future Steps

In this study, a framework for customer experience score is developed on the basis of two factors. The first one is customer satisfaction, and other one is customer dissatisfaction identified through customer churn. It is shown that churn probability of a customer can be used as the dissatisfaction index of a customer based on transaction pattern of customers. The framework has also captured how this quantified dissatisfaction can be discounted from satisfaction score to get the absolute customer experience score. The absolute customer experience shows the reliable experience score of a customer, and banks can use it to perform different types of retention programs according to the experience score of customers. This quantification mechanism is extremely important in the context of improvement of customer experience. Improvement of customer satisfaction can be done through understanding of customers' digital inclination, and then through cross-industry collaboration, the customers can be offered various products and services.

As a part of future scope of this research, a framework can be developed that would help to assess and quantify customer digital inclination. However, the quantification of customer experience is essential in order to justify the strategy. Also, the campaign strategy for improvement of customer satisfaction through cross-industry collaboration will require the framework to quantify the customer satisfaction both for test group and control group. The framework can be proposed to any retail bank for possible implementation in view of its unique benefits but rather easy to implement procedure. The framework is able to quantify the customer satisfaction by analyzing the transactional behavior of the customers. There is no separate survey or customer assessment to be conducted by the bank. Hence, for every customer in the bank's portfolio, the customer satisfaction score can be computed. In a similar manner, the churn propensity can also be computed, and the customer experience score generated using the two afore mentioned scores. The customer experience score will be of profound significance. Moreover, this score can be refreshed at a regular interval. The experience score *transition* analysis can also be performed. This will help retail banks to develop various cross-sell and up-sell strategies and plan their promotional campaign by identifying the trend of the portfolio.

References

1. Leninkumar, V.: The relationship between customer satisfaction and customer trust on customer loyalty. Int. J. Acad. Res. Bus. Soc. Sci. **7**(4), 450–465 (2017)
2. Hossain, M., Leo, S.: Customer perception on service quality in retail banking in Middle East: the case of Qatar. Int. J. Islamic Middle East. Finan. Manag. **2**(4), 338–350 (2009)
3. Mittal, V., Frennea, C.: Customer satisfaction: a strategic review and guidelines for managers. In: MSI Fast Forward Series. Marketing Science Institute, Cambridge (2010)
4. Zalatar, W.F.: Quantifying customers' gender effects on service quality perceptions of Philippine commercial banks. Soc. Behav. Sci. **57**, 268–274 (2012)

5. Pasha, M.A., Razashah, M.: Impact of service quality on customer satisfaction: an empirical study in selected public and private sector banks. Res. World **9**(1), 64–73 (2018)
6. White, C.: The impact of motivation on customer satisfaction formation: a self-determination perspective. Eur. J. Market. **49**(11/12), 1923–1940 (2015)
7. Raza, S.A., Umer, A., Qureshi, M.A., Dahri, A.S.: Internet banking service quality, e-customer satisfaction and loyalty: the modified e-SERVQUAL model. TQM J. **32**(6), 1443–1466 (2020)
8. Kaura, V.: Antecedents of customer satisfaction: a study of Indian public and private sector banks. Int. J. Bank Market. **31**(3), 167–186 (2013)
9. Dincer, H., Hacioglu, U.: Performance evaluation with fuzzy VIKOR and AHP method based on customer satisfaction in Turkish banking sector. Kybernetes **42**(7), 1072–1085 (2013)
10. Eren, B.A.: Determinants of customer satisfaction in chatbot use: evidence from a banking application in Turkey. Int. J. Bank Market. **39**(2), 294–311 (2021)
11. Muyeed, M.A.: Customer perception on service quality in retail banking in developing countries: a case study. Int. J. Market. Stud. **4**(1), 116 (2012)
12. Athanassopoulos, A., Gounaris, S., Stathakopoulos, V.: Behavioural responses to customer satisfaction: an empirical study. Eur. J. Market. **35**(5/6), 687–707 (2001)
13. Lee, Y.C., Wang, Y.C., Lu, S.C., Hsieh, Y.F., Chien, C.H., Tsai, S.B., Dong, W.: An empirical research on customer satisfaction study: a consideration of different levels of performance. Springerplus **5**(1), 1577 (2016)
14. Korkutata, A.: Determination of customer satisfaction by staff demographic properties: a research in recreation sports business. J. Educ. Train. Stud. **5**(6), 126–131 (2017)
15. Aimee, R.M.: A thorough literature review of customer satisfaction definition, factors affecting customer satisfaction and measuring satisfaction. Int. J. Adv. Res. **7**(9), 828–843 (2019)
16. Gilbert, G.R., Veloutsou, C.: A cross-industry comparison of customer satisfaction. J. Serv. Market. **20**(5), 298–308 (2006)
17. Rahi, S., Ghani, M.A.: Integration of DeLone and McLean and self-determination theory in internet banking continuance intention context. Int. J. Account. Inform. Manag. **27**(3), 512–528 (2018)
18. Chen, H.-G., Liu, J.Y.-C., Sheu, T.S., Yang, M.-H.: The impact of financial services quality and fairness on customer satisfaction. Manag. Serv. Qual. Int. J. **22**(4), 399–421 (2012)
19. Rahi, S., Ghani, M.A.: Integration of expectation confirmation theory and self-determination theory in internet banking continuance intention. J. Sci. Technol. Policy Manag. **10**(3), 533–550 (2019)

Chapter 4
Challenging the Dynamics of Time: Evaluating Real-World-Based Time Series Data Generation

Anna Valanzano, Luca Strazzera, Smita Gholkar, Kanhaiya Kumar, and Giacomo Veneri

Abstract Estimating accurate nitrogen oxides (NOx) emissions with advanced analysis tools is essential to monitor the pollution and the health condition of a turbomachine. Furthermore, providing advanced analysis tools that are compliant with environmental safety laws is fundamental. While machine learning models show promising solutions in estimating particle pollution levels through virtual sensor modeling of turbomachines, the scarcity and quality of real-world data pose significant challenges. To address these challenges, our study focuses on evaluating the performance of time series generative models on the turbomachinery sensor data generation quality using a comprehensive set of quantitative and qualitative metrics. To this end, we propose a task-specific metric to assess the effectiveness of the generated sensor data in supporting our industrial application. Moreover, we conduct a critical analysis regarding the evaluation metrics of the commonly proposed methods in the literature, for validating and assessing the effectiveness of the generative models. The analysis highlights the limitations of these metrics, as they do not consider the temporal correlations present in time series data, heavily rely on the specific implementation of the evaluation model, and do not provide a robust description of the overall model performance. By delving into those concerns, we believe that this work

A. Valanzano · L. Strazzera · G. Veneri
Baker Hughes, Firenze, Italy
e-mail: anna.valanzano@bakerhughes.com

L. Strazzera
e-mail: luca.strazzera2@bakerhughes.com

G. Veneri
e-mail: giacomo.veneri@bakerhughes.com

S. Gholkar (✉) · K. Kumar
Baker Hughes, Bangalore, India
e-mail: smita.anilgholkar@bakerhughes.com

K. Kumar
e-mail: kanhaiya.kumar@bakerhughes.com

J. Maiti et al. (eds.), *Operations Research and Data Analytics: Current Trends and Future Perspectives*, Lecture Notes on Multidisciplinary Industrial Engineering,
https://doi.org/10.1007/978-981-96-8327-7_4

contributes to the advancement of knowledge on temporal synthetic data generation, stimulating discussion in how it can be reliably evaluated in real-world settings.

4.1 Introduction

Our research originated from an industrial application that aims to estimate nitrogen oxides (NOx) emissions of a turbo machine. However, due to limited data availability, the performance of our regression model based on long short-term memory (LSTM) [1] falls short of the desired accuracy on unseen data. To address this issue, we propose to generate synthetic time series data that, according to our expectation, should increase the generalization ability of trained models by reducing overfitting. In this work, we explore three types of generative approaches based on Gaussian mixture (GM) [2], variational autoencoder (VAE) [1], and generative adversarial network (GAN) [3], and we compare them to a simple baseline model, which creates synthetic data adding random noise to the mean sensor value. We further investigate the metrics generally applied in research for evaluating time series generation [4], and we examine if these metrics can provide insights into the model's ability to consistently generate coherent samples over time, or if at the very least, generated time series exhibit essential characteristics comparable to real-world ones, such as mean and standard deviation. Moreover, we demonstrate that these metrics can be inadequate for evaluating the overall performance of our models, as our investigation revealed that some of these metrics, such as visual and statistical metrics, do not consider the temporal dependencies present in time series data. Additionally, these metrics may not adequately capture significant distinctions between complex models like VAE and GAN, compared to a basic baseline model. On the other hand, other metrics, based on the predictions of classification and regression models, may rely on the specific implementation of the evaluation model [5]. To overcome these limitations, we propose utilizing simpler metrics that offer deeper insights into the preservation of fundamental properties between synthetic and real data as mean, standard deviation, correlation, and autocorrelation. Furthermore, we have introduced task-specific metrics that assess the generated data's capacity to effectively support our industrial tasks. This assessment involves integrating synthetic data into our real dataset and evaluating the performance of the regression model. The main contribution of this paper lies in showcasing a practical application of time series generation within a business context, highlighting the limitations of evaluation metrics proposed in existing literature.

4.2 Methodology

Dataset The dataset comprises 300,000 time ordered data points for each sensor collected over a 2-year period from an operational gas turbine in an industrial environment. There were 35 sensors exploited to monitoring parameters such as temperatures, pressures, and emissions (measured in parts per million). The dataset features are labeled as $x_0,...,x_{34}$, and all values are appropriately normalized. During our analysis, we employ a feature selection technique, specifically the Boruta algorithm [2], to reduce the initial set of 35 features to a more manageable subset of 16 features. We reshape the two-dimensional dataset into a three-dimensional shape using a sliding window approach, where the windows size n is selected through a grid search assessment. At the end each sensor is divided into 24 sliding consecutive data points. We employ a holdout approach to split randomly the data into 80% for training and 20% for testing. While a train–test split is commonly used in machine learning research more broadly, we note that to the best of our knowledge, this division of data has not been a typical practice reported in the time series generation literature [6, 7].

We explore models based on four different architectures for generating synthetic time series data: GM, GAN, VAE, and a simpler baseline model. GM [2] involves representing a time series as a mixture of Gaussian distributions. GAN [1] generates synthetic data by training a generator network able to produce data that are close to the real one while a discriminator network tries to distinguish between real and fake data. VAE [8] uses an encoder to map input data to a latent representation and a decoder to map the latent representation back to the original space, exploiting the generation from the learnt latent space. The baseline model provides a comparison point for evaluating the performance of the more advanced models. It generates synthetic time series data by adding a random noise term to the mean of the sensor value in the training data. The noise term is added with a probability of 10% and has a value between 0 and the standard deviation of the sensor measurement for all the training timestamps. As for GAN and VAE, we selected TimeGAN [9] TimeVAE and [10] implementations, respectively, since at the best of our knowledge they got promising results in literature with time series data generation tasks.

Metrics to assess the effectiveness of our proposed models in generating synthetic time series data, we employed various techniques commonly used in the literature for evaluating the generation [11, 12]. We split them based on common definition of qualitative evaluation, where images are developed, and quantitative evaluation, where numeric scores are provided. About quantitative evaluation, we compute train on synthetic test on real (TSTR) and train on real test on synthetic (TRTS) metrics [13], which involve training a regression model on synthetic data generated by the model and testing it on the real data, and vice versa. The discriminative score (DS) [9] is calculated by training a post-hoc time series classification model to distinguish between sequences from the original and generated datasets. Lastly, the predictive score (PS) [9] is calculated by training a post-hoc sequence-prediction model on the generated data and testing it on the real data to predict next-step temporal vectors over each input sequence. It is important to note that lower values of these metrics (TRTS,

TSTR, DS, and PS) indicate a higher level of similarity between real and synthetic data, as they are directly related to the amount of error the classification/regression models produce. To obtaining further insights, beside the common metrics in literature, we show both the statistical properties of the real and generated data, as well as the LSTM regression score when trained on a new dataset composed of real and generated samples. To expose qualitative results, we picture a visual representation to evaluate the degree to which the synthetic data distribution resembles the original dataset when projected into a two-dimensional space. Specifically, principal component analysis (PCA) [14] and t-distributed stochastic neighbor embedding (t-SNE) [12] were used to project the high-dimensional time series data into two dimensions for visualization and comparison. We are aware that the cited data embedding methods suffer from several problems, and for this reason, we use them just for the qualitative evaluation [9, 15].

4.3 Results

Figure 4.1 presents t-SNE and PCA visualizations for both the real and the synthetic data generated by the baseline (top left), GM (top right), TimeVAE (bottom left), and TimeGAN (bottom right) models. In the visualization, red represents the original data points, while black represents the synthetic data points. According to Fig. 4.1a, the baseline model fails to accurately capture the complexity of the underlying distribution observed in the real data. Instead, it only produced an approximation of the mean value along with some random noise. Upon analyzing Fig. 4.1b–d, the TimeVAE model exhibits notably better results compared to other models, providing its ability to generate data that closest approximate the real data distribution.

We report the quantitative evaluation metrics for the four models in Table 4.1 revealing that different metrics may suggest different models as the best performer, suggesting that no single metric can express the effectiveness and the quality of the time series generator overall.

To point it out, TimeVAE achieves the best results in terms of DS, while TimeGAN achieves the best performance in terms of PS, TSTR, and TRTS instead suggest GM as the best model. Upon evaluating with all the presented metrics, the baseline model's performance overall is comparable to, or only slightly worse than, advanced models. This result is somewhat surprising since the generated data are simply based on the mean of each sensor plus random noise term. Therefore, we can conclude that these metrics can be helpful in measuring specific aspects of the performance of a generative model, but they may not always be reliable indicators neither of the model's overall power nor for its ability to learn temporal dependencies inside the data. Furthermore, we emphasize that the scores of metrics such as PS, DS, TSTR, and TRTS depend on the specific topology of the classification/regression model as well as the initialization of the parameters, and these facts make the metrics even less reliable and less comparable with results reported in other publications.

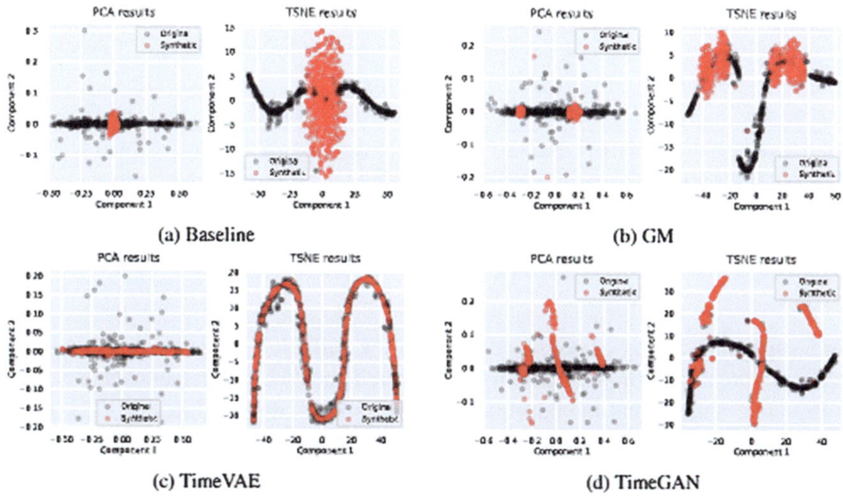

Fig. 4.1 t-SNE and PCA visualization for the baseline (top left), GM (top right), TimeVAE (bottom left), and TimeGAN (bottom right) models. Red denotes original data, and black denotes synthetic

Table 4.1 Comparison of quantitative metrics for the baseline, GM, TimeGAN, and TimeVAE models on the NOx dataset

Model	DS	PS	TSTR	TRTS
Baseline	0.50	0.04	0.68	0.34
GM	0.16	0.04	0.67	0.33
TimeVAE	0.01	0.13	1.65	0.68
TimeGAN	0.44	0.03	1.12	0.49

Best values are highlighted

Statistics to evaluate how well the generative models capture the distribution of the real data, we compare the mean and the confidence interval for four features randomly chosen, namely x_8, x_{10}, x_{11}, and x_{13}. By comparing these statistics, we can assess whether the generative models are able to learn these basic properties of the real data distribution. Specifically in Table 4.2 we report the sample mean with its corresponding 95% confidence interval. According to these results, TimeGAN performs the worst at matching the real data distribution, while the TimeVAE and GM perform similarly to the baseline. Clearly, the baseline looks promising but remember that its definition is chosen to fool evaluation when we exploit statistical metrics.

Autocorrelation represents the degree of similarity between a given time series and a lagged version of itself over successive time intervals. Figure 4.2 presents it for the single feature x_1, as a illustrative example, where can be observed that the autocorrelation values in the real dataset consistently hover around 1 for all lags. The pattern in Fig. 4.2e, observed also for all the other features but not shown

Table 4.2 Comparison of mean and 95% confidence interval for normalized features $x_8, x_{10}, x_{11}, x_{13}$ in the NOx dataset with real and synthetic data generated by: baseline, GM, TimeVAE, and TimeGAN models

Feature real		Baseline	GM	TimeVAE	TimeGAN
x_8	$0.675 \pm 7.4e^{-4}$	$0.671 \pm 2.3e^{-4}$	$0.626 \pm 3.0e^{-4}$	$0.670 \pm 6.8e^{-4}$	$0.660 \pm 7.3e^{-4}$
$\times$ 10	$0.708 \pm 5.8e^{-4}$	$0.708 \pm 1.8e^{-4}$	$0.671 \pm 5.4e^{-4}$	$0.706 \pm 5.6e^{-4}$	$0.712 \pm 5.9e^{-4}$
$\times$ 11	$0.851 \pm 3.7e^{-4}$	$0.850 \pm 1.2e^{-4}$	$0.830 \pm 3.8e^{-4}$	$0.850 \pm 3.6e^{-4}$	$0.854 \pm 3.9e^{-4}$
$\times$ 13	$0.957 \pm 1.2e^{-4}$	$0.956 \pm 3.8e^{-5}$	$0.956 \pm 4.9e^{-5}$	$0.957 \pm 8.0e^{-5}$	$0.960 \pm 9.0e^{-5}$

here, suggests a strong dependence between the variable's values with its own past values, indicating high predictability; however, TimeVAE and TimeGAN, as shown in Fig. 4.2c and d), and the baseline and GM models, depicted in Fig. 4.2a and b, show difficulties in capturing it.

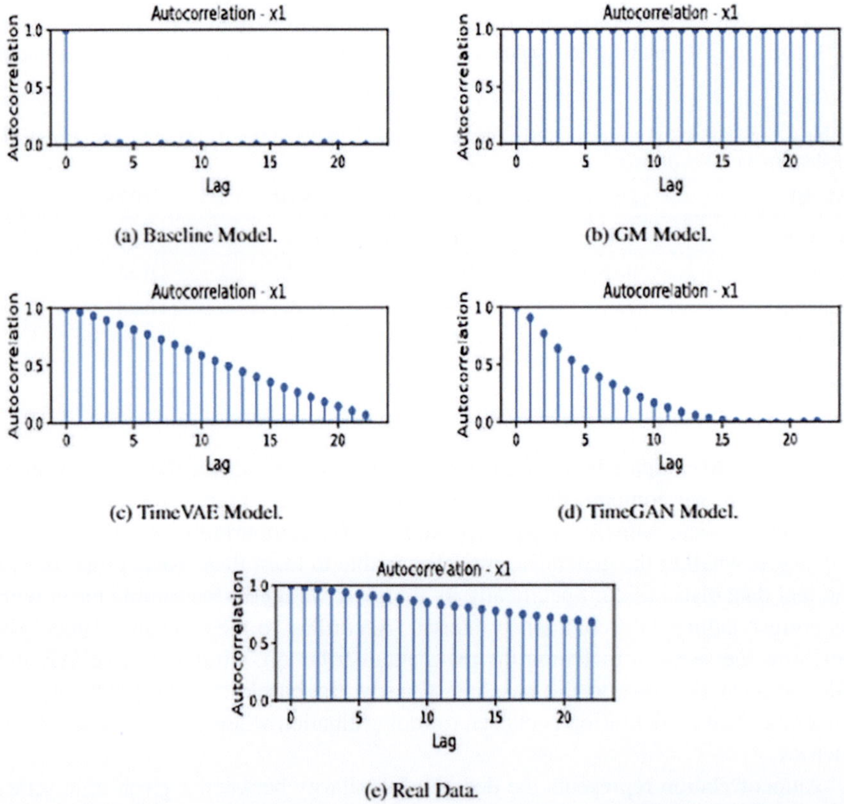

Fig. 4.2 Comparison of autocorrelation values for feature x_1 in real data and in synthetic data generated by the baseline, GM, and TimeGAN models

Table 4.3 Comparison of Pearson correlation coefficients between NOx emissions and variables $x_8, x_{10}, x_{11}, x_{13}$ observed in real data with those in synthetic data generated by the baseline, GM, and TimeGAN models

Variable	Real	Baseline	GM	TimeVAE	TimeGAN
x_8	−0.06	0.00	−0.01	0.01	−0.59
x_{10}	0.75	0.00	0.89	0.87	0.86
x_{11}	0.69	0.00	0.65	0.81	0.85
x_{13}	0.55	0.00	0.97	0.82	0.88

Finally, we analyze the Pearson correlation between NOx emissions and the above-mentioned 4 features, to manifest how well these correlations are preserved in the generated data. In Table 4.3, the correlation values for the data generated by the baseline model are all zero. This indicates the absence of a linear association between the variables being compared, specifically between NOx emissions and the features in the dataset. The lack of significant linear correlation suggests that the baseline model fails to capture and reproduce the same relationships observed in the real data. Therefore, the synthetic data generated by the baseline model does not accurately reflect the underlying correlations present in the original dataset. We can observe that the GM and TimeVAE models exhibit a higher degree of correlation preservation compared to the baseline model. Specifically, the highest correlation values consistently appear in the same three variables as in the real data: x_{10}, x_{11}, and x_{13}.This finding suggests that both the GM and TimeVAE models have a better ability to capture and preserve the underlying relationships between these specific variables and NOx emissions compared to the Baseline model. The fact that the highest correlation values align with those in the real data indicates a stronger fidelity in reproducing the correlations between these variables.

4.3.1 *Task-Specific Metrics*

To address the limitations of the evaluation metrics discussed previously, we decided to move toward a task-related metric that could better capture the performance of the evaluated generative models in our specific business use case. As described in Sect. 4.1, our main goal was to employ a time series generation model to improve actual scores given scarcity of data. We integrate the generated synthetic data into the original NOx dataset and evaluate the resulting LSTM model performance for estimating NOx emissions. To assess that, our task-related metric proposal is the mean absolute percentage error (MAPE) computed with the model trained on the real dataset combined with synthetic data with the expectation that in case the synthetic generated data is coherent, regression score should improve when dataset is combined. For reference, the LSTM trained on real data achieves a MAPE of 0.547 as mean over three attempts. Table 4.4 presents the MAPE for the LSTM model

Table 4.4 MAPE of the LSTM with varying percentages of synthetic data added to real data

Percentage (%)	MAPE			
	Baseline	GM	TimeVAE	TimeGAN
130	0.636	0.541	0.609	0.485
150	0.620	0.573	0.612	0.593
Mean	0.628	0.557	0.611	0.488

Bold results indicate better performance. Resulting values are the mean over three attempts

when increasing percentages of synthetic data generated from different generative models. The percentage refers to the amount of generated data added to the real dataset, assuming 100% it is the size of the real dataset.

The results indicate that augmenting real data with coherent synthetic data can improve MAPE, but the amount of percentage chosen impacts the results. While the baseline model yields no improvement in any case, understandably by the way the data is generated, TimeGAN gives better result only with 30% of data addition. Moreover, adding synthetic data from GM has negligible adverse effects, suggesting that the generated data closely matches the true distribution. As counterpart, TimeVAE looks not able to capturing time dependencies, despite looks promising in other quantitative metrics reported in Table 4.1. Experiments varying the percentage of additional data reveal that excess into adding synthetic data always decreases the performance since consistently better results are obtained adding the 30% instead of the 50%. These findings are coherent with respect to the literature metrics reported in Table 4.1, where no single model emerges as best. Last final open question regards the impact of embedding a good quality synthetic dataset inside our business use case, since an improvement of the regression score suggests that the underlying generator model is able to capture time dependencies, not a quantitative metric which everyone agrees on has been provided in literature yet, at the best of our knowledge.

4.4 Conclusions

We introduce a novel approach for monitoring NOx emissions through time series generative models, addressing the data scarcity challenge commonly encountered in industrial settings. Our study encompasses a comprehensive exploration of various generative techniques, assessing their performance within the context of a specific industrial dataset, chosen for its complexity and criticality. We highlight the limitations of conventional evaluation methods commonly employed in the literature, emphasizing that they may not consistently serve as reliable indicators of overall model performance. We advocate for the incorporation of additional, simpler comparison metrics to assess the fidelity of generated data. These metrics include correlation with NOx emissions, autocorrelation, and fundamental statistical measures such as mean and confidence interval. To cater the specific needs of the industrial applications, we put forth a task-specific regression evaluation metric. This metric underscores the potential of generative models while emphasizing the importance of developing evaluation metrics tailored to the unique requirements of the intended application. Notably, our findings reveal that the TimeGAN model outperforms the baseline but cannot be put on top of other models due to the scarcity of an effective metric to assess the quality of the data generated. This contradicts the outcomes obtained by conventional metrics commonly employed in the literature. In future work, the authors plan to explore the potential of synthetic data to improve the performance of other industrial models, particularly on anomaly detection tasks, and propose new evaluation metrics that can better handle the temporal quality assessment.

References

1. Hochreiter, S., Schmidhuber, J.: Long short-term memory. Neural Comput. **9**, 1735–1780 (1997). https://doi.org/10.1162/neco.1997.9.8.1735
2. Reynolds, D.A.: Gaussian mixture models. In: *Encyclopedia of Biometrics* (2009). https://api.semanticscholar.org/CorpusID:1063711
3. Goodfellow, I.J., Pouget-Abadie, J., Mirza, M., Xu, B., Warde-Farley, D., Ozair, S., Courville, A., Bengio, Y.: Generative adversarial networks (2014)
4. van der Maaten, L., Hinton, G.: Visualizing data using t-sne. J. Mach. Learn. Res. **9**, 2579–2605 (2008)
5. Valanzano, A., Strazzera, L., Lombardi, M., Maggio, A.: Challenging the dynamics of time: generate and evaluate real-world time series to estimate NOx emissions in a turbo machine [Master's Thesis submitted at the Department of Computer Science and Engineering of The University of Bologna]
6. Brophy, E., Wang, Z., She, Q., Ward, T.: Generative adversarial networks in time series: a survey and taxonomy (2021)
7. Brophy, E., Wang, Z., She, Q., Ward, T.: Generative adversarial networks in time series: a systematic literature review. ACM Comput. Surv. **55**(10), 540 (2023). https://doi.org/10.1145/3559540
8. Kingma, D.P., Welling, M.: Auto-encoding variational bayes (2013)
9. Yoon, J., Jarrett, D., van der Schaar, M.: Timeseries generative adversarial networks. In: Wallach, H., Larochelle, H., Beygelzimer, A., d'Alché-Buc, F., Fox, E., Garnett, R. (eds) Advances in Neural Information Processing Systems, vol. 32. Curran Associates, Inc. (2019). https://proceedings.neurips.cc/paper_files/paper/2019/file/c9efe5f26cd17ba6216bbe2a7d26d490-Paper.pdf.
10. Desai, A., Freeman, C., Wang, Z., Beaver, I.: Timevae: a variational auto-encoder for multivariate time series generation (2021)
11. Arnout, H., Bronner, J., Runkler, T.: Evaluation of generative adversarial networks for time series data. In: Proceedings of the 2021 International Joint Conference on Neural Networks (IJCNN), pp. 1–7 (2021). https://doi.org/10.1109/IJCNN52387.2021.9534373
12. Yadav, P., Gaur, M., Fatima, N., Sarwar, S.: Qualitative and quantitative evaluation of multivariate time-series synthetic data generated using mts-tgan: a novel approach. Appl. Sci. **13**(7), 4958 (2023)
13. Esteban, C., Hyland, S.L., Rätsch, G.: Real-valued (medical) time series generation with recurrent conditional gans (2017)
14. Bryant, F., Yarnold, P.: Principal-component analysis and exploratory and confirmatory factor analysis (2001)

15. Elhaik, E.: Why most principal component analyses (pca) in population genetic studies are wrong. BioRXiv **13**, 381 (2021). https://doi.org/10.1101/2021.04.11.439381
16. Kursa, M., Rudnicki, W.: Feature selection with boruta package. J. Stat. Softw. **36**, 1–13 (2010)
17. Yang, J.C.Z., Chen, Y.: T-sne is not optimized to reveal clusters in data (2021)

Chapter 5
Revolutionizing Sales Forecasting in Quick-Service Restaurants Using Hybrid Deep Learning Models

Rony Mitra, Abhiram Ramesh, Narasimha Kamath, and Manoj Kumar Tiwari

Abstract Effective management of inventory is one of the primary concerns for any industry or enterprise in Supply Chain Management. Sales Forecasting is one of the most critical tools in pursuing this for these industries. Researchers have employed a mixture of conventional and deep learning models. However, existing sales forecasting models need more comprehensiveness and flexibility to account for the dynamics and nonlinearities in re-tail-level sales time series. We have implemented a hybrid deep neural network model to overcome those challenges. This methodology allows the visualisation of product sales, promotional events, and weekend or holiday effects. A large volume of sales data from a fast food chain company is utilized to validate our model. This dataset contains multidimensional features—both indigenous to the dataset and de-rived. First, we perform exploratory data analysis to extract meaningful information from the sales data. Then, we infer seasonal effects and the impact of holidays, weekends, and marketing on sales. We applied the Parallel CNN-GRU algorithm, considering all the factors to predict the demand for the fast food chain company and minimizing its loss by reducing wastage. A comparative study is executed with various state-of-the-art sales forecasting techniques to showcase/confirm the significance of our model. We have also analyzed our proposed model's complexity compared with other existing models.

Keywords Sales forecasting · Regression model · Deep learning · Fast food chain

R. Mitra (✉) · M. K. Tiwari
Indian Institute of Management Mumbai, Mumbai, India
e-mail: mitrarony92@gmail.com

M. K. Tiwari
e-mail: mkt09@hotmail.com

A. Ramesh · N. Kamath
Research and Development, o9 Solutions, Bengaluru, India
e-mail: abhiram.ramesh@o9solutions.com

N. Kamath
e-mail: narasimha.kamath@o9solutions.com

J. Maiti et al. (eds.), *Operations Research and Data Analytics: Current Trends and Future Perspectives*, Lecture Notes on Multidisciplinary Industrial Engineering,
https://doi.org/10.1007/978-981-96-8327-7_5

5.1 Introduction

Demand planning and sales forecasting at the stock-keeping unit (SKU) level have been critical in many companies and business strategies and directly affect a company's profits. Especially for a fast food chain business, making accurate forecasting is an important and challenging problem. Companies can develop their capacity planning, production rate, and hiring budgets based on sales forecasts. Accurate sales forecasting reduces the cost of storing inventory, customer waiting time, and probability of stock-outs. Also, it will help the supply chain with material acquisitions and production capability, sales functions allocation planning, and domain sales strategy. Proper sales forecasting grants many advantages to businesses, including compaction of sales channels and risk prediction, reduced time in planning territory coverage and setting quota assignments, alignment of sales quotas and revenue expectations, and improved decision-making about the future. This emerging trend of data analytics, the rise of advanced tools, and broadly accessible sales data across multiple entities such as regions, domains, stores, items, manufacturers, distributors, and customers give new opportunities for improving product sales. Parallelly, it is challenging to identify the valuable data among the large volume and variety of collected data. One of the critical challenges for a fast food chain company is forecasting sales of multiple products across multiple stores. These multiple-product sales in different stores are affected by various factors, such as domain, selling price, purchasing behavior, seasonality, expected market conditions, promotional activities, festivals, and collaboration. How these factors impact product sales and what are the scopes to incorporate those details to enhance large-scale multidimensional sales forecasting are still challenging.

Advanced technology tools and historical data quality is crucial for accurate sales forecasting. Inaccurate data makes forecasting practices incomprehensible. One potential solution is leveraging sales data from regions, stores, and products. Earlier researchers have developed many traditional time series forecasting models based on statistical methods and machine learning algorithms for sales forecasting. Tehrani and Ahrens [1] used big data analytics for the fashion industry and applied the K-means clustering algorithm for sales prediction. Ma et al. [2] introduced a multistage LASSO regression model fitted with Granger causality for retail sales forecasting. Kechyn et al. [3] used WaveNet CNN model for a Grocery Sales Forecasting. The least absolute shrinkage and selection operator (LASSO) produced dependable forecasts for high-dimensional problems.

Sagaert et al. [4] applied multiple LASSO models to a supplier in the tire industry for sales forecasting. Loureiro et al. [5] applied deep NN to forecast sales and support purchasing operations in the fashion industry. Liang et al. [6] applied XGboost and LightGBM algorithms for Product marketing prediction. They showed that parameter tuning plays an essential role, and after adjusting the parameters, there is no significant difference based on accuracy and computational complexity between these two algorithms. Ganesan et al. [7] used Dynamic Artificial Neural Networks for food

Table 5.1 State-of-art methods for sales forecasting across various industries

Method	Industries	Year
k-means clustering	Fashion	2016
LASSO	Retail	2016
WaveNet CNN	Grocery	2018
LASSO, linear regression	Tire industry	2018
Deep NN	Retail	2018
XGboost and LightGBM	Product marketing	2019
Dynamic ANN	Food	2019
Stacking (NN, LASSO, RF, ARIMA, ExtraTrees)	Rossmann store	2019
CNN	E-commerce	2020
Deep-embedded LSTM	Pharmacies	2020
CNN-based meta-learner	Retail	2021
Aggregated LightGBM	Steel consumption	2022
Parallel CNN-GRU	Fast food chain	This study

sales forecasting in a Multiplex and helped them to reduce food waste by 170 units per day.

Pavlyshenko [8] introduced a stacking approach for building ML models and applied it to Rossmann Store Sales data. After that, the convolutional neural network is used to mine e-commerce data to predict commodity sales [9]. Kraus et al. [10] applied a deep-embedded LSTM algorithm for demand prediction for a Pharmaceutical Industry considering 1113 pharmacies. Ma and Fildes [11] presented a meta-learning framework based on deep CNN for retail sales forecasting at stores' SKU levels. Balamwar et al. [12] introduced an aggregated LightGBM algorithm for steel consumption prediction for a metal-forming industry (Table 5.1).

In this study, we introduce an advanced real-time sales forecasting model based on deep learning algorithms considering store-specific seasonality and product category-specific seasonality, holiday and weekend influences, regional impacts, product hierarchy levels, and marketing effects to capture the dynamic behavior of the product's sales.

5.2 Related Work and Proposed Methods

Today when every sector requires a practical and accurate insight into the future, forecasting methods have gained much importance. Forecasting helps to estimate the future behavior of the variables. These variables range from sales to price, or supply [13].

A. **Fully Connected Neural Networks**

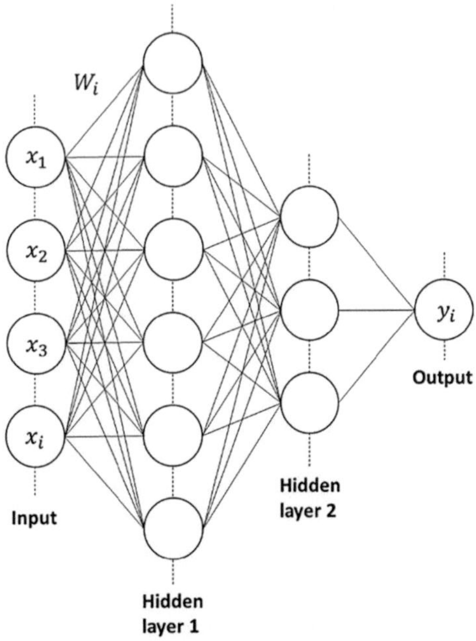

Fig. 5.1 Fully connected neural network

A Fully Connected Neural Network [14] is made up of several layers, each of which connects every neuron in one layer to every other layer, as shown in Fig. 5.1. A Neural Network consists of the input layer, hidden layers, and an output layer. Let $x_i \in \mathbb{R}$ be the i-th input and $y_i \in \mathbb{R}$ be the i-th output from the fully connected layer. Then y_i is computed as follows:

$$y_i = f(W_i x_i + b_i) \tag{5.1}$$

where f is an activation function like ReLU, Softmax, linear or sigmoid. W_i and b_i are the weight vector and bias for the the i-th layer. In each iteration, the neuron calculates a weightage average of the values of the vector x.

B. **Convolutional Neural Network**

A Convolutional Neural Network (CNN, or ConvNet) [15] is a class of artificial neural network (ANN). The ability to handle data with different dimensionalities and extract the relevant features makes it extremely useful in various areas like image processing and classification. In place of general matrix multiplication, CNNs use a mathematical operation called convolution in at least one of their layers. Natural Language Processing and Load Forecasting are the domains where CNN has become quite popular in recent years.

$$Y(i) = \sum_{i,j} x(i-j)w(j) \tag{5.2}$$

Y and x are the output and the input respectively and w represents the kernel. Convolutional neural networks use a filter to apply to an input to produce a feature map that lists the existence of features that were recognised in the input. CNNs architecture comprises convolution layers to extract the best features of, pooling layer for down sampling the input data. Finally, a fully connected layer is employed for the final prediction.

C. **Long Short-Term Memory**

Long Short-Term Memory (LSTM) [16] is an extension for recurrent neural networks which solves the problem of short-term memory. LSTMs were created in the 1990's to solve the problem of vanishing gradients. LSTM uses a mechanism called gates to learn long-term dependencies. Data is either deleted, filtered, or added for the next cells using three different gates i.e., Input gate, Output gate and Forget gate as shown in the Fig. 5.2. Each gate yields a number between 0 and 1 using a sigmoid unit.

$$\text{Forget gate}: f_t = \sigma\left(W_f\left[h_{t-1}, x_t\right] + b_f\right) \tag{5.3}$$

$$\text{Input gate}: i_t = \sigma\left(W_i\left[h_{t-1}, x_t\right] + b_i\right) \tag{5.4}$$

$$\text{Output gate}: o_t = \sigma\left(W_o\left[h_{t-1}, x_t\right] + b_o\right) \tag{5.5}$$

$$\text{Process input}: I_t = \tanh\left(W_C\left[h_{t-1}, x_t\right] + b_C\right) \tag{5.6}$$

$$\text{Cell update}: C_t = \sigma(f_t \times C_{t-1} + i_t \times I_t) \tag{5.7}$$

$$\text{Output}: h_t = \tanh(C_t) \times o_t \tag{5.8}$$

D. **Gates Recurrent Unit**

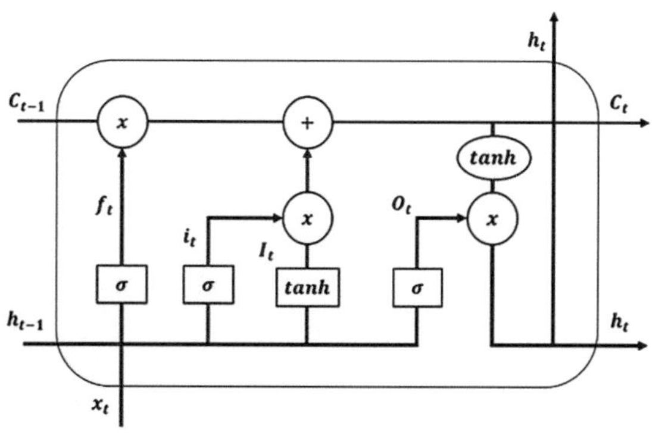

Fig. 5.2 Long short-term memory

The Gates Recurrent Unit Recurrent Neural Network [17] is similar to LSTM but with only two gates (i) an update gate z and (ii) a reset gate r. Hence it is easier to compute and implement. The function of reset gates is similar to forget gates of LSTM. We won't go into detail about the formula because GRU NNs are similar to LSTM NNs in various ways.

E. **CNN-LSTM**

The CNN output serves as the LSTM input in the sequential CNN-LSTM model. Although the forecasting results are accurate, the LSTM's training output is affected by the features derived from CNN layers, and its capacity to extract temporal information is not effectively utilized. The parallel CNN-LSTM model, on the other hand, uses LSTM to analyse time series data for the purpose of detecting temporal features, while CNN is employed for feature extraction that conveys spatial locality.

F. **Parallel CNN-GRU**

The CNN layers are designed to extract relevant features and filter out the noise from the input data while the GRU layers have the ability to extract sequence patterns. Previous researchers have used the CNN-GRU model [18] where the features of the input data are extracted using CNN, and the data is then fed into the GRU model. However, the issue is that the GRU's training is impacted by CNN features that were retrieved. A hybrid model comprising CNN and GRU layers is utilized to overcome the issue, in which input signals are first fed into two paths to be processed separately by GRU and CNN. These two layers then extract the data's features and time dependencies and prepare the input data for the final prediction. Outputs from both layers are concatenated and fed into the final dense layer. Figure 5.3 shows the working of Parallel CNN-GRU.

Fig. 5.3 Parallel CNN-GRU framework

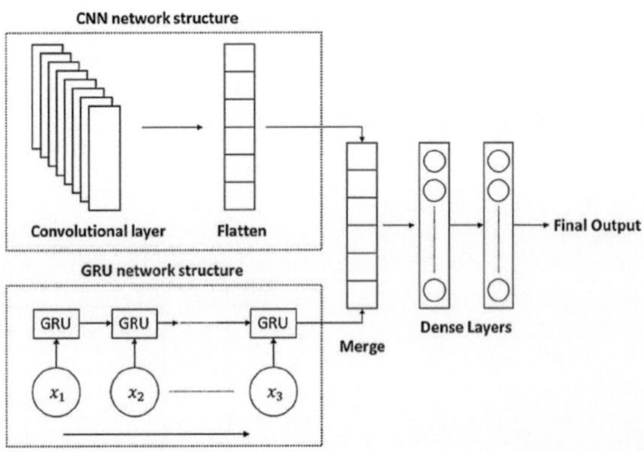

5.3 Data Modelling

A. Dataset Description

The dataset contains the sellout data of the 3866 items in 50 sales domains for 270 weeks (~5 years and 2 months). There are primarily four features that represent the sales data, i.e., Sold to whom (Sales domain), the Start date of the week (Week), Item name (Planning Item), and Sellout values (Sell out Stat).

This data is augmented with other features representing sales domain details like state (region), city, customer groups, and distribution channels. Features like Item levels (L1, L2, L3, L5), Promotion, and Initiative details were also included in the dataset. Finally, external factors like covid confirmed cases, vaccination status, disposable income, and holidays were included in this dataset. Due to the complexity of the data, it is difficult to accurately predict the sales of various products and make optimum decisions. However, effective feature engineering and advanced machine learning algorithms boosted prediction accuracy and speed.

B. Exploratory Data Analysis

Figure 5.4 shows the percentages of each sales domain in ascending order of their total sales across the timeline. It was observed that all 50 domains have a significant demand percentage. The highest demand percentage is 3.5%, while the lowest is 0.25%. As part of this domain analysis, the relation among Sales domains with distribution channels, state (region), and customer groups was studied using treemap. From Fig. 5.5, it was inferred that a hierarchical relationship existed amongst the mentioned features where sales domains were a subset of customer groups, which in turn were a subset of state (region). Finally, distribution channels were the superset of all the mentioned features. This insight reinforced the notion of exploring the data with tree-based algorithms to predict the demands.

Figure 5.6 shows the distribution of each item with respect to their availability for sale (here, referred to as time-step; in the number of weeks). It was observed that 2661 items were available for sale for only 39 weeks. Further, the only prominent time-step observed was of 270 weeks, which had 341 items available for sale for the mentioned time period. We augmented this analysis with the demand distribution

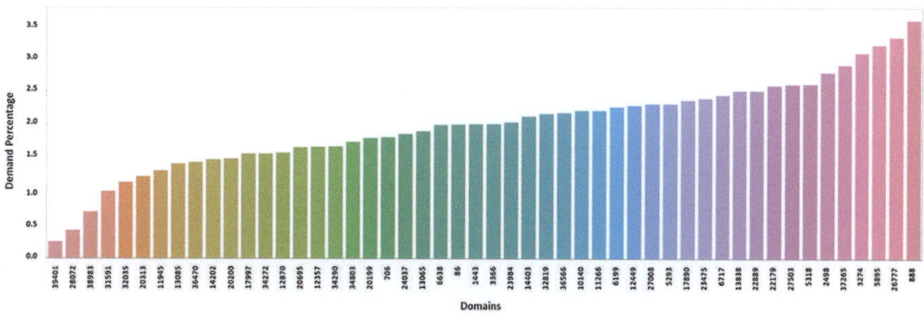

Fig. 5.4 Variation of demand percentage with respect to domains

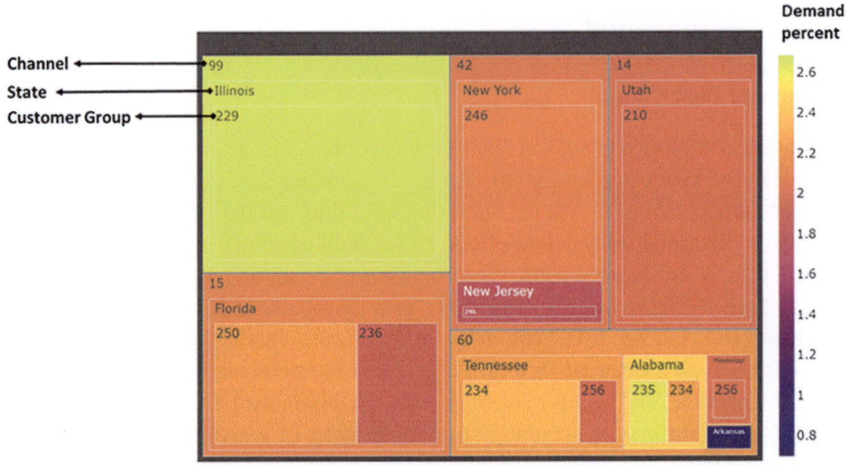

Fig. 5.5 Treemap describing the hierarchy between channels, states, and customer groups

of the respective time-steps to draw relevant insights and found that 95% of the total demand share is captured by these 341 items, which were available for sale for 270 weeks. This insight was also reinforced by the observation from normalized sales shown in Fig. 5.7. Here, the sales were normalized by taking the ratio of the total sale in that time-step by the product of time-step duration and the number of items in the time-step. This exploration was succeeded by Item analysis. As shown in Fig. 5.8, a hierarchical relationship was established among the different item levels by treemap, which implied L1 (level 1) being a subset of L2, L2 a subset of L3, and L3 a subset of L5. The cumulative demand was studied with respect to each item in ascending order of their sales percentage. It was inferred that a large number of items had negligible demands. The data was further analyzed according to the items, and we inferred many important insights from Fig. 5.9, which shows the pie chart of the distribution of items in accordance with their demand share. It was observed from this figure that 90% of the total demand is shared by only 116 items. Therefore, this insight helped in dramatically reducing the number of items needed for further analysis for better accuracy. In the beginning, we focused on the 341 items, which shared 95% of the demand share, and then to improve results further, we narrowed our analysis to the 116 items, which captured 90% of the demand share. This methodology has helped us with less uncertain and better predictions for our analysis.

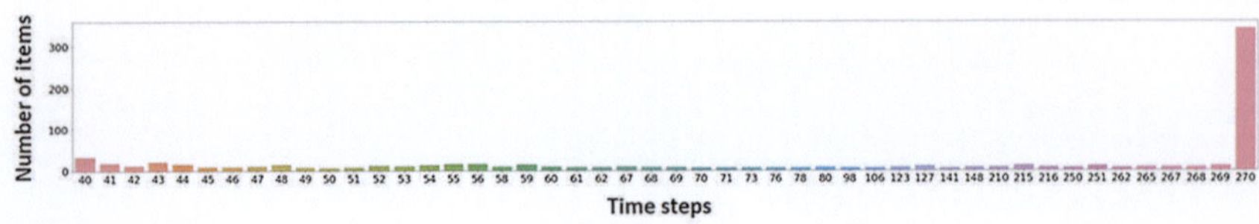

Fig. 5.6 Number of items in different time steps

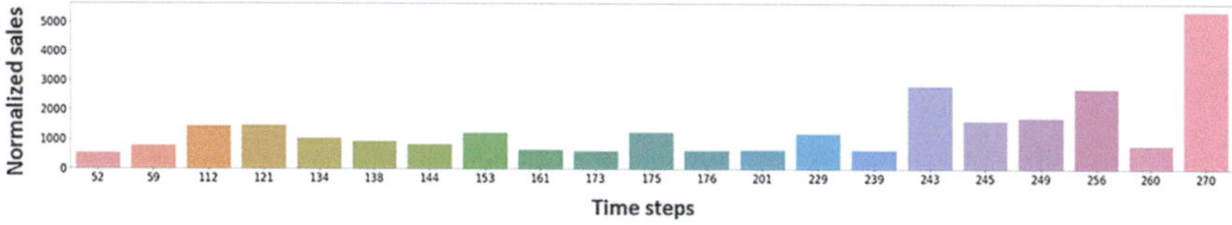

Fig. 5.7 Normalized sales in different time steps

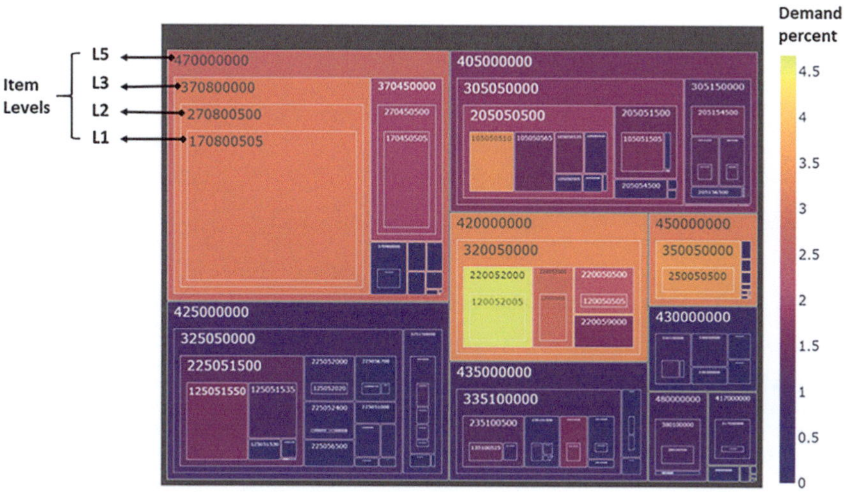

Fig. 5.8 Treemap describing hierarchy in item levels

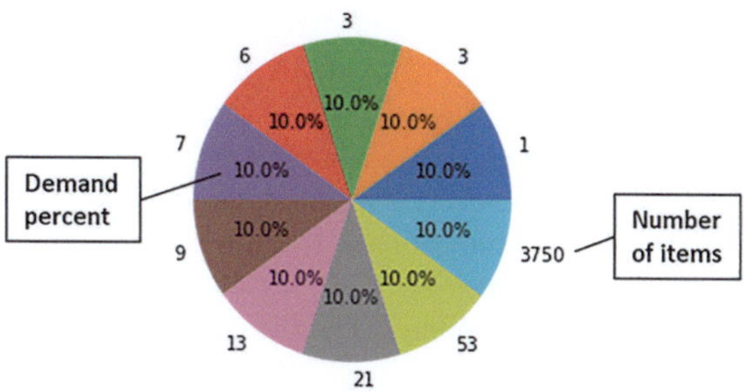

Fig. 5.9 Distribution of number of items according to the demand percentage

Figures 5.10 and 5.11 compare the sales of all the newly segregated items with the sales of all items. This is done to empirically check if there are any significant changes in the demand distribution across the entire time period of 270 weeks after removing the discarded items. As we can see from the above figures, the graphs show the same characteristics without any significant deviation, so we continued performing the test on the newly segregated 341 items.

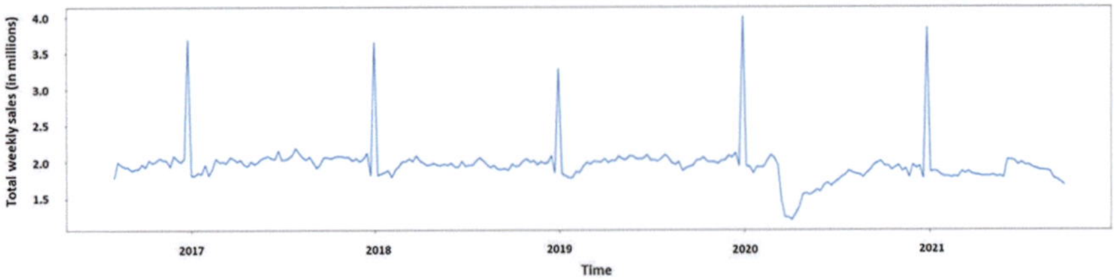

Fig. 5.10 Weekly Sales distribution of all items over the entire timeline

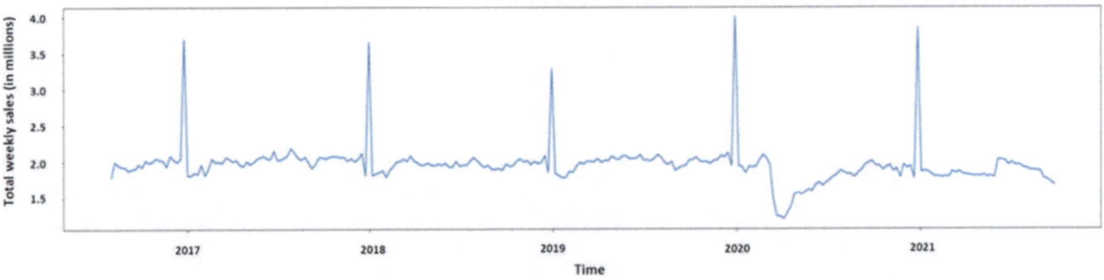

Fig. 5.11 Weekly Sales distribution of all newly segregated items over the entire timeline

5.4 Methodology and Experimental Setting

A. SARIMA and Exponential Smoothing

SARIMA includes 4 seasonal elements that are not part of ARIMA: P—Seasonal autoregressive order, D—seasonal difference order, Q—seasonal moving average order, S—number of time steps for a single seasonal period.

$$\text{ARIMA}(p, d, q) \times (P, D, Q)S \tag{5.9}$$

The S parameter has an impact on the P, D, and Q. parameters. For monthly data, a S of 12 indicates a yearlong seasonal cycle. $P = 1$ would utilize the model's initial seasonally offset observation, such as $t - (S \times 1)$ or $t - 12$. In our model, the corresponding values for P, D, and Q used are 2,1 and 0 respectively. Along with SARIMA we also applied exponential smoothing with a smoothing factor of 0.2.

B. Fully Connected Neural Network

A Fully Connected Neural Network has been used. The network has four hidden layers. The first layer is the input layer and the hidden layers consist of one flattened layer and three dense layers. ADAM optimizer has been used to learn the parameters in 100 epochs with a batch size of 32. ReLU activation function has been used.

C. Convolutional Neural Network

The Convolutional Neural Network (CNN) model uses three layers one convolutional layer and two dense layers. In layer 1, $1 - D$ CNN with ReLU activation function and filters = 32 and kernel size = 4 is implemented. After those two dense layers have been used to give the final output.

D. **Gated Recurrent Unit and Long Short-Term Memory**

The Gated Recurrent Unit (GRU) model studies the last 4 weeks' data to predict the next-week sales. It has two GRU layers and two dense layers. The activation function used is ReLU. The model is trained by ADAM optimizer in 100 epochs. ReLU (Rectified Linear Unit) acts like a linear function that outputs zero for negative input. Similarly, an LSTM has also been applied with the same features used in GRU.

E. **Sequential CNN-LSTM**

As discussed above, CNNs are widely used for feature extraction and LSTM to process time series data. CNN layer consists of one Conv1D layer with kernel size 4 and a number of filters 32. The activation used in ReLU. Then an LSTM layer has been applied with 32 units and ADAM optimizer followed by a dense layer.

F. **Parallel CNN-LSTM**

A Parallel CNN-LSTM model has been applied. One Conv1D layer with kernel size 4. and number of filters 32 and parallelly an LSTM layer with 32 units and ADAM optimizer has been applied. Then the concatenated output is passed to a dense layer with 32 units and 1 unit.

G. **Parallel CNN-GRU**

Another hybrid model of CNN and GRU has been applied where data is fed into both layers separately without any interaction between the two paths. In GRU, path input is fed into the GRU layer with 32 units and ADAM optimizer. Parallelly, input is sent to the CNN path with one Conv1D layer with kernel size 4 and 32 filters followed by a flattened layer. Then the output from both paths is concatenated and fed into a dense layer with 32 units followed by another dense layer of 1 unit.

5.5 Results and Analysis

The model was trained on the dataset from July 2016 to February 2021 (243 weeks). The comparison of actual data from March 2021 to September 2021 (27 weeks of data for testing) was done and the forecast efficiency was checked using rooted mean square error (RMSE), Mean Absolute Error (MAE), and Weighted Absolute Percentage Error (WAPE).

- The RMSE i.e., the Root Mean Squared error is the square root of Mean Squared Error. RMSE is not scaled to the demand and is given by,

$$\text{RMSE} = \sqrt{\left(\frac{1}{n}\right)\sum_{i=1}^{N}(A_i - F_i)^2} \tag{5.10}$$

- The Mean Absolute Error (MAE) is an excellent metric for determining forecast accuracy. It is the mean of the absolute error, as the name indicates. Mean Absolute Error (MAE) is given by,

$$\text{MAE} = \left(\frac{1}{n}\right)\sum_{i=1}^{N}|A_i - F_i| \tag{5.11}$$

- The Weighted Absolute Percentage Error is the sum of the absolute error normalized by the total demand. The WAPE penalizes both under-forecasting and over-forecasting equally, and neither situation is preferred. The absolute error is calculated using the average (mean) predicted value of projections. WAPE is given by,

$$\text{WAPE} = \frac{\sum_{i=1}^{N}|A_i - F_i|}{\sum_{i=1}^{N}A_i} \tag{5.12}$$

Every model has certain advantages as well as certain disadvantages. After multiple applications of each model, the average errors for every model are calculated and compiled. The results of the experiments are presented in the table given below. It can be observed from the table that the MAE, RMSE, and WAPE of our Dataset are the smallest for the Parallel CNN-GRU Model. For the parallel CNN-GRU Model, the WAPE is 0.094, RMSE is 1582.92, and MAE is 1303.72. It is followed by the GRU for which WAPE is 0.103, RMSE is 1744.44, and MAE is 1475.38. For the CNN Model, WAPE is 0.112, RMSE is 1831.66, and MAE is 1536.66.

The Mean Absolute Error and Root Mean Squared Error follow a similar order to the WAPE. Overall Parallel GRU-CNN model is forecasted most accurately, while the SARIMA Model is the least accurate. According to the error values of different models, the accuracy of the used models in the decreasing order follows the following order: Parallel CNN-GRU, GRU, Parallel CNN-LSTM, Fully Connected Neural Network, CNN, LSTM, Exponential Smoothing, Sequential CNN-LSTM, and SARIMAX.

Table 5.2 shows that the CNN and GRU models performed well, but the Parallel CNN-GRU Model proved better than the individual implementation of both the hybrid models. The performance of the Parallel CNN-GRU is the best among all the models, with RMSE 1582.922, MAE 1303.725, and WAPE 0.09434. Therefore, the Parallel CNN-GRU Model proved the best for sales forecasting with multidimensional sales data.

Table 5.2 Results

Model	RMSE	MAE	WAPE
SARIMA	2602.706	2302.952	0.19697
Exponential smoothing	1779.258	1505.490	0.13214
Fully connected neural network	1731.045	1438.928	0.11247
CNN	1831.661	1536.666	0.11283
LSTM	1959.832	1689.350	0.12322
GRU	1744.440	1475.381	0.10306
Sequential CNN-LSTM	2479.874	2192.821	0.13803
Parallel CNN-LSTM	1820.556	1524.888	0.10622
Parallel CNN-GRU	1582.922	1303.725	0.09434

5.6 Summary and Conclusion

This paper demonstrates various methods to forecast future sales of a fast food chain company and proposes a parallel CNN-GRU model. Firstly, the feature engineering on the data has been described, and then various methods have been used, which involve statistical, deep learning, and hybrid models. Statistical models work better with predictable sales and stable data, whereas deep learning models work better for unstable and unpredictable sales. The hybrid models are implemented to acquire the features of the various individual models like CNN and LSTM. The hybrid models developed are then tested on the fast food chain's Sales data, and results show that the hybrid model of CNN and LSTM in two different paths outperform the individual models, as the MAE, RMSE, and WAPE of the hybrid model were found to be less than the CNN and the LSTM. If, however, the features extracted by the CNN model are used by the LSTM model, that might also affect the training of the LSTM model and result in poor accuracy. To rectify the above problem, the models followed different paths such that there is no relation between the two. Ultimately, to get more precise sales forecasting, a hybrid model of CNN and GRU is developed, which performs better than the previous models. The CNN model can capture the features of the data, and the GRU model can capture the long-term dependency in the data. Given the data in the structured format with multiple hierarchies, the proposed parallel CNN-GRU model proved to be the best for sales forecasting of the items.

Acknowledgements This research work has been supported by the Indian Institute of Management (IIM) Mumbai and o9 Solutions, Inc.

References

1. Tehrani, A.F., Ahrens, D.: Improved forecasting and purchasing of fashion products based on the use of big data techniques. Supply Manag. Res. **14**, 293–312 (2016)
2. Ma, S., Fildes, R., Huang, T.: Demand forecasting with high dimensional data: the case of SKU retail sales forecasting with intra-and inter-category promotional information. Eur. J. Oper. Res. **249**(1), 245–257 (2016)
3. Kechyn, G., Yu, L., Zang, Y., Kechyn, S.: Sales forecasting using WaveNet within the framework of the Kaggle competition. arXiv preprint arXiv:1803.04037 (2018)
4. Sagaert, Y.R., Aghezzaf, E.-H., Kourentzes, N., Desmet, B.: Temporal big data for tactical sales forecasting in the tire industry. Interfaces **48**(2), 121–129 (2018)
5. Loureiro, A.L.D., Migueis, V.L., da Silva, L.F.M.: Exploring the use of deep neural networks for sales forecasting in fashion retail. Decis. Supp. Syst. **114**, 81–93 (2018)
6. Liang, Y., Wu, J., Wang, W., Cao, Y., Zhong, B., Chen, Z., Li, Z.: Product marketing prediction based on XGboost and LightGBM algorithm. In: Proceedings of the 2nd International Conference on Artificial Intelligence and Pattern Recognition (2019)
7. Adithya Ganesan, V., Divi, S., Moudhgalya, N.B., Sriharsha, U., Vijayaraghavan, V.: Forecasting food sales in a multiplex using dynamic artificial neural networks. Sci. Inform. Confer. **35**, 69–80 (2019)
8. Pavlyshenko, B.M.: Machine-learning models for sales time series forecasting. Data **4**, 5485 (2019)
9. Pan, H., Zhou, H.: Study on convolutional neural network and its application in data mining and sales forecasting for E-commerce. Electr. Commer. Res. **20**(2), 297–320 (2020)
10. Kraus, M., Feuerriegel, S., Oztekin, A.: Deep learning in business analytics and operations research: models, applications and managerial implications. Eur. J. Oper. Res. **281**(3), 628–641 (2020)
11. Ma, S., Fildes, R.: Retail sales forecasting with meta-learning. Eur. J. Oper. Res. **288**(1), 111–128 (2021)
12. Balamwar, A., Mitra, R., Tiwari, M.K., Verma, P.: Prediction and analysis of seasonal dynamic metal consumption using aggregated LightGBM: a case study. IFAC PapersOnLine **55**(10), 725–730 (2022)
13. Tsay, R.S.: Time series and forecasting: brief history and future research. J. Am. Stat. Assoc. **95**, 638–643 (2000)
14. Tonello, A.M., Letizia, N.A., Righini, D., Marcuzzi, F.: Machine learning tips and tricks for power line communications. IEEE Access **7**, 82434–82452 (2019)
15. Gu, J., Wang, Z., Kuen, J., Ma, L., Shahroudy, A., Shuai, B., Liu, T., Wang, X., Wang, G., Cai, J.: Recent advances in convolutional neural networks. Pattern Recogn. **77**, 354–377 (2018)
16. Hochreiter, S., Schmidhuber, J.: Long short-term memory. Neural Comput. **9**, 1735–1780 (1997)
17. Kim, J., Kim, H.: Classification performance using gated recurrent unit recurrent neural network on energy disaggregation. In: Proceedings of the 2016 International Conference on Machine Learning and Cybernetics (ICMLC) (2016)
18. Li, C., Li, G., Wang, K., Han, B.: A multi-energy load forecasting method based on parallel architecture CNN-GRU and transfer learning for data deficient integrated energy systems. Energy **259**, 1134 (2022)

Part II
Operations Research in Transportation and Logistics

Chapter 6
Air Cargo Transportation Considering Load Compatibility Issues

Deepak Kumar Kushwaha, Arnav Aakash, and Goutam Sen

Abstract In emergency scenarios, such as natural disasters or public health crises, the prompt and effective transportation of multiple loads to different destinations emerges as a crucial element of a well-coordinated response. This study delves into the logistical challenges of transportation planning, addressing the compatibility of various load types among themselves, and with the aircraft in a one-to-many and many-to-one transportation network (1-M/M-1). The investigation encompasses fleet selection, cargo loading, routing, and the efficient utilization of available cargo aircraft. The proposed model finds application in scenarios like military movements during crises and humanitarian missions, offering a strategic approach to plan the transportation of sensitive cargo, equipment, and supplies. The paper outlines vital facets of emergency transportation, supported by illustrative examples and the computational performance of the solver on a smaller-scale problem. Through a comparative analysis of load compatibility issues, the study provides valuable insights into how the model can optimize the logistics of emergency response efforts, ensuring the timely and efficient delivery of essential resources to critical locations.

Keywords Cargo compatibility · Pickup-delivery · Transportation · Branch and cut

---

D. K. Kushwaha (✉) · G. Sen
Department of Industrial and Systems Engineering, IIT Kharagpur, Kharagpur, India
e-mail: deepak78888@iitkgp.ac.in

G. Sen
e-mail: gsen@iem.iitkgp.ac.in

A. Aakash
Department of Chemical Engineering, IIT Bombay, Mumbai, India

J. Maiti et al. (eds.), *Operations Research and Data Analytics: Current Trends and Future Perspectives*, Lecture Notes on Multidisciplinary Industrial Engineering,
https://doi.org/10.1007/978-981-96-8327-7_6

6.1 Introduction

Effective logistics management becomes a crucial necessity in times of crisis, addressing various situations such as natural disasters, public health emergencies, conflicts, and humanitarian aid efforts. Successfully navigating the complexities of facilitating a smooth distribution of relief supplies amid limited resources and evolving demands requires a sophisticated comprehension of logistical intricacies [1–3]. A significant obstacle in these circumstances lies in efficiently organizing and enhancing the collection and delivery of aid, especially when dealing with multiple depots and vehicles operating under capacity constraints [4–6].

In aviation, a thorough evaluation of load compatibility, encompassing the cargo's internal compatibility and suitability with the aircraft type, is a crucial determinant in fleet selection and assignment during an emergency where mission completion time is vital. The compatibility of load types within themselves refers to the harmonious arrangement and combination of different cargo items within an aircraft [7]. Group incompatibility involves restrictions or rules that apply to combinations within specified groups of items or entities. For instance, an aircraft may need to transport a mix of perishable medical supplies, equipment, and personnel in medical emergencies. Ensuring these loads are incompatible regarding weight distribution, temperature sensitivity, and spatial constraints is crucial. An illustrative example could involve organizing temperature-sensitive medical supplies in a way that does not compromise the integrity of other essential equipment during transport. In aviation, it is crucial to address group incompatibility to safeguard against cargo transportation with diverse requirements or attributes in a manner that could jeopardize the quality or integrity of the goods. Up-holding conditions specific to each group aid in averting spoilage, damage, or malfunction of delicate cargo items during air transit. Moreover, it facilitates effective planning and utilization of cargo space onboard the aircraft, thereby enhancing the efficiency of the transportation process.

Secondly, ensuring the compatibility of loads with the aircraft type entails aligning the attributes of the cargo with the specifications and capabilities of the selected aircraft. For instance, prioritizing an aircraft with adequate payload capacity is crucial in responding to a disaster relief operation necessitating the transportation of heavy machinery and construction materials. An illustrative scenario could involve selecting a cargo plane equipped to efficiently manage the weight and volume of construction materials, thereby ensuring the secure and efficient delivery of resources to the disaster-affected area.

The work presented is the extension of the study on the emergency transportation planning problem studied by Kushwaha et al. [8] in the 1-M/M-1 transportation network with the consideration of cargo and aircraft compatibility issues for fleet selection and route planning. The research examines how relief agencies can best use air transport to meet transportation requirements during emergencies, considering the intricate interplay between load in-compatibility and aircraft specifications during fleet selection, especially in the 1-M/M-1 transportation network.

6.2 Literature Survey

The vehicle routing problem stands as a pivotal aspect within the realm of transportation and logistics. In recent years, there has been a notable surge in contributions to the combination of routing and loading predicaments, underscoring the increasing attention accorded to this area. The incorporation of loading constraints into vehicle routing constitutes a substantial focus in research [9]. This issue spans diverse transportation modes, including road, maritime, and air, alongside automated guided vehicles, emphasizing the intricate and vital nature of proficient route planning in logistics and the substantial challenges associated with real-world scenarios.

The investigation conducted by Bortfeldt and Wascher [10] extensively categorizes loading constraints, encompassing container-related, item-related, and cargo or load-related constraints. However, there appears to be a scarcity of literature addressing the integration of loading constraints within the realm of fleet selection and air cargo transportation. Addressing operational load planning issues in air cargo carriers during fleet selection and route planning reveals the preference for air transportation despite higher costs than surface transport, particularly for goods with attributes such as urgency, perishability, high value, and hazardous nature.

Wiebernei's study [11] on service network design for freight transportation delves into numerous pivotal inquiries concerning the enhancement of freight operations. These inquiries encompass identifying the optimal route, service timetable, and the most advantageous frequency for its provision. Furthermore, the study examines decisions regarding vehicle selection, taking into account their respective capacities to effectively operate these services. Another study [12] introduces an integrated planning approach encompassing two primary aspects. Firstly, it proposes a practical planning paradigm for optimally adjusting existing schedules. Secondly, it integrates additional crucial planning considerations for cargo airlines, such as available capacity on external flights, cargo handling costs, constraints, and aircraft maintenance regulations. Fenge and Shan's [7] study highlights the importance of considering specific limitations, including the limited availability of certain Unit Load Device (ULD) types at departure airports, three-dimensional packing constraints, and airport-specific limitations. Moreover, recent literature on the air cargo transport problem emphasizes the influence of cargo compatibility on operational efficiency, emphasizing the necessity of determining the optimal configuration of ULDs to maximize cargo volume while minimizing handling effort in accordance with airline preferences [13].

Furthermore, efficient cargo transportation networks are crucial in optimizing cargo movement across various supply chain scenarios. In a one-to-many and many-to-one (1-M/M-1) transportation network, where goods originate from a single source and are destined for multiple locations, the focus lies on centralized distribution, cost minimization, and resource utilization [5]. This strategy ensures a structured flow of goods, reducing redundancy and enhancing overall logistics efficiency. Conversely, in a many-to-one cargo transportation network, resources from multiple origins are consolidated to deliver goods to a single destination. This consolidation simplifies logistics processes, optimizes route planning, and increases transport capacity utilization, resulting in streamlined operations and reduced complexity. Whether distributing resources from one source to many destinations or consolidating goods from numerous sources to a single location, the selection of the transportation network depends on specific supply chain requirements, aiming to achieve optimal delivery efficiency and cost-effectiveness [5, 14].

The research conducted in [15] and [16] addressed a significant logistical challenge involving cargo aircraft, aiming to address multi-commodity, multi-depot, and heterogeneous vehicle pickup and delivery issues. However, these studies did not encompass the compatibility issues inherent in cargo transportation during an emergency where the time-sensitive nature of operations further complicates the fleet selection and assignment decisions. In their study, Kushwaha and sen [8] attempted to investigate the logistical challenges considering the multi-depot, multi-trip aircraft transportation planning problems to minimize mission completion time during emergencies. The study examined the 1-M/M-1 transportation network for relief distribution and highlighted the adaptability of network structure in emergency logistics. Nevertheless, the research addressed a significant logistical obstacle concerning cargo aircraft, aiming to resolve multi-commodity, multi-depot, and heterogeneous vehicle pickup and delivery challenges. However, it did not encompass compatibility issues in cargo transportation.

Moreover, while challenges such as service network design, flight schedule problems, and constraints related to Unit Load Devices (ULDs) are significant, the issue of load compatibility in air cargo carriers stands out as particularly crucial during fleet selection and route planning stages. It becomes imperative to address the specific attributes of urgency, perishability, high value, and hazardous nature to ensure the efficient and timely delivery of essential supplies and aid during critical situations. Additionally, strategically utilizing air cargo services while considering the compatibility of aircraft with cargo and the compatibility of cargo itself has significantly contributed to the success of emergency response efforts, underscoring the vital role of air transportation in mitigating the impact of disasters and emergencies.

In the context of emergency scenarios, the study focuses on integrating load compatibility constraints into the fleet selection and scheduling processes to enhance the flexibility of air cargo transportation during emergencies. The study contributes to the state of the art by delving into cargo compatibility within groups and with aircraft to promote operational efficiency, safety, and regulatory adherence in air cargo transportation. Moreover, adhering to cargo compatibility standards enhances customer satisfaction by ensuring punctual and secure deliveries, meeting customer expectations, and fostering trust in air cargo services for various cargo types, such as hazardous materials or perishable goods, which is crucial for maintaining legal compliance and avoiding operational disruptions. Furthermore, integrating incompatibility factors during fleet selection and route planning establishes an organized air cargo transportation plan within the 1-M/M-1 service network and embeds a novel approach.

6.3 Problem Description

This problem deals with cargo transportation, considering cargo aircraft. The fleet selection and route plan for an aircraft is based on the compatibility of an aircraft with the cargo loads and the earliest completion of the overall transportation requirements. The cargo's compatibility with an aircraft appears to be the most critical criterion during fleet selection. For instance, if a plane transports one load type from one airport to another, it cannot simultaneously carry any quantity of the other incompatible load type on the same journey. In transportation, it is imperative to transport incompatible load types separately, either by utilizing distinct fleets or by taking into account multiple flight legs, ensuring that the integrity of the load types is not compromised. Another challenge arises when dealing with the compatibility between load types and aircraft.

Different aircraft may have varying capacities, weight restrictions, and other specifications in transportation logistics or aviation. Likewise, different types of loads may have specific requirements or constraints. The challenge arises when matching these load types with suitable aircraft to ensure a smooth and efficient transportation process. For example, specific aircraft may be better suited for carrying heavy or bulky loads, while others may have limitations in terms of weight or size. Suppose there is a mismatch between the characteristics of the load and the capabilities of the aircraft. For instance, an oversized industrial crate might not fit through the cargo door of a Boeing 737, even if its weight falls within the aircraft's payload capacity. Similarly, temperature-sensitive pharmaceuticals may require specialized cold storage compartments that some aircraft may not possess. These are just two examples, but compatibility issues can arise due to various factors such as specific hazardous materials requiring special handling or conflicting with other cargo onboard; certain high-value cargo needing security features unavailable on certain aircraft. In all that case, fleet selection and load planning can lead to operational challenges, potential safety concerns, or inefficiencies in the transportation process.

Addressing these challenges may involve careful planning, considering compatibility factors, and potentially using different aircraft or transportation methods for specific loads. It emphasizes the need for effective coordination and logistics management to optimize the transportation of diverse loads using the appropriate aircraft.

It is more or less possible that an aircraft may start its journey from the home airbase to the source to get loaded if the requested load type is unavailable at the existing location where the aircraft is parked. A plane can make multiple visits to fulfill the load requirement of a specific demand location. There is another possibility that an aircraft can deliver its complete load at a single-demand airbase and serve only one demand airbase in a route.

6.3.1 Mathematical Model

The airport close to the affected region in the humanitarian network makes transportation requests related to the movement of different cargo types. All these requests are fulfilled by cargo aircraft positioned at various locations across the region. In formulation, the subscripts $i, j \in N$ are used to denote the airports or airbases, and $k \in K$ represents the aircraft assumed to be available at the time of mission. The cargo aircraft must return to their home bases after the tour. The superscript $l \in L$ denotes the load types moving from source to destination. To ensure that incompatible load types are not carried together, a subset of load types $\overline{L} \subseteq L$ introduced in the model.

Multiple cargo aircraft are accessible within the network, yet some may lack the capability to transport specific types of cargo indicated by the parameter δ_k^l. The logistical challenge of planning routes and schedules for cargo planes revolves around a dedicated 1-M/M-1 flight network. The feasibility of reaching an airport from various locations is denoted by a finite distance measurement indicated by d_{ij}.

Set:

N {Set of total airbases in the network}
L {Set of load types}
$\overline{L}$ {subset of incompatible load types $\overline{L} \subseteq L$}.
K {Set of aircraft K available in the network}.

Model Parameters:

q_k Cargo capacity of plane type $k \in K$;
spe_k Cruise velocity for plane *type* $k \in K$;
g_k Flight range of plane type $k \in K$;
β_k Ground service time for plane type $k \in K$ in minutes;
e_{ki} Auxiliary binary parameter ensuring availability of plane $k \in K$ in airbase $i \in N$ at the start;
d_{ij} Haversine distance between airbase i and j; $i, j \in N$;
γ Maximum number of flight legs in a tour;
r_{ij}^l The demand of load type $l \in L$ to transfer from airbase $i \in N$ to airbase $j \in N$;
wl^l Unit weight of load type $l \in L$;
δ_k^l Auxiliary binary parameter indicating compatibility of load $l \in L$ with aircraft type $k \in K$;
M A big number.

Decision variable:

The decision variables used in the proposed model are described as follows:

T Operation/mission completion time;
x_{ijk} A binary variable such that $x_{ijk} = 1$ indicating that an aircraft $k \in K$ departs from node $i \in N$ towards node $j \in N, i \neq j$; 0 otherwise;
y_{ijkl} A binary variable such that $y_{ijkl} = 1$ indicating that load type $l \in L$ tranferring from node $i \in N$ to node $j \in N$ on aircraft $k \in N$; 0 otherwise;

c_{ijk}^{l} Quantity of load type $l \in L$ carried by an aircraft $k \in K$ while traversing from $i \in N$ to $j \in N$;

u_{ijk}^{l} Quantity of load type $l \in L$ sent from $i \in N$ to $j \in N$ on aircraft $k \in K$.

Assumtions

The proposed model is based on the following operating assumptions:

- Demand for load type at different airfields might be larger than the airplane's cargo capacity; therefore, split delivery and pickups are allowed.
- Demand airfield will not store excess load type, i.e., it will not work as a transshipment node.
- Pickup and delivery demands are known deterministically.
- The problem assume disruption-free transportation and does not consider flight delays, diversion, or cancellations caused by severe weather conditions such as strong crosswinds, low visibility, and turbulence.
- The distance matrix is known and satisfies the triangular inequality.

Formulation

$$\text{Min.} \sum_{k \in K} T \tag{6.1}$$

where

$$T = \sum_{i \in N} \sum_{j \in N} \left(\frac{d_{ij}}{spe_k} x_{ijk} \right) + \sum_{i \in N} \sum_{j \in N} \beta_k x_{ijk}$$

$$\text{s.t.} \quad \sum_{i \in e_{ki}=1} \sum_{j \in N} x_{ijk} \leq 1 \quad \forall k \in K. \tag{6.2}$$

$$M * \sum_{i \in A} \sum_{j \in N} x_{ijk} \geq \sum_{i \in N} \sum_{j \in N} x_{ijk} \quad \forall k \in K \tag{6.3}$$

$$\sum_{\substack{j \in N \\ j \neq i}} x_{ijk} = \sum_{j \in N} x_{jik} \quad \forall i \in N, k \in K \tag{6.4}$$

$$\sum_{i \in N} \sum_{j \in N} x_{ijk} \leq \gamma \quad \forall k \in K \tag{6.5}$$

$$\sum_{i \in N} \sum_{i \in N} x_{iik} = 0 \quad \forall k \in K \tag{6.6}$$

$$d_{ij}.x_{fijtr} \leq g_k \quad \forall i \in N, j \in N, k \in K \tag{6.7}$$

$$\sum_{l\in L} c_{ijk}^{l} \leq M * x_{ijk} \quad \forall i, j \in N, k \in K \tag{6.8}$$

$$c_{ijk}^{l} \leq M * y_{ijkl} \quad \forall i, j \in N, k \in K, l \in L \tag{6.9}$$

$$\sum_{l\in L} wt^{l}.\delta_{k}^{l}.c_{ijk}^{l} \leq m_{k} \quad \forall i, j \in N, k \in K \tag{6.10}$$

$$\sum_{l,l'\in \overline{L}} y_{ijkl} = 1 \quad \forall i, j \in N, k \in K, l, l' \in \overline{L}, l \neq l' \tag{6.11}$$

$$\sum_{k\in K} \delta_{k}^{l}.u_{ijk}^{l} = r_{ij}^{l} \quad \forall i, j \in N, l \in L. \tag{6.12}$$

$$\sum_{p,i\in N} (c_{ijk}^{l} - u_{ijk}^{l} + u_{jpk}^{l} - c_{jpk}^{l}) = 0; \quad \forall j \in N, k \in K, l \in L \tag{6.13}$$

$$\sum_{j\in N} (c_{ijk}^{l} - u_{ijk}^{l}) \geq 0 \quad \forall i \in N, k \in K, l \in L \tag{6.14}$$

$$x_{ijk} \in \{0, 1\}; y_{il} \in \{0, 1\}; c_{ijk}^{l} \geq 0; u_{ijk}^{l} \geq 0; \forall i, j \in N, k \in K, l \in L \tag{6.15}$$

Objective (6.1) tries to minimize the total operation time of the mission. The overall mission completion time is computed by adding traveling time and ground service time at all the airbases from where the plane lands or takes off. Constraints (6.2) ensure that each aircraft begins its journey from its initial parking location. Constraint (6.3) ensures that if a plane does not take off from its initial parking airbase, it will not be included in the solution. Constraint (6.4) will ensure that the aircraft can finish its route in its home airbase. Moreover, constraint (6.4) provides that if an aircraft departs from node i towards node j, there must be a corresponding arrival at node i from node j. This essentially allows for a recurring path in the solution, meaning that the aircraft can visit the same node multiple times. Constraint (6.5) limits the number of flight legs in a tour, and (6.6) eliminates the possibility of loop formation. Constraint (6.7) will ensure that the distance traversed by each aircraft will not exceed its flight range (taken in km) limit.

The second set of constraints (6.8)–(6.15) are load planning constraints. In all these relations, multiple visits to a node by different planes are assumed to be permitted when partial delivery of load is made. Constraint (6.8) links the loading decision to aircraft routes. Constraint (6.9) links the loading decision to the load transported within the network. Constraints (6.10) limit the load type carried by the plane using its cargo carrying limitations and respect the compatibility of the aircraft with the load types. Constraints (6.11) ensure that for each pair of nodes i and j, and each aircraft k, only one load type (l) can be transferred from i to j. This formulation works for any number of incompatible load types, avoids conflicts in load transfers, and maintains the integrity of the load type. The variable y_{ijkl}, making it impossible

Table 6.1 Aircraft technical attributes

Parameters	Aircraft			
	C-130 ($k1$)	C-17 ($k2$)	IL-76 ($k3$)	AN-32 ($k4$)
q_k(kg)	19,051	77,519	48,000	6700
spe_k(kmph)	644	830	780	470
g_k(km)	3300	4480	4400	2500
β_k(min)	90	90	90	90

for incompatible load types to be carried together on the same aircraft. Constraint (6.12) will ensure that the picked-up and delivered load type always respects the transportation demand of the load type by compatible aircraft. Constraints (6.13) and (6.14) are recursive load relations balancing the loaded and unloaded quantities as the aircraft moves from one airport to another. Constraint (6.15) defines the search space.

6.4 Computational Analysis and Example Illustration

The proposed MIP model's computational illustration was evaluated using CPLEX Concert Technology version 20.1.0 on small synthetic datasets. A script written in an optimization programming language (OPL) was executed on a system equipped with a processor running at 2.38 GHz with 8GB (7.23 usable) of RAM capacity.

Test problems were generated using the data presented in Tables 6.1 and 6.2. The data sets considered for model testing are not random in their totality since some of the parameters (Table 6.1) indicate the fundamental attributes (except β_k) of fleets under consideration. The fleet attributes shown in Table 6.1 have been compiled from respective aircraft flight manuals from available sources.[1] The difficulty in overseeing actual data for ground service time (β_k) emerges due to its reliance on airport ground resources. Consequently, we have presumed a consistent value for this parameter across all fleets to eliminate ambiguity in the fleet selection process for the computational illustration of the proposed model. It is assumed that these dependencies remain consistent across all fleets. The upper bound for the number of flight legs in a trip is limited to $b = 5$. Table 6.2 shows the haversine distance between the geographical location of the point of interest in the study. Transportation requests for test instances are randomly generated in the interval [0, 1000]. The in-compatibility within load types and with the aircraft is considered for result comparison for different test instances.

[1] http://www.axonaviation.com/commercial-aircraft/aircraft-data/aircraft-specifications.

Table 6.2 Distance matrix (in km)

d_{ij}	$i1$	$i2$	$i3$
$i1$	0	810	790
$i2$	810	0	1198
$i3$	790	1198	0

6.4.1 Test Instance 1

In the first instance, the problem is defined for 3-node and 2-aircraft with 2-load types. We examined the model with the initial location (e_{ki}) of aircraft $k1$ at $i1$ and $k2$ at $i2$, respectively. Two load types, $l1$, and $l2$, were considered, without in-compatibility with themselves in instance-1 and with in-compatibility in the test instance-1. The aircrafts $k1$ and $k2$ are compatible with both load types in instance-1. Load $l1$ weighed 30 kgs, and $l2$ weighed 76 kgs. The transportation requests between locations are presented in Table 6.3. The outcomes depicted in Fig. 6.1 illustrate the transportation specifics for a 3-node, 2-aircraft problem, for instance: 1.

- **Nodes:** The circles in Fig. 6.1 represent nodes, which are pickup or delivery locations for the aircraft.

Table 6.3 Transportation request

N	N	L	r_{ij}^{l}
$i2$	$i1$	$l1$	300
$i3$	$i1$	$l1$	300
$i1$	$i2$	$l1$	200
$i2$	$i3$	$l2$	1000
$i3$	$i1$	$l2$	600

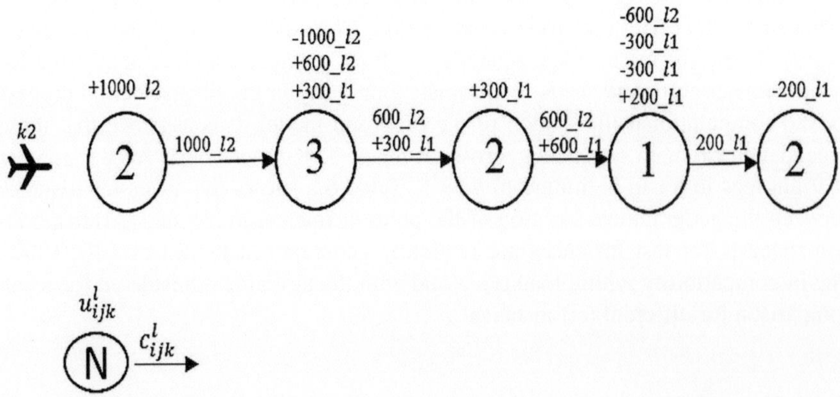

Fig. 6.1 Transportation plan: Instance 1

- **Paths:** The arrows connecting the nodes indicate the path of the aircraft. The value aligned with the arrow indicates the total amount carried (c^{l}_{ijk}) by the aircraft while moving across the locations.
- **Pickups and deliveries:** The values adjacent to each node indicate the pickup and delivery amounts $\left(u^{l}_{ijk}\right)$ at that location. A positive value (+) indicates a pickup, while a negative value (-) indicates a delivery.

Based on the figure, the breakdown of the transportation details for aircraft $k2$ is as follows:

Node 2: Aircraft $k2$ starts at node 2 with 1000 units of load type $l2$.
Node 3: It then flies to node 3, delivering 1000 units of load type $l2$. It picks up 600 units of load type $l2$ and 300 units of load type $l1$, leaving node 3 with 600_ $l2$ and 300 _$l1$.
Node 2: From node 2, the aircraft flies to node 1 with a total amount carrying 1200 units, i.e., (600_$l2$ + 600_$l1$). It picks 300_$l1$ at this location.
Node1: Next, the aircraft flies to node 1, delivering 6 00_$l2$; 300_$l1$; 300_ $l1 units$. It picks 200 units of load type $l1$.
Node 2: Finally, the aircraft flies to node 2, delivering the remaining 200_$l1$ units.

The results of the test instance-1 demonstrate the existence of three feasible solutions, achieving an optimal mission time of 10.839 h through 60 iterations within an elapsed time of 0.03 s.

6.4.2 Test Instance 2

In the second test case, Fig. 6.2a, b showcases the transportation details in the presence of aircraft load compatibility ($\delta^{l1}_{k1} = 1$; $\delta^{l2}_{k1} = 0$; $\delta^{l1}_{k2} = 0$; $\delta^{l2}_{k1} = 1$). Figure 6.2a, b concisely represents the transportation plan. The optimal mission time in the second scenario was determined to be 16.715 h, accompanied by the assignment of $k1$ and $k2$ through 13 iterations with an elapsed time of 0.09 s. The number of feasible solutions found in test instance 2 was three.

A comparative analysis of the computational matrices of instances 1 and 2 indicates that the elapsed time in the second scenario is twice as long as in the previous scenario, where aircraft-load compatibility was absent.

6.4.3 Test Instances 3 and 4

In the third scenario, the problem was defined for a 3-node and 4-aircraft system with two load types. The model was investigated with initial aircraft locations (e_{ki}), specifically $k1$ at $i1$, and $k2$ at $i2$, $k3$ at $i3$, and $k4$ at $i4$. In instance 3, the model

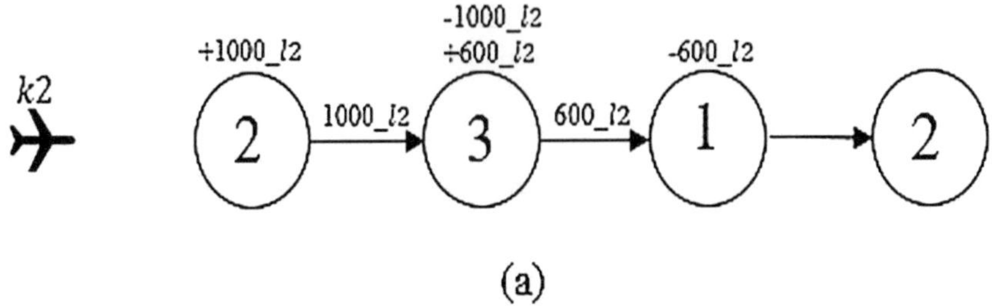

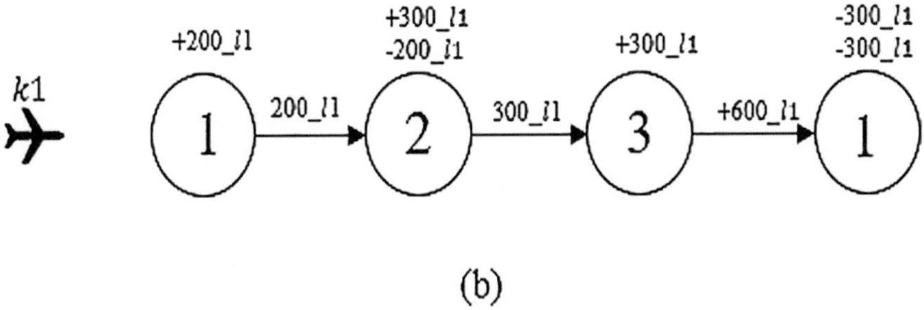

Fig. 6.2 Transportation plan: Instance 2

was executed without load type in compatibility within the load groups and with the aircraft. The solution obtained for this case matched the solution found for test instance -1 and illustrated in Fig. 6.1. The computational performance of CPLEX was observed to extend with increased fleet size. The elapsed time was recorded as 0.05 s to reach the optimality through 82 iterations. The number of feasible solutions found was four for test instance 3.

While ensuring compatibility with all aircraft, introducing load type in-compatibility within the load groups resulted in CPLEX finding four feasible solutions for test instance-4. The solution found indicates the assignments of aircraft $k2$ and $k3$. The optimal solution for this scenario is presented in Fig. 6.3, with an optimal mission time of 15.864 h and an elapsed time of 0.11 s through 119 iterations.

In the initial phase of the flight, aircraft $k2$ transports load type $l2$, while for all subsequent flight segments, the aircraft exclusively transports load type $l1$ to maintain load in-compatibility.

6.4.4 *Test Instances 5*

In the fifth scenario, the transportation details are illustrated in Fig. 6.4a, b, highlighting the presence of aircraft load compatibility ($\delta_{k1}^{l1} = 0$; $\delta_{k1}^{l2} = 1$; $\delta_{k2}^{l1} = 1$; $\delta_{k2}^{l2} = 1$; $\delta_{k3}^{l1} = 1$; $\delta_{k3}^{l2} = 0$; $\delta_{k4}^{l1} = 1$; $\delta_{k4}^{l2} = 1$) in a 3-node and 4-aircraft system with two

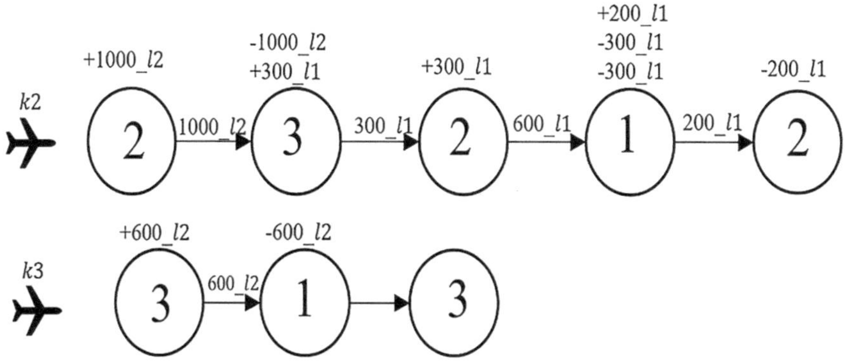

Fig. 6.3 Transportation plan: Instance 4

load types. The model was examined with initial aircraft locations (e_{ki}) identical to those specified for instance-3. The model was executed without load type incompatibility within the load groups. The solution for this case was obtained in a mere 0.02 s, involving the assignment of k1 and k3 through 81 iterations. The computational performance of CPLEX demonstrated rapid processing, particularly when the largest fleet size was available for transportation. Test instance-5 yielded three feasible solutions.

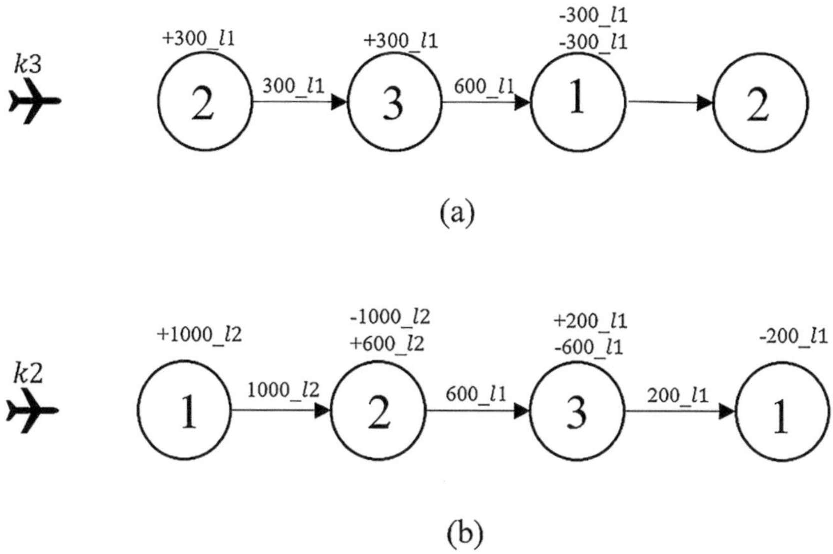

Fig. 6.4 Transportation plan: Instance 5

6.5 Application in Relief Delivery

In air cargo transportation for relief delivery following a natural disaster, the formulation and execution of a framework for optimizing cargo loading and handling procedures are very crucial. This process is essential to ensure that the proposed model aligns with a humanitarian mission's real-world requirements and complexities. The examination of the model's relevance involves an assessment of the group compatibility challenges arising from the procedures implemented in humanitarian missions. For example, the relief delivery mission considers diverse relief items categorized into three distinct load types based on their nature and storage needs. The objective is to create an optimized air cargo transportation plan that considers group incompatibility among relief aids such as medical supplies ($l1$), food and perishable goods ($l2$), and shelter and construction materials ($l3$).

The first step in transportation planning involves assessing the practical relevance of the categorized relief items and their specific storage requirements. It is assumed that load type $l1$, consisting of temperature-sensitive medical supplies, should not be loaded together with load type $l2$, which has standard storage conditions. Furthermore, in a humanitarian mission, one can recognize that the heavy and bulky items in Group C could potentially compromise the integrity of temperature-sensitive medical supplies in Group A.

The proposed study suggests a strategic approach to group-based segregation, with designated cargo compartments for each load type to prevent direct contact and potential damage during transportation. This organizational structure is intended to mirror the reality of humanitarian air cargo operations where efficient grouping and segregation are essential for the safety and integrity of relief items.

The integrated transportation of temperature-sensitive load $l1$, followed by type $l2$ and type $l3$, aligns with the humanitarian mission's logistical requirements and reflects the applicability of the developed model for ensuring group incompatibility. The choice of an aircraft to meet the specific transportation requirement reflects the model's adaptability to real-world constraints and the need for load and aircraft compatibility. To validate the model, a comparison of the proposed loading plan with humanitarian air cargo operations in similar contexts is essential. Moreover, the real-world data on relief item characteristics, storage requirements, and logistical challenges will provide empirical evidence of the model's effectiveness. Regular assessments and adjustments based on feedback from humanitarian organizations engaged in relief missions will contribute to the ongoing validation and refinement of the model.

In summary, the model for optimizing humanitarian air cargo transportation, with a focus on group incompatibility, requires continuous validation against real-world scenarios. By aligning with the complexities and challenges of actual relief delivery missions, the model can be refined and validated through ongoing collaboration with humanitarian organizations, ensuring its relevance, effectiveness, and contribution to the success of relief operations.

6.6 Conclusion

The model proposed in this paper represents a specific instance of the pickup and delivery vehicle routing problem, focusing on cargo compatibility challenges encountered by mission planners during emergencies. The formulated cargo transportation problem is expressed as a mixed-integer linear program for a 1-MM-1 transportation network, ensuring fleet selection and route planning that takes into account the compatibility of load types both among themselves and with the aircraft. The model, solved using the branch-and-cut algorithm in commercial off-the-shelf optimization software, demonstrates solution times for small instances and underscores the need to test the model on large-scale problems involving diverse fleet types and load categories. Future investigations could be oriented towards developing a methodology for solving the model with user-specified constraints, such as prioritizing the transportation of urgent requirements, considering actual service times for fleets, accounting for delays in cargo loading and unloading, and evaluating the impact of risky environments on fleet selection and routing. While the paper treats demand as static, recognizing that real-world scenarios involve dynamic demands suggests that further research might explore stochastic demand models. Additionally, extending the problem could involve the development of heuristic and metaheuristic approaches with more realistic constraints, such as aircraft queuing before cargo loading, pre and post-landing delays, and the influence of ground resources on mission completion time. Furthermore, performing sensitivity analysis to assess the robustness of the proposed model under varying input parameters such as fleet capacities, travel times, or demand distribution would definitely provide valuable insights on model's stability and reliability.

References

1. Cordeau, J.F., Laporte, G., Potvin, J.Y., Savelsbergh, M.W.P.: Transportation on demand. Handbooks Oper Res Manage Sci 14:429–466. https://doi.org/10.1016/S0927-0507(06)14007-4
2. Bertsimas, D., Chang, A., Miši'miši'c, V.V., Mundru, N.: The airlift planning problem. Transp. Sci. **53**(3), 773–795 (2016)
3. Nance, R.L., Roesener, A.G., Moore, J.T.: An advanced tabu search for solving the mixed payload airlift loading problem. J. Oper. Res. Soc. **62**, 337–347 (2011). https://doi.org/10.1057/jors.2010.119
4. Hansen, M., Li, A.: Multiple airport regions based on inter-airport temporal distances. Transp. Res. E Logist. Transp. Rev. **101**, 84–98 (2017). https://doi.org/10.1016/J.TRE.2017.03.002
5. Koç, Ç., Laporte, G., Tükenmez, İ: A review of vehicle routing with simultaneous pickup and delivery. Comput. Oper. Res. **122**, 104987 (2020). https://doi.org/10.1016/J.COR.2020.104987
6. Sherali, H.D., Bish, E.K., Zhu, X.: Airline fleet assignment concepts, models, and algorithms. Eur. J. Oper. Res. **172**, 1–30 (2006). https://doi.org/10.1016/J.EJOR.2005.01.056
7. Feng, B., Li, Y., Shen, Z.J.M.: Air cargo operations: literature review and comparison with practices. Transp. Res. Part C Emerg. Technol. **56**, 263–280 (2015). https://doi.org/10.1016/j.trc.2015.03.028

8. Kushwaha, D.K., Sen, G.: Multi-trip pickup and delivery problem in one to many and many to one (1-M/M-1) transportation network. In: 2023 IEEE International conference on industrial engineering and engineering management (IEEM), Singapore, Singapore, pp 1588–1592. https://doi.org/10.1109/IEEM58616.2023.10406505
9. Pollaris, H., Braekers, K., Caris, A., Janssens, G.K., Limbourg, S.: Vehicle routing problems with loading constraints: state-of-the-art and future directions. OR Spectrum **37**, 297–330 (2015). https://doi.org/10.1007/s00291-014-0386-3
10. Bortfeldt, A., Wäscher, G.: Constraints in container loading-A state-of-the-art review. Eur. J. Oper. Res. **229**, 1–20 (2013). https://doi.org/10.1016/j.ejor.2012.12.006
11. Wieberneit, N.: Service network design for freight transportation: a review. OR Spectrum **30**, 77–112 (2008). https://doi.org/10.1007/s00291-007-0079-2
12. Derigs, U., Friederichs, S.: Air cargo scheduling: Integrated models and solution procedures. OR Spectrum **35**, 325–362 (2013). https://doi.org/10.1007/S00291-012-0299-Y/METRICS
13. Brandt, F., Nickel, S.: The air cargo load planning problem—a consolidated problem definition and literature review on related problems
14. Lyu, Z., Yu, A.J.: The pickup and delivery problem with transshipments: critical review of two existing models and a new formulation. Eur. J. Oper. Res. **305**, 260–270 (2023). https://doi.org/10.1016/J.EJOR.2022.05.053
15. Kim, S.-K., Lee, M.-G., Lee, H.-S.: Effective route scheduling for military cargo-plane operation. Kor. Manage. Sci. Rev. **33**, 89–99 (2016). https://doi.org/10.7737/kmsr.2016.33.1.089
16. Aksoy, Ö., Kapanoglu, M., Kapanoğlu, M.: Multi-commodity, multi-depot, heterogeneous vehicle pickup and delivery problem for air transportation in the Turkish Air Force. J. Aeronautics Space Technol. **5**(4), 43–57 (2012)

Chapter 7
Optimizing Packaging Efficiency in the Fast-Moving Consumer Goods (FMCG) Industry: Multivariate Approach

Jai Prakash Kumar, Rony Mitra, and Manoj Kumar Tiwari

Abstract The paper focused on enhancing packaging efficiency in the Fast-Moving Consumer Goods (FMCG) sector, specifically for powder-form products. The primary challenge faced by FMCG companies is the precise filling of product packets during machine-based packaging. This paper explored the factors contributing to underfilling and overfilling and proposed techniques and strategies to improve packaging efficiency. The study employed causality analysis, multivariate regression, and statistical methods to uncover the root causes of packaging variations. The findings also highlighted the importance of reducing overfilling to minimize material wastage and environmental impact. Two key propositions were presented and examined in the paper in order to investigate and support the objective. The first proposition examined the impact of external environmental factors on packaging efficiency, including lump formation, temperature, humidity, and rainfall. The second proposition investigated the effect of time difference on packaging efficiency. The reduction of overfilling not only reduces material wastage but also aligns with sustainability practices, portrays companies as responsible environmental stewards and increases revenue for the company. This paper discussed some actionable recommendations for improving the packaging process, reducing waste, and enhancing industry efficiency. The findings have implications for similar products and contribute to the broader goal of sustainability.

Keywords FMCG · Packaging efficiency · Multivariate · Underfilling · Overfilling · Sustainability

J. Prakash Kumar (✉) · R. Mitra
Indian Institute of Technology Kharagpur, Kharagpur, India
e-mail: imjai.p9@gmail.com

M. K. Tiwari
Indian Institute of Management Mumbai, Mumbai, India

J. Maiti et al. (eds.), *Operations Research and Data Analytics: Current Trends and Future Perspectives*, Lecture Notes on Multidisciplinary Industrial Engineering,
https://doi.org/10.1007/978-981-96-8327-7_7

7.1 Introduction

Accurate packaging is a major challenge for the Fast-Moving Consumer Goods (FMCG) sector [1], especially for powder and liquid products. This paper discussed the importance of precise filling during machine-based packaging and the regulations that govern this process. This paper uncovered various in-house and external factors like humidity, rainfall, and temperatures that affect packaging efficiency. This paper emphasized products in powder form, such as detergents [2], Horlicks, Bournvita, milk powder, and other similar products which are frequently used in our daily usage. The primary challenge faced by FMCG companies in this context is the precise filling of product packets during machine-based packaging [3]. Following government regulations, these companies must adhere to error margins, encompassing lower and upper permissible limits beyond the target fill weight. These regulations play a crucial role in managing variances introduced by packaging machinery.

For companies involved in packaging millions of units daily, the significance of even a single additional gram of product cannot be overstated. The cumulative effect of these slight variances can result in significant financial losses and the wastage of materials for a company. A multitude of factors affect underfilling and overfilling while packaging [4], and this paper focused on elucidating some of these factors while also delving into the techniques and strategies that can be employed to enhance packaging efficiency. In optimizing packaging efficiency in the FMCG sector [5], it is essential to acknowledge the intricate interplay of environmental factors [6]. While overfilling in packaging primarily impacts a company's finances, its consequences extend to the environment on a significant scale. This issue is closely related to the excessive use of non-biodegradable materials, as seen in detergent packaging, and it carries cascading environmental effects.

Reducing overfilling per packet serves a dual purpose. Firstly, it directly reduces material wastage, eliminating non-biodegradable [7], thus diminishing the environmental impact of the FMCG industry. Excessive use of nonbiodegradables (like detergents) can lead to problems such as soil erosion, water pollution, and harm to natural ecosystems. Secondly, this reduction aligns seamlessly with modern sustainability practices [8], portraying companies as responsible environmental stewards and earning goodwill from environmentally conscious consumers. Sustainability has become a hallmark of ethical business practices, influencing consumer choices and safeguarding corporate reputations. This paper embarked on a comprehensive exploration of the complex challenge of improving packaging efficiency within the FMCG sector. It aimed to uncover the intricate factors contributing to variations in machine-based packaging processes, resulting in persistent issues like underfilling and overfilling. The paper employed a multifaceted approach to achieve this objective, utilizing techniques such as causality analysis, multivariate regression [9], and various statistical methods. The paper also discussed two important propositions.

This paper aimed to provide a holistic understanding of the challenges and opportunities surrounding optimizing packaging efficiency in the FMCG sector by amalgamating these techniques and strategies. While it focused on powder packaging

[10], the findings can be applied to similar products, extending the relevance of the research.

Furthermore, this paper underscored the broader global context of sustainability challenges. The reduction of overfilling served as a microcosm of the FMCG sector's commitment to sustainability, illustrating how conscientious packaging practices can significantly alleviate the industry's environmental impact. Essentially, it emphasized the pivotal role played by the FMCG sector in advancing sustainable and environmentally responsible practices on a global scale [11].

7.2 Overview: Imperatives and Significance

In today's dynamic and competitive landscape, the Fast-Moving Consumer Goods (FMCG) sector finds itself at the crossroads of multiple pressing concerns. The quest for enhanced packaging efficiency has taken canter stage, responding to various exigencies. Chief among them is the necessity of adhering to stringent government compliance and industry standards. Maintaining regulatory compliance is not just a requisite; it is synonymous with earning and sustaining consumer trust. Furthermore, it's closely intertwined with the fabric of the FMCG industry, where the meticulous observance of standards forms the bedrock of success. In tandem with this, financial considerations take the spotlight. The industry's pursuit of cost reduction and revenue enhancement compels businesses to innovate and optimize their packaging processes. Finally, the global call for sustainability resonates loudly in this context. Efficient packaging practices are not just about reducing waste; they are a tangible commitment to environmental responsibility, aligning with broader sustainability goals.

7.2.1 Government Compliance and Industry Standards

Government regulations wield significant influence in the FMCG sector, stipulating precise packaging standards. For instance, a 1000 g (1 kg) packet must adhere to strict guidelines, permitting a maximum allowable variation of 1.5%, equivalent to 15 g. Consequently, companies must ensure a minimum fill weight of 985 g.

Companies exercise their discretion in establishing upper limits, typically falling within the range of 1020–1030 g, a measure aimed at mitigating machine-related variances and minimizing the incidence of packet failure.

Any variance falling below 985 g or exceeding 1030 g is classified as a packaging failure, necessitating costly repackaging efforts. Adhering to these parameters is of paramount importance. The overarching goal is to reduce the occurrence of failed packets and achieve packaging as close to the desired weight as possible. Striking this equilibrium is essential, as deviations result in financial losses and affect operational efficiency, customer satisfaction, and legal compliance.

Table 7.1 Government rules for company (maximum permissible error in excess)

Serial No.	Declared quantity (g or ml)	As % of declared quantity	Grams or ml
1	up to 50	9	–
2	50–100	–	4.5
3	100–200	4.5	–
4	200–300	–	9
5	300–500	3	–
6	500–1000	–	15
7	1000–10,000	1.5	–
8	10,000–15,000	–	150
9	More than 15,000	1.0	–

Table 7.1 overviews the government's regulations governing Fast Moving Consumer Goods (FMCG) companies and other companies. It delineates the prescribed tolerance levels for different package sizes within this context. Notably, for packages within the range of 1000 –10,000 g, a maximum permissible error of 1.5% is allowed, while for packages weighing between 500 and 1000 g, the allowable variance is 15 g. This policy strictly dictates that companies are obligated to refrain from selling products that fall below the defined minimum weight in the market.

7.2.2 *Efficiency-Driven Cost Savings and Income Amplification*

Even the most minute excess in packet weight takes on considerable significance when extrapolated across high-volume production lines. For instance, consider a company that packages millions of units daily; every additional gram results in substantial material and operational expenses. The principles of economies of scale underscore the imperative nature of precise packaging, as each fraction of an ounce conserved directly augments the company's financial bottom line.

Moreover, going beyond immediate cost reduction, increasing packaging accuracy serves as a catalyst for enhancing revenue streams [12]. By consistently packaging products to their specified weight, companies prevent customer dissatisfaction, returns, and potential legal liabilities and cultivate a deep trust among customers regarding the quantity they receive. This trust, in turn, nurtures brand loyalty and repeat business, ultimately propelling the company's sales and profitability.

7.2.3 Environmental Insights and Impact Analysis: Towards Sustainability

Elevating packaging efficiency in the Fast-Moving Consumer Goods (FMCG) sector, the imperative to delve comprehensively into environmental factors becomes even more pronounced. This multifaceted exploration extends far beyond mere financial considerations, reaching into the profound implications for our broader ecosystem. It underscores the need to balance optimizing business practices and ensuring they are ecologically responsible. This endeavor fosters more efficient and cost-effective packaging and aligns with the global call for sustainability, acknowledging the critical role the FMCG sector plays in safeguarding the environment and our collective future.

7.2.3.1 Minimizing the Use of Non-biodegradable Materials

A substantial environmental concern [13] revolves around the overuse of non-biodegradable materials, notably exemplified by products such as detergents. The excessive reliance on these materials contributes to pressing issues like soil erosion, water pollution, and the degradation of natural ecosystems. Since non-biodegradable substances, including detergent residues, linger in the environment, they threaten the delicate balance of ecosystems and the quality of water bodies.

7.2.3.2 Sustainable Waste Management Strategies

Another critical environmental facet pertains to the management of waste. Overfilled packaging gives rise to surplus waste, a significant portion of which is non-biodegradable. Disposing of this excess often entails incineration or landfilling non-recyclable materials, exacerbating challenges related to waste management and environmental pollution.

7.2.3.3 Eco-conscious Packaging

The quest for packaging efficiency closely aligns with broader sustainability objectives [14]. Prioritizing precision in packaging epitomizes responsible business conduct and resonates with environmentally conscious consumers. Companies that champion sustainability by curbing overfilling contribute to the larger movement promoting environmentally responsible packaging practices.

These environmental factors underscore the intricate nexus between packaging efficiency and our ecosystem. The reduction of overfilling transcends financial considerations, representing a conscientious step toward mitigating environmental degradation, resource conservation, waste reduction, and the cultivation of a sustainable future for the FMCG sector.

7.3 Methodology

7.3.1 Exploratory Data Analysis (EDA)

The initial primary objective was to develop a profound understanding of the data, focusing intently on the packaging processes associated with various machines and products. To achieve this, various statistical metrics like standard deviation and mean calculations for the data collected from each machine (a total of 5 machines) and for each product.

These statistical techniques were used to identify underlying patterns, variances, and trends in the packaging process. As the mean made it easier to spot central tendencies, the standard deviation helped to evaluate the degree of variability and dispersion in the data. This analysis laid the groundwork for future data-driven packaging optimization techniques and decision-making by providing insightful information about the accuracy and consistency of packaging operations across various machines and goods.

Plotting standard deviations against month-product combinations on the *x*-axis, the five images (Fig. 7.1) performed an analysis unique to each machine.

Identifying differences in each machine's performance was the primary objective. The supplied data highlights notable differences and variations between the various machines. Lower standard deviation values in the analytical framework suggest more consistency and accuracy in the packaging process. The outcomes show that every machine performs at different precision levels. After thoroughly examining inter-machine variations, the analysis was expanded to include a closer look at the differences when various machines packed identical products. The dataset illustrates the notable differences between seven items packaged across several months on different machines.

In the seven images (Fig. 7.2), the *y*-axis displays the standard deviation, while the *x*-axis shows machine-month combinations. The pronounced differences in standard deviations emphasize the notable variations present.

Significant differences between various machines and products were revealed by a preliminary study. These differences became quite noticeable when comparing the standard deviations linked to different machines and products. For example, differences in standard deviations were noted when a single product was packaged using three different machines. Moreover, month-to-month variances appeared when the identical product was packaged for several months. The findings were examined in greater detail in the next areas of the analysis, along with their consequences and possible approaches to rectifying these discrepancies to improve packaging accuracy and productivity.

(a) Comparative analysis for machine M_1

(b) Comparative analysis for machine M_2

(c) Comparative analysis for machine M_3

(d) Comparative analysis for machine M_4

(e) Comparative analysis for machine M_5

Fig. 7.1 Comparative analysis of month-products and machines

7.3.2 *Defining the Acceptable Range*

An acceptable range is discussed in the paper as the basis for the accuracy analysis. The precision of the packaging operations was assessed and measured by the model using this range as a reference. This acceptable range was defined, providing a crucial reference point for assessing the packing procedures' correctness. The details of this acceptable range, its ramifications, and how it fits into the larger picture were explored in detail in the ensuing sections. It is important to note that this foundational model is based on empirical techniques, including the widely accepted 3 sigma rule. This one is one of the accuracy matrices to assess accuracy in addition to the total failure percentage compared to the current failure percentage.

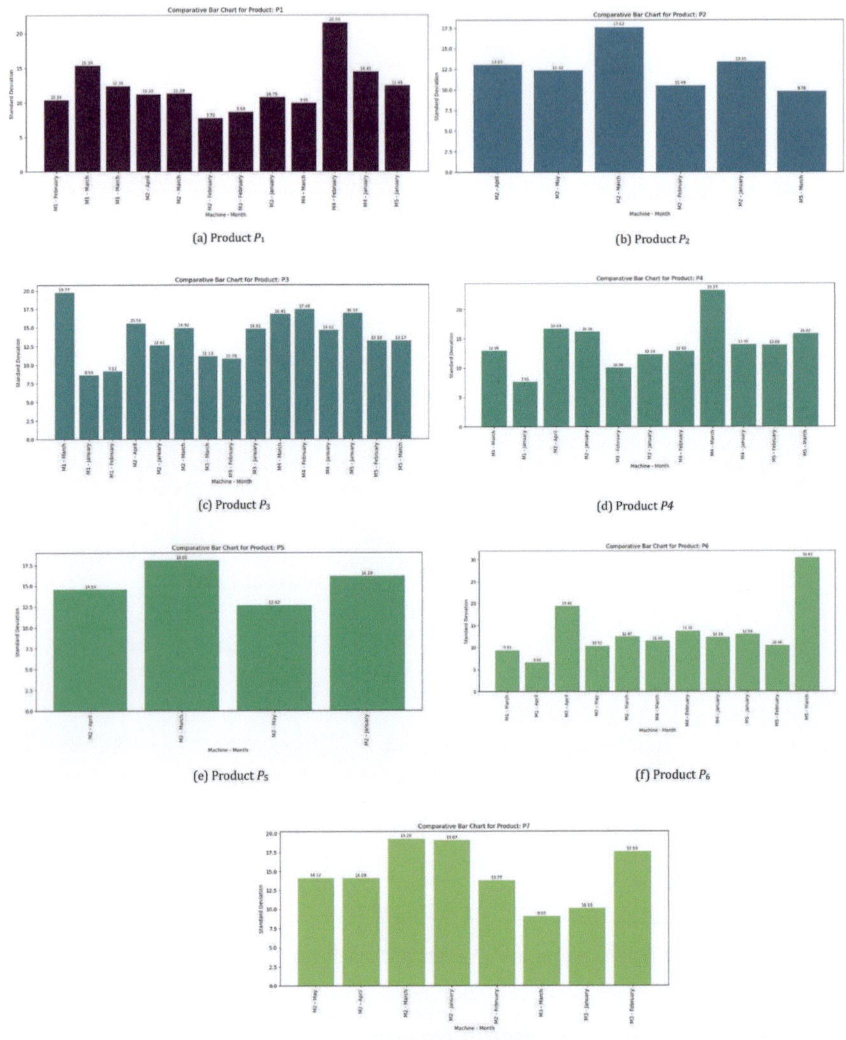

(a) Product P_1 (b) Product P_2 (c) Product P_3 (d) Product $P4$ (e) Product P_5 (f) Product P_6 (g) Product P_7

Fig. 7.2 Comparative analysis of machine-months and products

In statistics, the 68–95–99.7 rule (3 sigma), also known as the empirical rule, is a shorthand used to remember the percentage of values (Eq. 7.1) that lie within an interval estimate in a normal distribution. After substituting $n = 1,2,3$ in Eq. 7.1, we analyze that 68.27%, 95%, and 99.7% of the values lie within one, two, and three standard deviations of the mean, respectively.

$$\Pr(\mu - n\sigma \le X \le \mu + n\sigma) = \int_{\mu-n\sigma}^{\mu+n\sigma} \frac{1}{2\pi\sigma^2} \exp\left[-\frac{(x-\mu)^2}{2\sigma^2}\right] \mathrm{d}x \tag{7.1}$$

$$\Pr(\mu - \sigma \le X \le \mu + \sigma) = \int_{-1}^{1} \frac{1}{2\pi\sigma^2} \exp\left[-\frac{(x-\mu)^2}{2\sigma^2}\right] \mathrm{d}x \approx 68 \tag{7.2}$$

$$\Pr(\mu - 2\sigma \le X \le \mu + 2\sigma) = \int_{-2}^{2} \frac{1}{2\pi\sigma^2} \exp\left[-\frac{(x-\mu)^2}{2\sigma^2}\right] \mathrm{d}x \approx 95.45\% \tag{7.3}$$

$$\Pr(\mu - 3\sigma \le X \le \mu + 3\sigma) = \int_{-3}^{3} \frac{1}{2\pi\sigma^2} \exp\left[-\frac{(x-\mu)^2}{2\sigma^2}\right] \mathrm{d}x \approx 99.73\% \tag{7.4}$$

The analysis centered on evaluating the proportion of data points falling within the $\pm\, 0.5\sigma$ range relative to the target weight, determined by packet size. This range, the Acceptable Range, served as a crucial reference point. This analysis considered the packaging process within this range to operate optimally, meeting the desired fill weight. Conversely, any data points exceeding or falling short of this $\pm\, 0.5\sigma$ range raised concerns, signifying potential overfilling or underfilling issues (Fig. 7.3).

Essentially, applying the 3-sigma rule provided a robust statistical framework for assessing the consistency and accuracy of the packaging process. Upon delving deeper into the analysis of the acceptable range, the results showed that the percentage of products within this acceptable range typically fell between 40 and 50% (see images below). This meant that a substantial portion of the product packets either fell

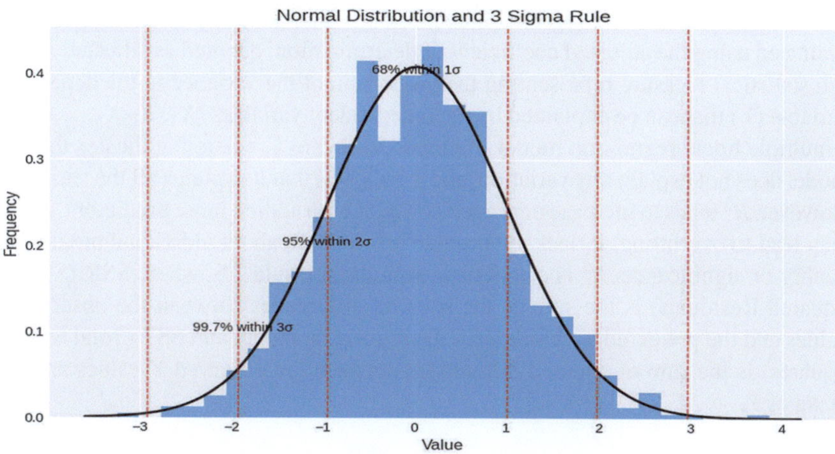

Fig. 7.3 Data points falling within $\pm\, 3\sigma$ range

outside this range, categorized as failed, or experienced overfilling and underfilling without being classified as failed.

Two specific focus areas were identified. Firstly, addressed the Upper Data that exceeds the $+ 0.5\sigma$ threshold but remains inside the defined upper limit and the "Lower Data" below the $- 0.5\sigma$ threshold. Bringing upper and lower data points closer to the acceptable range became crucial for minimizing overfilling and underfilling issues and fostering a more precise and cost-effective packaging process.

7.3.3 Multiple Linear Regression

Regression analysis is a statistical technique for estimating the relationship among variables that have a cause-and-effect relationship. Univariate regression analyses the relationship between a dependent variable and a single independent variable to formulate a linear equation. Regression models with multiple independent variables are known as multiple regression or multivariate regression analysis. The multivariate regression analysis model is formulated based on Eq. 7.5, where

$$Y = \beta_0 + \beta_1 X_1 + \beta_2 X_2 + \beta_3 X_3 + \cdots + \beta_n X_n + \varepsilon \quad (7.5)$$

Y is the dependent variable, $X_1, X_2, X_3, \ldots, X_n$ are the independent variables, β_0 is the intercept (the value of Y when all Xs are zero), $\beta_1, \beta_2, \beta_3, \ldots, \beta_n$ are the coefficients, representing the change in Y for a one-unit change in each corresponding X, while holding all other variables constant, ϵ is the error term, representing the unexplained variance (Fig. 7.4).

Prediction models are developed by estimating the regression coefficients and the constant term through multiple linear regression using data. In this study, Python was used to estimate the regression coefficients. The accuracy of the models was evaluated using the adjusted coefficient of determination, denoted as adjusted R^2. R^2 is a statistical measure representing the proportion of the variance in the dependent variable (Y) that can be explained by the independent variables ($X_1, X_2, X_3, \ldots, X_n$) in a multiple linear regression model. It ranges from 0 to 1, where 0 indicates that the model does not explain any variation, and 1 indicates that it explains all the variation. However, R^2 tends to increase or remain constant with adding more predictors, which may lead to overfitting. It does not provide information about individual predictors' quality or significance. R^2 is calculated using the formula 7.6, where SSR (Sum of Squared Residuals) is the sum of the squared differences between the observed Y values and the predicted Y values from the regression model and SST (Total Sum of Squares) is the sum of squared differences between the observed Y values and the mean of Y.

$$R^2 = 1 - \frac{\text{SSR}}{\text{SST}} \quad (7.6)$$

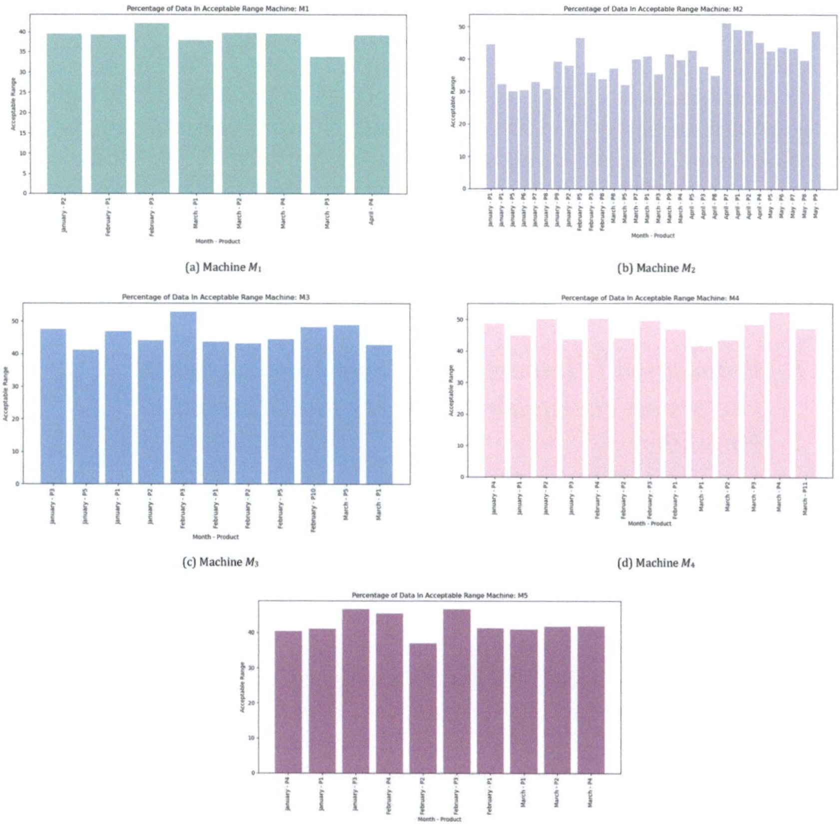

(a) Machine M_1 (b) Machine M_2

(c) Machine M_3 (d) Machine M_4

(f) Machine M_5

Fig. 7.4 Percentage of data in acceptable range with respect to month-product and machines

Adjusted R-squared, denoted as ${R_{\text{adj}}}^2$, is a modification of R-squared that considers the number of predictors (independent variables) in the model. It provides a more balanced measure of model fit, penalizing the inclusion of irrelevant predictors. Adjusted R-squared, denoted as ${R_{\text{adj}}}^2$, is calculated using the formula 7.7, where R^2 is the standard R-squared value, n is the number of observations (data points), and k is the number of predictors (independent variables) in the model.

$$R_{\text{adj}}^2 = 1 - \left[\frac{\left(1 - R^2(1-1)\right)}{n-k-1} \right] \tag{7.7}$$

It is important to note that Adjusted R-squared is always less than or equal to R-squared. A higher Adjusted R-squared is preferred as it indicates that the model provides a good fit while penalizing the addition of unnecessary predictors. It accounts for model complexity and helps to avoid overfitting.

7.4 Propositions

This section introduced and discussed two important propositions. The first proposition examined the impact of external environmental factors on packaging efficiency, including lump formation, temperature, humidity, and rainfall. The second proposition investigated the effect of time difference on packaging efficiency and how variations in time difference, including production schedules and packaging packet time delays, influence the efficiency of packaging operations.

7.4.1 Proposition I: Impact of External Environmental Factors on Packaging Efficiency and Materials Humidity

Humidity levels have a significant role in the packaging efficiency of powder materials. Elevated humidity can induce lump formation within powder materials like detergent, primarily due to the propensity of certain detergent components to absorb moisture from the atmosphere. This absorption leads to agglomeration, resulting in inconsistent product quality and suboptimal packaging. The water-absorbing properties of detergent components, such as surfactants and hygroscopic additives, make them susceptible to moisture induced clumping. Therefore, it is imperative to mitigate humidity-related challenges to maintain consistent packaging precision. The relationship between humidity and lump formation can be expressed based on the Eq. 8, where'f' is a function incorporating the moisture-absorption characteristics of materials.

$$\text{Lump formation } = f\,(\text{humidity}) \tag{7.8}$$

In-House Temperature:

The temperature differential between the in-house environment and external conditions profoundly influences packaging materials and efficiency. We observed that maintaining a stable in-house temperature is essential for packaging efficiency. Fluctuations in temperature can cause thermal expansion or contraction of materials, affecting their physical properties. For instance, temperature variations can alter the flow characteristics of powder particles, impacting their uniform distribution within packaging containers. Consequently, precise packaging becomes contingent on meticulously controlling in-house temperature, ensuring that materials retain their intended properties during packaging. Temperature differentials between the in-house environment (T_{in}) and external conditions (T_{ext}) can be quantified by Eq. 7.9.

$$\text{Temperature difference } = T_{\text{in}} - T_{\text{ext}} \tag{7.9}$$

Rainfall:

Rainfall events introduce a unique set of challenges to packaging efficiency and material integrity. We discerned that the occurrence of rain correlates with increased humidity and a concurrent decrease in temperature. This atmospheric dynamic has a direct bearing on packaging outcomes. We isolated data sets on days marked by rainfall to examine their impact. Notably, we observed a higher rate of packaging failures on rainy days. The elevation in humidity levels, combined with the cooling effect of rainfall, compounds the challenges posed by high humidity alone. The combined influence of these factors heightens the risk of material clumping and altered flow characteristics, leading to packaging errors. Rainfall introduces variations in humidity (ΔH) and temperature (ΔT), which are defined by the Eqs. 7.10 and 7.11, respectively.

$$\Delta H = H\text{rain} - H\text{dry} \tag{7.10}$$

$$\Delta T = T\text{rain} - T\text{dry} \tag{7.11}$$

These expressions quantify the differences in humidity and temperature between rainy and dry conditions. The impact of rainfall on packaging efficiency can be modeled by considering the change in humidity (ΔH) and temperature (ΔT) during rainfall events. The combined effect can be expressed by Eq. 7.12, where f is a function that incorporates the relationship between humidity, temperature, and packaging efficiency.

$$\text{Rainfall impact } = f(\Delta H, \Delta T) \tag{7.12}$$

Incorporating the above three factors and verifying the assumptions, we noted the dates when rainfall occurred and then extracted the data corresponding to those days. The analysis was particularly insightful when the data was compared to the failure percentage on these rainy days with the previous non-rainy days. The revelations were startling and confirmed the initial assumptions. According to the statistics, the failure rate was found to be significantly higher on rainy days than on dry ones. This result suggests that external variables, such as humidity and precipitation, have a substantial impact on the packing procedure and the flow of materials within the manufacturing facility.

In the above five images, we presented bar graphs for each of the five months, illustrating the failure percentages across all machines. Notably, these percentages represent the instances falling outside the upper and lower bands, thus reflecting the failed outcomes. The graphs did not encompass overfilling and underfilling within the lower and upper ranges.

7.4.2 Proposition II. Impact of Time Difference on Packaging Efficiency

The time difference (Δt) between the time of production (T_{p}) and the time of dumping (T_{d}) played a crucial role in packaging efficiency. A greater (Δt) implied a longer duration between production and packaging. When (Δt) exceeded a certain threshold, often set at 2 h (in this case, a range of time 0–2 h, 2–4 h, 4–6 h, and so on), there was a notable increase in packaging failures. Furthermore, with a hold time ($T_{\mathrm{h}} > 0$), the fail ratio increased even further, introducing an additional factor (T_{h}), into the equation, resulting in $T_{\mathrm{d}} - T_{\mathrm{p}} + T_{\mathrm{h}}$.

One key observation was that lumps formed in the materials as (Δt) become significantly larger than expected. This phenomenon became even more impactful in high humidity and rainfall conditions. The presence of lumps disrupted the uniform flow and distribution of powder materials like detergent within packaging containers. In conjunction with elevated humidity levels and rainfall, discussed earlier, this compounded the challenges associated with packaging efficiency.

The time difference (Δt) between the time of production (T_{p}) and the time of dumping (T_{d}) was a critical factor influencing packaging efficiency. A mathematical representation of the relationship between Δt, fail ratio (F_{r}), and hold time (H_{t}) can be expressed as Eq. 7.13, where f represents a function that considers both Δt and H_{t} (Fig. 7.5).

$$F_r = f(\Delta t, H_t) \tag{7.13}$$

In Figs. 7.6 and 7.7, time horizons and their corresponding failure percentages across various products and machines were comprehensively depicted. The recurring trend in this representation was prominent: the lowest failure percentages were consistently observed in the 0–2 and 2–4 time intervals.

This compelling observation demonstrated a consistent pattern across multiple machines and products. As time progressed beyond these initial intervals, a corresponding increase in the failure percentages was readily discerned. This trend reinforces the link between time and packaging efficiency.

7.5 Results

One pivotal aspect of the study was examining external factors and their collective impact on packaging efficiency. To signify this influence, we conducted a rigorous regression analysis with the dependent variable being the percentage change (P_c).

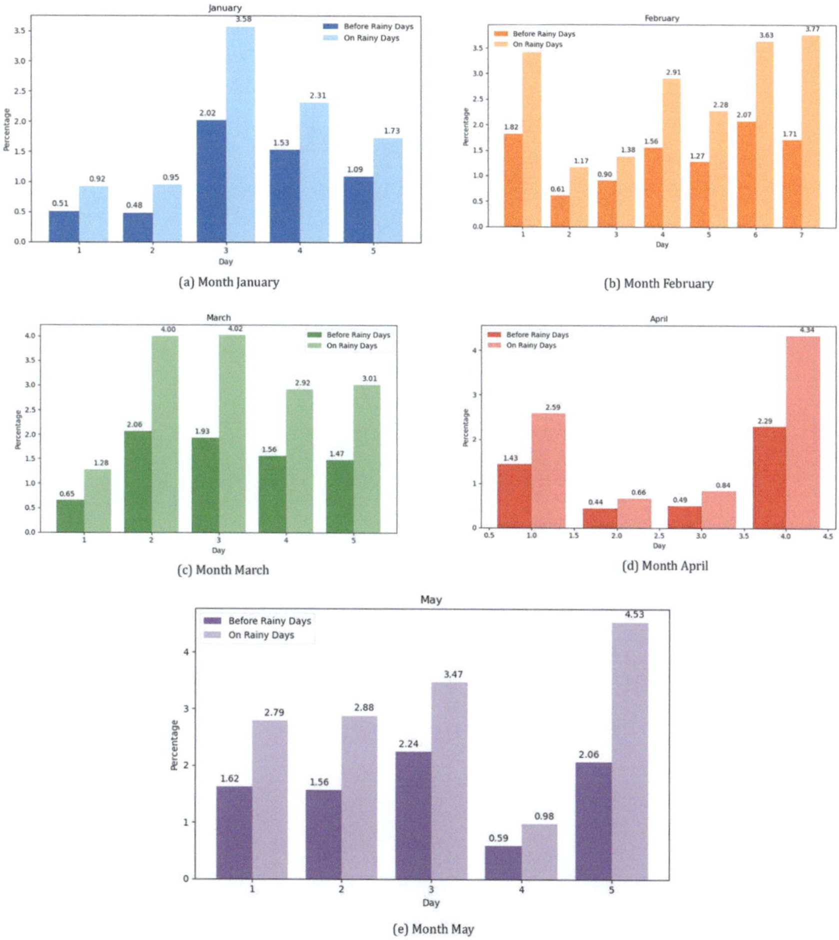

(a) Month January

(b) Month February

(c) Month March

(d) Month April

(e) Month May

Fig. 7.5 Monthly failure percentage with respect to rainy days

7.5.1 Correlation Analysis

The correlations among the variables were notably low, suggesting minimal to no linear relationships between the factors examined. This indicates that multicollinearity is likely not a major issue in this dataset, which is beneficial for conducting regression analysis (Figs. 7.8 and 7.9).

The objective was to understand how various external factors, such as humidity, rainfall, temperature, time differences between production and packaging, seasonal

	Product	P1				P2				P3			
	Month	April	February	January	March	April	February	January	March	April	February	January	March
Machine	time_interval												
M1	0-2hrs	0.009907	0.094178	0.015174	0.102114	0.028596	0.106946	0.128472	0.089479	0.087196	0.181018	0.376981	0.455227
	2-4hrs	0.011966	0.028652	0.117112	0.253607	0.056444	0.131657	0.038144	0.050169	0.110298	0.066483	1.018357	0.027779
	4-6hrs	0.022569	0.037777	0.209065	0.206269	0.04768	0.088883	0.217584	0.46558	2.773246	0.072446	0	0.050079
	greater than 6hrs	0.071498	0.100365	0.189178	0.104081	0.1491	0.523797	0.336859	0.09029	0.022103	0.230363	0.200807	0.294966

	Product	P1				P2				P3			
M2	Month	April	February	January	March	April	February	January	March	April	February	January	March
	0-2hrs	0.765983	1.913609		7.425001	1.111823	1.175538	1.063847	1.549692	0.860088			1.029837
	2-4hrs	1.162709	1.016801		2.038969	1.146335	1.319604	1.258845	1.3441	0.938875			1.523813
	4-6hrs	1.184141	6.043046		8.166189	0.892211	1.179068	1.284268	0.976505	0.942727			1.223364
	greater than 6hrs	1.37214	1.556439		3.271228	0.980279	1.198364	1.701042	1.631988	1.812806			1.581952

Fig. 7.6 M1 and M2

M3	Product	P1				P2				P3				P4				P5			
	Month	April	February	January	March	April	February	January	March	April	February	January	March	April	February	January	March	April	February	January	March
	0-2hrs		2.19007	3.27307	2.86765		3.12824	3.89624	3.65413		1.3946	3.98671	1.343		1.61975		1.74106		0.90233	1.00227	0.39867
	2-4hrs		3.83687	3.8681	3.51654		5.08927	5.67458	4.28427		1.75308	0.58447	1.26258		1.36549		2.34567		1.49289	1.14709	0.99472
	4-6hrs		6.21536		13.8754				4.48113		0.28195	2.47728	1.73659		1.62933		1.20284		1.08243	1.89966	0.98079
	greater than 6hrs		3.97271		3.42303			2.12479	5.27867		2.32144	0.97245	0.7794		1.68435		2.74873		0.98269	2.00289	1.09156

M4	Product	P2				P3				P4				P5				P6				P7		
	Month	April	February	January	March	April	February	January	March	April	February	January	March	April	February	January	March	April	February	January	March	April	February	January
	0-2hrs	0.04819	0.0245	0.1436	0.10771	1.22833		1.07945	0.08214	1.10386	1.54321	0.13475	0.79821	0.04303	0.18249	0.00229	0.01439	1.01713	1.18165	0.86836	0.52026	0.37402	1.10181	1.14877
	2-4hrs	0.11868	0.15656	0.21397	0.09821	3.61038		0.12862	1.31167	1.53312		1.40845	1.07644	0.32329	0.598	0.24931	0.30889	1.08441	0.93928	1.69652	3.27213	0.65171	2.28288	2.51139
	4-6hrs	0.06314	0.3955	0.17268	0.19626	1.56163		0.25988	1.50794	0.33501			0	0.05228	1.80849	0.22063	0.03345	0	2.21247	1.44806	4.56955	0.59256	0.49214	1.54623
	greater than 6hrs	0.21737	0.20158	0.17846	0.1224	1.26295		0.22262	0.42933		4.27934	0.63005	0.37906	0.07017	0.46947	0.02046	0.09557	1.53581	1.21331	1.33984	1.18811	0.90204	7.47457	1.46101

M5	Product	P1				P2				P5				P6				P7			
	Month	April	February	January	March	April	February	January	March	April	February	January	March	April	February	January	March	April	February	January	March
	0-2hrs	1.63512	1.99566	1.41789	1.36264	1.10236	0.50058	0.51356	0.4036	0.49785	1.69938	0.45484	1.36102	0.60868	0.41004	0.73632	0.28726	0.91793	0	0.34556	0.73226
	2-4hrs	3.27345	2.44578	2.93513	1.39468	1.17475	0.50562	0.98144	1.31885	0.88496	8.44875	1.39757	3.43807	1.42719	1.14443	0.81582	1.16618	0.63871	4.375	0.35717	0.7343
	4-6hrs	1.71403	2.35576	3.15552	1.60091	1.58873	0.59697	1.36719	1.60436	4.9292	0.15275	1.80869	5.85932		0.88638	0.42208	0.72341	1.70983	1.21355	1.61638	0.50708
	greater than 6hrs	2.2687	2.26849	1.8784	1.49402	1.34755	0.54502	2.46914	0.57645	8.83759	3.34844	1.13374	1.78649	0.69065	2.53889	1.3023	2.67784	0.76324	1.502	0.64061	0.41076

Fig. 7.7 M3,M4, and M5

	humidity	rainfall	temperature	timedifference	season_summer	season_winter	weather_moderate
humidity	1.000000	0.001073	-0.001406	-0.004202	-0.000754	0.002780	-0.002249
rainfall	0.001073	1.000000	0.011225	0.000975	0.000660	0.003573	-0.006657
temperature	-0.001406	0.011225	1.000000	-0.000751	0.001685	0.000759	0.001784
timedifference	-0.004202	0.000975	-0.000751	1.000000	0.001705	-0.007635	0.001099
season_summer	-0.000754	0.000660	0.001685	0.001705	1.000000	-0.003550	-0.001738
season_winter	0.002780	0.003573	0.000759	-0.007635	-0.003550	1.000000	-0.002625
weather_moderate	-0.002249	-0.006657	0.001784	0.001099	-0.001738	-0.002625	1.000000

Fig. 7.8 Correlation matrix of variables

	coef	std err	t	P>|t|	[0.025	0.975]
const	-58.3444	0.112	-520.828	0.000	-58.564	-58.125
rainfall	2.2427	0.009	236.774	0.000	2.224	2.261
timedifference	1.1147	0.001	1537.561	0.000	1.113	1.116
season_summer	0.9111	0.053	17.277	0.000	0.808	1.014
humidity	0.0233	3.18e-05	731.570	0.000	0.023	0.023
temperature	0.5011	0.002	320.964	0.000	0.498	0.504
weather_moderate	-0.0311	0.055	-0.567	0.570	-0.138	0.076
season_winter	-0.0511	0.055	-0.935	0.350	-0.158	0.056

Fig. 7.9 Statistical significance of multivariate regression model

variations, and weather conditions, collectively contribute to fluctuations in packaging efficiency. To this end, we employed a multivariate regression model, incorporating these external factors as independent variables. The multivariate equations defined by Eq. 7.14, where P_c, represents percentage change (Eq. 7.15) and $v_1, v_2, v_3, v_4, v_5, v_6, v_7$, represent humidity, rainfall, temperature, time difference (Eq. 7.16), summer season, winter season, and moderate weather, respectively.

$$Pc = \beta 0 + \beta 1v1 + \beta 2v2 + \beta 3v3 + \beta 4v4 + \beta 5v5 + \beta 6v6 + \beta 7v7 \tag{7.14}$$

$$P_{\mathrm{c}} = \frac{\text{target weight} - \text{actual weight}}{\text{target weight}} \times 100 \tag{7.15}$$

$$v_4 = T_{\mathrm{p}} - T_{\mathrm{d}} + T_{\mathrm{h}} \tag{7.16}$$

The results showed that the model explained a significant amount of the variation in the percentage change in weight. The adjusted R-squared value of 0.871 indicated that the model accounts for 87.1% of the data variation.

The multivariate model analysis indicated that the coefficients corresponding to the independent variables exhibited statistical significance, indicating a substantive influence on the percentage change (P_{c}). The coefficient associated with rainfall manifested a positive direction, signifying that an increased rainfall correlated with an augmented percentage change in weight.

This phenomenon could be attributed to the propensity of rainfall and humidity to induce moisture absorption, consequently forming lumps in the product.

Moreover, an elevation in humidity, temperature, or time-difference was positively linked to an increased percentage change in weight. The coefficient analysis further revealed that the season summer variable exhibited a positive coefficient, suggesting that the summer season was associated with a higher percentage change in weight. Conversely, the coefficients for the weather moderate and season winter variables displayed a negative orientation, indicating that moderate weather conditions and the winter season were linked to a lower percentage change in weight. A formerly overlooked variable became pivotal in ensuring precision by incorporating time difference considerations into the workflow. Most notably, efforts to reduce the failure ratio have been highly successful, as timely packaging curtailed material irregularities and deviations, fortifying the overall robustness of the processes. This optimization ensured that more of the packages met the specified weight criteria.

Simultaneously, we addressed the impact of external environmental factors, particularly humidity, on the packaging operations. Stabilizing the inhouse environment through temperature control led to heightened environmental resilience. The packaging process displayed remarkable consistency and predictability, even under challenging external conditions. These results underscored the approach's effectiveness in optimizing packaging efficiency and ensuring reliable outcomes.

The work is based on a proprietary dataset from a top multinational FMCG company operating in India. Due to confidentiality agreements, this dataset is not available for public disclosure. As outlined in the paper's introduction section, the findings can be generalized to similar products. Specifically, it applies to powder-like FMCG products, including detergents, Bournvita, Horlicks, and other comparable powdered goods.

This study uncovered notable fluctuations in machine performance that were associated with different months and product types, affecting packaging efficiency. The variability in machine efficiency was influenced by several environmental factors, such as temperature, rainfall, and humidity. These elements impacted the operational dynamics of the machines, leading to variations in performance. Consequently, the company faced challenges in maintaining consistent packaging efficiency. Further investigation into the interplay between these variables offered valuable insights into how to optimize machine settings and maintenance schedules, thereby improving efficiency regardless of environmental changes. This research was pivotal for stabilizing production processes and minimizing inefficiencies along the packaging line.

7.6 Conclusions

Optimizing packaging efficiency within the FMCG sector yielded results and transformative outcomes. We addressed challenges, introduced changes, and harnessed cutting-edge tools to shape a future of packaging precision and sustainability. This research underscored the dynamic interplay of environmental factors, time management, machine variability, and data-driven insights in the scope of packaging. Beyond mere efficiency gains, it painted a picture of responsible corporate practices, where sustainability and environmental stewardship are paramount. The impact of the findings extends beyond the confines of the FMCG sector. In an era where every gram matters, the research points the way toward a more efficient, sustainable, and responsible future in packaging practices.

7.7 Dataset

The analysis is based on a proprietary dataset from a top multinational FMCG company operating in India. Due to confidentiality agreements, this dataset is not available for public disclosure.

References

1. Simms, C., Trott, P.: Conceptualising the management of packaging within new product development: a grounded investigation in the UK fast moving consumer goods industry. Eur. J. Mark. **48**(11/12), 2009–2032 (2014)

2. Coughlin, R.: Detergent packaging. In: Second World Conference on Detergents: Looking Towards the 90's: Proceedings, vol. 2, p. 197. The American Oil Chemists Society (1987)
3. Fu, X., Mao, Y., Wang, N., Li, G., Sun, R., Lu, J.: Efficient numerical and ANN models for optimization of filler gradation of particulate-filled composites. Powder Technol. **426**, 118669 (2023)
4. Engel, G., Wilke, S., König, O., Harris, K., Leusen, F.: Powdersolve—a complete package for crystal structure solution from powder diffraction patterns. J. Appl. Crystallogr. **32**(6), 1169–1179 (1999)
5. Sharma, M., Luthra, S., Joshi, S., Kumar, A., Jain, A.: Green logistics driven circular practices adoption in industry 4.0 era: a moderating effect of institution pressure and supply chain flexibility. J. Cleaner Prod. **383**, 135284 (2023)
6. Xue, M., Chai, X.-S., Li, X., Chen, R.: Migration of organic contaminants into dry powdered food in paper packaging materials and the influencing factors. J. Food Eng. **262**, 75–82 (2019)
7. Ma, X., Park, C., Moultrie, J.: Factors for eliminating plastic in packaging: the European FMCG experts' view. J. Clean. Prod. **256**, 120492 (2020)
8. Jain, P., Hudnurkar, M.: Sustainable packaging in the FMCG industry. Cleaner Respons. Consump. **7**, 100075 (2022)
9. Eberly, L.E.: Multiple linear regression. Topics Biostat. 165–187 (2007)
10. Abu-Lebdeh, T., Damptey, R., Lamberti, V., Hamoush, S.: Powder packing density and its impact on slm-based additive manufacturing. In: TMS 2019 148th Annual Meeting & Exhibition Supplemental Proceedings, pp. 355–367. Springer, Berlin (2019)
11. Ma, X., Moultrie, J., et al.: Understand sustainable packaging design in practice. In: DS 92: Proceedings of the DESIGN 2018 15th International Design Conference, pp. 2693–2704.
12. Robalo, K., Soldado, E., Costa, H., do Carmo, R., Alves, H., Jùlio, E.: Efficiency of cement content and of compactness on mechanical performance of low cement concrete designed with packing optimization. Construct. Build. Mater. **266**, 121077 (2021)
13. Shaikh, S., Yaqoob, M., Aggarwal, P.: An overview of biodegradable packaging in food industry. Curr. Res. Food Sci. **4**, 503–520 (2021)
14. Feber, D., Granskog, A., Lingqvist, O., Nordigarden, D.: Sustainability in packaging: inside the minds of us consumers, McKinsey & Company (2020). Available online: https://www.mckinsey.com/industries/paper-forestproducts-and-packaging/our-insights/sustainability-in-packaginginside-the-minds-of-us-consumers. Accessed on 21 Oct 2020

Chapter 8
A Spherical Fuzzy-MARCOS-Based Decision-Making Model for Optimizing UAV Landing Zone Selection and Payload Delivery

Saptadeep Biswas, Rajdip Mahajan, Abhijit Baidya, and Uttam Kumar Bera

Abstract This research presents a sophisticated method for optimizing Unmanned Aerial Vehicle (UAV) landing zone selection and payload delivery in flood disaster responses. We propose the Spherical Fuzzy-Measurement of Alternatives and Ranking according to the COmpromise Solution (SF-MARCOS) technique for addressing the decision-making uncertainties. Criteria, including safety, visibility, communication range, emergency landing options, surface and terrain conditions, wind factors, altitude, and proximity to the disaster, undergo linguistic assessments with decision-makers. The method, tailored for multiple decision-makers, carefully balances ideal and anti-ideal trade-offs using the Spherical Weighted Arithmetic Mean (SWAM) operator. Resulting in a comprehensive ranking of landing zones and payload delivery strategies, this study underscores stakeholder perspectives' significance and criteria importance. Sensitivity analyses ensure adaptability to diverse disaster scenarios, enhancing reliability. This research contributes significantly to disaster management, logistics efficiency, and data-driven decision-making.

Keywords Unmanned aerial vehicles · Spherical fuzzy-MARCOS · Disaster response · Spherical weighted arithmetic mean · Decision-making

---

S. Biswas (✉) · R. Mahajan · A. Baidya · U. K. Bera
Department of Mathematics, National Institute of Technology Agartala, Agartala 799046, Tripura, India
e-mail: saptadeepbiswas7@gmail.com; saptadeepmath.sch@nita.ac.in; sbiswas3@gitam.edu

S. Biswas
Department of Operations and Supply Chain, GITAM School of Business, GITAM University, Visakhapatnam 530045, Andhra Pradesh, India

J. Maiti et al. (eds.), *Operations Research and Data Analytics: Current Trends and Future Perspectives*, Lecture Notes on Multidisciplinary Industrial Engineering,
https://doi.org/10.1007/978-981-96-8327-7_8

8.1 Introduction

The global landscape is witnessing an alarming surge in natural and man-made disasters, a trend evident in the EM-DAT database (Fig. 8.1a). From 2016 to 2020, a staggering 708.35 million people were directly impacted by the ravages of natural disasters, underscoring the escalating frequency and magnitude of catastrophic events globally. Floods, constituting 44% of all disasters and causing 31% of economic losses, stand out as one of the most prevalent and economically impactful natural disasters. From 2001 to 2022, the primary natural disaster was floods, representing the highest proportion at 66.5%, as indicated by the data from the EM-DAT database (Fig. 8.1b). Their severity varies from localized incidents to large-scale catastrophes triggered by heavy rainfall, snowmelt, storm surges, or dam failures. The severity and recurrence of floods are influenced by factors such as rainfall intensity, terrain, flood defences, and climate patterns, with climate change exacerbating their frequency and intensity. This surge in disaster occurrences emphasizes the urgent need for innovative disaster response and humanitarian logistics approaches.

Traditional relief transport, relying on trucks and rail, encounters challenges like inadequate transportation networks, traffic, route navigation difficulties, and adverse environmental conditions during the response stage. Integrating drone technology with conventional relief transport methods holds significant potential to overcome these challenges [1]. Drones collaborating with existing transportation modes address concerns about human errors and time consumption, providing a solution for efficient deliveries, especially in areas where conventional transport faces limitations [31]. Drones play a crucial role in disaster response by rapidly deploying to areas with limited or damaged infrastructure, a challenge for traditional transport. This is especially vital in disaster-affected regions where immediate and precise deliveries are essential for saving lives. The efficiency of drone technology in relief transport is evident in its ability to bypass road congestion and navigate challenging terrains, ensuring swift and targeted deliveries. Drones excel at transporting lightweight, time-sensitive supplies with speed and precision, including medical aid, food, and communication equipment. Integrating drones with traditional transport enhances the overall effectiveness of relief efforts and establishes a more resilient and adaptable response

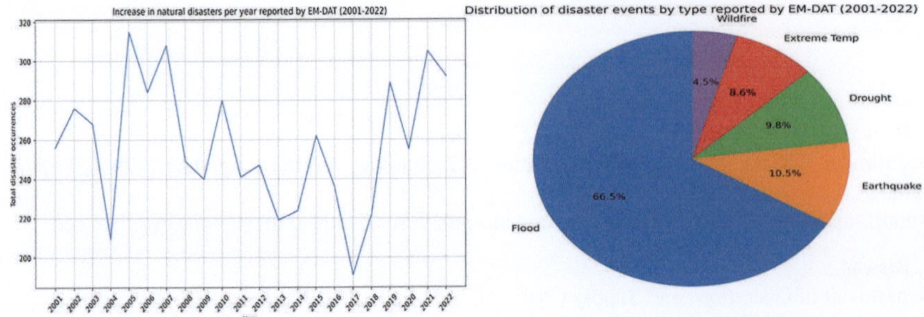

Fig. 8.1 Natural disasters reported by EM-DAT (2001–2022)

system for dynamic and unpredictable disaster scenarios. This collaborative approach optimizes the distribution process, fostering a coordinated and streamlined strategy in relief operations that leverages both technologies' strengths, overcoming their limitations.

In introducing technological advancements like drones, theoretical foundations are crucial in understanding their acceptance and feasibility [21]. The risk theory, a noteworthy conceptual foundation, explores public acceptance relying on perceived risk as a critical indicator [6]. Researchers and practitioners systematically address potential safety, privacy, and reliability concerns by adopting the risk theory for the feasibility study of drone-based deliveries [26]. This theoretical framework assists in cultivating public trust, ensuring regulatory compliance, and facilitating the successful integration of drone technology into conventional relief transport systems, providing valuable insights into the dynamics of public acceptance.

Unmanned Aerial Vehicles (UAVs) emerge as indispensable tools in natural and man-made disaster scenarios, offering various applications to enhance emergency response efforts [27]. These include establishing temporary communication networks, conducting visual inspections for structural stability, aiding in search and rescue missions [7], monitoring crowds for effective control, managing wildfires through detection, detecting chemical and radiation hazards, contributing to the crime scene and accident reconstruction, capturing imagery for anomaly detection, enhancing security through threat detection, and deploying rescue materials to specific locations [10, 12]. The adaptability of UAVs, equipped with advanced sensors and imaging technologies, enables them to provide valuable aerial perspectives and real-time data, significantly contributing to disaster response and management initiatives' efficiency, safety, and success.

Our research addresses critical challenges within disaster response scenarios, specifically optimizing UAV landing site selection and payload delivery efficiency during emergencies. Recognizing the potential synergy between UAVs and traditional relief transport methods, we aim to develop a comprehensive framework that leverages Multi-Criteria Decision-Making (MCDM) methodologies tailored for UAV landing site selection in disaster response scenarios. Integrating the Spherical Fuzzy Measurement of Alternatives and Ranking according to the COmpromise Solution (SF-MARCOS) MCDM method into our decision-making model is a key feature aiming to bridge the identified gaps in the existing literature. Our research scope extends beyond conventional approaches by emphasizing dynamic decision models that adapt to real-time changes in disaster scenarios. Applying the SF-MARCOS MCDM method will yield an efficient and adaptive framework, improving safety and overall operational effectiveness in UAV-based disaster response scenarios. Our research endeavours to fill a crucial gap by offering a comprehensive framework and practical tools that contribute valuable insights for Decision-Makers (DMs) engaged in disaster management and relief operations.

8.2 Literature Review

The UAV landscape in disaster management has witnessed remarkable advancements, with ground-breaking studies showcasing diverse applications and innovations [17]. Their versatility, mobility, and ability to access remote and disaster-stricken areas make them indispensable in these critical phases of disaster management. UAVs equipped with cameras and sensors can swiftly conduct aerial surveys of disaster-affected areas, enabling rapid assessment of damage, identification of hazards, and assessment of the needs of affected populations [28]. In search and rescue operations, drones equipped with thermal imaging cameras and infrared sensors can quickly cover large areas, increasing the chances of locating and rescuing distressed individuals. Furthermore, drones are pivotal in delivering medical supplies, monitoring logistics, conducting damage assessments, and ensuring connectivity in communication-deprived areas [11, 35]. Drone relief delivery has transformed disaster response and humanitarian logistics, offering unprecedented speed, access, and efficiency in aid distribution. Drones bridge logistical challenges, providing rapid response, reaching inaccessible areas, and ensuring last-mile delivery to remote populations. They are critical in transporting medical supplies, food, water, and vital equipment, reducing human risk and enabling data-driven decision-making.

Griffin [13] emphasizes the strategic importance of UAVs in disaster management, highlighting the need to navigate variables influencing operational capabilities, ensure safety and regulatory compliance, and address community concerns. Verykokou et al. [41] contribute significantly by evaluating both commercial and open-source software for 3D modelling of disaster scenes using UAV imagery, specifically tailored for Urban Search and Rescue (USaR) operations. Nimilan et al. [30] present the pioneering Drone-Aid, a Multipurpose Medical Drone that leverages GPS for swift responses to emergency locations, demonstrating versatility with life-saving components and telemetric transmission of vital signs. Zhao et al. [44] introduce a comprehensive UAV-aided earthquake emergency system, addressing disaster condition collection, on-site rescue, emergency communication, and image return. Liu and You [22] integrate an integer programming algorithm with the Dijkstra algorithm to optimize medical supply distribution by rescue drones in disaster areas, demonstrating efficiency in resource allocation post-Hurricane Maria. Yamazaki et al. [43] present innovative human detection system using UAVs in disaster sites, incorporating audio-processing-based approaches for enhanced search and rescue operations. Mathew et al. [24] explore consumer attitudes toward drone food delivery services, providing valuable insights for food delivery companies. Rajan et al. [34] categorize UAV applications in disaster management, offering a comprehensive overview of algorithms, sensing systems, communication architectures, and hardware considerations. Rottondi et al. [36] propose heuristic solutions for optimizing emergency tasks in post-disaster scenarios, introducing multitask mission planning for UAV fleets.

Pinto and Lagorio [32] provide a network design perspective for extending the operational range of drones through intermediate charging stations, addressing challenges in last-mile logistics. Lyu et al. [23] offer a comprehensive overview of UAVs in Search and Rescue operations, synthesizing research trends, exploring sensor integrations, and assessing various applications and methodologies. Yakushiji et al. [42] highlight the novel application of UAVs in short-range transportation for medical supplies during disasters in Japan, overcoming logistical challenges in urban areas. Edwards et al. [9] analyse adoption barriers to delivery drones for humanitarian operations among logistics service providers, shedding light on operational, supply chain, and behavioural factors influencing adoption. Dukkanci et al. [8] address the relief distribution problem using drones after earthquakes, considering uncertainties in demand and road networks and providing a comprehensive approach to optimizing relief logistics.

Recent research has introduced innovative methodologies in the context of decision-making under uncertainty and multiple criteria. Jaller and Otay [15] address the challenges of evaluating sustainable vehicle technologies, utilizing AHP and TOPSIS with Spherical Fuzzy Sets (SFSs). Kutlu Gu¨ndo˘gdu and Kahraman [19] extend classical AHP with SF-AHP, showcasing superior performance in renewable energy applications. Buyuk and Temur [4] propose a novel framework for food waste management using Spherical Fuzzy AHP, addressing linguistic and uncertain aspects comprehensively.

Lin et al. [20] present a MULTIMOORA-based MCDM model for car-sharing station site selection under a picture-fuzzy environment. Büyüközkan and Güler [5] introduce a combined hesitant fuzzy MCDM approach for evaluating Supply Chain Analytics tools. Simić et al. [39] extend the WASPAS method under the picture fuzzy environment for Last-Mile Delivery mode selection. Boyacı and Şişman [3] propose a GIS-based MCDM approach for pandemic hospital site selection using Pythagorean fuzzy sets. Nguyen et al. [29] pioneer a hybrid PF-AHP and PF-CoCoSo method for AI technology adoption, emphasizing managerial capabilities. Sotoudeh-Anvari [40] conducted a comprehensive review of the applications of MCDM methods in addressing COVID-19 challenges, emphasizing the prevalent use of fuzzy sets. Rahnamay Bonab et al. [33] present a decision support model for logistic autonomous vehicle assessment using a Spherical Fuzzy set integrated Choquet Integral approach. Memarpour Ghiaci et al. [25] utilized MARCOS methodology within Spherical Fuzzy framework and simulation modelling to improve the emergency departments amid COVID-19 pandemic. Jawad et al. [16] introduce an automatic spherical fuzzy AHP algorithm for portfolio selection, emphasizing the significance of addressing uncertainty. Jafarzadeh Ghoushchi et al. [14] integrate SWARA with MARCOS under a Spherical Fuzzy set environment for road safety assessment and risk prioritization. Sharaf [38] introduces novel approaches for spherical fuzzy TOPSIS and VIKOR, showcasing versatility in complex evaluations.

The literature has elucidated the robust capabilities of UAVs in disaster management, emphasizing their rapid deployment, access to remote areas, and collection of real-time data. Concurrently, the strength of MCDM methodologies, particularly in handling uncertainty, has been discussed. However, a notable research gap has been identified within the existing literature—specifically, the absence of substantial work on MCDM-based solutions tailored for selecting optimal landing sites for UAVs in disaster response scenarios. While individual criteria for UAV landing site selection have been explored, a holistic decision-making framework that integrates various factors simultaneously is lacking. Moreover, an absence of research focuses on the adaptability of decision models to real-time changes in disaster scenarios and the explicit prioritization of safety criteria and emergency landing options for UAVs. This research gap underscores the need for innovative studies that leverage MCDM approaches to enhance UAV operations' efficiency, safety, and overall effectiveness in disaster management.

In addressing the notable research gap in existing literature, our study presents a pioneering framework for optimizing UAV landing site selection and payload delivery during disaster response through the application of the SF-MARCOS MCDM method. Recognizing the limited focus on MCDM-based solutions tailored for UAV landing site selection in disaster scenarios, our research bridges this gap by integrating a comprehensive set of criteria and considering the uncertainties inherent in dynamic environments. This innovative approach contributes to the academic discourse surrounding disaster management and offers a practical tool for DMs, enabling them to navigate the complexities of disaster response operations effectively.

Our research endeavours to answer the following key questions:

- How can MCDM methodologies seamlessly integrate into the UAV landing site selection process for enhanced disaster response scenarios?
- What criteria are indispensable for the optimization of UAV landing sites in disaster-affected areas, giving due consideration to safety, efficiency, and adaptability?
- In what ways can the SF-MARCOS MCDM method contribute to the development of dynamic decision models capable of adapting to real-time changes in disaster scenarios?
- What is the impact of prioritizing safety criteria and emergency landing options on the overall effectiveness of UAV operations during disaster response?

Our research aspires to serve as a catalyst for practical advancements in disaster response operations. The interdisciplinary nature of our approach, combining UAV technology, MCDM methodologies, and a safety-centric focus, positions our framework as a valuable asset for stakeholders involved in mitigating the impact of disasters and safeguarding communities. Through this holistic perspective, we aim to inspire a new wave of research and application in the realm of disaster management, reinforcing the vital role of technology and decision-making in creating resilient and adaptive response systems.

8.3 Prelamination of Spherical Fuzzy Sets

The Spherical Fuzzy Set (SFS) theory, introduced by Kutlu Gundo˘gdu and Kahraman [18], represents a relatively recent extension of fuzzy set theory. This section delves into the characteristics, arithmetic operations, and fundamental principles that underlie SFSs.

Definition 1 According to Kutlu Gu¨ndo˘gdu and Kahraman [18], an SFS Z *of* the universe of discourse X is given by.

$$\boldsymbol{Z} = \{(\boldsymbol{x} \cdot (\boldsymbol{\mu}_Z(\boldsymbol{x}), \boldsymbol{\nu}_Z(\boldsymbol{x}), \boldsymbol{\pi}_Z(\boldsymbol{x})))|\boldsymbol{x} \in \boldsymbol{X}\} \tag{8.1}$$

where

$\mu_Z : X \to [0, 1]$: the membership degree for every $x \in X$ in the SFS Z,

$\nu_Z : X \to [0, 1]$: the non-membership degree for every $x \in X$ in the SFS Z,

$\pi_Z : X \to [0, 1]$: the hesitance degree for every $x \in X$ in the SFS Z.
Also, the following constraint holds:

$$0 \leq (\boldsymbol{\mu}_Z(\boldsymbol{x}))^2 + (\boldsymbol{\nu}_Z(\boldsymbol{x}))^2 + (\boldsymbol{\pi}_Z(\boldsymbol{x}))^2 \leq 1. \tag{8.2}$$

Definition 2 [18]

Suppose $Z_1 = (\mu_1, \nu_1, \pi_1)$ and $Z_2 = (\mu_2, \nu_2, \pi_2)$ are two spherical fuzzy numbers (SFNs) and let $k \geq 0$. The mathematical operations that govern Z_1 and Z_2 are presented below:

$$Z_1 \oplus Z_2 = \left(\sqrt{\mu_1^2 + \mu_2^2 - \mu_1^2\mu_2^2}, \nu_1\nu_2, \sqrt{(1 - \mu_2^2)\pi_1^2 + (1 - \mu_1^2)\pi_2^2 - \pi_1^2\pi_2^2}\right) \tag{8.3}$$

$$Z_1 \otimes Z_2 = \left(\mu_1\mu_2, \sqrt{\nu_1^2 + \nu_2^2 - \nu_1^2\nu_2^2}, \sqrt{(1 - \nu_2^2)\pi_1^2 + (1 - \nu_1^2)\pi_2^2 - \pi_1^2\pi_2^2}\right) \tag{8.4}$$

$$k \cdot Z = \left(\sqrt{1 - (1 - \mu_Z^2)^k}, \nu_Z^k, \sqrt{(1 - \mu_Z^2)^k - (1 - \mu_Z^2 - \pi_Z^2)^k}\right) \tag{8.5}$$

$$Z^k = \left(\mu_Z^k, \sqrt{1 - (1 - \nu_Z^2)^k}, \sqrt{(1 - \nu_Z^2)^k - (1 - \nu_Z^2 - \pi_Z^2)^k}\right) \tag{8.6}$$

Definition 3 [19]

SFS represented by $Z_1 = (\mu_1, \nu_1, \pi_1)$ and $Z_2 = (\mu_2, \nu_2, \pi_2)$ adhere to the following rules under the conditions $k, k_1, k_2 > 0$:

$$Z_1 \oplus Z_2 = Z_2 \oplus Z_1 \tag{8.7}$$

$$Z_1 \otimes Z_2 = Z_2 \otimes Z_1 \tag{8.8}$$

$$k(Z_1 \oplus Z_2) = kZ_1 \oplus kZ_2 \tag{8.9}$$

$$k_1 Z_1 + k_2 Z_1 = (k_1 + k_2) Z_1 \tag{8.10}$$

$$(Z_1 \otimes Z_2)^k = Z_1^k \otimes Z_2^k \tag{8.11}$$

$$Z^{k_1} \otimes Z^{k_2} = Z^{(k_1+k_2)} \tag{8.12}$$

Definition 4 [18]

Consider a SFN denoted as $Z = (\mu_Z, \nu_Z, \pi_Z)$. The Score Value (SV) and Accuracy Function (AF) of Z are determined through the following expressions:

$$\text{Score}(Z) = \mu_Z^2 + \nu_Z^2 + \pi_Z^2 \tag{8.13}$$

$$\text{Accuracy}(Z) = \mu_Z^2 + \nu_Z^2 + \pi_Z^2 \tag{8.14}$$

$Z_1 < Z_2$ is true if and only if

$$\begin{aligned} &\text{score}(Z_1) < \text{score}(Z_2) \text{ or} \\ &(\text{score}(Z_1) = \text{score}(Z_2) \text{ and Accuracy}(Z_1) < \text{Accuracy}(Z_2)) \end{aligned} \tag{8.15}$$

Definition 5 [18]

The calculation of Single-Valued Spherical Weighted Arithmetic Mean (SWAM) corresponding to the weight parameter w is based on the given equation:

If there are $'n'$ no's of SFS such that $Z_i, \forall i = 1, 2, \ldots, n$

$$\begin{aligned} &SWAM_w(Z_1,\ Z_2, \ldots, Z_n) \\ &= w_1\tilde{Z}_1 + w_2\tilde{Z}_2 + \ldots + w_n Z_n \\ &= \left(\left[1 - \prod_{i=1}^{n}(1-\mu_{Z_i}^2)^{w_i}\right]^{1/2}, \prod_{i=1}^{n}\nu_Z^{w_i}, \left[\prod_{i=1}^{n}(1-\mu_{Z_i}^2)^{w_i} - \prod_{i=1}^{n}(1-\mu_{Z_i}^2-\pi_{Z_i}^2)^{w_i}\right]^{1/2}\right) \end{aligned} \tag{8.16}$$

where $w_i \in [0, 1], \sum_{i=1}^{n} w_i = 1$

Definition 6 [18]

The calculation of Single-Valued Spherical Weighted Geometric Mean (SWGM) relative to the weight parameter w is determined by the following equation:

If there are $'n'$ no's of SFS such that $Z_i, \forall i = 1, 2, \ldots, n$

$$SWGM_w(Z_1, Z_2, \ldots, Z_n) \\ = Z_1^{w_1} + Z_2^{w_2} + \cdots + Z_n^{w_n} \\ = \left(\prod_{i=1}^{n} \mu_{Z_i}^{w_i}, \left[1 - \prod_{i=1}^{n} (1 - \nu_{Z_i}^2)^{w_i} \right]^{1/2}, \left[\prod_{i=1}^{n} (1 - \nu_{Z_i}^2)^{w_i} - \prod_{i=1}^{n} (1 - \nu_{Z_i}^2 - \pi_{Z_i}^2)^{w_i} \right]^{1/2} \right) \tag{8.17}$$

where $w_i \in [0, 1], \sum_{i=1}^{n} w_i = 1$

Definition 7 (*Distance between Two SFNs and Normalized Spherical Distance*)

$$\text{dis}(A, B) = \arccos\left(1 - \frac{1}{2}\left((\mu_A - \mu_B)^2 + (\nu_A - \nu_B)^2 + (\pi_A - \pi_B)^2 \right) \right) \tag{8.18}$$

$$\text{dis}(A, B) = \frac{2}{\pi} \arccos\left(1 - \frac{1}{2}\left((\mu_A - \mu_B)^2 + (\nu_A - \nu_B)^2 + (\pi_A - \pi_B)^2 \right) \right) \tag{8.19}$$

$$\text{dis}^n(A, B) = \frac{2}{n\pi} \sum_{i=1}^{n} \arccos\left(1 - \frac{1}{2}\left((\mu_A - \mu_B)^2 + (\nu_A - \nu_B)^2 + (\pi_A - \pi_B)^2 \right) \right) \tag{8.20}$$

This preliminary discussion introduces the foundational concepts of SFSs, providing clarity on their definitions, arithmetic operations, and key properties.

8.4 Problem Statements

The challenge of selecting optimal landing sites for UAVs is critical, especially in disaster response scenarios such as flood management. This multifaceted problem involves navigating various considerations to ensure UAV operations' safety, efficiency, and success. Safety is paramount, necessitating the identification of landing locations devoid of obstacles and hazards, ranging from trees and buildings to debris in flood-affected areas. The adaptability of UAVs to diverse terrains, including urban environments and rugged landscapes, is crucial for successful landings in different conditions. Anticipating unforeseen circumstances, such as emergency landings, requires strategic planning and the prioritization of suitable emergency landing zones.

Moreover, the selected landing sites must fall within the UAV's communication range to maintain seamless connectivity with operators and ground control, ensuring effective control and real-time data transmission. In disaster response scenarios, landing sites often serve as crucial points for payload delivery. Therefore, optimizing landing sites to accommodate safe and efficient payload drop-offs becomes vital, contributing significantly to the overall success of relief efforts. Considering adverse weather conditions common in disaster scenarios, evaluating the feasibility of safe and effective landings is essential. Integration into broader humanitarian logistics operations is a key aspect, emphasizing the coordination of landing sites with logistics activities for the seamless delivery of aid and essential supplies to affected areas. Advanced navigation systems and precision landing technologies ensure accurate UAV landings, especially in challenging conditions. Addressing the UAV landing site selection problem requires a holistic approach, combining strategic planning, technological innovation, and a deep understanding of the dynamic and unpredictable nature of disaster-affected environments to enhance the overall effectiveness of UAV operations.

8.4.1 Criteria Considerations in UAV Landing Site Selection During Relief Delivery

In the context of UAV landing in disaster response scenarios, the criteria considerations play a pivotal role in ensuring the safety and effectiveness of operations. Each criterion is strategically designed to address specific aspects of the landing site selection process, contributing to the success of UAV activities.

After a comprehensive literature review, we have selected a total of ten criteria. Ensuring safe arrangements (C_1) is paramount as it directly influences the well-being and security of the UAV and its surroundings [2]. This encompasses hazard identification, clearance of landing zones, risk mitigation, emergency preparedness, and adherence to safety protocols. Prioritizing safe arrangements is fundamental for UAV operators and stakeholders to minimize risks, prevent accidents, and uphold safety standards. Visibility and Line of Sight (C_2) ensure unobstructed communication and operational control during the landing phase. This involves assessing visibility conditions for clear communication between the UAV and operators, facilitating precise navigation, and immediate response to potential challenges. Emphasizing line of sight contributes to overall safety by preventing collisions and accidents, particularly in disaster response scenarios where optimal visibility is critical. Communication Range (C_3) underscores the importance of maintaining seamless connectivity between the UAV and operators or ground control during the landing phase. This ensures the selected landing site falls within the UAV's communication range, allowing continuous and reliable data transmission. Prioritizing the Communication Range is imperative for DMs seeking to optimize the success and safety of UAV operations, especially in disaster response scenarios where effective communication is pivotal. Emergency Landing Option (C_4) enhances UAV operations' safety and preparedness during the critical landing phase. This criterion emphasizes anticipating unforeseen circumstances and establishing suitable emergency landing zones. Offering operational flexibility, predetermined emergency landing zones enable the UAV to adapt to changing conditions without compromising safety, contributing to the overall effectiveness of emergency response efforts. Surface Condition (C_5) focuses on the critical assessment and adaptability of the landing surface to ensure successful and safe landings. This becomes paramount in disaster response scenarios, encompassing the evaluation of the UAV's ability to adapt to various surfaces and the stability of the landing surface to prevent accidents.

Distance from Disaster (C_6) strategically addresses the proximity of the landing site to the disaster-affected area. This criterion is crucial for optimizing response efforts by minimizing the distance between the landing site and affected areas, ensuring swift UAV deployment for crucial activities such as assessment, surveillance, and payload delivery. Terrain and Topography Conditions (C_7) direct attention to the evaluation of natural features influencing the operational capabilities of the UAV. This criterion is essential for assessing topographic complexity, elevation changes, and the presence of vegetation or obstacles in the landing area, ensuring the effective adaptation of UAV operations to various landscapes. Wind Conditions (C_8) underscore the importance of evaluating factors related to wind that could affect the safety and stability of the landing. This includes assessing the speed and direction of the wind in the landing area and considering its impact on the UAV's stability during the landing phase. Altitude and Obstructions (C_9) emphasize assessing altitude-related factors and potential obstacles that could influence the safety and effectiveness of the landing. DMs ensure that UAVs can safely navigate obstacles and effectively land at varying altitudes, contributing to the overall success and safety of landing operations. Distance from People and Property (C_{10}) places paramount importance on assessing the proximity of the landing site to people and property. DMs evaluate safety considerations to minimize the risk of accidents or damage to the UAV and the surrounding environment during the landing phase.

The comprehensive consideration of these criteria in UAV landing site selection for disaster response ensures operations' safety, effectiveness, and success. The strategic integration of these factors contributes to the overall optimization of UAV activities in critical scenarios, aligning with the evolving landscape of disaster management.

8.4.2 Landing Sites

The study draws inspiration from Guwahati's urban flood hazard zones, as [37] classified, encompassing factors like topography, hydrology, historical flood data, and vulnerability assessments. Guwahati's division into five urban flood hazard zones aids in assessing and categorizing flood risk levels across different city areas. While the information does not explicitly outline specific UAV landing sites, it underscores the significance of considering various factors when identifying suitable landing alternatives for flood response operations. These factors include safety arrangements, visibility, communication range, emergency landing options, surface conditions, distance from people and property, terrain and topography, wind conditions, altitude, obstructions, and distance from the disaster. The implication is that the classification of flood hazard zones informs decision-making processes related to resource deployment, such as the strategic use of UAVs, in flood response operations within a city. Considering these factors ensures a comprehensive and informed approach to UAV landing site selection, aligning with the broader goal of optimizing disaster response efforts in urban flood-prone areas.

8.5 Methodology

8.5.1 DMs' Weight

Experts in real-world group decision-making scenarios may possess varying knowledge, skills, and experience, resulting in different interpretations of attribute importance. In our study, DMs were carefully chosen based on a comprehensive skill set encompassing expertise in disaster response, proficiency in UAV technology, awareness of geospatial considerations, emergency management experience, regulatory compliance knowledge, risk assessment capabilities, a humanitarian perspective, and adaptability to innovative solutions. This meticulous selection process ensures that the data considered for evaluating UAV landing site selection during the disaster response phase is both reliable and appropriate, allowing for effective navigation of the complex and dynamic landscape of disaster response using UAVs.

8.5.2 Criteria Weight

Criteria play a crucial role in decision-making. The SWAM operator is employed to aggregate criteria weights, offering a comprehensive approach that considers uncertainties associated with SFSs. The SWAM operator is defined as follows:

$$\mathrm{SWAM}_w(Z_1, Z_2, \ldots, Z_n) = w_1 Z_1 + w_2 Z_2 + \cdots + w_n Z_n$$
$$= \left(\left[1 - \prod_{i=1}^{n} (1 - \mu_{Z_i}^2)^{w_i} \right]^{1/2}, \prod_{i=1}^{n} \nu_Z^{w_i}, \left[\prod_{i=1}^{n} (1 - \mu_{Z_i}^2)^{w_i} - \prod_{i=1}^{n} (1 - \mu_{Z_i}^2 - \pi_{Z_i}^2)^{w_i} \right]^{1/2} \right)$$

Here, $Z_1, Z_2, \ldots, Z_n$ represents the SFSs associated with different criteria, and $w_1, \ldots, w_n$ denote the corresponding weights assigned to these criteria.

To apply the SWAM operator for aggregating criteria weights:

Table 8.1 Correspondence of linguistic terms to SFNs

Linguistic variable	μ	ν	π
Absolutely more importance (AMI)	0.9	0.1	0.1
Very high importance (VHI)	0.8	0.2	0.2
High importance (HI)	0.7	0.3	0.3
Slightly more importance (SMI)	0.6	0.4	0.4
Equally importance (EI)	0.5	0.5	0.5
Slightly low importance (SLI)	0.4	0.6	0.4
Low importance (LI)	0.3	0.7	0.3
Very low importance (VLI)	0.2	0.8	0.2
Absolutely low importance (ALI)	0.1	0.9	0.1

1. Assign SFSs $Z_1, Z_2, \ldots, Z_n$ to represent the weights associated with each criterion.
2. Assign weights $w_1, \ldots, w_n$to denote the importance of each criterion.
3. Use the SWAM operator formula to calculate the aggregated weight, resulting in a new SFS representing the combined weight, considering the uncertainties associated with each criterion.

This aggregated weight comprehensively represents the overall importance of criteria in a Spherical Fuzzy context and can be utilized in decision-making processes.

8.5.3 Proposed Spherical Fuzzy MARCOS Method

Step 1: Conversion of Linguistic Variables Expert expressions of linguistic variables are transformed into Spherical Fuzzy Numbers (SFNs) using Table 8.1.

Step 2: Construction of the Decision Matrix Assess each alternative against each criterion using SFNs. Employ the SWAM operator and aggregated weight to create the group decision matrix for alternative performance for each criterion.

Step 3: Weighted Decision Matrix Form the weighted decision matrix v_{ij} by multiplying matrix elements with corresponding criteria weight coefficients based on Eq. (8.21).

$$
\begin{aligned}
v_{ij} &= w_j \otimes x_{ij} = \left(\mu_{\mathrm{j}}, \nu_{\mathrm{j}}, \pi_{\mathrm{j}}\right) \otimes \left(\mu_{\mathrm{ij}}, \nu_{\mathrm{ij}}, \pi_{\mathrm{ij}}\right) \\
&= \left(\mu_{\mathrm{ij}}^{v}, \nu_{\mathrm{ij}}^{v}, \pi_{\mathrm{ij}}^{v}\right) \\
&= \left(\mu_{\mathrm{j}}\mu_{\mathrm{ij}}, \sqrt{\nu_j^2 + \nu_{ij}^2 - \nu_{\mathrm{j}}^2\nu_{ij}^2}, \sqrt{\left(1 - \nu_{\mathrm{ij}}^2\right)\pi_j^2 + \left(1 - \nu_{\mathrm{j}}^2\right)\pi_{ij}^2 - \pi_{\mathrm{j}}^2\pi_{ij}^2}\right)
\end{aligned} \tag{8.21}
$$

Step 4: Ideal and Anti-Ideal Values Determine the ideal ($\tilde{A}_{id}$) and anti-ideal ($\tilde{A}_{ai}$) values.

$$\tilde{A}_{ai} = \min_{1 \le i \le m} x_{ij}, \quad j \in B^{\max}, \quad \max_{1 \le i \le m} x_{ij}, \quad j \in C^{\min} \tag{8.22}$$

$$\tilde{A}_{id} = \max_{1 \le i \le m} x_{ij}, \quad j \in B^{\max}, \quad \min_{1 \le i \le m} x_{ij}, \quad j \in C^{\min} \tag{8.23}$$

Step 5: Utility Degrees of Alternatives

Calculate utility degrees (K_i^+ and K_i^-) of alternatives using Eqs. (8.24) and (8.25).

$$K_i^+ = \frac{2}{n\pi} \sum_{i=1}^{n} \arccos\left(\mu_{ij}^{v}.\mu_{idj} + v_{ij}^{v}.v_{idj} + \pi_{ij}^{v}.\pi_{idj}\right) \tag{8.24}$$

$$K_i^- = \frac{2}{n\pi} \sum_{i=1}^{n} \arccos\left(\mu_{ij}^{v}.\mu_{aij} + v_{ij}^{v}.v_{aij} + \pi_{ij}^{v}.\pi_{aij}\right) \tag{8.25}$$

Step 6: Utility Functions and Ranking

Calculate utility functions ($f\left(K_i^+\right)$ and $f\left(K_i^-\right)$) and rank alternatives using Eq. (8.26).

$$F = \frac{K_i^+ + K_i^-}{1 + \frac{1-f\left(K_i^+\right)}{f\left(K_i^+\right)} + \frac{1-f\left(K_i^-\right)}{f\left(K_i^-\right)}} \tag{8.26}$$

where, $f\left(K_i^+\right) = \frac{K_i^+}{K_i^+ + K_i^-}$ and $f\left(K_i^-\right) = \frac{K_i^-}{K_i^+ + K_i^-}$

Step 7: Ranking

Rank the alternatives according to F_i in descending order (Fig. 8.2).

8.6 Results

This section presents the outcomes of the devised methodology for optimizing UAV landing site selection and enhancing payload delivery efficiency in emergencies. The study introduces a novel approach employing SF-MARCOS MCDM methods within an uncertain environment.

We establish a linguistic framework for evaluating the significance of criteria in the context of UAV landing zone selection during disaster response scenarios. Table 8.1 presents a set of linguistic terms and their corresponding Spherical Fuzzy Numbers (SFNs) utilized by DMs to express the importance of each criterion. These linguistic variables include terms such as Absolutely More Importance (AMI), Very

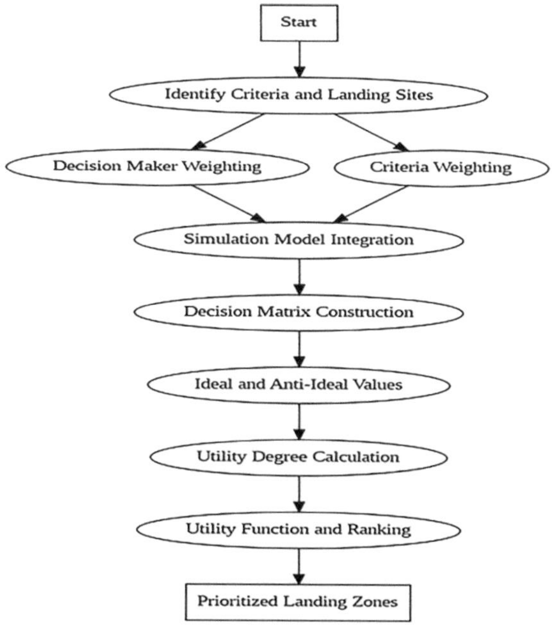

Fig. 8.2 Flowchart of spherical fuzzy MARCOS method for optimizing UAV landing zone selection

High Importance (VHI), High Importance (HI), Slightly More Importance (SMI), Equally Importance (EI), Slightly Low Importance (SLI), Low Importance (LI), Very Low Importance (VLI), and Absolutely Low Importance (ALI). The SFNs associated with each linguistic variable, denoted by μ, v, and π, capture the degree of membership, non-membership, and hesitancy, respectively, in a comprehensive and flexible manner.

In Table 8.2, we observe evaluations provided by three DMs ($D1$, $D2$, and $D3$) for each criterion (C_1 through C_{10}). The DMs express their perspectives using the linguistic terms established in Table 8.1. For instance, D1 considers C_1 to be of Absolutely More Importance (AMI), while D2 and D3 perceive it as Very High Importance (VHI). These diverse evaluations highlight the subjectivity in decision-making processes, particularly in complex tasks such as UAV landing zone selection.

To systematically integrate these linguistic expressions into a quantitative framework, we employ the SWAM operator, as defined in Eq. (8.16). This operator facilitates the aggregation of linguistic evaluations, considering the uncertainties associated with SFSs. The calculation for the aggregated SF weights of Criterion C_1 is provided as an illustrative example.

The calculation on the SWAM operator for criteria C_1 is as follows:

$$\mathrm{SWAM}_w((0.9, 0.1, 0.1), (0.8, 0.2, 0.2), (0.8, 0.2, 0.2))$$
$$= [\sqrt{1-\left(1-0.9^2\right)^{0.5}.\left(1-0.8^2\right)^{0.3}.\left(1-0.8^2\right)^{0.2}}, 0.1^{0.5}.0.2^{0.3}.0.2^{0.2},$$

Table 8.2 Significance of criteria using spherical fuzzy linguistic variables

Criteria	*D*1	*D*2	*D*3
C_1	AMI	VHI	VHI
C_2	HI	SMI	HI
C_3	HI	VHI	HI
C_4	VHI	HI	HI
C_5	VHI	VHI	HI
C_6	LI	SLI	VLI
C_7	VHI	HI	VHI
C_8	SMI	SLI	SMI
C_9	SLI	LI	LI
$C10$	SLI	SLI	VLI

$$\sqrt{\begin{array}{c}\left(1-0.9^2\right)^{0.5}.\left(1-0.8^2\right)^{0.3}.\left(1-0.8^2\right)^{0.2}-\\ \left(1-0.9^2-0.1^2\right)^{0.5}.\left(1-0.8^2-0.2^2\right)^{0.3}.\left(1-0.8^2-0.2^2\right)^{0.2}\end{array}}$$

$$= \left[\sqrt{1-0.436.0.736.0.815}, 0.316.0.617.0.725,\right.$$

$$\left.\sqrt{0.436.0.736.0.815-0.424.0.710.0.796}\right]$$

$$= \left[\sqrt{1-0.262}, 0.141, \sqrt{0.262-0.240}\right]$$

$$= [0.859, 0.141, 0.147]$$

This process, illustrating the aggregation of linguistic evaluations using the SWAM operator, is integral to our approach for deriving aggregated Spherical Fuzzy weights. Similar calculations are performed for all criteria and the aggregated SF weights for C_1 through C_{10} are presented in Table 8.3. These weights reflect a comprehensive and nuanced understanding of the significance of each criterion, incorporating the varied perspectives of DMs. The SF approach enhances the representation of uncertainty and subjectivity in decision-making, contributing to a more robust and reliable framework for UAV landing zone selection in disaster response scenarios.

Tables 8.4, 8.5 and 8.6 outline the assessments made by three DMs (D1, D2, and D3) on different criteria (C_1 to C_{10}) for each alternative (R_1 to R_5). These evaluations employ linguistic terms to convey each criterion's perceived importance or suitability for the respective alternatives. In Table 8.8.4, D1's subjective judgments are reflected in the assessments. For example, alternative R_1 is evaluated with Very High Importance (VHI) for C_1, High Importance (HI) for C_2, Slightly More Importance (SMI) for C_3, and so forth. Similarly, Table 8.5 presents *D*2's evaluations for the same set of alternatives and criteria. *D*2 assigns High Importance (HI) to C_1 for R_1, Very High Importance (VHI) to C_2, and so on, representing *D*2's perspective on the importance of each criterion for the given alternatives. Table 8.6 captures *D*3's

Table 8.3 Aggregate weight of criteria using SWAM operator

Criteria	μ	v	π
C_1	0.859	0.141	0.147
C_2	0.674	0.327	0.330
C_3	0.735	0.266	0.269
C_4	0.756	0.245	0.249
C_5	0.784	0.217	0.219
C_6	0.319	0.686	0.324
C_7	0.775	0.226	0.229
C_8	0.553	0.452	0.402
C_9	0.355	0.648	0.357
$C10$	0.370	0.636	0.375

assessments for the alternatives and criteria, expressed through linguistic terms like High Importance (HI) and Very High Importance (VHI).

The calculation of the aggregated group decision matrix in Table 8.7 involves the utilization of the SWAM operator. This operator plays a crucial role in integrating the linguistic evaluations provided by DMs, as outlined in Tables 8.4, 8.5, and 8.6. These tables capture the diverse perspectives of DMs regarding the importance or suitability of criteria (C_1 through C_{10}) for each alternative (R_1 through R_5). The SWAM operator, defined in Eq. (8.16), systematically processes the linguistic evaluations represented as SFNs. Combining these SFNs, the SWAM operator produces aggregated weights of μ, v, and π for each criterion-alternative pair. These aggregated weights are then

Table 8.4 Criteria-alternative evaluations for $D1$

$D1$	C_1	C_2	C_3	C_4	C_5	C_6	C_7	C_8	C_9	$C10$
R_1	VHI	HI	SMI	SLI	SMI	SMI	LI	VLI	SLI	SLI
R_2	VLI	SMI	HI	VLI	SLI	VLI	HI	LI	SMI	SLI
R_3	HI	HI	VHI	SMI	VHI	HI	VLI	VLI	VLI	SMI
R_4	LI	SMI	VLI	LI	SLI	LI	VHI	SMI	HI	VHI
R_5	HI	SMI	SLI	SMI	HI	SMI	SMI	LI	SLI	HI

Table 8.5 Criteria-alternative evaluations for $D2$

$D2$	C_1	C_2	C_3	C_4	C_5	C_6	C_7	C_8	C_9	$C10$
R_1	HI	VHI	HI	SMI	HI	SMI	VLI	LI	LI	VLI
R_2	SLI	SMI	SLI	VLI	LI	SLI	VHI	HI	HI	LI
R_3	VHI	AMI	HI	HI	VHI	SMI	VLI	LI	LI	SLI
R_4	SLI	LI	LI	ALI	SLI	SLI	HI	VHI	SMI	HI
R_5	HI	HI	SMI	SLI	SMI	HI	SLI	VLI	SLI	SMI

Table 8.6 Criteria-alternative evaluations for *D*3

*D*3	C_1	C_2	C_3	C_4	C_5	C_6	C_7	C_8	C_9	$C10$
R_1	HI	HI	HI	SMI	SMI	HI	LI	VLI	LI	LI
R_2	LI	SLI	LI	LI	SLI	SMI	HI	SMI	SMI	LI
R_3	HI	VHI	VHI	VHI	HI	VHI	ALI	VLI	VLI	SLI
R_4	VLI	SLI	VLI	VLI	LI	SLI	SMI	HI	SMI	HI
R_5	SMI	HI	SMI	SLI	HI	SMI	SLI	SLI	LI	HI

compiled into the aggregated group decision matrix, which serves as a comprehensive representation of the collective preferences and priorities of the DMs.

Table 8.8 represents the Weighted Decision Matrix, a crucial step in the decision-making process for UAV landing zone selection during disaster response scenarios. The entries in the table are the weighted scores assigned to each criterion—alternative pair. These weights are derived by combining the aggregated weights from the SWAM operator with the original linguistic evaluations provided by the DMs. Table 8.8 serves as the foundational matrix for the subsequent steps in the SF-MARCOS procedure.

Table 8.9 provides a clear representation of the extreme values for each criterion, offering a basis for evaluating the performance of each alternative in the subsequent steps of the decision-making process. The Anti-Ideal and Ideal values are used to calculate the proximity of each alternative to these extreme solutions. The Anti-Ideal solution represents each criterion's least desirable or acceptable conditions. For example, in C_1, a low membership value (0.254) indicates a lower value is less desirable for this criterion. The ideal solution represents the most desirable or acceptable conditions for each criterion. For instance, in C_1, a high membership value (0.650) indicates a higher value is more desirable for this criterion.

Table 8.10 plays a pivotal role in the UAV landing zone selection decision-making amid disaster response scenarios. The Positive Deviation ($K+$) column assesses how well each alternative performs relative to the Ideal solution for the specified criteria. In contrast, the Negative Deviation ($K-$) column gauges how far each alternative deviates from the least desirable conditions based on the criteria. The Normalized Positive and Negative Deviation columns ($F(K+)$ and $F(K-)$) provide scaled measures of proximity to the Ideal and Anti-Ideal solutions, respectively. The Comprehensive Evaluation Index (Fi) combines these normalized deviations, offering a holistic evaluation for each alternative. The final ranking is determined by Fi, with a lower rank indicating a more favourable alternative. This systematic approach aids DMs in selecting the most suitable UAV landing zone alternative in the complex landscape of disaster response, considering both desirable and undesirable criteria.

Table 8.7 Aggregated group decision matrix

	C_1			C_2			C_3			C_4			C_5		
	μ	v	π	μ	v	π	μ	v	π	μ	v	π	μ	v	π
R_1	0.756	0.245	0.249	0.735	0.266	0.269	0.655	0.346	0.350	0.517	0.490	0.402	0.634	0.367	0.370
R_2	0.296	0.715	0.302	0.569	0.434	0.401	0.579	0.438	0.332	0.224	0.779	0.225	0.374	0.628	0.376
R_3	0.735	0.266	0.269	0.804	0.199	0.213	0.775	0.226	0.229	0.683	0.319	0.328	0.784	0.217	0.219
R_4	0.319	0.686	0.324	0.499	0.513	0.383	0.235	0.769	0.237	0.238	0.775	0.241	0.383	0.619	0.384
R_5	0.683	0.318	0.320	0.655	0.346	0.350	0.517	0.490	0.402	0.517	0.490	0.402	0.674	0.327	0.330
	C_6			C_7			C_8			C_9			C_{10}		
	μ	v	π	μ	v	π	μ	v	π	μ	v	π	μ	v	π
R_1	0.623	0.378	0.380	0.274	0.729	0.276	0.235	0.769	0.237	0.355	0.648	0.357	0.334	0.675	0.340
R_2	0.387	0.639	0.336	0.735	0.266	0.269	0.536	0.485	0.332	0.634	0.367	0.370	0.355	0.648	0.357
R_3	0.701	0.302	0.309	0.185	0.819	0.185	0.235	0.769	0.237	0.235	0.769	0.237	0.517	0.490	0.402
R_4	0.355	0.648	0.357	0.743	0.259	0.268	0.697	0.307	0.317	0.655	0.346	0.350	0.756	0.245	0.249
R_5	0.634	0.367	0.370	0.517	0.490	0.402	0.300	0.706	0.304	0.383	0.619	0.384	0.674	0.327	0.330

Table 8.8 Weighted decision matrix

	C_1			C_2			C_3			C_4			C_5		
	μ	ν	π	μ	ν	π	μ	ν	π	μ	ν	π	μ	ν	π
R_1	0.650	0.281	0.313	0.495	0.412	0.463	0.481	0.427	0.391	0.390	0.534	0.262	0.497	0.419	0.396
R_2	0.254	0.721	0.299	0.384	0.524	0.374	0.426	0.499	0.337	0.169	0.794	0.387	0.293	0.651	0.299
R_3	0.632	0.299	0.334	0.542	0.377	0.442	0.570	0.343	0.278	0.517	0.395	0.275	0.614	0.303	0.404
R_4	0.275	0.694	0.343	0.336	0.585	0.438	0.173	0.787	0.440	0.180	0.791	0.435	0.300	0.642	0.376
R_5	0.587	0.345	0.147	0.441	0.463	0.330	0.380	0.542	0.269	0.390	0.534	0.249	0.528	0.386	0.219
	C_6			C_7			C_8			C_9			C10		
	μ	ν	π	μ	ν	π	μ	ν	π	μ	ν	π	μ	ν	π
R_1	0.199	0.739	0.331	0.213	0.745	0.337	0.130	0.821	0.440	0.126	0.815	0.415	0.124	0.822	0.374
R_2	0.124	0.829	0.368	0.570	0.343	0.219	0.296	0.626	0.319	0.225	0.706	0.278	0.131	0.809	0.425
R_3	0.224	0.721	0.339	0.143	0.829	0.337	0.130	0.821	0.459	0.083	0.873	0.409	0.191	0.740	0.400
R_4	0.113	0.833	0.385	0.575	0.339	0.430	0.385	0.528	0.374	0.232	0.700	0.381	0.280	0.663	0.418
R_5	0.203	0.736	0.324	0.400	0.528	0.229	0.166	0.775	0.402	0.136	0.801	0.357	0.250	0.684	0.375

Table 8.9 Anti ideal and ideal values

	C_1			C_2			C_3			C_4			C_5		
	μ	ν	π	μ	ν	π	μ	ν	π	μ	ν	π	μ	ν	π
Anti-ideal	0.254	0.281	0.147	0.336	0.377	0.330	0.173	0.343	0.269	0.169	0.395	0.249	0.293	0.303	0.219
Ideal	0.650	0.721	0.343	0.542	0.585	0.463	0.570	0.787	0.440	0.517	0.794	0.435	0.614	0.651	0.404

	C_6			C_7			C_8			C_9			C10		
	μ	ν	π	μ	ν	π	μ	ν	π	μ	ν	π	μ	ν	π
Anti-ideal	0.113	0.721	0.324	0.575	0.829	0.430	0.385	0.821	0.459	0.232	0.873	0.415	0.280	0.822	0.425
Ideal	0.224	0.833	0.385	0.143	0.339	0.219	0.130	0.528	0.319	0.083	0.700	0.278	0.124	0.663	0.374

Table 8.10 Final evaluations matrix

	$K+$	$K-$	$F(K+)$	$F(K-)$	Fi	Ranking
R_1	0.514	0.552	0.518	0.482	0.355	4
R_2	0.513	0.593	0.536	0.464	0.366	2
R_3	0.522	0.549	0.513	0.487	0.357	3
R_4	0.489	0.581	0.543	0.457	0.353	5
R_5	0.549	0.594	0.519	0.481	0.380	1

8.7 Comparative and Sensitivity Analysis

In this section, we present the results of a comparative analysis among three methods: SF-MARCOS, SF-TOPSIS, and PF-TOPSIS, followed by a sensitivity analysis specifically focusing on SF-MARCOS. The goal is to identify the most acceptable landing site for UAVs under varying decision-maker weights. The comparative analysis, as illustrated in Fig. 8.3, indicates that R_5 emerges as the most favourable landing site for UAVs when applying SF-MARCOS, SF-TOPSIS, and PF-TOPSIS methods. This conclusion is drawn from the collective evaluation of multiple criteria, incorporating the expertise and preferences of DMs.

Sensitivity analysis was conducted to assess the robustness and stability of SF-MARCOS under different weight scenarios assigned to DMs (D1, D2, D3). Four distinct weight scenarios denoted as Scenario 1, Scenario 2, Scenario 3, and Scenario 4, with weight distributions (0.4, 0.3, 0.3), (0.4, 0.4, 0.2), (0.3, 0.4, 0.3), and (0.2, 0.5, 0.3), respectively, were considered. The outcomes of the sensitivity analysis, as depicted in Fig. 8.4, consistently indicate that Landing Site R_5 consistently retains its top-ranking position for UAVs across all scenarios. This consistency in ranking highlights the stability of SF-MARCOS outcomes, emphasizing the robustness of R_5 as the preferred choice for UAV landings. The sensitivity analysis reinforces the

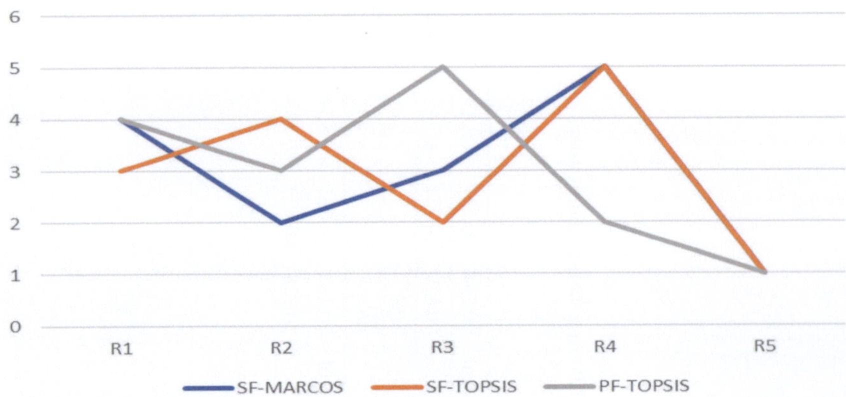

Fig. 8.3 Visual representation of comparative analysis

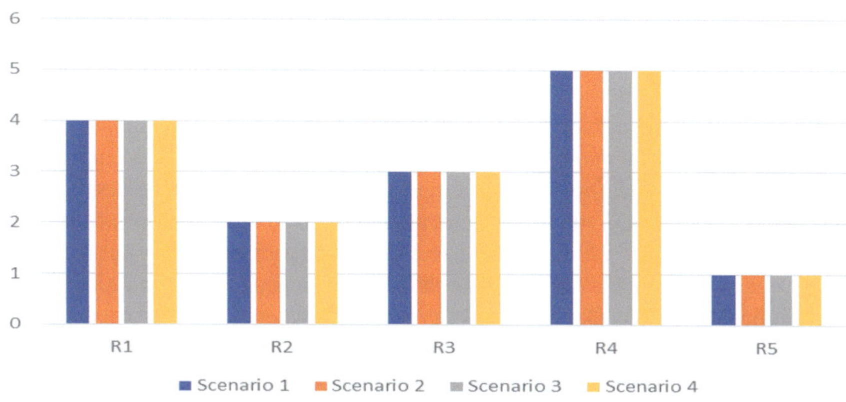

Fig. 8.4 Sensitivity analysis for different weights of decision-makers

reliability of SF-MARCOS in UAV landing site selection. Regardless of variations in decision-maker weights, R_5 consistently exhibits the highest utility and is consistently ranked as the optimal landing site. This consistency instils confidence in the stability of the decision-making process, emphasizing R_5 as a reliable and preferred choice for UAV landings.

The comparative and sensitivity analyses collectively affirm that SF-MARCOS effectively identifies R_5 as the most suitable landing site for UAVs across various decision-making scenarios. These findings contribute to the robustness and reliability of SF-MARCOS in decision-making for UAV operations in disaster response scenarios.

8.8 Discussions and Limitations

The significance of this study lies in its contribution to optimizing UAV landing site selection in disaster response scenarios, particularly in the context of urban flood management. By integrating the SWAM operator and a simulation model, the proposed SF-MARCOS methodology offers a systematic and robust approach to decision-making. This study addresses complex and dynamic disaster response operations by considering diverse criteria, such as safety, communication, emergency preparedness, adaptability, and more. The comparative and sensitivity analyses demonstrate the reliability and stability of SF-MARCOS in consistently identifying the most suitable landing site for UAVs, specifically R_5, in the case of Guwahati's flood hazard zones. The findings contribute to MCDM, disaster management, and UAV applications by providing a comprehensive and adaptable methodology that can enhance the effectiveness of UAV operations in critical scenarios. This study's outcomes can inform DMs, emergency responders, and researchers, fostering advancements in disaster response strategies and technologies.

The study and methodologies presented have certain limitations. Assigning weights to DMs and criteria involves subjectivity, simplifying the representation of decision-maker preferences. Assumptions about the effectiveness of SFNs in capturing linguistic variables may vary in practicality. The simulation model, while valuable, makes assumptions about UAV capabilities and disaster scenarios that might not entirely reflect real-world complexities. Findings are context-specific to Guwahati's flood hazard zones and generalizing them to other regions or disasters may require additional validation. The exclusion of specific criteria and the static nature of assumed weights may limit the study's comprehensiveness and adaptability to dynamic conditions. Decision-making accuracy depends on the chosen DM's expertise, introducing potential biases. While the study demonstrates effectiveness within its scope, further validation across diverse scenarios is needed. Data quality and technological assumptions, such as UAV capabilities, also impact the study's reliability. Recognizing these limitations is crucial for refining methodologies and enhancing their applicability in real-world disaster response contexts.

8.9 Conclusions and Future Scope

This study has introduced the SF-MARCOS methodology as an effective approach for optimizing UAV landing site selection in disaster response, particularly in the context of urban flood management in Guwahati. Integrating the SWAM operator, a simulation model, and MCDM has demonstrated consistent and reliable outcomes, with R_5 identified as the most suitable landing site across various decision-making scenarios. The comparative and sensitivity analyses have underscored the stability of SF-MARCOS, highlighting its potential for informed decision-making in dynamic disaster response environments. Future research should explore the adaptability of the methodology to different disaster types, validate it in real-world scenarios, and consider dynamic adjustments to decision-maker weights. Additionally, efforts should focus on community engagement, technological advancements, integration with existing systems, criteria refinement, and cross-disciplinary collaboration to enhance the methodology's versatility and applicability in diverse disaster response contexts. Addressing these future directions will contribute to the continued evolution and effectiveness of SF-MARCOS as a decision-making tool for optimizing UAV landing site selection.

References

1. Aggarwal, S., Gupta, P., Mahajan, N., Balaji, S., Singh, K.J., Bhargava, B., Panda, S.: Implementation of drone based delivery of medical supplies in North-East India: experiences, challenges and adopted strategies. Front. Public Health 11 (2023)

2. Balestrieri, E., Daponte, P., De Vito, L., Picariello, F., Tudosa, I.: Sensors and measurements for UAV safety: an overview. Sensors **21**(24), 8253 (2021)
3. Boyacı, A., Şişman, A.: Pandemic hospital site selection: a GIS-based MCDM approach employing Pythagorean fuzzy sets. Environ. Sci. Pollut. Res. **29**(2), 1985–1997 (2022)
4. Buyuk, A.M., Temur, G.T.: A framework for selection of the best food waste management alternative by a spherical fuzzy AHP based approach. In: Intelligent and Fuzzy Techniques: Smart and Innovative Solutions Proceedings of the INFUS 2020 Conference, Istanbul, Turkey, 21–23 July 2020, vol 1197, pp 151–159. Springer, Cham (2021)
5. Bu¨yu¨k¨ozkan, G., Güler, M.: A combined hesitant fuzzy MCDM approach for supply chain analytics tool evaluation. Appl. Soft Comput. **112**, 107812 (2021)
6. Chamata, J., Winterton, J.: A conceptual framework for the acceptance of drones. Int. Technol. Manage. Rev. **7**(1) (2018)
7. Chowdhury, S., Shahvari, O., Marufuzzaman, M., Li, X., Bian, L.: Drone routing and optimization for post-disaster inspection. Comput. Ind. Eng. **159**, 107495 (2021)
8. Dukkanci, O., Koberstein, A., Kara, B.Y.: Drones for relief logistics under uncertainty after an earthquake. Eur. J. Oper. Res. **310**(1), 117–132 (2023)
9. Edwards, D., Subramanian, N., Chaudhuri, A., Morlacchi, P., Zeng, W.: Use of delivery drones for humanitarian operations: analysis of adoption barriers among logistics service providers from the technology acceptance model perspective. Ann. Oper. Res. (2023)
10. Erdelj, M., Natalizio, E., Chowdhury, K.R., Akyildiz, I.F.: Help from the sky: leveraging UAVs for disaster management. IEEE Pervasive Comput. **16**(1), 24–32 (2017)
11. Flemons, K., Baylis, B., Khan, A.Z., Kirkpatrick, A.W., Whitehead, K., Moeini, S., Schreiber, A., Lapointe, S., Ashoori, S., Arif, M., Berenger, B., Conly, J., Hawkins, W.: The use of drones for the delivery of diagnostic test kits and medical supplies to remote First Nations communities during Covid-19. Am. J. Infect. Control **50**(8), 849–856 (2022)
12. Glantz, E.J., Ritter, F.E., Gilbreath, D., Stager, S.J., Anton, A., Emani, R.: UAV use in disaster management. In: Proceedings of the International ISCRAM Conference, vol. 2020 (2020)
13. Griffin, G.F.: The use of unmanned aerial vehicles for disaster management. Geomatica **68**(4) (2015)
14. Jafarzadeh Ghoushchi, S., Bonab, S.R., Ghiaci, A.M.: A decision-making framework for COVID-19 infodemic management strategies evaluation in spherical fuzzy environment. Stochastic Environ. Res. Risk Assess. **37**(4), 1635–1648 (2023)
15. Jaller, M., Otay, I.: Evaluating sustainable vehicle technologies for freight transportation using spherical fuzzy AHP and TOPSIS. In: Intelligent and Fuzzy Techniques: Smart and Innovative Solutions Proceedings of the INFUS 2020 Conference, Istanbul, Turkey, 21–23 July 2020, volume 1197, pp. 118–126. Springer, Cham (2021)
16. Jawad, M., Naz, M., Muqaddus, H.: A multi-criteria decision-making approach for portfolio selection by using an automatic spherical fuzzy AHP algorithm. J. Oper. Res. Soc. 1–14 (2023)
17. Khan, A., Gupta, S., Gupta, S.K.: Emerging UAV technology for disaster detection, mitigation, response, and preparedness. J. Field Robot. **39**(6), 905–955 (2022)
18. Kutlu Gu¨ndo˘gdu, F., Kahraman, C.: Spherical fuzzy sets and spherical fuzzy TOPSIS method. J. Intell. Fuzzy Syst. **36**(1), 337–352 (2019)
19. Kutlu Gu¨ndo˘gdu, F., Kahraman, C.: A novel spherical fuzzy analytic hierarchy process and its renewable energy application. Soft Comput. **24**(6), 4607–4621 (2020)
20. Lin, M., Huang, C., Xu, Z.: MULTIMOORA-based MCDM model for site selection of car sharing station under picture fuzzy environment. Sustain. Cities Soc. **53**, 101873 (2020)
21. Lin Tan, L.K., Lim, B.C., Park, G., Low, K.H., Seng Yeo, V.C.: Public acceptance of drone applications in a highly urbanized environment. Technol. Soc. **64**, 101462 (2021)
22. Liu, L., You, Z.: Drone transports medical supplies to Puerto Rico based on shortest path. IOP Conf. Ser. Earth Environ. Sci. **512**(1), 012140 (2020)
23. Lyu, M., Zhao, Y., Huang, C., Huang, H.: Unmanned aerial vehicles for search and rescue: a survey. Remote Sens. **15**(13), 3266 (2023)
24. Mathew, A.O., Jha, A.N., Lingappa, A.K., Sinha, P.: Attitude towards drone food delivery services—role of innovativeness, perceived risk, and green image. J Open Innov. Technol. Market Complexity **7**(2), 144 (2021)

25. Memarpour Ghiaci, A., Garg, H., Jafarzadeh Ghoushchi, S.: Improving emergency departments during COVID-19 pandemic: a simulation and MCDM approach with MARCOS methodology in an uncertain environment. Comput. Appl. Math. **41**(8), 368 (2022)
26. Merkert, R., Bushell, J.: Managing the drone revolution: A systematic literature review into the current use of airborne drones and future strategic directions for their effective control. J. Air Transp. Manag. **89**, 101929 (2020)
27. Mohd Daud, S.M.S., Mohd Yusof, M.Y.P., Heo, C.C., Khoo, L.S., Chainchel Singh, M.K., Mahmood, M.S., Nawawi, H.: Applications of drone in disaster management: a scoping review. Sci. Justice **62**(1), 30–42 (2022)
28. Mohsan, S.A.H., Othman, N.Q.H., Li, Y., Alsharif, M.H., Khan, M.A.: Unmanned aerial vehicles (UAVs): practical aspects, applications, open challenges, security issues, and future trends. Intell. Serv. Robot. (2023)
29. Nguyen, T.M.H., Nguyen, V.P., Nguyen, D.T.: A new hybrid Pythagorean fuzzy AHP and COCOSO MCDM based approach by adopting artificial intelligence technologies. J. Experim. Theoret Artif. Intell. 1–27 (2022)
30. Nimilan, V., Manohar, G., Sudha, R., Stanley, P.: Drone-aid: an aerial medical assistance. Int. J. Innov. Technol. Exploring Eng. **8**(11S), 1288–1292 (2019)
31. Olatunji, G., Isarinade, T.D., Emmanuel, K., Olatunji, D., Aderinto, N.: Exploring the transformative role of drone technology in advancing healthcare delivery in Africa; a perspective. Ann. Med. Surg. **85**(10), 5279–5284 (2023)
32. Pinto, R., Lagorio, A.: Point-to-point drone-based delivery network design with intermediate charging stations. Transport. Res. Part C: Emerg. Technol. **135**, 103506 (2022)
33. Rahnamay Bonab, S., Jafarzadeh Ghoushchi, S., Deveci, M., Haseli, G.: Logistic autonomous vehicles assessment using decision support model under spherical fuzzy set integrated Choquet Integral approach. Expert Syst. Appl. **214**, 119205 (2023)
34. Rajan, J., Shriwastav, S., Kashyap, A., Ratnoo, A., Ghose, D.: Disaster management using unmanned aerial vehicles. In: Unmanned Aerial Systems, pp. 129–155. Elsevier, Amsterdam (2021)
35. Rejeb, A., Rejeb, K., Simske, S.J., Treiblmaier, H.: Drones for supply chain management and logistics: a review and research agenda. Int J Log Res Appl **26**(6), 708–731 (2023)
36. Rottondi, C., Malandrino, F., Bianco, A., Chiasserini, C.F., Stavrakakis, I.: Scheduling of emergency tasks for multiservice UAVs in post-disaster scenarios. Comput. Netw. **184**, 107644 (2021)
37. Sarmah, T., Das, S., Narendr, A., Aithal, B.H.: Assessing human vulnerability to urban flood hazard using the analytic hierarchy process and geographic information system. Int. J. Disaster Risk Reduct. **50**, 101659 (2020)
38. Sharaf, I.M.: A new approach for spherical fuzzy TOPSIS and spherical fuzzy VIKOR applied to the evaluation of hydrogen storage systems. Soft. Comput. **27**(8), 4403–4423 (2023)
39. Simić, V., Lazarević, D., Dobrodolac, M.: Picture fuzzy WASPAS method for selecting last-mile delivery mode: a case study of Belgrade. Eur. Transp. Res. Rev. **13**(1), 43 (2021)
40. Sotoudeh-Anvari, A.: The applications of MCDM methods in COVID-19 pandemic: a state of the art review. Appl. Soft Comput. **126**, 109238 (2022)
41. Verykokou, S., Doulamis, A., Athanasiou, G., Ioannidis, C., Amditis, A.: UAV-based 3D modelling of disaster scenes for Urban Search and Rescue. In: 2016 IEEE International Conference on Imaging Systems and Techniques (IST), pp. 106–111. IEEE (2016)
42. Yakushiji, K., Fujita, H., Murata, M., Hiroi, N., Hamabe, Y., Yakushiji, F.: Short-range transportation using unmanned aerial vehicles (UAVs) during disasters in Japan. Drones **4**(4), 68 (2020)
43. Yamazaki, Y., Premachandra, C., Perea, C.J.: Audio-processing-based human detection at disaster sites with unmanned aerial vehicle. IEEE Access **8**, 101398–101405 (2020)
44. Zhao, J., Yang, W., Wang, F., Zhang, C.: Research on UAV aided earthquake emergency system. IOP Conf. Ser. Earth Environ. Sci. **610**(1), 012018 (2020)

Part III
Analytics in Healthcare

Chapter 9
Social Media Sentiment Analysis of Impacts of COVID-19 on Mothers and Children

Abhishek Kumar, Amrita, and Sobhan Sarkar

Abstract The COVID-19 pandemic, declared by the World Health Organization in 2020, not only brought about a global health crisis but also unleashed a flood of information and misinformation on social media platforms. This study delves into the profound effects of the pandemic on mothers and children by employing sentiment analysis on social media data. Analyzing over 30,000 posts from platforms, namely Twitter and Reddit, this research discerns the emotional undercurrents amidst the pandemic chaos. We have used sentiment analysis to classify posts into positive and negative. Further, we have analyzed the classified posts using different sets of machine and deep learning classifiers. To display and examine the current public discourse of worry about the effects of COVID-19 on mothers and children, social media posts are modeled on different topics. Our findings demonstrate that although the participants followed recommendations and safety precautions, their opinions occasionally fluctuated, resulting in a negative impact on society due to misinformation.

Keywords COVID-19 effect · Mothers and children · Social media · Sentiment analysis · Machine learning models

A. Kumar (✉)
Chemical Engineering, IIT Kharagpur, Kharagpur 721302, West Bengal, India
e-mail: abhishekkumartrn@gmail.com

Amrita
Coggin College of Business, University of North Florida, Jacksonville 32224, Florida, USA

S. Sarkar
Information Systems & Business Analytics, IIM Ranchi, Ranchi 835303, Jharkhand, India

J. Maiti et al. (eds.), *Operations Research and Data Analytics: Current Trends and Future Perspectives*, Lecture Notes on Multidisciplinary Industrial Engineering,
https://doi.org/10.1007/978-981-96-8327-7_9

9.1 Introduction

COVID-19 is a contagious illness resulting from the severe acute respiratory syndrome coronavirus 2 (SARS-CoV-2), which causes severe acute respiratory syndrome. They are members of a big viral family. SARS-CoV-2 is the source of the COVID-19 health catastrophe, which was initially reported in December 2019. This is a new viral strain that evolved from a family of previous strains. COVID-19 is an infectious illness resulting from the severe acute respiratory syndrome coronavirus 2 (SARS-CoV-2). It emerged in December 2019 in Wuhan, China, catching the world by surprise [1, 2]. Since then, the virus has reached numerous countries, prompting global efforts to curb its spread, even in regions with minimal or no reported cases. On March 11, 2020, the World Health Organization (WHO) officially declared it a pandemic, marking a critical moment in the ongoing battle against the disease, an organization that is working tirelessly to contain it [3]. The virus had already been circulating in various populations in various nations around the world, resulting in 701,684,455 confirmed cases and 6,968,394 deaths worldwide. As of now, 672,690,394 patients have recovered from this disease.[1] The prevalence of COVID-19, as well as the shutdown of several enterprises, would lower GDP. According to research, the global spread of the disease will result in a monthly loss in global GDP of 3.4.[2]

COVID-19 has evolved beyond being solely an infectious disease transmitted through physical contact and tiny droplets produced during coughing, sneezing, or talking. It has transformed into a cause of distress, concern, and unease due to the dissemination of misinformation on social media. Because of the undesired conspiracies surrounding COVID-19, Social media platforms such as Twitter, Facebook, Reddit, and Instagram are actively engaged in scrutinizing and fact-checking to combat the dissemination of misinformation. With the number of coronavirus cases increasing by the day, many are diligently striving to develop innovative point-of-care tests swiftly to curb the transmission. The unpredictability and fluctuation of this adverse situation keep people on edge, wondering what will happen next. This circumstance can cause anxiety and make people feel terrified and helpless. And in these cases, the most affected sections of the world are children and mothers. Due to the use of social media mothers can share their opinions, affect, and updates regarding this situation. That's why we have used social media posts of different mothers and children to analyze how they respond to COVID-19. Many research efforts have primarily concentrated on examining preventative care, recovery, healthcare, social networks, and economic data. Consequently, there has been limited exploration into the evaluation of trends in conspiratorial communication on social media and the accumulation of personal-level data related to its impact on children and mothers. That is why, methods of analysis that can be rapidly implemented to comprehend the dynamics of information dissemination and evaluate the evolution of public sentiment during pandemic situations are essential.

[1] https://www.worldometers.info/coronavirus/.

[2] https://www.statista.com/topics/6139/covid-19-impact-on-the-global-economy/#topicOverview.

Particularly in crisis situations like pandemics, social media becomes a crucial tool for crisis communication, providing a platform for information exchange, socialization, and emotional support. Social media has become integrated in our day-to-day lives. It creates a link between people and the outside world. Social media enables individuals to share their personal experiences privately, conveniently, and at their own discretion. People increasingly depend on the content they share through posts and tweets on various social media platforms like Twitter, Facebook, Reddit, and similar platforms.

Individuals in crises have greater needs for information and social assistance. In high-risk scenarios, including pandemics, natural catastrophes, and terrorist attacks, social media is an important tool for crisis communication. Unprecedented occurrences, uncertainty, a lack of information, and the need for sense-making are all characteristics of crises. When individuals are trying to cope with uncertainty and deal with disasters creatively, they turn to social media. According to the interpretive paradigm, people, particularly mothers, utilize social media to meet a variety of needs, including knowledge, socialization, and emotional support. Audiences may also choose to interact with material that reinforces their present thoughts and mindsets. But nowadays individuals are not utilizing social media as a means to navigate the challenges posed by the pandemic. Instead, the tweets and viewpoints related to COVID-19 are proving to be detrimental and are raising concerns. It is imperative to address this issue to effectively manage the spread of inaccurate information originating from various sources.

Increased use of social media during pandemics, a knowledge repository is created by individuals sharing information and updates, and this information is stored in the form of text, making effective communication platforms and social media an important information source for researching issues involving quickly changing public sentiment. As a result, this article focuses on an in-depth examination of the sentiments expressed in social media posts about the novel COVID-19 from the beginning of the year. Sentiment analysis involves gathering individuals' perspectives on real life events. Recognizing people's emotions is particularly important in the current global circumstances. The challenging situation where individuals are confined to their homes highlights the need to explore authentic sentiments about the situation, making sentiment analysis the most effective approach.

Therefore, the present study aims to discover the COVID-19-related issues and community sentiment dynamics expressed on social media to assess the COVID-19 impact on children and mothers. The two main issues that we addressed in this paper are first, how people express their sentiments on social media regarding COVID-19 whether it affects society positively or negatively. And secondly, analyze users commonly expressed feelings about COVID-19 by discussing prevalent topics. The study begins with the collection of posts from various COVID-19-related social media groups, which are then thoroughly cleaned. We collected posts from Twitter and Reddit that are related to mothers and children using specific keywords like "child COVID", "pregnancy pandemic", etc. The issues outlined earlier were investigated by compiling diverse social media posts shared on Twitter and Reddit between January 2020 and April 2022. This process resulted in the creation of a dataset comprising approximately 30,000 posts for analysis. Further, we labeled the dataset using the Valence Aware Dictionary and Sentiment Reasoner (VADER) sentiment lexicon as positive and negative. We extracted indicative topics using Latent Dirichlet Allocation (LDA) [4, 5] and Non-negative Matrix Factorization (NMF) topic modeling to show and discuss the prevailing discourse of public concern regarding COVID-19's impact on children and mothers. To extract the true sentiment of social media users during the pandemic, we used a variety of deep learning and machine learning classifiers on feature-extracted text data.

The rest of this paper will be organized as follows: The related work is presented in Sect. 9.2. The proposed technique for this article is explained in Sect. 9.3. A case study has been presented in Section IV. In Sect. 9.4, the findings are reported, and finally, in Sect. 9.5, conclusions are drawn with limitations and future research areas.

9.2 Related Works

This article is informed by a variety of academic fields, and as a result, the literature review for the impact of COVID-19 on children, as well as sentiment analysis and textual analytics, is presented in this part. Because COVID-19 is a huge health threat to the global population, all governments and citizens must work together to tackle it. Media and social media platforms should strive to provide unbiased and positive information, presenting accurate news ethically. This approach aims to inspire and motivate the general public rather than contributing to division and anxiety, as biased or negative coverage tends to do.

COVID-19 impacts on pregnant mothers are well studied by Talbot et al. [6]. The research examined 192 tweets from a global sample that discussed pregnant women's mental health in the early COVID-19 pandemic months. This study revealed that insights into the challenges faced by women during the pandemic, as well as the available resources and support systems, can be gained by utilizing the Twitter platform as a means to observe the current experiences of women. Sentiment analysis results indicated that individuals and firms tended to employ the same negative tonality whereas those from organizations and health advisors were neutral. This study highlights the challenges faced by women, shedding light on the obstacles they encounter and the available resources. Unequal expectations regarding domestic chores, barriers to digital access to online education, heightened risks of girl child marriage, adolescent pregnancy, and sexual exploitation, exacerbated by the pandemic, are factors contributing to the educational setbacks. Moreover, indications suggest that these negative consequences may disproportionately impact girls. A study by Dey et al. [7] highlights concerns related to gender equity in education, international education organizations raised issues during the COVID-19 pandemic. We discover that the problems brought up on Twitter highlight the negative effects of closing schools on girls' educational opportunities. The study by Yukich et al. [8] projects the potential impact of COVID-19 on child marriage in the five countries in which the burden of child marriage is the largest: Bangladesh, Brazil, Ethiopia, India, and Nigeria.

The primary difficulty faced by pregnant women during the pandemic, according to the results of the thematic analysis, was stress and coping with it. The government issued a national lockdown order to prevent the spread of the virus. This resulted in house confinement, home (smart) working, and the temporary closure of nonessential businesses and schools. A study by Di Giorgio [9] examined the change in behavioral patterns of mothers and preschool children in their care as a result of these restrictive restrictions. Data demonstrated that these restricted restrictions had adverse impacts, as was to be expected. In general, it appeared that smart working methods or unplanned job interruptions had a detrimental impact on routines and sleep quality. There is an increase in social media use after the pandemic hit the study done by Gao et al. [10] showed a negative impact on mental health problems due to social media exposure. According to this study's findings, 82.0% of individuals regularly use social media, and frequent social media exposure is connected with increased probabilities of anxiety which is somewhat consistent with other studies [11]. Osakwe [12] research study shows a better understanding of the public concerns and reactions during the COVID-19 pandemic on social media. The results of this study help to improve knowledge of how text mining of Twitter data is used to gauge public response to the COVID-19 epidemic. As these have significant implications for public health initiatives, public health institutions need to be more involved in monitoring and evaluating the information, queries, and trends within the public's tweets. Amid the recent COVID-19 pandemic, social media channels have played a role in spreading inaccurate information and conspiracy theories. Analyzing the spread of false health-related information reveals that the phenomenon of an infodemic significantly impacts individuals globally, fostering distrust in authorities, scholars, and medical professionals. This lack of trust can have adverse consequences on individuals' well-being and overall health. Yasmim et al. [13] study shows that the media rose in the backdrop of the epidemic to learn more about the illness. Nevertheless, a considerable number of these occurrences turned out to be false information presented as reliable strategies for disease prevention and control, contributing to a rise in the dissemination of deceptive content. As a result of meddling with people's behavior and health during this period, there was social unrest characterized by violence, mistrust, social upheavals, and attacks on health professionals. Their findings show that some age groups are more likely than others to engage with the false news spread through social media, especially the elderly. Keeping mental health in consideration for children Nazish et al. [14] did research and their study shows the impact of increased screen time on children's mental health. A study found that after 9/11, people who watched too much television experienced more cases of Post-Traumatic Stress Disorder (PTSD) and other mental health conditions [15]. A lot of healthcare professionals (HCPs) are involved in the pandemic response and are among the most crucial factors in maintaining coping capacities. It is important to also analyze the state of mind of HCPs. To prevent fatigue, burnout, and mental health disorders during the postpandemic period, it may be crucial to provide emotional support for HCPs [16]. There are worries that youngsters who use technology and social media excessively and in a way that would expose others to harmful information could acquire comparable illnesses. Mothers are conversing about the secondary impacts of the COVID pandemic on their daily lives, emphasizing the potential repercussions of disturbances in regular healthcare services and reduced availability of food. The study by Roberton [17] focuses on how shortages of competent health professionals and a rise in women's resistance to using the healthcare system may result in less funding for antenatal care, postpartum care, and facility- and community-based lactation assistance and counselling. Aligned with the Global Breastfeeding Collective's plea for action amid the COVID-19 pandemic, this examination underscores the crucial need for governments to persist in offering assistance to encourage and protect breastfeeding.

9.2.1 Research Issues

Based on the above-mentioned discussion, some research issues have been identified, which are given below.

(i) Existing research has mainly addressed preventive care, recovery, healthcare, social networks, and economic data. However, there is a limited exploration of trends in conspiratorial communication on social media and its impact on mothers and children at the personal level.
(ii) In the past, various studies have been done on text analysis and sentiment analysis of tweets related to COVID-19 but mostly these are limited to Twitter.
(iii) Reddit is a social media platform where individuals tend to share more realistic scenario updates. An analysis of data extracted from Reddit was missing.

9.2.2 Research Contributions

To address the above-mentioned research issues, our present study endeavors to contribute by addressing the main issues as follows:

(i) Main focus on social media posts related to mothers and children to analyze the direct impact of COVID-19 on them, and understand the behaviors and sentiments of individual mothers during the pandemic.
(ii) Determining the most common topics expressed by Twitter users when they share their feelings about COVID-19.
(iii) The results of this study have the potential to assist governments globally in formulating effective public health strategies for children and mothers. Additionally, they may offer valuable insights into understanding how individuals respond to a pandemic such as COVID-19.

9.3 Proposed Methodology

The following provides an overview of our suggested framework. It begins with data collection and labeling, followed by analysis and classification into various sentiments.

Our database contains 30,000 unique social media posts related to the impact of the COVID-19 pandemic on children and mothers. Some of the keywords used for the collection of posts are coronavirus pregnancy, COVID-19, pregnant, maternity, SARS-CoV-2, baby, Pandemic birth, child COVID-19, toddler, etc. For example, “the coronavirus will cause child poverty to soar. So what can we do about it?”. We have maintained our data selection criteria like it must be limited to English and the keywords of the corpus focus most on the impact of COVID-19 on children and mothers.

9.3.1 Data Pre-processing and Labeling

The datasets that we collected were raw data i.e. it contains lots of irrelevant texts and emoticons that must be cleaned using various text-cleaning methods. And, for classifying the posts, they should be labeled. Further, in this section, we have described the methods used for text cleaning and labeling of text data (Fig. 9.1).

9.3.1.1 Text Cleaning

Unstructured, unprofessional, and messy tweets are common, the first step in sentiment analysis is to pre-process the data. The following steps are followed to pre-process the text [18]:

- *Step 1*: The hashtags #COVID-19, #StayHomeChild, #pregnancy, and #Coronavirus are widely employed across various social media platforms to symbolize themes. In many cases, they are deemed unnecessary for expressing sentiment and can reduce overall performance. As a result, in the first phase, we cleaned up the text by deleting any extraneous hashtags not only characters but also text.
- *Step 2*: In various social media posts there are mentions(@) which are not necessary so we remove mentioned also. Social media contains various emoticons which after converting into text begin with “&” and are also removed to make the text better. The hyperlinks are also removed.
- *Step 3*: We convert all uppercase letters to lowercase to prevent confusion and ensure consistent interpretation of the same word, eliminating the risk of misreading due to capitalization differences.
- *Step 4*: The subsequent step involves stemming, where morphological analysis is performed to revert words to their base or dictionary form. Through the utilization of the PorterStemmer tool from the Natural Language Toolkit (NLTK) library, we

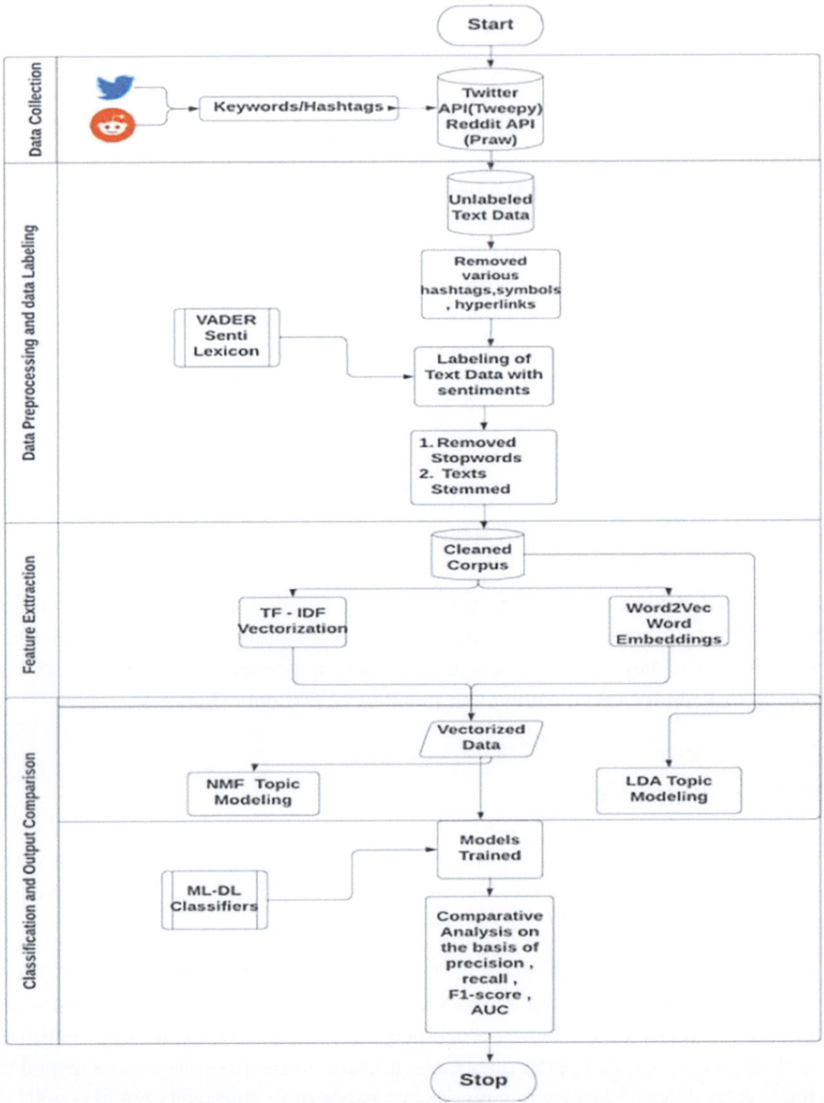

Fig. 9.1 Overview of the proposed framework

undertake the process of stemming to transform various forms of a word into its fundamental root form. Stop words are commonly removed from textual data to reduce noise. Stop words have little effect on interpreting the sentiment valence of a sentence. We excluded stopping words (words typically present in a document with minimal semantic value) from all but sentiment analysis.

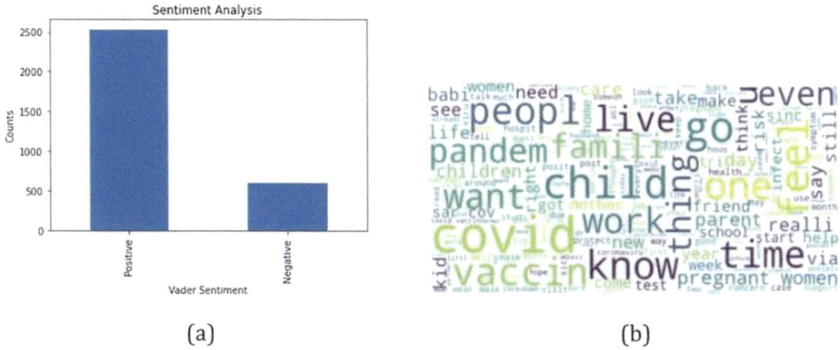

Fig. 9.2 **a** Vader sentiment analysis, and (b) word cloud of the extracted social media posts

9.3.1.2 Data Labeling

We have collected unlabeled datasets, for classification they must be labeled. After comparing the three Rule-Based Approaches for sentiment analysis namely Senti-Word Net [19] VADER [20], and Text-Blob [19] we labeled each text using the VADER approach into negative and positive tweets. We used VADER because it is specially customized to social media sentiments and also works on texts from other domains.

VADER sentiment analysis essentially uses a lexicon to attribute lexical features to sentiment scores. Each token was then put through a rule-based checker to determine how strongly they felt. In addition to lexical aspects, several heuristics also influence the polarity's intensity. Particularly, punctuation is crucial to VADER because it views sentences with punctuation, like exclamation points, as being more emotionally powerful. The employment of degree modifiers in a sentence is another heuristic. Depending on their proximity to the word they modify, degree modifiers change the emotional intensity of a text. The sentiment score of each token in a tweet is added up to determine the total sentiment score of that tweet. The calculation of a word's sentiment score uses a scale from -4, which is the most negative, to + 4, which is the most positive. The neutral sentiment is represented by a sentiment score of 0. However, a sentence's sentiment score can be anywhere between -1 and + 1. Sentiment Analysis on a small batch of the dataset and its distribution can be seen in Fig. 9.2a.

9.3.2 Exploratory Analysis

We do exploratory analysis in this section to get a more complete picture of our data set.

(i) *Word cloud*: Basic text analysis is demonstrated through word clouds, which visually depict key or commonly used words in a given text (Fig. 9.2b). These visual representations highlight the most significant terms within a body of text.

Algorithm 1 Data Labeling using VADER Lexicon

```
Initialize empty lists: Vader, Sentiment
for each t in Text Corpus T do
    Search for t in Vader Lexicon A
    if t is an adjective, verb, adverb, noun, or pronoun in A then
        S ← S-SWN-Score(t)
        if t is in the positive word list then
            pos ← pos + S
        else if t is in the negative word list then
            neg ← neg + S
        else
            neu ← neu + S
        end if
    end if
end for
for each t in Text Corpus T do
    if Vader[i].compound > 0 then
        Sentiment ← Positive
    else
        Sentiment ← Negative
    endif
endfor
```

From Word cloud, we can say that people are mostly talking about children, vaccines, and pandemics.

(ii) *Topic Modeling*: We use LDA and NMF to explore the topic distributions in our data set to analyze the topics.

(a) LDA: To extract topics from a corpus, LDA is a popular topic modeling technique [5, 21]. In the case of LDA, we decided on a total of ten topics.

After LDA training, topics represented by a distribution of words and document topic distributions are learned. LDA does two tasks: it locates topics in a corpus and allocates these subjects to documents within that corpus. Topics with their associated word can be seen in Table 9.2.

(b) NMF: Non-Negative Matrix Factorization serves as a statistical method for diminishing the dimensions of datasets. It utilizes factor analysis to assign lower weights to words with limited coherence, aiding in the modeling of topics. The Top-10 words in the top 10 topics generated using the NMF model are given in Table 9.3.

9.3.3 Feature Extraction

Vectorization methods and word embedding are employed to extract features in this research. For vectorization, the term frequency-inverse document frequency (TF-IDF) method was applied and for word embedding, we have used the Word2Vec technique.

9.3.4 Classification

We employed Machine Learning (ML) and deep learning (DL)-based classifiers to assess performance in the sentiment classification task to give a comprehensive study. In this study, we used ML-based classifiers including the Support vector machine (SVM) [22], Naive Bayes (NB), Decision tree (DT), and Random forest (RF), K-Nearest Neighbors (KNN) on both the feature extraction technique (i.e., TF-IDF and Word2Vec). In addition to classical machine learning, we used Bidirectional longshort-term memory (Bi-LSTM), a DL-based classifier. Three BiLSTM cells with varied numbers of hidden nodes were employed in our tests. A dropout of 0.3 is utilized for regularization after the three BiLSTM layers. The BiLSTM cell's output is linked to dense layers, employing the Rectified Linear Unit (ReLU) for activation, succeeded by a subsequent activation function using sigmoid.

9.4 Results and Discussion

This section explains the experimental procedure for assessing performance on data sets and presenting findings for comparison. We have used various ML and DL methods to obtain the results. The results are summarized in Table 9.1 and plots below.

In our study, we employed a comprehensive approach to feature extraction and baseline establishment by utilizing traditional methods such as TF-IDF and Word2Vec. Employing TF-IDF with SVM, Naive Bayes, Random Forest, K-Nearest Neighbor, Decision Tree, and Bi-LSTM yielded accuracy scores of 0.69, 0.53, 0.79, 0.76, 0.72, and 0.71 respectively. Notably, TF-IDF with RF, demonstrated superior performance compared to other classifiers. Subsequently, when we applied Word2Vec embeddings with ML classifiers (SVM, DT, RF, and Naive Bayes), the

Table 9.1 Comparative results of text classification using ML and DL models

Models	SVM	NB	KNN	DT	RF	Bi-LSTM
TF-IDF	0.69	0.53	0.76	0.72	0.79	0.71
Word2Vec	0.77	0.77	0.71	0.69	0.78	0.886

resulting accuracy scores were 0.77, 0.69, 0.78, 0.71, and 0.77 respectively. Remarkably, Random Forest (RF) outperformed the other ML-based classifiers with an accuracy score of 0.79, although models utilizing TF-IDF generally exhibited better performance than those using Word2Vec embeddings.

In contrast, for deep learning-based classifiers, we leveraged word embeddings like Word2Vec and applied a Bidirectional Long Short-Term Memory (BiLSTM) model, achieving an accuracy score of 0.886. Transitioning to topic extraction using LDA and NMF models, we observed a rich landscape of discussions on social media. The LDA model revealed that approximately 37% of the total discussions centered around topics related to "Advice and Measures," with other major topics including "Impact on pregnant mothers," "Society efforts and discussion," and "Support and Health care Facilities". Similarly, the NMF model highlighted discussions on topics such as "Impact and Protective measures for Child," "COVID-19 update with Advices", and "Support for pregnant mothers" as the predominant subjects, illustrating the versatility of our topic modeling method in identifying diverse and internally coherent subjects within tweet content (Table 9.2).

9.5 Conclusion

Since the rise of COVID-19 conspiracy theories, social media has been widely used to provide knowledge and dispel myths. The sentiment relations of COVID-19-related social media posts are discussed in this article. In this study, we compare sentiment analysis approaches. Our results show that the people followed guidance and precautionary measures; yet, their opinions varied occasionally, resulting in a detrimental impact on society as a result of misinformation; hence, a proactive and agile public health presence is needed to counteract the spread of fake news. This study also sheds light on the differences in topic content provided by citizens on social media about the COVID-19 pandemic's impact on children and mothers. We use an unsupervised text analysis approach, LDA, and NMF topic modeling approach to analyze social media postings created by citizens to see how disruptive technology may aid with the research. In this study, substantial contributions on both theoretical and practical implications are revealed. Several topics on the COVID-19 pandemic are recovered from the LDA and NMF approach, in line with prior studies of social media postcontent analysis. Citizens' posts fall under a variety of categories, including Advice and Protective Measures, Impact on Pregnant Mothers, Support and Health Care, Society efforts, Express Feelings, and many others. This research adds to the debate over whether the general population uses social media for emotional or information needs in times of crisis. The outcomes of this study add to the theoretical literature on disruptive technology analysis using COVID-19. The study uses an unsupervised machine learning approach in textual data analysis to uncover citizens' interest in the COVID-19 problem as they use social media (Fig. 9.3; Table 9.3).

Table 9.2 Topic and top words by LDA topic modeling

Topic 1	Safe	Know	Duty	Need	Take	Change	Care	Expect	Get	Us
Topic 2	Pregnant	Women	COVID	Pregnancy	Risk	Birth	New	Infect	Pandemic	Maternal
Topic 3	Social	Test	Distance	Full	Positive	Tip	Continue	Question	Reason	Account
Topic 4	Health	Inform	Antenatal	Appointment	Team	Visit	Support	Million	Remove	Address
Topic 5	Worker	Make	Role	Obligation	Doctor	Worry	Alternative	Still	Adjust	Suspend
Topic 6	Coronavirus	Help	Doctor	Toward	Research	News	Article	Impact	Pandemic	Healthy
Topic 7	Mom	Look	Mother	Please	Share	Concern	Check	Contact	Fear	Introduce
Topic 8	Baby	Blog	Plan	Affect	Business	Well	Fertility	Reproductive	Ultrasound	Insight
Topic 9	Exercise	Egzersizleri	Kbakani	nefe	pm	Grow	Silent	Service	Session	fb
Topic 10	Pay	Around	Automate	Breastfeed	blood	Interview	md	de	Injury	Gynecology

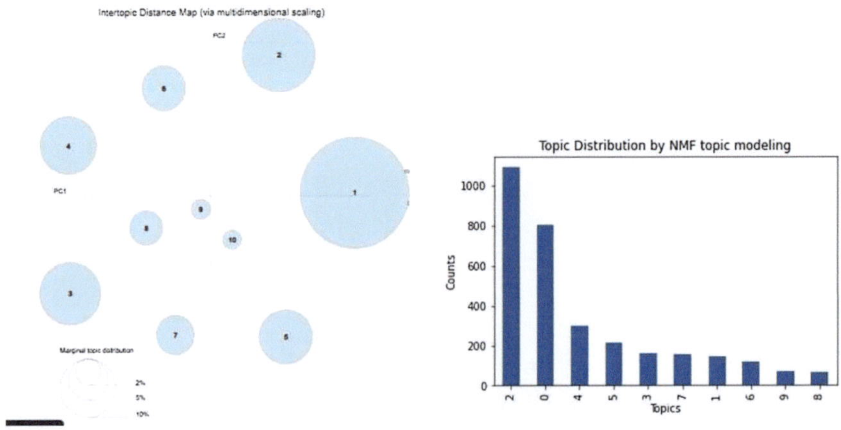

(a) (b)

Fig. 9.3 **a** LDA topic modeling, **b** NMF topic distribution

Despite the valuable insights gained from our exploration of sentiment relations in COVID-19-related social media posts, it is crucial to acknowledge certain limitations in our study. Firstly, the dynamic nature of social media platforms makes it challenging to capture real-time sentiments comprehensively, as opinions may evolve rapidly. Additionally, our analysis focused on publicly available content, potentially excluding sentiments expressed in private forums or messaging app. This study only focuses on content in English, which limits its ability to include a more diverse population. It is also important to highlight potential future avenues for research. Firstly, the incorporation of advanced modeling techniques, such as fuzzy modeling [23], and methodologies like rough set theory [24] and new clustering approaches [25, 26], could enhance the depth and precision of our sentiment analysis. The utilization of these techniques may uncover more nuanced sentiments and patterns within social media discussions. Additionally, the development of a Decision Support System (DSS) [27] holds promise for improving the decision-making process based on the sentiments expressed in social media. Integrating DSS could provide a more structured and informed approach to addressing public health challenges arising from misinformation. Furthermore, the potential application of Kernel-free SVM [28, 29] in sentiment analysis could be explored to enhance the accuracy of classification and better capture the nuances of opinions expressed in COVID-19-related social media content. While our study contributes valuable insights, the integration of these advanced modeling techniques and methodologies, along with addressing the highlighted limitations, presents exciting opportunities for future research. Exploring these avenues could not only refine our understanding of sentiment dynamics but also contribute to the development of more effective strategies for managing public health crises in the digital age.

Table 9.3 Topics and top words by NMF topic modeling

Topic 0	Women	Pregnant	COVID	Vaccine	Risk	Severe	Study	Ill	Increase	Among
Topic 1	Learn	News	Regain	Compound	Resilience	Loss	Impact	Major	Student	Great
Topic 2	Get	Child	Go	Know	Feel	Year	One	Like	Work	Mask
Topic 3	Coronavirus	Pandemic	Mom	Stress	Anytime	Healthy	Tip	Stay	Possible	Help
Topic 4	Pregnancy	COVID	egzersizleri	kbakanl	nefe	Exercise	Vaccine	Latest	Please	Helpline
Topic 5	COVID	SARS	Infect	Vertical	Study	Immune	Association	Newborn	Maternal	Antibody
Topic 6	Safe	Worker	Make	Role	Duty	Pregnant	Adjust	Alternative	Find	Obligation
Topic 7	via	Face shield	Birth	Due	Child	Healthday	February	May	March	Women
Topic 8	Doctor	Share	Oximeter	Pulse	Buy	May	Explain	Tip	Worry	Want
Topic 9	com	Application	Accept	Fall	Childcare	Higherheight	Fun	Question	Today	Detail

References

1. Das, S., Sarkar, S.: News media mining to explore speed-crash-traffic association during covid-19. Transport. Res. Record (2022). https://doi.org/10.1177/03611981221121261
2. Sarkar, S., Pramanik, A., Maiti, J., Reniers, G.: Covid-19 outbreak: a datadriven optimization model for allocation of patients. Comput. Ind. Eng. **161**, 107675 (2021)
3. Organization, W.H., et al.: Who director-general's opening remarks at the media briefing on covid-19 (2020)
4. Sarkar, S., Vinay, S., Raj, R., Maiti, J., Mitra, P.: Application of optimized machine learning techniques for prediction of occupational accidents. Comput. Oper. Res. **106**, 210–224 (2019). https://doi.org/10.1016/j.cor.2018.02.021
5. Sarkar, S., Pramanik, A., Maiti, J., Reniers, G.: Predicting and analyzing injury severity: a machine learning-based approach using class-imbalanced proactive and reactive data. Safety Sci. **125**, 104616 (2020). https://doi.org/10.1016/j.ssci.2020.104616
6. Talbot, J., Charron, V., Konkle, A.T.: Feeling the void: lack of support for isolation and sleep difficulties in pregnant women during the covid-19 pandemic revealed by twitter data analysis. Int. J. Environ. Res. Public Health **18**(2), 393 (2021)
7. Dey, A., Dehingia, N., Raj, A., Saseendrakumar, B.R., Tripathi, T.: Topic modeling of twitter data to identify issues impacting women and girls' education during the covid-19 pandemic (2021)
8. Yukich, J., Worges, M., Gage, A., Hotchkiss, D., Preaux, A., Murray, C., Cappa, C.: Projecting the impact of the covid-19 pandemic on child marriage. J. Adolesc. Health. **69**, 23–30 (2021). https://doi.org/10.1016/j.jadohealth.2021.07.037
9. Di Giorgio, E., Di Riso, D., Mioni, G., Cellini, N.: The interplay between mothers' and children behavioral and psychological factors during covid-19: an Italian study. Eur. Child Adolesc. Psychiatry **30**(9), 1401–1412 (2021)
10. Gao, J., Zheng, P., Jia, Y., Chen, H., Mao, Y., Chen, S., Wang, Y., Fu, H., Dai, J.: Mental health problems and social media exposure during covid-19 outbreak. PLoS ONE **15**(4), 0231924 (2020)
11. Neria, Y., Sullivan, G.M.: Understanding the mental health effects of indirect exposure to mass trauma through the media. JAMA **306**(12), 1374–1375 (2011)
12. Osakwe, Z.T., Ikhapoh, I., Arora, B.K., Bubu, O.M.: Identifying public concerns and reactions during the covid-19 pandemic on Twitter: a text-mining analysis. Public Health Nurs. **38**(2), 145–151 (2021)
13. Rocha, Y.M., Moura, G.A., Desid´erio, G.A., Oliveira, C.H., Lourenco, F.D., Figueiredo Nicolete, L.D.: The impact of fake news on social media and its influence on health during the covid-19 pandemic: a systematic review. J. Public Health 1–10 (2021)
14. Imran, N., Zeshan, M., Pervaiz, Z.: Mental health considerations for children & adolescents in covid-19 pandemic. Pak. J. Med. Sci. **36**(COVID19-S4), 67 (2020)
15. Garfin, D.R., Silver, R.C., Holman, E.A.: The novel coronavirus (covid-2019) outbreak: amplification of public health consequences by media exposure. Health Psychol. **39**(5), 355 (2020)
16. Elyashar, A., Plochotnikov, I., Cohen, I.-C., Puzis, R., Cohen, O., et al.: The state of mind of health care professionals in light of the covid-19 pandemic: text analysis study of twitter discourses. J. Med. Internet Res. **23**(10), 30217 (2021)
17. Busch-Hallen, J., Walters, D., Rowe, S., Chowdhury, A., Arabi, M.: Impact of covid-19 on maternal and child health. Lancet Glob. Health **8**(10), 1257 (2020)
18. Sarkar, S., Vinay, S., Djeddi, C., Maiti, J.: Text mining-based association rule mining for incident analysis: a case study of a steel plant in India. In: Pattern Recognition and Artificial Intelligence, pp. 257–273 (2021). https://doi.org/10.1007/978-3-030-71804-6_19
19. Onyenwe, I., Samuel, N., Orji, R., Gr, O., Onyenwe, I., Nwafor, C., Agbata, O.: Location-based sentiment analysis of 2019 Nigeria Presidential Election using a voting ensemble approach. https://doi.org/10.21203/rs.3.rs-2104488/v1

20. Hutto, C.J., Gilbert, E.: Vader: A parsimonious rule-based model for sentiment analysis of social media text (2015)
21. Sarkar, S., Sammangi, V., Raj, R., Maiti, J., Mitra, P.: Application of optimized machine learning techniques for prediction of occupational accidents. Comput. Oper. Res. **106**, 210–224 (2019)
22. Pramanik, A., Nande, V., Pradhan, A.S., Sarkar, S., Maiti, J.: Dynamic functional bandwidth kernel-based SVM: an efficient approach for functional data analysis. In: Emerging Technologies in Data Mining and Information Security, vol. 1286, pp. 673–681. Springer, Berlin (2021)
23. Sarkar, S., Lakha, V., Ansari, I., Maiti, J.: Supplier selection in uncertain environment: a fuzzy MCDM approach, pp. 257–266 (2016). https://doi.org/10.1007/978-981-10-2035-3_27
24. Sarkar, S., Baidya, S., Maiti, J.: Application of rough set theory in accident analysis at work: a case study. In: 2017 Third International Conference on Research in Computational Intelligence and Communication Networks (ICRCICN), pp. 245–250 (2017). https://doi.org/10.1109/ICRCICN.2017.8234514
25. Pramanik, A., Sarkar, S., Maiti, J., Mitra, P.: RT-GSOM: rough tolerance growing self-organizing map. Inform. Sci. **566**, 19–37 (2021). https://doi.org/10.1016/j.ins.2021.01.039
26. Sarkar, S., Lodhi, V., Maiti, J.: Text-clustering based deep neural network for prediction of occupational accident risk: a case study. In: iSAI-NLP 2018, Thailand (2018)
27. Sarkar, S., Chain, M., Nayak, S., Maiti, J.: Decision support system for prediction of occupational accident: a case study from a steel plant. In: Emerging Technologies in Data Mining and Information Security, vol. 813, pp. 787–796 (2019). https://doi.org/10.1007/978-981-13-1498-8_69
28. Saha, S., Das, M., Mondal, B.S., Sarkar, S., Maiti, J.: DiPSVM: a polynomial kernel-free support vector machine. In: 2021 International Conference on Data Analytics for Business and Industry (ICDABI). IEEE (2021)
29. Pradhan, S., Kumar, S., Sarkar, S., Maiti, J.: A kernel-free support vector machine with q-margin. In: 2021 International Conference on Data Analytics for Business and Industry (ICDABI). IEEE (2021)

Chapter 10
Alzheimer's Disease Detection Using Gaussian–Bayesian Parameter Optimization-Based Deep Convolution Neural Network

Ansh Varshney, Ishita Luharuka, Sobhan Sarkar, and Indranil Bose

Abstract This study addresses Alzheimer's disease detection through MRI scans, employing the renowned ResNet50, CNN with its distinctive 50 (48 convolutional layers, one MaxPool layer, and one average pool layer). ResNet50 is navigated using the Gaussian-based Bayesian Parameter Optimization (GBPO) technique for hyperparameter optimization. This innovative approach yields optimal configurations for the network. To validate our model, we plan to compare its performance with the default parameter that optimizes are set to. When applied with the best parameters obtained from GBPO, these outcomes underscore the effectiveness and robustness of our optimized ResNet50 in Alzheimer's disease detection. This paper serves as a concise illustration of the application of the proposed technique within a broader context, showcasing its adaptability to other deep convolutional neural networks (CNNs) such as Inception V3 and VGG19, DenseNet-169, LeNet-5, and AlexNet.

Keywords ResNet50 · Deep convolution neural network (DCNN) · Gaussian-based Bayesian parameter optimization (GBPO) · Alzheimer's detection · Inception-V3 · VGG19

A. Varshney (✉) · I. Luharuka
Chemical Engineering, IIT Kharagpur, Kharagpur, West Bengal, India
e-mail: ansh.v0606@gmail.com

S. Sarkar
Information Systems and Business Analytics, IIM Ranchi, Ranchi, Jharkhand, India
e-mail: sobhan.sarkar@iimranchi.ac.in

I. Bose
NEOMA Business School, Reims, France
e-mail: indranil.bose@neoma-bs.fr

J. Maiti et al. (eds.), *Operations Research and Data Analytics: Current Trends and Future Perspectives*, Lecture Notes on Multidisciplinary Industrial Engineering,
https://doi.org/10.1007/978-981-96-8327-7_10

10.1 Introduction

Alzheimer's Disease (AD) is a reversible, progressive brain illness that first causes short-term memory loss before gradually impairing thinking, memory, behavior, and other general abilities that interfere with a person's capacity to function as an individual [1]. This is the common cause of dementia that is linked to memory loss, reasoning impairment, problem-solving, and cognitive differences. The typical form of dementia, known as AD, is more prevalent in those over 60 [2]. Every five years, the chance of acquiring AD doubles, particularly after age 65. The causes of AD, a progressive and irreversible brain disease, are yet unknown. Between 2000 and 2006, there was a 47.1% rise in AD-related deaths. Early application is when medication for AD is most effective [3]. Alzheimer's disease currently has no known cure, yet there are several strategies to slow the illness's course [4]. Aging baby boomers will have significant social and economic effects on society as they go closer to the risk of rising AD. For this reason, it is important to diagnose the illness as soon as possible. Memory loss, poor judgment, feelings, and changes in nature are a few of the early signs of the condition [4]. This stage is known as mild cognitive impairment (MCI). The stage between being in good physical health and having Alzheimer's disease is known as MCI. Reduced cognitive abilities, such as memory and reasoning capacity, may result from MCI. Though slight, the decline in certain abilities is noticeable. Patients with mild cognitive impairment are more likely to develop AD [5], but only 15% of MCI patients do so. To assist physicians in improving the accuracy of MCI diagnosis, minimizing needless examinations, and lowering the misdiagnosis rate of patients, an increasing number of scholars advocate diagnosing MCI using artificial intelligence such as machine learning [6].

Convolutional Neural Networks (CNN) have been utilized successfully in medical image analysis, including MRI, X-ray, CT scans, and Ultrasonography. CNN has also seen significant progress in natural language processing, computer vision, audio recognition, and speech recognition [7]. Deep learning algorithms have been utilized for image analysis and interpretation, as well as to assist physicians and radiologists in making diagnostic judgments [8]. Convolutional Neural Network (CNN) applied on MRI images achieves good accuracy in the early diagnosis and classification of Alzheimer's Disease, showcasing the superior performance of deep learning over traditional machine learning approaches in handling large medical datasets [9]. In the paper [10], the authors explored diverse CNN-based transfer learning methods, achieving remarkable accuracy, with DenseNet outperforming other models and presenting significant advancements in specificity, sensitivity, and accuracy compared to previous works on the ADNI dataset. The paper [10] suggested a unique and lightweight binary classifier to differentiate AD from CN. They pre-processed the data using a typical procedure, including spatial normalization, skull stripping, and stationary wavelet entropy, using just the axial slice from the 3D MRI volume. A single-layer neural network was used, and a particle swarm optimizer was used to train the network parameters, yielding an accuracy of about 93%. Another paper [11] introduced a deep convolutional neural network-based pipeline utilizing

MRI scans for the accurate multiclass classification of Alzheimer's disease stages (AD, MCI, LMCI) and healthy individuals, achieving a notable prediction accuracy of 98.8% on the ADNI dataset, surpassing previous studies and highlighting the effectiveness of the proposed method. A pivotal facet of our methodology revolved around hyperparameter optimization, a critical determinant of deep neural network performance. Using GBPO, costly black box tasks like creating a new product of the highest caliber or fine-tuning a machine learning algorithm's hyperparameters may be efficiently optimized [12]. Departing from conventional approaches, we embraced the dynamism of Gaussian Bayesian optimization (GBPO) to fine-tune hyperparameters. GBPO employs Gaussian processes to intricately model the complex objective functions entwined with neural network training. This not only expedites the search for optimal parameters but also endows our methodology with a probabilistic framework for comprehending the uncertainties associated with diverse configurations. The dynamic nature of this optimization process, in stark contrast to static methods, enables our approach to adapt and iteratively learn from preceding experiments, thereby markedly amplifying efficiency and effectiveness. Central to our methodology is an iterative refinement process, a departure from traditional static hyperparameter tuning. Through successive iterations, our approach refines the surrogate model, ushering in improvements in sample efficiency. This adaptive learning paradigm ensures that the optimization process becomes increasingly effective over time, assimilating insights garnered from prior experiments to augment overall robustness and performance.

Our proposition extends beyond a conventional framework; it marks a pioneering stride in reshaping perspectives on Alzheimer's disease (AD). Rooted in the utilization of ResNet50 and advanced transfer learning techniques, our research holds the promise of significantly advancing the landscape of AD diagnosis. Picture a future where the application of this methodology allows for precise detection of Alzheimer's in its early stages, ushering in timely interventions and improving overall patient outcomes. This forward-looking vision serves as the impetus behind our research, propelling us to continually refine our framework and push the boundaries of early AD detection. Within this study, we delve into the intricacies of our proposed methodology, meticulously dissecting each component—from the utilization of ResNet50 and transfer learning to the intricacies of hyperparameter optimization through GBPO. Our exploration is a journey of enlightenment, meticulously analyzing results and uncovering the inherent strengths and possibilities within our framework. Ultimately, we take a moment to reflect on the road ahead, recognizing and addressing potential limitations while strategically charting a course for future advancements in Alzheimer's detection methodologies.

The inherent probabilistic nature of Gaussian processes assumes a central role in our innovative methodology, contributing significantly to precision-informed decision- making. Beyond conventional predictive modeling, our approach leverages Gaussian processes to furnish nuanced uncertainty estimates—a crucial aspect, especially in medical applications like Alzheimer's detection, where precision and reliability are paramount considerations. In the context of our methodology, the utilization of Gaussian processes extends beyond the mere provision of predictions.

It introduces a sophisticated layer of understanding that goes beyond deterministic outcomes. By offering detailed uncertainty estimates, our approach enhances the reliability of diagnostic outcomes. In the intricate landscape of Alzheimer's detection, where subtle patterns and complex interactions prevail, the ability to provide predictions and nuanced insights into the uncertainty surrounding those predictions is invaluable. This incorporation of uncertainty estimates elevates our methodology beyond the realm of traditional predictive models, transforming it into a more comprehensive and reliable diagnostic tool. It facilitates precise decision-making and establishes a foundation of transparency and trust in the diagnostic process, ultimately contributing to the advancement of medical applications and reinforcing the significance of probabilistic modeling in pushing the boundaries of diagnostic accuracy.

In a departure from conventional techniques, our methodology places paramount importance on rigorous comparative analysis. This involves a meticulous evaluation of its effectiveness in contrast to a scenario where GBPO is not applied. This comparative scrutiny serves as a robust validation mechanism, spotlighting the superior performance and efficacy of our methodology. Our approach seamlessly integrates advanced neural network architectures, hyperparameter optimization techniques, and probabilistic modeling, thereby presenting a comprehensive and pioneering paradigm for advancing the state-of-the-art in Alzheimer's detection. We also apply GBPO to different architectures to find the increase in the robustness of our approach. By systematically applying GBPO to each architecture and quantifying the resultant improvements in robustness, we validate its effectiveness across the ensemble and fine-tune the individual models for enhanced performance. This thorough exploration further solidifies our pioneering paradigm, reinforcing the significance of advanced techniques in pushing the boundaries of Alzheimer's detection methodologies.

The rest of this paper will proceed as follows: The proposed methodology for this article is explained in Sect. 10.2. Results and discussions are given in Sect. 10.3. The concluding part along with future scope and limitations is given in Sect. 10.4.

10.2 Methodology

In this section, we have discussed the proposed methodology in detail. The entire methodology has been divided into two phases: (i) *Phase I:* Data collection and pre-processing, and (ii) *Phase II:* Prediction model building with evaluation metrics.

10.2.1 Phase I: Dataset and Pre-processing

In the conducted study, the Alzheimer MRI Preprocessed Dataset was a central component, comprising 6400 MRI images used for dementia classification. The dataset, depicted in Fig. 10.1 and obtained from Kaggle,[1] encompassed four distinct classes: Mild Demented, Moderate Demented, Non-Demented, and Very Mild Demented. A rigorous preprocessing pipeline was implemented to ensure the images' consistency and quality. This preprocessing involved standardization procedures, enhancing the images, and establishing a uniform format conducive to effective neural network training. Subsequently, a stratified split of 7:2:1 was applied, allocating the data into training, validation, and testing sets, respectively. Rotation and zoom variations were introduced to further augment the dataset and improve the model's generalization. The augmentation process was implemented using the Keras Sequential API, incorporating two key operations: random rotation with a factor ranging from − 0.15 to 0.15 and random zoom with a height factor between − 0.3 and − 0.1. These variations served to enhance the model's robustness by introducing diverse perspectives of the input data during training. In addition to data augmentation, the study addressed potential class imbalance issues through the computation of class weights for the multi-class classification problem. The computation of these weights considered both the total number of images and their distribution across different classes. This class weight calculation was crucial during training to ensure that the model assigns appropriate importance to each class, particularly in scenarios where there was a significant difference in the number of images per class. The computed class weights were printed for each class, revealing their inverse proportions to the number of images per class in the dataset. This weighting mechanism played a pivotal role in achieving a balanced learning process during the training of neural networks for image classification tasks. In summary, the utilization of the Alzheimer MRI Preprocessed Dataset, coupled with rigorous preprocessing, stratified data splitting, and the incorporation of data augmentation techniques, formed a robust foundation for training neural networks for dementia classification. The introduction of class weights further ensured that the model is equipped to handle potential imbalances in class distribution, fostering a balanced and effective learning process. These meticulous steps collectively contributed to the overall success and reliability of the neural network in accurately classifying Alzheimer's disease across different severity levels.

10.2.2 Phase II: Prediction Model Building

In this section, we have discussed how we trained our model using diverse CNN architectures, with subsequent evaluation using various metrics. After that, a GBPO approach was implemented to determine an optimized learning rate. The model was then retrained using the optimized learning rate, and the results were compared between the two scenarios.

[1] https://www.kaggle.com/datasets/sachinkumar413/alzheimer-mri-dataset/data.

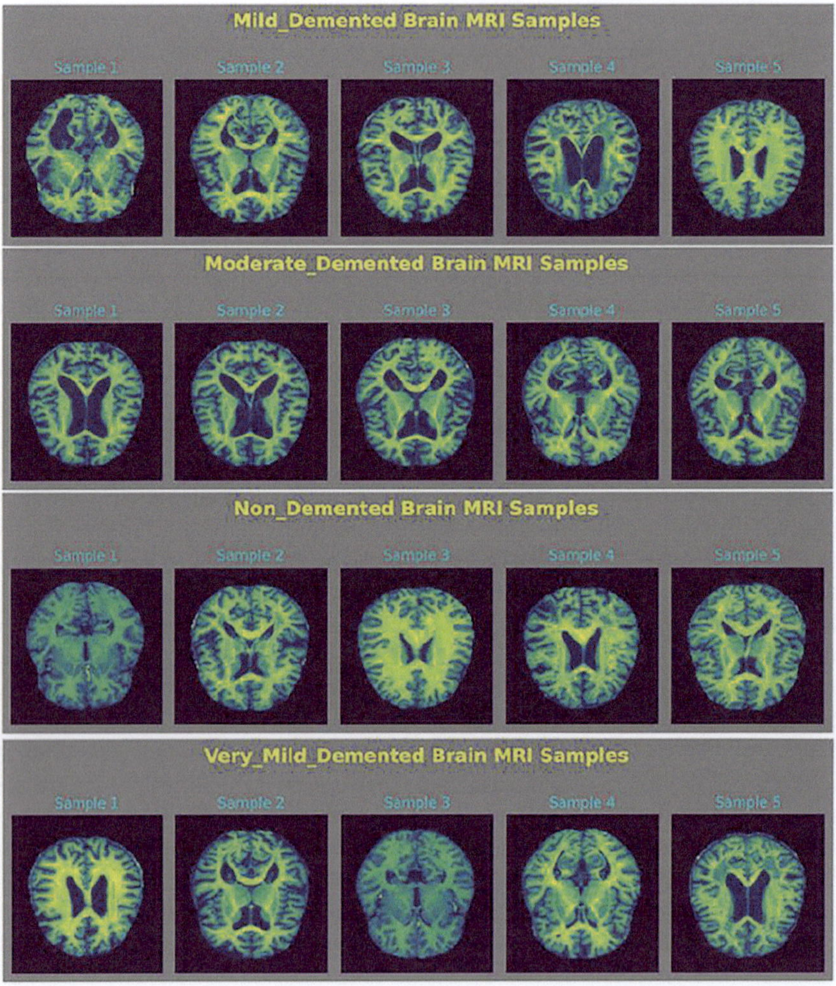

Fig. 10.1 Dataset of MRI scans

10.2.2.1 Transfer Learning-Based CNN Architecture

In the realm of Alzheimer's disease (AD) detection, leveraging state-of-the-art deep learning architectures is crucial for achieving high accuracy. There are several hidden qualities in various forms that are rarely investigated by conventional machine learning techniques [13]. Transfer learning, a paradigm in deep learning, is employed to harness the capabilities of pre-trained models on large-scale datasets and adapt them to specific tasks. ResNet-50, a prominent convolutional neural network (CNN) architecture, is chosen for its ability to handle complex feature hierarchies effectively. The ResNet- 50 model is initially trained on ImageNet, a massive image dataset, where it learns to recognize a diverse range of visual patterns. The architecture of ResNet-50 is characterized by its deep structure, featuring 50 layers, and its distinctive residual blocks, which facilitate the training of deep networks. In transfer learning, the pre-trained ResNet-50 model serves as a feature extractor. The convolutional layers act as filters that automatically learn hierarchical features, starting from simple edges and textures to complex high-level patterns. These learned features are crucial for image classification tasks. To adapt ResNet-50 for Alzheimer's detection, the final fully connected layer was modified to suit the binary classification nature of the task—distinguishing between Alzheimer's and normal brain images. Additional layers were incorporated into the model to enhance its capabilities. Global average pooling was applied to reduce the spatial dimensions of the feature maps, effectively summarizing the learned features. Batch normalization was introduced to stabilize and accelerate training by normalizing each layer's inputs. The modified architecture extended with densely connected layers, introducing non-linearities through activation functions like ReLU (Rectified Linear Unit). These additional layers further refined the features extracted by ResNet-50, enabling the model to capture intricate patterns specific to Alzheimer's disease. Fine-tuning, a crucial step in transfer learning, involved training the modified ResNet-50 on a dataset specifically curated for Alzheimer's detection. This dataset included brain images representative of both Alzheimer's-affected and normal cases. The process allowed the model to adapt its learned features to the nuances of the target task, enhancing its discriminatory power. During the training phase, the model optimized its parameters using the Adam optimizer and categorical cross-entropy loss function.

Metrics such as accuracy, area under the curve (AUC), precision, and recall were monitored to comprehensively assess the model's performance. The resulting architecture, born from the fusion of transfer learning, feature extraction, and fine-tuning, emerged as a powerful tool for accurate Alzheimer's disease detection. By utilizing the abstract features learned by ResNet-50 from diverse visual patterns in the original dataset, the model effectively distinguished between Alzheimer's and normal brain images. This intricate approach combined the strengths of a pre-trained deep learning model with task-specific adaptations, showcasing a holistic methodology for advancing Alzheimer's disease detection using cutting-edge neural network architectures.

10.2.2.2 Optimization Using Gaussian-Based Bayesian Process Optimization

Optimizing the hyperparameters using typical optimization procedures might take an eternity if each function evaluation is costly, such as in the case of a neural network where the parameters are the hyperparameters and the function evaluation is the mean cross-validation score over ten folds. If we deploy the traditional method for hyperparameter tuning, i.e., GridSearch CV or RandomSearch CV, then it would take infinite time as our model would have a lot of parameters to be learned, and learning those parameters again and again with different hyperparameters would be time-consuming. Making better decisions was made possible by using past information or opinions on the goal function. Exploration and exploitation were also balanced. While exploitation concentrated on using the identified promising regions to maximize the currently best solution, exploration permitted a more thorough examination of the search space and led to the discovery of superior solutions. Finding better answers and honing the best solution should balance these two factors. The optimization process of tuning hyperparameters for Convolutional Neural Networks (CNNs) using Gaussian- based Bayesian Process Optimization (GBPO) systematically explored the learning rate space. The Gaussian Process (GP) is a probabilistic model employed in this context, serving as a surrogate for the validation accuracy of the CNN with different learning rates. At the outset, the GP was initialized with random samples of learning rates, and its predictions were updated as the optimization proceeds based on actual validation accuracies. The Expected Improvement (EI) acquisition function was integral to the decision-making process in selecting the next learning rate to evaluate. The mathematics underlying the GP involved modeling the distribution of the objective function (negative validation accuracy) across the space of learning rates. Mathematically, the GP predicted the mean of the objective function and its variance, capturing the uncertainty associated with different learning rates. The EI function then calculated the expected improvement in the objective function, considering both the predicted mean and variance. By balancing exploration (favoring regions with high uncertainty) and exploitation (considering regions with potentially higher objective values), EI guided the optimization towards regions likely to yield the optimal learning rate. The optimization loop iterated through updating the GP, selecting the next learning rate to evaluate, and refining its understanding of the objective function. This iterative process efficiently navigated the hyperparameter space, ensuring that the search was directed toward regions that were likely to yield the optimal learning rate. The convergence of the optimization occurred when the specified maximum number of iterations was reached, providing the learning rate that maximized the CNN's validation accuracy. In terms of a visual representation, the Gaussian-based Bayesian Process Optimization flowchart is a helpful depiction of the step-by-step procedure (as shown in Fig. 10.2). The flowchart outlines the sequential stages of updating the GP, selecting learning rates, and optimizing the CNN's validation accuracy. This systematic approach to hyperparameter tuning showcased the power of Gaussian-based Bayesian optimization in efficiently exploring and exploiting the hyperparameter space for enhanced CNN performance. This methodology encapsulated a comprehensive and mathematically grounded approach to hyperparameter tuning, ensuring that the learning rate was optimized to maximize the validation accuracy of the CNN. The combination of Gaussian Process modeling, Expected Improvement acquisition function, and iterative optimization loop formed a robust framework for hyperparameter tuning, contributing to the overall efficacy of CNN.

Gaussian processes as a prior for functions

We require a prior probability model for $f(x)$ and a likelihood model for the samples from $f(x)$ in order to construct a posterior expectation. We assume a normal likelihood

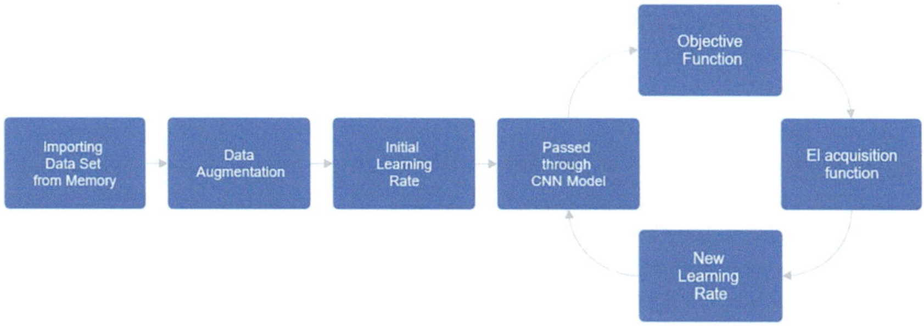

Fig. 10.2 Proposed methodological flowchart for GBPO

with noise in Bayesian search.

$$y = f(x) + \varepsilon, \quad \varepsilon \sim N(0, \sigma^2) \tag{10.1}$$

We suppose that a Gaussian process (GP) can explain the loss function $f(x)$ for the prior distribution. A generalized Gaussian distribution (GP) is a Gaussian distribution over functions rather than random variables. Similar to how the mean and variance of a Gaussian distribution fully characterize it, the **mean function** $m(x)$ and **covariance function** $k(x,x)$ fully characterize a GP. We suppose that a multivariate Gaussian distribution may represent the values of the loss function $f_{1:n} = \{f(x_1), \ldots, f(x_n)\}$ for a collection of data points $x_{1:n} = \{x_1, \ldots, x_n\}$.

$$f_{1:n} \sim N(m(\mathbf{x}_{1:n}), K), \tag{10.2}$$

where $m(\mathbf{x}_{1:n}) = [m(x_1), \ldots, m(x_n)]^T$, and the $n \times n$ kernel matrix K has entries given by $[K]_{ij} = k(x_i, x_j)$ A general purpose function (GP) can be conceptualized as a function that yields the mean and variance of a normal distribution over the range of potential values of f at x, as opposed to a scalar $f(x)$. Because it produces an analytically tractable posterior distribution over the loss function, a generalized probability model (GP) is a widely used probability model. This allows us to compute the loss for a fresh set of hyperparameters and update our views about what f looks like.

Acquisition functions

We shall select the point that maximizes an acquisition function in order to determine the optimal location from which to sample f next. This function characterizes the utility for every value of the hyperparameters and is dependent on the posterior distribution over f. The values for which we compute the loss next are those that have the highest utility. Although the literature has presented a number of acquisition functions, the anticipated improvement (EI) function appears to be the most often used. It is described as

$$\mathrm{EI}(x) = E\big[\max\{0, f(x) - f(x^*)\}\big], \tag{10.3}$$

where x^* is the current best combination of hyperparameters. We shall get at the point that, hopefully, improves upon f the most when we maximize this quantity. The wonderful thing about the projected improvement is that it can be calculated using integration by parts under the GP model, where the (multivariate) standard normal distribution's cumulative distribution and probability density function are denoted by the symbols $\Phi(z)$ and $\phi(z)$. We can see from this closed-form solution what kinds of values lead to a larger predicted improvement.

$$\mathrm{EI}(x, Z) = \begin{cases} (\mu(x) - f(x^*))\Phi(Z) + \sigma(x)\phi(Z) & \text{if } \sigma(x) > 0 \\ 0 & \text{if } \sigma(x) = 0 \end{cases} \tag{10.4}$$

10.2.2.3 Evaluation Metrics

The evaluation of ResNet-50, VGG-19, Inception-V3, DenseNet-169, LeNet-5, and AlexNet models before and after employing Gaussian Bayesian Process Optimization (GBPO) involved a comprehensive set of metrics to assess their performance in detecting Alzheimer's disease. Initially, these models were trained and evaluated using standard hyperparameters, and the metrics included accuracy, precision, recall, F1-score, and area under the receiver operating characteristic curve (AUC-ROC). These metrics provided a holistic view of a model's classification performance, considering aspects such as true positive rate, false positive rate, and overall accuracy. Before GBPO, the models were trained using default hyperparameters, and their evaluation metrics were recorded on the validation set. Subsequently, GBPO was employed to optimize hyperparameters, specifically learning rates, for each model individually. The hyperparameter search space and the objective function were adapted for each model's architecture. GBPO leveraged a probabilistic model to explore the hyper- parameter space efficiently, and the acquisition function guided the search toward promising regions. After GBPO, the models were retrained with the optimized hyper-parameters, and their evaluation metrics were once again computed on the validation set. The comparison involved assessing how the optimization process influenced the models' accuracy, precision, recall, F1-score, and AUC-ROC performance. Improvements or changes in these metrics indicated the efficacy of GBPO in fine-tuning the hyper-parameters and enhancing the models' overall classification capabilities. This comprehensive evaluation strategy not only quantified the impact of GBPO on the models' overall performance but also shed light on the nuances of improvements or adjustments in specific aspects of classification. The comparison results contributed valuable insights into the efficacy of GBPO in optimizing hyperparameters for diverse CNN architectures, offering a nuanced understanding of the interplay between hyperparameter tuning and the models' classification capabilities in Alzheimer's disease detection (Fig. 10.3).

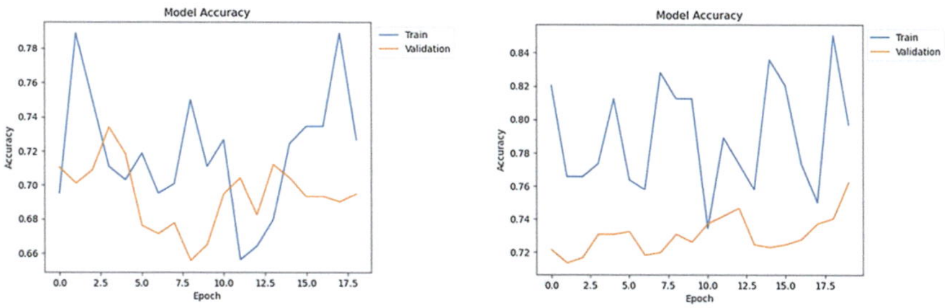

(a) at default learning rate. (b) optimised learning rate of 0.00034

Fig. 10.3 Extra 20 epochs for Inception-V3

10.3 Results and Discussion

In this section, the results of ResNet-50, VGG-19, Inception-V3, DenseNet-169, LeNet- 5, and AlexNet are compared and discussed with the model having a default learning rate.

By using the GBPO technique, there was an increase in the performance of the model. This was due to its capacity to use and explore the parameter space efficiently. Through the use of acquisition functions and a clever balance between exploration and exploitation GBPO effectively navigated the parameter space to find configurations that result in enhanced model performance. This feature was especially helpful in complicated optimization environments where it might be difficult for standard optimization approaches to identify the global optimum [14]. Furthermore, GBPO used Gaussian process models to capture forecast uncertainty, which helped it decide where to sample next and concentrate computing resources on areas that were most likely to improve performance. Additionally, GBPO's adaptive sampling technique maximized the use of computing resources by directing assessments toward regions of the parameter space with the greatest potential for improvement [15]. Additionally, the method repeatedly improved model configurations for improved performance since GBPO automated the hyperparameter tuning process, which minimized the requirement for operator involvement [16]. Overall, GBPO's efficacy in improving machine learning model performance across various evaluation metrics could be attributed to its capacity to effectively explore the parameter space, manage uncertainty, carry out global optimization, adaptively sample points, and automate hyperparameter tuning.

10.3.1 Results of ResNet-50

ResNet-50, short for Residual Network with 50 layers, is a DCNN architecture renowned for its success in image classification tasks. It addresses the vanishing gradient problem by introducing residual blocks, which utilize skip connections to facilitate the flow of information. ResNet-50 comprises a series of building blocks containing convolutional layers and shortcut connections. The residual blocks enable the network to learn residual functions, making it highly effective for training very deep networks. In our Alzheimer's detection project, ResNet-50 is applied to the dataset of medical images. The network learnt to discern intricate patterns indicative of Alzheimer's disease through extensive pre-training and subsequent fine-tuning. After integrating Gaussian-based Bayesian Process Optimization (GBPO) into the optimization process, we observed notable improvements in performance metrics. The loss decreased from 0.58 to 0.5, showcasing enhanced convergence and model fitting. Accuracy saw a substantial increase from 70 to 76%, indicating improved overall model correctness. The area under the curve (AUC) increased from 0.93 to 0.95, underscoring heightened discriminative power. Precision witnessed a boost from 0.73 to 0.81, reflecting improved positive predictive value, while recall increased from 0.67 to 0.72, indicating the model's ability to capture more instances of true positives. These results highlighted the effectiveness of GBPO in fine-tuning the ResNet-50 architecture for Alzheimer's detection, enhancing both sensitivity and specificity. However, the optimization process posed challenges, including the need for careful hyperparameter tuning and computational resources. Training a deep network like ResNet-50 demanded substantial computational power, and optimizing hyper-parameters required a delicate balance to prevent overfitting [17].

10.3.2 Results of VGG-19

VGG-19 is a convolutional neural network (CNN) architecture renowned for its simplicity and effectiveness in image classification tasks. Comprising 19 layers, VGG-19 is characterized by its uniform design, featuring 3×3 convolutional layers and max-pooling layers. The convolutional layers are organized into blocks, and the repetitive structure allows the network to learn hierarchical features of increasing complexity. In each block, multiple 3×3 convolutions are followed by max-pooling, culminating in three fully connected layers and a softmax activation for classification. This architecture has demonstrated strong performance in various image classification tasks, including Alzheimer's disease detection. On our dataset, VGG-19 effectively extracted features indicative of Alzheimer's pathology. After applying Gaussian-based Bayesian Process Optimization (GBPO) for hyperparameter tuning, we observed notable improvements in performance metrics. The loss decreased from 0.37 to 0.33, accuracy increased from 85 to 89%, AUC increased from 0.98 to 0.98, precision increased from 0.86 to 0.89, and recall increased from 0.84 to 0.88. These enhancements underscored the efficacy of GBPO in refining the hyperparameters, leading to a more optimized VGG-19 model for Alzheimer's detection with improved accuracy, precision, and recall metrics. The training of the model is shown in Fig. 10.5b. The optimized learning rate after applying GBPO was found to be 0.0063. The backstopping was done, taking validation accuracy as the parameter (Fig. 10.4).

10.3.3 Results of Inception-V3

Inception-V3 is a convolutional neural network (CNN) architecture developed by Google, primarily designed for image classification and object recognition tasks. Its architecture is characterized by the innovative use of inception modules, which incorporate parallel convolutional operations with varying kernel sizes to capture features at multiple scales. The network consists of stem layers followed by multiple inception modules, culminating in fully connected layers for classification. Inception-V3 employs factorized convolutions and auxiliary classifiers to enhance gradient flow during training. In our dataset of Alzheimer's images, Inception-V3 excelled at capturing intricate patterns indicative of the disease, leveraging its ability to learn from diverse scale information. The model's architecture, comprising approximately 48 layers, facilitated effective hierarchical feature learning. Regarding the results, before applying Gaussian-based Bayesian Process Optimization (GBPO), the Inception-V3 model exhibited a loss of 0.69 and an accuracy of 0.69. After applying GBPO, there was a notable improvement, with the loss decreasing to 0.61 and the accuracy increasing to 0.74. Additionally, crucial performance metrics such as AUC-ROC, precision, and recall also saw improvements. The AUC-ROC increased from 91 to 94%, precision rose from 0.72 to 0.77, and recall improved from 0.64 to 0.70. These results underscored the efficacy of GBPO in optimizing the hyper-parameters of Inception-V3 for our specific Alzheimer's detection task, leading to a substantial enhancement in model performance across various evaluation metrics. The improvements suggested that the tailored optimization facilitated by GBPO enhanced the model's capacity to discern and classify Alzheimer's-related features in our dataset. The training of the model before and after applying the optimizer is illustrated in Fig. 10.3b.

10.3.4 Results of DenseNet-169

DenseNet-169 is a deep convolutional neural network renowned for its densely connected architecture, comprising 169 layers. DenseNet-169 features densely connected blocks within its structure, where each layer receives direct inputs from all preceding layers. This interconnectedness fosters efficient feature reuse, mitigating the vanishing gradient problem and enabling the network to learn intricate patterns. In the context of our Alzheimer's detection dataset, DenseNet-169 excels in capturing subtle features indicative of the disease. Through extensive pre-training on diverse datasets and fine-tuning for Alzheimer's detection, the architecture is optimized to discern relevant patterns in our specific image data. Applying Gaussian-Based

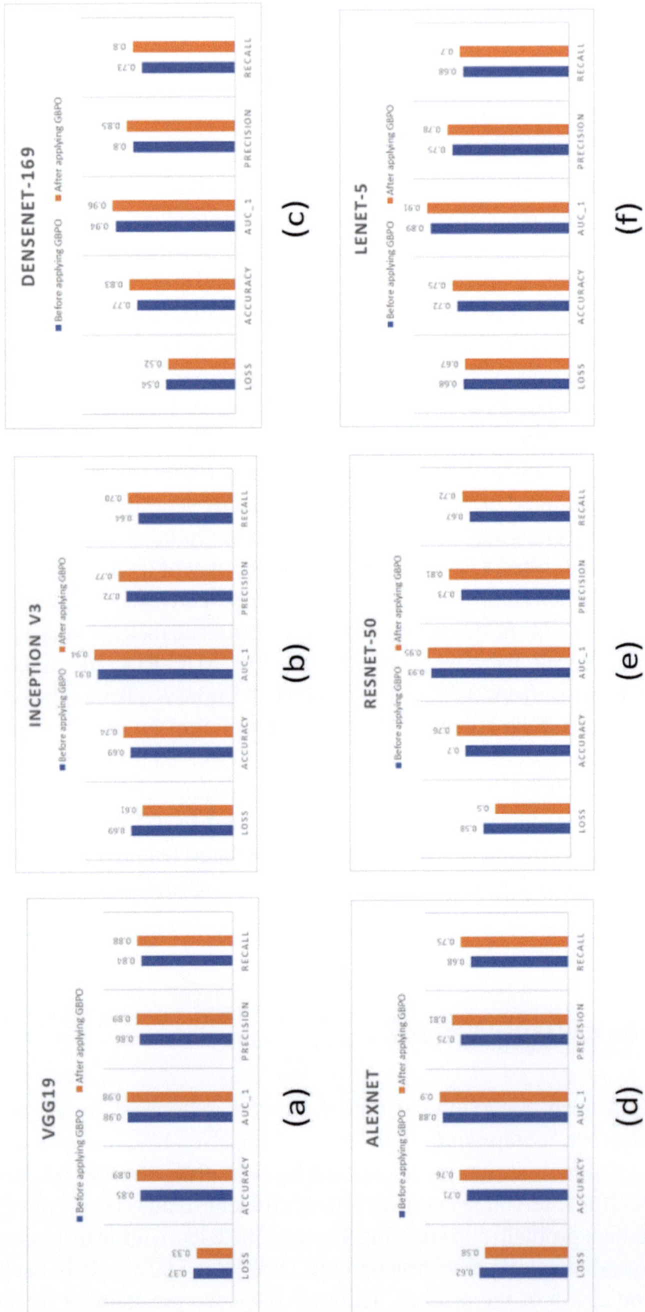

Fig. 10.4 Increase in performance metric for various CNN architectures after applying GBPO for **a** VGG-19 **b** Inception-V3 **c** DenseNet-169 **d** AlexNet **e** ResNet-50 **f** LeNet-5

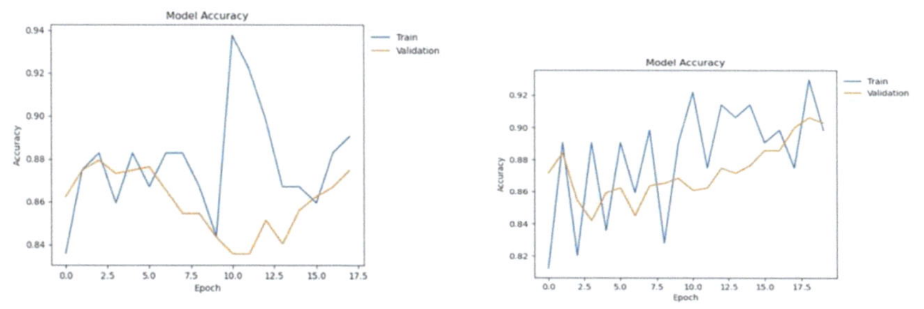

(a) at default learning rate. (b) optimised learning rate of 0.0063

Fig. 10.5 Extra 20 epochs for VGG-19

Bayesian Process Optimization (GBPO) to DenseNet-169 yielded notable improvements in performance metrics. Before GBPO, the model exhibited a loss of 0.54, accuracy of 77%, AUC of 0.94, precision of 0.8, and recall of 0.73. Post-GBPO, these metrics improved, demonstrating a reduction in loss 0.52, increased accuracy 0.83%, elevated AUC (0.96), enhanced precision (0.85), and improved recall (0.8). The application of GBPO facilitated a more refined and effective optimization of hyper-parameters, showcasing the potential of this technique in boosting the performance of DenseNet-169 for Alzheimer's detection. These results underscored the significance of tailoring the model's parameters for the specific task, ultimately contributing to a more robust and accurate diagnostic tool for Alzheimer's disease. While the highly dense structure of DenseNet-169 offered numerous advantages, it also introduced considerations related to the potential for overfitting. The extensive connectivity within the dense blocks allowed for rich feature interactions and enhanced the model's capacity to learn intricate patterns. However, this increased model complexity might lead to a higher risk of overfitting, especially when the dataset is limited. In the context of Alzheimer's detection, the highly dense structure of DenseNet-169 became particularly valuable. It enabled the model to effectively capture intricate patterns and subtle features associated with the disease. Our methodology needed to work well with this highly dense network. We achieved a significant increase of 7.79% in accuracy.

10.3.5 Results of LeNet-5

LeNet-5, designed by Yann LeCun and his colleagues, is a pioneering convolutional neural network (CNN) architecture primarily developed for handwritten digit recognition. It comprises seven layers, including three convolutional layers, two sub-sampling (pooling) layers, and two fully connected layers. The architecture begins with the input layer, followed by the first convolutional layer, which extracts low-level features using a 5×5 kernel. Subsequently, sub-sampling layers reduce spatial dimensions, aiding in feature selection. The second convolutional layer further extracts higher- level features, followed by additional sub-sampling. The fully connected layers process these features for classification. LeNet-5 is characterized by its efficient use of convolutional and sub-sampling operations, enabling hierarchical feature extraction. In our Alzheimer's detection project, we adopted LeNet-5 for image classification. Before applying Gaussian-Based Bayesian Process Optimization (GBPO), the model showed a loss of 0.68, accuracy of 72%, AUC of 0.89, precision of 0.75, and recall of 0.68. After applying GBPO, we observed improvements with a reduced loss of 0.67, increased accuracy to 75%, elevated AUC to 0.91, improved precision of 0.78, and enhanced recall of 0.7. These results underscored the efficacy of GBPO in optimizing the hyperparameter space of LeNet-5, contributing to superior performance metrics and reinforcing the value of fine-tuning for enhanced Alzheimer's disease detection.

10.3.6 Results of AlexNet

AlexNet, a groundbreaking convolutional neural network (CNN) architecture, was introduced by Krizhevsky et al. in 2012 and played a pivotal role in popularizing deep learning. AlexNet comprises eight learned layers, including five convolutional layers and three fully connected layers. The convolutional layers leverage rectified linear unit (ReLU) activation functions and local response normalization, enhancing feature extraction and non-linearity. The fully connected layers contribute to high-level feature representation and decision-making. Importantly, AlexNet incorporates techniques like dropout and data augmentation to prevent overfitting. In our study, AlexNet was applied to our dataset for Alzheimer's detection. The architecture demonstrated notable performance improvements before and after the application of Gaussian-based Bayesian Process Optimization (GBPO). The initial results showed a loss of 0.62, accuracy of 71%, AUC of 0.88, precision of 0.75, and recall of 0.68. Post-GBPO, the performance metrics exhibited significant enhancement, with the loss reduced to 0.58, accuracy increased to 76%, AUC improved to 0.9, precision rose to 0.81, and recall reached 0.75. The application of GBPO not only optimized the hyper-parameters of AlexNet but also refined its architecture for improved specificity and sensitivity in Alzheimer's disease detection. This outcome underscored the efficacy of GBPO in fine-tuning deep neural networks for enhanced performance on our specific dataset.

In summary, the combination of diverse model architectures and meticulous hyperparameter tuning has positively impacted the detection accuracy of Alzheimer's disease. The increase in evaluation metrics on various CNN architecture models is summarized in Fig. 10.6.

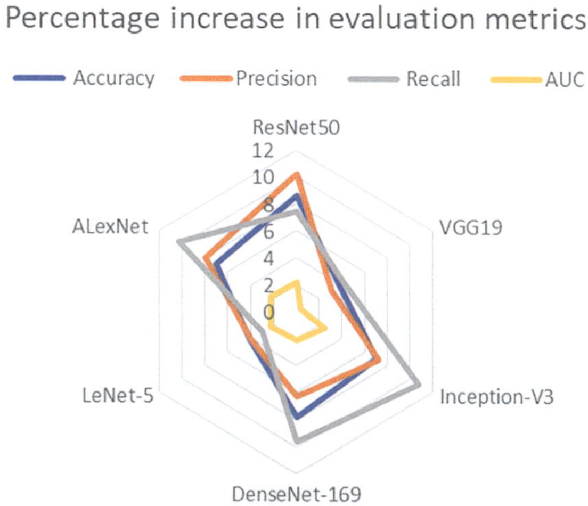

Fig. 10.6 Percentage increase in performance metric after applying GBPO

10.4 Conclusions

With the recent developments in biomedical engineering, one of the main fields of research is now the analysis and study of medical scans [18, 19]. The implementation of various models for Alzheimer's disease detection, coupled with the optimization of the default learning rate using Gaussian-based Bayesian process optimization, has yielded positive improvements in the evaluation metrics across different models. The increase in accuracy, precision, recall, and AUC values indicates enhanced performance in the ability of these models to classify instances of Alzheimer's disease correctly. Specifically, when comparing the models, it is evident that Inception-V3 and AlexNet exhibited notable improvements across multiple metrics, showcasing their effectiveness in detecting Alzheimer's disease. The increase in accuracy, precision, recall, and AUC for these models suggested that the fine-tuning of the learning rate positively impacted their overall performance. It is crucial to acknowledge that the choice of model architecture played a significant role in the success of Alzheimer's disease detection. Moreover, the successful optimization of the learning rate using Gaussian-based Bayesian process optimization highlighted the importance of hyperparameter tuning in the performance of machine learning models. This approach proved to be effective in enhancing the models' ability to learn and generalize from the data, leading to improved diagnostic capabilities for Alzheimer's disease. Notably, the average accuracy of all models witnessed a substantial boost, reaching 6.44%. Specifically, the Inception-V3 model exhibited the most significant improvements, with a remarkable 10.78% increase in recall and a 7.19% enhancement in precision. While other models also showcased positive outcomes, it is essential to acknowledge the varying degrees of impact across architectures. In summary, the combination of diverse model architectures and meticulous hyperparameter tuning positively impacted the detection accuracy of Alzheimer's disease. These findings contributed to the ongoing efforts to leverage machine learning for medical diagnostics, particularly in the early detection of neurodegenerative diseases. Furthermore, although helpful, fine-tuning hyperparameters complicated the model building process and might not necessarily result in appreciable performance improvements in actual clinical scenarios. Lastly, there was still a worry about how interpretable deep learning models would be as it was critical to comprehend how these models made decisions in order to win the confidence of medical professionals and guarantee the accuracy of diagnoses.

We anticipate and strive to support further efforts in the future. Therefore, deep convolutional neural networks, which have recently demonstrated their potential in neuroimaging research, might be used to further improve the outcome. Consequently, the algorithm's capacity to identify AD would be greatly enhanced by the employment of deep CNN with large MRI scan pictures. Additionally, this deep learning technique not only benefits the patient, physician, and radiologist suffering from this ailment, but it also gives the researcher invaluable information to diagnose other types of diseases. Future studies should validate these models' performance across various demographics and imaging modalities in order to increase their generalizability. Another crucial area of research is improving the interpretability of deep learning models. There are continuous efforts to create explainable AI methods that can clarify how these intricate systems make decisions. Additionally, merging MRI scans with other forms of medical data (such as genetic information or patient history) or integrating multi modal data might result in a more thorough diagnostic tool. Last but not least, in order to guarantee that technological developments are successfully incorporated into clinical practice, researchers, engineers, and medical professionals will need to work together continuously. This will improve patient outcomes and further our understanding of neurodegenerative diseases like Alzheimer's.

References

1. Liu, Y., Spulber, G., Lehtimäki, K.K., Könönen, M., Hallikainen, I., Gröhn, H., Kivipelto, M., Hallikainen, M., Vanninen, R., Soininen, H.: Diffusion tensor imaging and tract-based spatial statistics in Alzheimer's disease and mild cognitive impairment. Neurobiol Aging **32**(9), 1558–1571 (2011)
2. Albright, J.: Forecasting the progression of Alzheimer's disease using neural networks and a novel preprocessing algorithm. Alzheimer's Dementia: Transl. Res. Clin. Intervent. **5**, 483–491 (2019). https://doi.org/10.1016/j.trci.2019.07.001
3. Iyengar, S., Dauwels, J., Varshney, P.K., Cichocki, A.: Quantifying EEG synchrony using copulas, pp. 505–508 (2010). https://doi.org/10.1109/ICASSP.2010.5495664
4. Moutinho, S.: The long road to a cure for Alzheimer's disease is paved with failures. Nat. Med. **28**(11), 2228–2231 (2022). https://doi.org/10.1038/s41591-022-02062-0
5. Li, F., Tran, L., Thung, K.-H., Ji, S., Shen, D., Li, J.: A robust deep model for improved classification of ad/mci patients. IEEE J. Biomed. Health Inform. **19**(5), 1610–1616 (2015). https://doi.org/10.1109/JBHI.2015.2429556
6. Shen, T., Jiang, J., Li, Y., Wu, P., Zuo, C., Yan, Z.: Decision supporting model for one-year conversion probability from mci to ad using CNN and SVM. In: 2018 40th Annual International

Conference of the IEEE Engineering in Medicine and Biology Society (EMBC), pp. 738–741 (2018). https://doi.org/10.1109/EMBC.2018.8512398
7. Hussain, E., Hasan, M., Hassan, S.Z., Hassan Azmi, T., Rahman, M.A., Zavid Parvez, M.: Deep learning based binary classification for Alzheimer's disease detection using brain MRI images. In: 2020 15th IEEE Conference on Industrial Electronics and Applications (ICIEA), pp. 1115–1120 (2020). https://doi.org/10.1109/ICIEA48937.2020.9248213
8. Al-Adhaileh, M.H.: Diagnosis and classification of Alzheimer's disease by using a convolution neural network algorithm. Soft. Comput. **26**(16), 7751–7762 (2022)
9. Salehi, W., Baglat, P., Upadhya, A., Sharma, B., Gupta, G.: A CNN model: earlier diagnosis and classification of Alzheimer disease using MRI (2020)
10. Ashraf, A., Naz, S., Shirazi, S.H., Razzak, I., Parsad, M.: Deep transfer learning for Alzheimer neurological disorder detection. Multimedia Tools Appl. 1–26 (2021)
11. Farooq, A., Anwar, S., Awais, M., Rehman, S.: A deep CNN based multi-class classification of Alzheimer's disease using MRI. In: 2017 IEEE International Conference on Imaging Systems and Techniques (IST), pp. 1–6 (2017). https://doi.org/10.1109/IST.2017.8261460
12. Rana, S., Li, C., Gupta, S., Nguyen, V., Venkatesh, S.: High dimensional Bayesian optimization with elastic Gaussian process. In: Precup, D., Teh, Y.W. (eds.) Proceedings of the 34th International Conference on Machine Learning. Proceedings of Machine Learning Research, vol. 70, pp. 2883–2891. PMLR (2017). https://proceedings.mlr.press/v70/rana17a.html
13. Sarkar, S., Vinay, S., Raj, R., Maiti, J., Mitra, P.: Application of optimized machine learning techniques for prediction of occupational accidents. Comput. Oper. Res. **106**, 210–224 (2019). https://doi.org/10.1016/j.cor.2018.02.021
14. Paul, D., Goswami, A.K., Chetri, R.L., Roy, R., Sen, P.: Bayesian optimization- based gradient boosting method of fault detection in oil-immersed transformer and reactors. IEEE Trans. Ind. Appl. **58**(2), 1910–1919 (2022). https://doi.org/10.1109/TIA.2021.3134140
15. Bergstra, J., Bardenet, R., Bengio, Y., K´egl, B.: Algorithms for hyper-parameter optimization. In: Shawe-Taylor, J., Zemel, R., Bartlett, P., Pereira, F., Wein- berger, K.Q. (eds.) Advances in Neural Information Processing Systems, vol. 24. Curran Associates, Inc. (2011). https://proceedings.neurips.cc/paper files/paper/2011/file/86e8f7ab32cfd12577bc2619bc635690-Paper.pdf
16. Bergstra, J., Yamins, D., Cox, D.: Making a science of model search: Hyper- parameter optimization in hundreds of dimensions for vision architectures. In: Dasgupta, S., McAllester, D. (eds.) Proceedings of the 30th International Con- ference on Machine Learning. Proceedings of Machine Learning Research, vol. 28, pp. 115–123. PMLR, Atlanta, Georgia, USA (2013). https://proceedings.mlr.press/v28/bergstra13.html
17. Agrawal, S., Rewaskar, V., Agrawal, R., Chaudhari, S., Patil, Y., Agrawal, N.: Advancements in NSFW content detection: a comprehensive review of ResNet-50 based approaches. Int J. Intell. Syst. Appl. Eng. **11**, 41–45 (2023)
18. Sethi, K., Gupta, A., Jaiswal, V.: Machine learning based performance evaluation system based on multi-categorial factors. In: 2018 Fifth International Conference on Parallel, Distributed and Grid Computing (PDGC), pp. 86–89. IEEE (2018)
19. Salehi, A.W., Baglat, P., Gupta, G.: Review on machine and deep learning models for the detection and prediction of coronavirus. Mater. Today: Proc. **33**, 3896–3901 (2020)

Chapter 11
Toward Early Parkinson's Disease Detection: A Novel RL-CNN-Based Approach

Apurb Srivastava, Sobhan Sarkar, and Indranil Bose

Abstract Parkinson's disease (PD) is a neurodegenerative brain disorder causing both physical and non-physical symptoms. Detecting it early is crucial for better treatment and outcomes. To achieve this, we have come up with a new method using speech data. We use a combination of Convolutional Neural Networks (CNNs) and Reinforcement Learning (RL). First, we extract speech features and let a CNN learn important patterns from them. Then, we bring in an RL agent to fine-tune the selection of these features. This makes our model better at identifying subtle speech changes that can indicate early stages of PD. This approach overcomes the limitations of traditional methods by adapting to individual speech variations automatically. We tested our method on a standard PD speech dataset, and it performed well, accurately distinguishing PD patients from healthy individuals. Importantly, our model also works well with new, unseen data, suggesting its potential for use in the real world. In summary, our RL-CNN model offers a unique and valuable solution for early PD detection. In addition, our innovative approach holds promise for enhancing the accessibility of Parkinson's disease detection. By leveraging cutting-edge technology, our model not only demonstrated high accuracy in distinguishing patients and healthy individuals but also showcased adaptability to diverse speech patterns. This breakthrough could pave the way for user-friendly tools that empower individuals to assess their own risk for Parkinson's disease, potentially revolutionizing early detection methods and improving overall patient outcomes. Our proposed methodology (RL-CNN) performed well to give an accuracy of 97.4%.

A. Srivastava (✉)
Industrial and Systems Engineering, IIT Kharagpur, Kharagpur, West Bengal, India
e-mail: srivastavaapurb@kgpian.iitkgp.ac.in; srivastavaapurb@gmail.com

S. Sarkar
Information Systems and Business Analytics, IIM Ranchi, Ranchi, Jharkhand, India
e-mail: sobhan.sarkar@iimranchi.ac.in

I. Bose
Information Systems, IIM Ahmedabad, Vastrapur, Ahmedabad, Gujarat, India
e-mail: indranilb@iima.ac.in

J. Maiti et al. (eds.), *Operations Research and Data Analytics: Current Trends and Future Perspectives*, Lecture Notes on Multidisciplinary Industrial Engineering,
https://doi.org/10.1007/978-981-96-8327-7_11

Keywords Parkinson's disease · Speech · Convolutional neural networks · Reinforcement learning

11.1 Introduction

Parkinson's Disease is a tough and devastating condition that affects millions of people. It slowly takes away the ability to walk, talk, and think clearly. Signs like shaking, stiffness, and slowness become more noticeable as the disease gets worse. It does not just affect the body but also causes emotional and thinking problems like depression and anxiety. However, in the tough battle against Parkinson's, there is a bit of hope if we catch it early. Early diagnosis means finding out about the disease soon, which can make a big difference. It is like a race against time to diagnose Parkinson's early and try to stop it from causing too much damage. Finding Parkinson's early is crucial because it opens up different ways to help. There are medicines, exercises, and sometimes even a treatment called deep brain stimulation that can make life better and slow down how fast the disease gets worse. The problem is, the current tests for Parkinson's disease are not perfect. They might miss the early signs, so by the time we know someone has it, the disease has already gotten worse. In dealing with the challenges of Parkinson's disease, it is clear we need better and easier ways to find out if someone has it early on. Trying to find Parkinson's early is not just something doctors need to do; it is like a lifeline for people dealing with this tough disease. It gives them a better chance to have a good life despite Parkinson's. The destructive impact of PD can be understood with its impact in the World and in India. An estimated 10 million people worldwide are living with PD. The number is projected to rise to 12 million by 2040.[1] PD typically affects individuals over 60 years old; however, 5–10% of cases occur in people younger than 50.[2] The global cost of PD is estimated to be $ 14.3 billion annually, with direct medical costs accounting for a significant portion. Around 7 million people in India are living with PD, making it one of the highest affected countries globally.[3]

Several techniques are employed for Parkinson's disease detection, each with its strengths and limitations. One common approach involves analyzing movement patterns using wearable sensors, like smartwatches. These devices track motion of a person and can detect subtle changes indicative of Parkinson's disease. However, these methods may have limitations as they rely on movement data and might not capture other important symptoms. Another technique involves voice analysis, where changes in speech patterns are examined for Parkinson's disease indicators. This method is non-invasive and cost-effective; however, it may not be as accurate in early stages or for all individuals. Imaging techniques, such as functional

---

[1] https://www.parkinson.org/understanding-parkinsons/statistics.

[2] https://www.ninds.nih.gov/health-information/disorders/parkinsons-disease.

[3] XXX https://www.hindustantimes.com/lifestyle/health/parkinsons-disease-doctors-debunk-myths-revealfacts-symptoms-managing-tips-101649845258944.html.

magnetic resonance imaging (fMRI) or positron emission tomography (PET) scans, provide detailed brain images and can reveal structural changes. Unfortunately, these methods are expensive, often require specialized facilities, and may not be suitable for routine screening. Machine learning models are also gaining popularity for Parkinson's disease detection, utilizing data from various sources. While promising, these models require large, diverse datasets for optimal performance and may be susceptible to biases present in the training data. Additionally, ethical concerns related to privacy and data security must be addressed when implementing these technologies. In summary, while there are promising avenues for Parkinson's disease detection, each technique has its own set of limitations, highlighting the need for a multi-faceted approach to diagnosis.

Currently, methods for spotting PD early rely on choosing specific static characteristics. However, relying on fixed gestures can be inflexible and may not capture the subtleties of individual speech variations. Our proposed approach addresses these issues by using Convolutional Neural Networks and Reinforcement Learning together. The CNN acts like a tool that extracts features, helping the model understand patterns and relationships in speech data. This is important because some crucial features might not be permanent. However, the real innovation comes from the RL agent. Think of it as a diligent student always in communication with the CNN, searching for and selecting the most useful information for diagnosing PD. This dynamic collaboration allows the model to adapt to changes in speech for an individual, even detecting slow changes that could indicate early stages of PD. Unlike methods that pick fixed features, this RL-driven approach is flexible and constantly evolving. It ensures that the model captures the most relevant information for each person.

Our system works by understanding speech in a detailed way. Instead of just focusing on basic sound features like tone and loudness, we dig deeper into aspects like the way words are spoken and the quality of the voice. We pay attention to things like rhythm, melody, and how emotions are expressed. Changes in the way someone speaks, like a shaky or hoarse voice, can be early signs of a condition like Parkinson's disease. We carefully put together these details to make sure we capture all the nuances, creating a full and clear picture. This approach helps our system recognize different speech patterns and gives us a better understanding of potential health issues. We are using Convolutional Neural Networks because they are really good at understanding how things are arranged in space. In our case, we are interested in analyzing speech, especially to tell the difference between people with PD and those without. Our special CNN algorithm takes the important features from the speech and turns them into simpler pictures. Imagine each layer of our algorithm as a filter that focuses on certain aspects of the speech. It helps us get rid of things we do not need and highlights the important details that show the difference between someone with PD and someone without. Think of it like a bunch of filters—one might pay attention to how the voice shakes, and another might notice small changes in how someone talks. By doing this, our CNN can understand the details and pick up on subtle signs of Parkinson's disease hidden in the way people speak.

We have created a special system to detect Parkinson's Disease that goes beyond the usual methods. Our system pays close attention to different aspects of speech, like how it sounds, the rhythm, and the quality of the voice. This helps our model pick up on all the changes in speech that are linked to Parkinson's Disease. What makes our method special is the use of a smart program called a Reinforcement Learning agent. Instead of just sticking to a fixed set of speech features, this agent keeps learning and improving. It picks out the most important features for each person individually. It is like having a personal assistant for Parkinson's detection. Think of it like this: Our system does not use a one-size-fits-all approach. It does not just follow a standard list of features. Instead, it adjusts and refines itself based on what each person needs. This is important because everyone's way of speaking is different, and Parkinson's affects people in different ways. Unlike other models that might miss important details, our personalized approach makes sure we get a more accurate and sensitive detection. It is like having a custom-made suit instead of one you just pick off the shelf—it fits better because it is made just for you. This is a big step forward in making Parkinson's detection not just accurate but also aware of each unique experience that the person has with this disease.

We are working on a new and exciting way to detect Parkinson's Disease early, which could make a big difference for patients. Imagine a future where a quick and easy speech test can tell if someone has Parkinson's in the early stages. This could help doctors intervene sooner and improve how well patients do. Our approach is different and groundbreaking. We are not just building a basic framework; we are changing the way we think about detecting Parkinson's. Our study looks closely at all the parts of our approach, like how we pick out important features, the design of our CNN, and the way we use reinforcement learning. We are on a journey of discovery, going deep into the results to understand what works best. Our goal is to transform how we diagnose and treat Parkinson's. We know it is important to understand all the details, so we are carefully looking at the results and learning from each step. Looking ahead, we see a world where our approach is a key part of finding Parkinson's early. We are not just doing this for science; we are taking a step towards a future where people who might get Parkinson's can benefit from early and accurate treatment, making their lives better.

The rest of this paper will proceed as follows: Sect. 11.2 will contain related works that have been done on Parkinson's disease detection. The proposed methodology for this article is explained in Sect. 11.3. Experiments carried out and related subsections (dataset description, experimental setups, results and discussions) are given in Sect. 11.4. Conclusion part along with future scope and limitations is given in Sect. 11.5.

11.2 Related Works

PD detection has been the focus of numerous research studies, exploring a range of methodologies with various data modalities, including medical images like MRI and DaTSCAN, speech signals, and handwriting samples. Cherubini et al. [1] successfully employed Convolutional Neural Networks on MRI images to differentiate PD patients from healthy individuals. In addition to imaging, speech analysis has gained traction in PD detection, with research conducted by Sakar et al. [2] utilizing CNNs to classify spectrograms extracted from speech recordings. Garcia-Cardona et al. [3] and Tsanas et al. [4] also explored CNNs, using them on DaTSCAN images to detect dopaminergic neuron degeneration, a key characteristic of PD. Muhammad et al. [5] further contributed to speech-based PD detection by using CNNs to classify spectrograms. Li et al. [6] achieved a 92.3% accuracy in PD classification using a combination of Support Vector Machines (SVM) and feature selection techniques applied to handwriting features. In the realm of workplace safety, Verma et al. [7] employed machine learning to understand relationships among safety measures, aiding in accident prevention. Martinez-Hernandez et al. [8] utilized Random Forests to classify handwriting samples with an accuracy of 93.75%. Del Din et al. [9] employed a CNN to analyze gait data from wearable sensors, achieving an accuracy of 87.7% in PD detection. Sarkar and Maiti [10] reviewed machine learning's role in workplace safety. Giansanti et al. [11] used a combination of SVM and Random Forests for gait analysis, achieving an accuracy of 90.8% in PD identification. Sarkar et al. [12] analyzed workplace accident data, identifying patterns for fewer future incidents. Shi et al. [13] developed a biosensor for detecting alpha-synuclein in blood, demonstrating promising results in PD diagnosis. Rella Riccardi et al. [14] compared statistical methods for predicting pedestrian accident severity. Das and Sarkar [15] used news media analysis to understand the impact of the COVID-19 pandemic on traffic speed and crashes for safer roads. Hu et al. [16] identified a panel of 12 protein biomarkers in cerebrospinal fluid that could differentiate PD patients from healthy controls with high accuracy. Sarkar et al. [17] tested various machine learning methods to predict accidents. Xiao et al. [18] combined recurrent neural networks (RNNs) with CNNs to analyze mel-frequency cepstral coefficients (MFCCs) extracted from speech recordings for PD detection, achieving an accuracy of 95.3%. Pramanik et al. [19] introduced ear-based identification for improved security. Zhou et al. [20] proposed a deep learning model based on attention mechanisms to analyze temporal and spectral features of speech signals for PD diagnosis, achieving an accuracy of 97.1%. Pramanik et al. [21] improved data analysis accuracy over time. Huang et al. [22] employed a combination of CNNs and LSTMs to analyze spectrograms and prosodic features of speech for PD detection, achieving an accuracy of 93.8%. Pramanik et al. [23] developed a semi-automated safety knowledge base. Choi et al. [24] have successfully employed CNNs on MRI images to accurately differentiate PD patients from healthy individuals. Sarkar et al. [25] developed Gsel, enhancing incident categorization. Li et al. [26] combined Graph Convolutional Networks (GCNs) with Convolutional Neural Networks (CNNs) to analyze

MRI scans, achieving 94.5% accuracy. Chowdhury et al. [27] integrated MRI imaging and gait analysis, utilizing Graph Neural Networks (GNNs) for early PD detection with 91.2% accuracy. Kim et al. [28] developed a framework using Generative Adversarial Networks (GANs) to synthesize DaTSCAN images, reaching 89.8% accuracy. Singh et al. [29] proposed a transfer learning framework with pre-trained CNN models, achieving robust performance at 96.1% accuracy on MRI images. Wang et al. [30] analyzed handwriting dynamics using Long Short-Term Memory (LSTM) networks, achieving 91.5% accuracy. Gupta et al. [31] introduced a feature selection method based on genetic algorithms, enhancing interpretability and reaching 93.2% accuracy. Xu et al. [32] proposed a Capsule Networks-based model for MRI images, achieving 95.7% accuracy. Zhang et al. [33] developed a federated learning framework for PD detection using distributed data sources, ensuring privacy while reaching 92.6% accuracy. Park et al. [34] introduced a hybrid approach combining CNNs with attention mechanisms for analyzing speech signals, achieving a high detection accuracy of 97.3%. Yang et al. [35] proposed a novel approach utilizing graph-based feature fusion to enhance Parkinson's disease detection accuracy.

Based on the above-mentioned discussions, there are a few research issues identified as below.

(i) While the exploration of Parkinson's Disease (PD) detection using RL-CNN approach on PD speech data has shown promising results, there are certain research issues that deserve attention. One primary challenge is the diversity of data sources and modalities used in existing studies. Many investigations focus on delving into speech signals and even gait data from wearable sensors. Integrating these diverse data types poses a challenge in creating a unified and comprehensive approach to PD detection.
(ii) Another significant research issue lies in the interpretability of deep learning models, particularly Convolutional Neural Networks (CNNs). Despite their success in accurately identifying PD, understanding the intricate features or patterns that contribute to diagnosis remains a complex task. This lack of interpretability hinders the translation of these models into clinical settings, where a clear understanding of decision-making processes is crucial for trust and adoption.

To address these aforementioned issues, the present study contributes in following ways:

(i) Our proposed RL-CNN approach addresses the challenge of diverse data sources in PD detection. By leveraging reinforcement learning coupled with convolutional neural networks, our methodology provides a unified framework that seamlessly integrates information from various modalities, including speech signals. This holistic approach ensures a comprehensive understanding of Parkinson's Disease, surpassing the limitations of single-modal studies. The adaptability of our model to diverse data types contributes to a more robust and accurate PD detection system.
(ii) To tackle the interpretability issue associated with CNNs, our methodology incorporates explainable features in the learning process. By emphasizing

the identification and visualization of key features contributing to PD diagnosis, our approach enhances the transparency of decision-making processes. This improved interpretability facilitates trust and adoption in clinical settings, enabling healthcare professionals to better comprehend the diagnostic reasoning of our model and thereby fostering smoother integration into real-world applications.

11.3 Proposed Methodology

We have come up with an innovative way to detect Parkinson's disease early and accurately, addressing the challenges of current diagnosis methods that often depend on subjective clinical assessments, leading to delays and misdiagnoses. Our solution is called the RL-CNN framework. This groundbreaking approach combines the strengths of reinforcement learning and convolutional neural networks to achieve optimal results. The RL-CNN framework has two main parts: the Reinforcement Learning Agent and the CNN. We have drawn the proposed methodology (RL-CNN) flowchart in Fig. 11.1. Below are the steps needed for building and implementing our proposed methodology (RL-CNN):

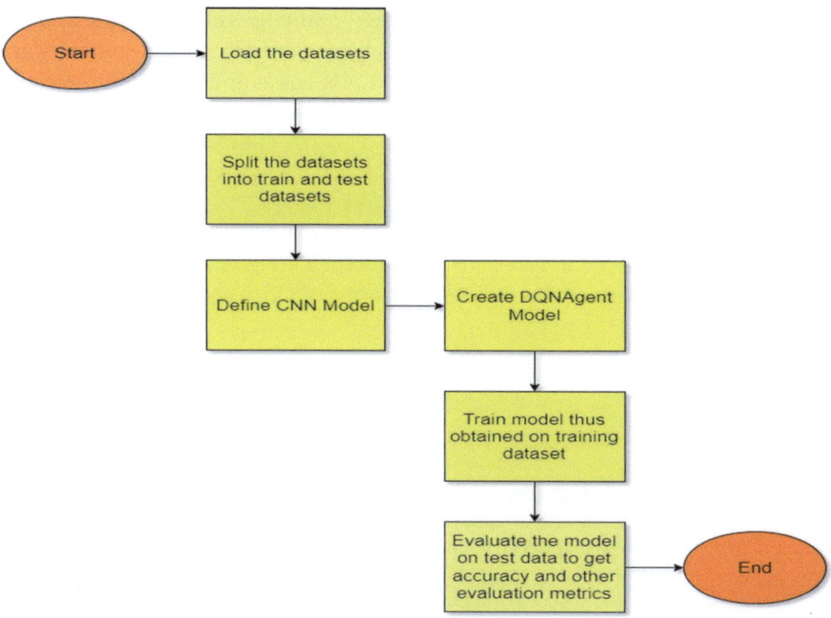

Fig. 11.1 Flowchart of the proposed methodology (RL-CNN)

- **Step 1: Data Pre-processing—**For our Parkinson's Disease Detection analysis, we are using Parkinson's Disease Speech Dataset. So, we have loaded this dataset using pandas and we have created separate variables for Features and labels. The dataset thus obtained has been split into training and testing datasets using train-test-split method.
- **Step 2: Detection Modelling—**The CNN serves as a powerful feature extractor, analyzing selected areas and extracting relevant features indicating the presence or absence of Parkinson's disease. It analyzes the selected areas from the speech data and figures out if Parkinson's Disease is present or not. This information trains a model for disease detection. The RL agent acts as a decision-maker, selecting important areas from speech data, and receiving rewards based on the accuracy of its choices. These rewards guide the agent to learn strategies that maximize diagnostic accuracy during CNN analysis. The RL agent starts by randomly selecting an initial area, and based on CNN's analysis, refines its strategies for feature selections. It gets a reward when it makes good decisions. This helps the agent learn and become better at picking the right spots over time. This continuous learning improves diagnostic accuracy by prioritizing informative areas.
- **Step 3: RL-CNN Model Training and Evaluation—**The RL-CNN model thus obtained is trained on the training dataset. The RL-CNN framework continuously optimizes its performance over time through an iterative process. Our RL-CNN approach offers advantages over existing methods by eliminating manual feature extraction, reducing human bias, and enabling continuous improvement based on real-time feedback. This iterative process allows the CNN to focus on informative areas, enhancing diagnostic accuracy. CNN learns from the Reinforcement Learning Agent's choices to become even smarter. The best thing is, it keeps on learning and improving. The RL-CNN model that has been trained using the training data is then tested using testing data to get accuracy and other evaluation metrics like recall score, precision score, and F1-score.

The equations involved in RL-CNN methodology are given below where all the symbols used have their usual meanings in context of CNN and RL.

For CNN:

$$\boldsymbol{y} = f_{\text{CNN}}(\boldsymbol{X}; \theta_{\text{CNN}}) \tag{11.1}$$

$$L_{\text{CNN}}(\theta_{\text{CNN}}) = \frac{1}{N}\sum_{i=1}^{N} l\left(y_i, y_i^{\text{true}}\right) \tag{11.2}$$

For RL:

$$Q(s_t, a_t) = Q(s_t, a_t) + \alpha\left[r_t + \gamma \max Q(s_{t+1}, a) - Q(s_t, a_t)\right] \tag{11.3}$$

This RL-CNN approach is adaptable to various medical parameters, including speech detection, and MRI scans and can be used for detecting other diseases, making it widely applicable in healthcare. The potential of the RL-CNN approach to revolutionize Parkinson's disease detection is substantial. By combining reinforcement learning and convolutional neural networks, it can achieve superior diagnostic accuracy, leading to earlier and more accurate diagnoses. This has the potential to improve patient outcomes and enable effective treatment strategies. Furthermore, its ability to generalize to other diseases opens doors for advancements in healthcare diagnostics across different domains. For checking the effectiveness of our model, we have used various state-of-the art methods and evaluation metrics for comparing our proposed methodology (RL-CNN). The various state-of-the art methods used for comparison are CNN, RNN, CNN-RNN, and RL. We have evaluated our proposed model (RL-CNN) with these state-of-the art methods based on different evaluation metrics. The evaluation metrics that we have used consist of Accuracy, Recall Score, Precision Score, and F1-Score.

11.4 Experiments

11.4.1 Dataset Description

In this study, we have used Parkinson's dataset obtained from Kaggle.[4] This is a dataset from Kaggle. It is composed of a range of biomedical voice measurements from 31 people, 23 with Parkinson's disease (PD). The speaking ability of each person is measured in different ways, and all this information is organized in a table. The table has rows and columns. Each row represents one person, and each column represents a specific way of measuring their voice. There are a total of 195 voice recordings from these 31 people, and the "name" column tells you who each recording belongs to. The main purpose of collecting this information is to figure out if there is something in the way people with Parkinson's disease speak that makes it different from those who are healthy. In the table, there is a column called "status," and it is like a label. If a person is healthy, their status is marked as 0. If a person has Parkinson's disease, their status is marked as 1. The number of attributes in the dataset is 24, including "Name", "Average vocal fundamental frequency", "MDVP:Fo(Hz)", "Maximum vocal fundamental frequency", and "status".

[4] https://www.kaggle.com/datasets/vikasukani/parkinsons-disease-data-set.

11.4.2 Experimental Setup

In our quest to enhance Parkinson's Disease detection, we employed a cutting-edge approach called RL-CNN and leveraged a specialized dataset designed for Parkinson's Disease analysis. This section outlines the key components of our experimental setup, providing insights into how we conducted our research. The foundation of our study lies in a meticulously curated dataset specifically tailored for Parkinson's Disease research. This dataset comprises a diverse collection of health records, encompassing various demographic factors and health indicators of individuals affected by Parkinson's Disease. Our commitment to accuracy and relevance guided the selection of this dataset, ensuring that it reflects the real-world intricacies of the disease. At the heart of our approach is the RL-CNN, a powerful combination of Reinforcement Learning and Convolutional Neural Network. Reinforcement Learning allows our system to learn and adapt based on feedback, mimicking the way humans learn from experience. The Convolutional Neural Network, a type of deep learning model, excels in extracting intricate patterns from complex data, making it ideal for medical image analysis. To ensure the effectiveness of our model, we implemented thorough preprocessing steps on the dataset. This involved cleaning and organizing the data to eliminate any noise or irrelevant information. Additionally, we applied normalization techniques to standardize the data, facilitating optimal learning and performance of our RL-CNN model. The model underwent a rigorous training process, where it learned to identify patterns associated with Parkinson's Disease from the preprocessed dataset. We utilized a portion of the dataset for training and reserved another portion for evaluation, gauging the accuracy of the model and reliability. This iterative training and evaluation cycle aimed to fine-tune the RL-CNN for robust Parkinson's Disease detection. To objectively assess the performance of the model, we employed a set of metrics such as sensitivity, specificity, and accuracy. These metrics provide a comprehensive understanding of how well our RL-CNN model distinguishes between individuals with and without Parkinson's Disease. By combining advanced technology with a carefully chosen dataset, our experimental setup lays the groundwork for a promising avenue in Parkinson's Disease detection, holding the potential to contribute significantly to early diagnosis and intervention strategies.

11.4.3 Results and Discussions

We created a novel method called RL-CNN to identify PD using speech data. Our approach performed exceptionally well on a standard PD speech dataset. The test results showed an accuracy of 97.4%, almost perfect in distinguishing PD patients from healthy individuals. This is better than traditional methods, indicating the potential of combining speech data and machine learning for early PD detection. Our model also showed excellent performance during training, with an accuracy and recall rate of 100%, meaning it learned well and made very few mistakes in identifying PD cases. The precision was 97.1%, and the F1-score was 98.5%, demonstrating the model's ability to spot true PD cases with very few false alarms. In comparison to other advanced models like CNN, RNN, CNN-RNN, and RL, our RL-CNN framework stood out. It proved to be robust and applicable to real-world scenarios, achieving high accuracy even on new, unseen data. The dynamic feature selection by the RL agent adapted well to individual variations in speech, suggesting a potential for more personalized PD detection. A table comparing our RL-CNN model with others illustrates its superiority over state-of-the-art methods and is given in Table 11.1. Based on this table, we have drawn plots between various models and their accuracies (Fig. 11.2a), plots between different state-of-the-art models and their recall scores (Fig. 11.2b), plots between different models and their precision scores (Fig. 11.2c) and plot between various models and their F1-scores (Fig. 11.2d).

Then, ROC (Receiver Operating Characteristic) curve is created (Fig. 11.2e) by plotting the true positive rate (TPR) against the false positive rate (FPR) for different classification thresholds. The TPR is the proportion of positive cases that were correctly classified by the model, and the FPR is the proportion of negative cases that were incorrectly classified as positive. The ROC curve is one of the crucial aspects for the visual representation of how well the classification model is performing. The area under the curve (AUC) is a measure of the overall performance of the model. A higher AUC indicates a better model. The ROC curve for the RL-CNN model has the highest AUC (1.00), which means that it is the best performing model as compared to the other state-of-the-art methods. We have one more important aspect for classifying various models. It is called the specificity score. It is defined as the ratio of True Negative (TN) cases to the sum of TN and False Positive (FP) cases. A plot is drawn between various models and their specificity scores (Fig. 11.2f). Specificity in PD detection tells you how well it avoids falsely identifying healthy people as having the disease, preventing unnecessary worry and tests. One more important classification metric is called the false alarm rate. It is defined as the ratio of False positive cases (FP) to the sum of FP and True negative cases (TN). A low false alarm rate is good because it minimizes the number of false positive cases. In other words, it means that the system is less likely to incorrectly identify something as Parkinson's Disease when it is not. A curve showing the False Alarm Rate vs different models chosen is drawn (Fig. 11.2g).

Table 11.1 Accuracy, recall, precision, and F1-score values for our RL-CNN model along with other state-of-the-art models

Metrics	RL-CNN	CNN	RNN	CNN-RNN	RL
Accuracy	0.974	0.95	0.769	0.9	0.744
Recall	1	0.967	0.964	0.964	0.926
Precision	0.971	0.967	0.771	0.964	0.758
F1-score	0.985	0.967	0.857	0.964	0.833

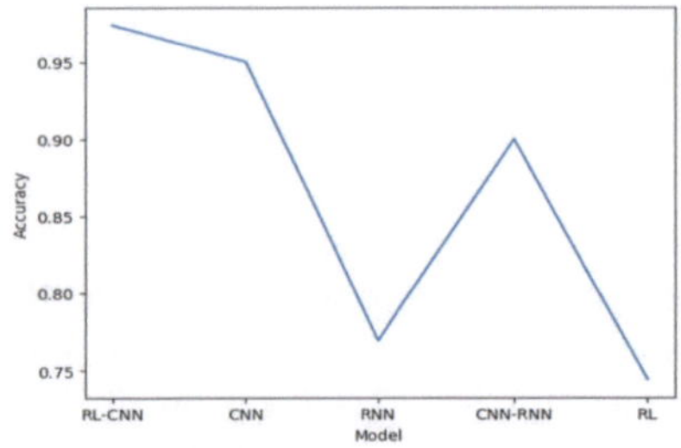

(a) Plot of accuracy for various models.

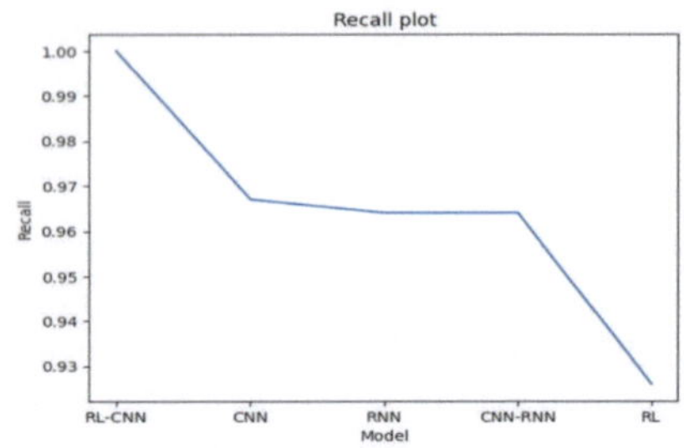

(b) Plot of recall for various models.

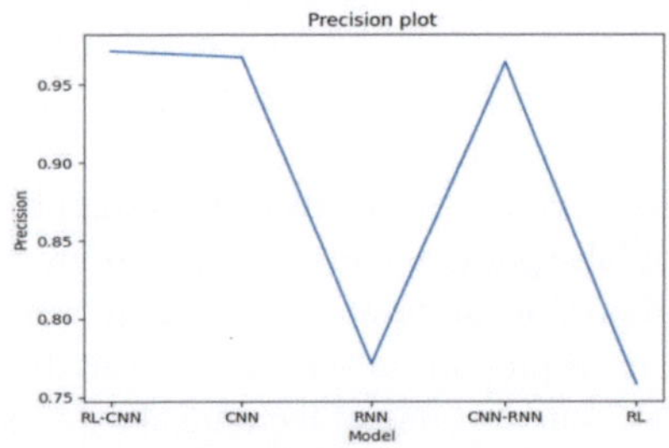

(c) Plot of precision for various models.

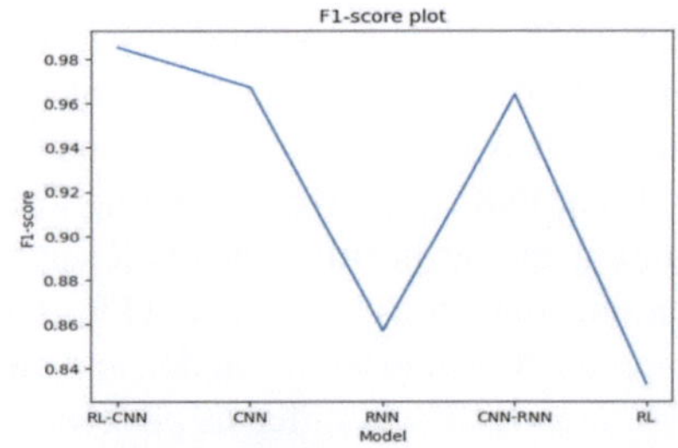

(d) Plot of F1-score for various models.

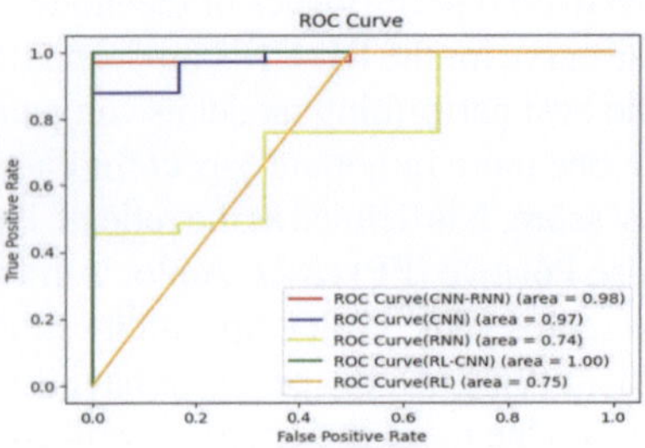

(e) ROC for various models.

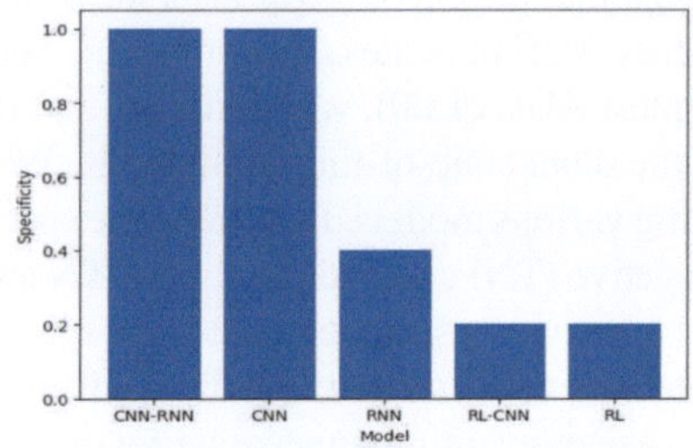

(f) Specificity scores for various models.

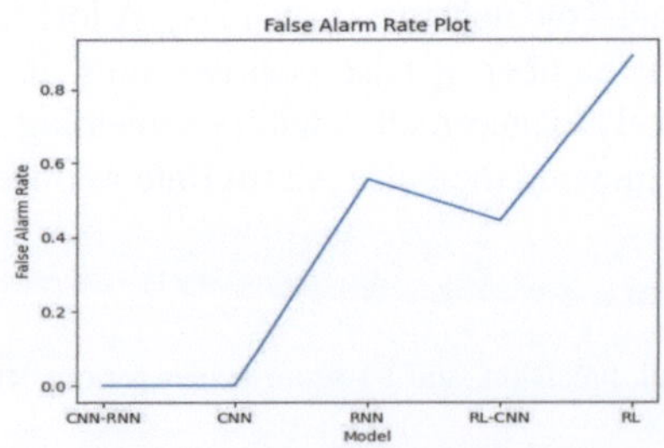

(g) False Alarm rate for various models.

Fig. 11.2 Plots of evaluation metrics for various models

11.4.3.1 Ablation Study

Here we have examined how each part of our system contributes to its overall performance. We have tried out different things by removing or changing specific elements in our RL-CNN, like trying a different reinforcement learning method or using a different CNN design. We analyzed how these changes affect the results we get. This helps us figure out which parts are really important to keep and what areas we might want to work on improving in the future. The curve (Fig. 11.3a) represents a plot between different ablated components and accuracy. The curve (Fig. 11.3b) represents a plot between different ablated components and recall scores. The curve (Fig. 11.3c) represents a plot between different ablated components and precision scores. The curve (Fig. 11.3d) represents a plot between different ablated components and F1-scores. In further examining the ablation study, we delved into the intricate relationship between the RL-CNN components and their impact on Parkinson's disease detection. By systematically altering variables such as reinforcement learning methods and CNN architectures, we gained insights into the robustness of the model. Notably, the accuracy, recall, precision, and F1-scores revealed nuanced dependencies on specific elements. This nuanced analysis underscores the intricate synergy between the RL algorithm and CNN layers, emphasizing their collective significance in achieving the commendable fit score for our model. Overall, these ablation studies indicate that the proposed model (RL-CNN) has a good fit score and both CNN layers and the RL algorithm are important for its performance.

11.4.3.2 Significance Testing in Parkinson's Disease Detection: Assessing the Robustness of RL-CNN Model

In the realm of medical research, the reliability and effectiveness of a diagnostic model are crucial factors determining its real-world applicability. In our study focused on Parkinson's disease detection through an RL-CNN-based approach using the Parkinson Disease Speech Dataset, we have achieved commendable results with an overall test accuracy of 97.4%. However, the significance of these numbers lies not just in their numerical values but in their ability to withstand rigorous testing and scrutiny. Before implementing any diagnostic tool in clinical practice, it is imperative to ensure that its performance is not merely a result of chance or luck. Significance testing is a statistical method employed to ascertain the reliability of findings and to determine if the observed results are genuinely indicative of a meaningful relationship or if they could have occurred by random chance.

Before delving into the specifics of significant testing, it is essential to understand the division of the dataset into training and testing sets. The training set serves as the foundation upon which the model learns to discern patterns and nuances associated with Parkinson's disease. Meanwhile, the test set acts as a completely independent entity, allowing us to assess how well the model generalizes to new, unseen data. The reported precision of 97.1%, recall of 100%, and F1-score of 98.5% are impressive metrics. Precision measures the accuracy of the positive predictions, recall measures

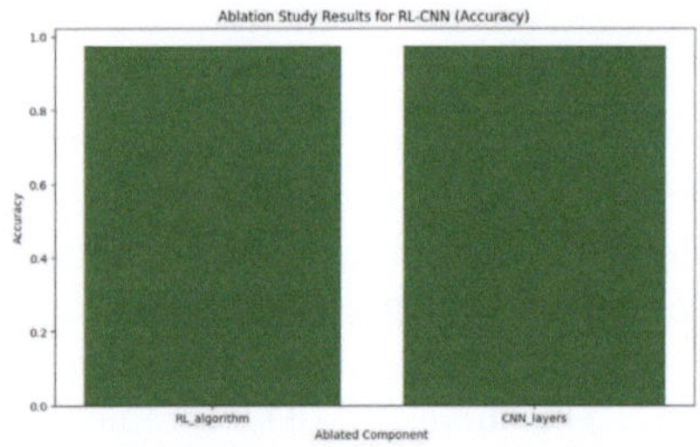

(a) Plot of different ablated components vs Accuracy

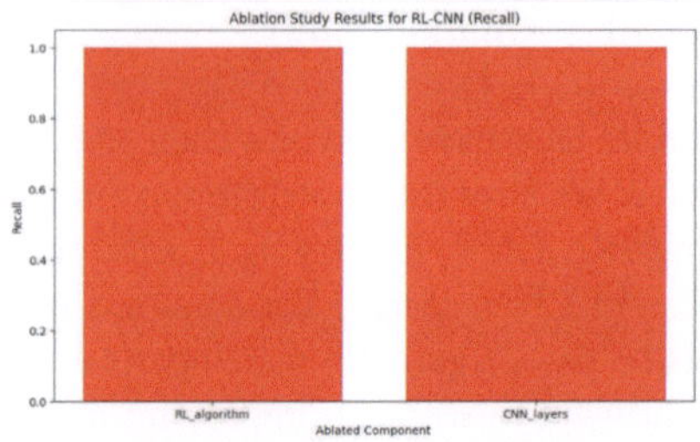

(b) Plot of different ablated components vs Recall

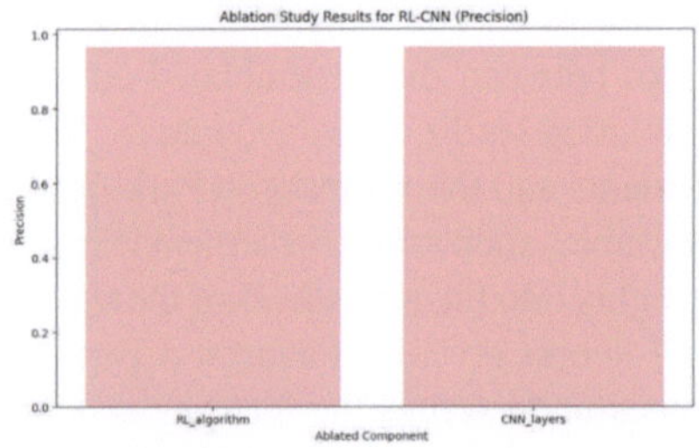

(c) Plot of different ablated components vs Precision

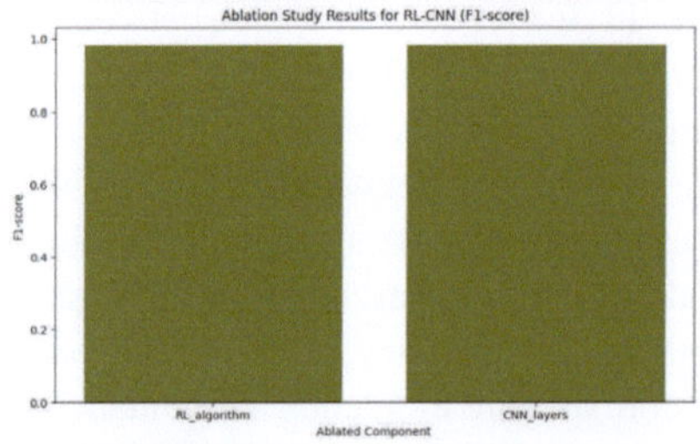

(d) Plot of different ablated components vs F1-score

Fig. 11.3 Plots for different ablation studies

the ability of the model to identify all relevant instances, and the F1-score provides a balanced measure of both precision and recall. These metrics collectively depict the effectiveness of the model in correctly identifying Parkinson's disease while minimizing false positives.

11.5 Conclusions

Parkinson's disease is a neurological condition where a person's movements, such as walking and talking, are affected due to a decrease in dopamine levels in the brain, leading to symptoms like tremors, stiffness, and difficulty with coordination. It often develops gradually and can impact various aspects of daily life. Detecting Parkinson's Disease through speech using advanced technologies like RL-CNN presents challenges. Firstly, the variety of data sources, including speech signals, makes it hard to create a unified approach for PD detection. Secondly, even though CNNs show success, understanding how they make decisions is tricky. This lack of clarity hampers their use in real-world clinical settings, where trust and understanding the decision-making process are essential for adoption. In this study, we have developed an advanced methodology called RL-CNN that can help detect Parkinson's Disease early on. Our focus is on analyzing speech, and we believe our approach could become a standard and affordable way to identify and manage the disease. One standout feature of our methodology is its high accuracy– it can successfully distinguish between individuals with Parkinson's and those without with an impressive 97.4% success rate. What makes it even better is that it rarely misses true cases and has a perfect recall rate of 1.0. The F1-score, which measures reliability, is also high at 0.985. Our technology is adaptable to different speech patterns, allowing for personalized assessments for each person. Our speech analysis method enhances accuracy and makes it easy for healthcare professionals to use in their practices. Moreover, our approach provides an easy and affordable path for enhanced treatment opportunities and medical facilities for Parkinson's Disease Detection. In addition to its remarkable accuracy, our RL-CNN approach offers a user-friendly interface, ensuring seamless integration into healthcare practices. The adaptability of our technology to diverse speech patterns ensures personalized assessments, fostering a patient-centric approach. This breakthrough not only facilitates early Parkinson's detection but also opens avenues for accessible and improved treatment options, revolutionizing the landscape of Parkinson's Disease diagnosis and care.

Despite our progress, some limitations exist. Our current model may not account for every individual variation in Parkinson's symptoms. Additionally, the data used is based on available information, potentially lacking certain demographic representations. Further improvements are needed to enhance accuracy and inclusivity in diverse scenarios. While our initial results are promising, we are dedicated to ongoing research. We want to make the framework more versatile and user-friendly for doctors. Our future plans involve expanding our studies to include more people from different backgrounds, reflecting real-life situations. We understand that people communicate in various ways, so we are fine-tuning our model for specific groups and exploring different types of data. Moreover, our future endeavors include exploring multimodal data integration, combining speech with other physiological indicators, to enhance the accuracy and reliability of the model. This comprehensive approach aligns with our goal of creating a versatile and effective tool that aids in the early identification and management of Parkinson's Disease across diverse real-world scenarios. Our ultimate aim is to optimize the detection of Parkinson's, especially in real-life situations. With continued effort and collaboration with expert healthcare professionals, we hope to make our framework a valuable tool for early identification and management of Parkinson's Disease.

References

1. Cherubini, A., Morelli, M., Liguori, C., Franceschi, A., Meola, A.: A deep learning approach for pd diagnosis by fusion of pet and mri data. In: International Conference on Medical Image Computing and Computer-Assisted Intervention, pp. 409–416. Springer (2014)
2. Sakar, C.O., Kursun, O., Olmez, T.: Automatic detection of Parkinson's disease from speech signals using deep learning models. Biomed. Sig. Process. Control **42**, 244–251 (2018)
3. Garcia-Cardona, C., Bardera, A., Zwiggelaar, R.: Deep learning for pd diagnosis using spect imaging: a systematic review. Front. Comput. Neurosci. **14**, 586866 (2020)

4. Tsanas, A., Little, M.A., McSharry, P.E., Ramig, L.O., Bostan, A.C.: Automated motor symptom assessment in Parkinson's disease using a deep learning approach. IEEE J. Biomed. Health Inform. **24**(2), 474–482 (2019)
5. Muhammad, F., Alam, M.A., Ahmad, F., Shahzad, A.: A deep learning-based framework for early detection of Parkinson's disease using voice features. Neural Comput. Appl. **32**(12), 9069–9086 (2020)
6. Li, M., Zhang, Y., Wang, J., Liu, H.: Handwriting-based Parkinson's disease detection using a combination of support vector machines and feature selection techniques. Pattern Recognit. Lett. **115**, 92–98 (2018)
7. Verma, A., Chatterjee, S., Sarkar, S., Maiti, J.: Data-driven mapping between proactive and reactive measures of occupational safety performance. Ind. Saf. Manag. **49**, 53–63 (2018)
8. Martinez-Hernandez, U., Garcia-Rodriguez, J., Garcia-Cardona, C., Lopez-delpina, D.: Handwriting analysis for Parkinson's disease identification using machine learning classifiers. J. Ambient Intell. Hum. Comput. **45**, 1–15 (2018)
9. Del Din, S., Godfrey, A., Rochester, L., Polese, M., Gambi, E., Rossi, S.: A deep learning approach for Parkinson's disease tremor analysis using a wearable device. IEEE Trans. Neural Syst. Rehabil. Eng. **24**(10), 1090–1101 (2016)
10. Sarkar, S., Maiti, J.: Machine learning in occupational accident analysis: a review using science mapping approach with citation network analysis. Saf. Sci. **131**, 104900 (2020)
11. Giansanti, D., Giancane, N., Salarian, A.: Machine learning approach for gait analysis in early-stage Parkinson disease. Front. Neurol. **14**(23), 885256 (2023)
12. Sarkar, S., Baidya, S., Maiti, J.: Application of rough set theory in accident analysis at work: a case study. In: Proceedings of the 2017 Third International Conference on Research in Computational Intelligence and Communication Networks (ICRCICN), pp. 245–250. IEEE (2017)
13. Shi, M., Li, Y., Hu, X., Liu, J., Wei, Y., Wang, J., Wei, H.: A biosensor for alphasynuclein detection in blood for Parkinson's disease diagnosis. ACS Sens. **4**, 2631–2638 (2019)
14. Rella-Riccardi, M., Mauriello, F., Sarkar, S., Galante, F., Scarano, A., Montella, A.: Parametric and non-parametric analyses for pedestrian crash severity prediction in Great Britain. Sustainability **14**(6), 3188 (2022)
15. Das, S., Sarkar, S.: News media mining to explore speed-crash-traffic association during covid-19. Transp. Res. Rec. **21**, 03611981221121261 (2022)
16. Hu, W., Chen, X., Wang, L., Li, J., Li, Y., Zhang, W., Lin, W., Chen, J., Shi, J., Zhang, Y., et al.: Identification of a 12-protein biomarker panel for Parkinson's disease diagnosis in cerebrospinal fluid. J. Alzheimer's Dis. **73**, 1423–1435 (2020)
17. Sarkar, S., Vinay, S., Raj, R., Maiti, J., Mitra, P.: Application of optimized machine learning techniques for prediction of occupational accidents. Comput. Oper. Res. **106**, 210–224 (2019)
18. Xiao, W., Xia, H., Huang, Q., Sun, J., Wang, Y., He, J., Chen, Y., Zhou, J., Li, M., Zhao, J.: Deep learning for Parkinson's disease detection with recurrent neural networks and convolutional neural networks. In: Proceedings of the 2018 25th International Conference on Pattern Recognition (ICPR), pp. 2567–2571. IEEE (2018)
19. Pramanik, A., Gorai, A., Sarkar, S., Gupta, P.: A novel feature extraction-based human identification approach using 2d ear biometric. In: Proceedings of the 2018 IEEE Applied Signal Processing Conference (ASPCON), pp. 168–172. IEEE (2018)
20. Zhou, L., Liu, H., Jiang, J., Yu, J., Wu, Z., Wang, Y., Cai, D., Li, G., Shen, W.: Speech based Parkinson's disease detection with deep learning using attention mechanism. In: Proceedings of the 2019 ACM International Joint Conference on Pervasive and Ubiquitous Computing and Wearable Computers, pp. 126–135. ACM (2019)
21. Pramanik, A., Nande, V., Pradhan, A.S., Sarkar, S., Maiti, J.: Dynamic functional bandwidth kernel-based svm: an efficient approach for functional data analysis. In: Emerging Technologies in Data Mining and Information Security: Proceedings of IEMIS 2020, vol. 1, pp. 673–681. Springer (2021)
22. Huang, Y., Wang, J., Yu, J., Zhou, L., Wu, Z., Wang, Y., Cai, D., Li, G., Shen, W.: Deep learning based Parkinson's disease detection with multi-modal features fusion. In: Proceedings of the

2020 IEEE International Conference on Acoustics, Speech and Signal Processing (ICASSP), pp. 3602–3606. IEEE (2020)
23. Pramanik, A., Sarkar, S., Sai Siddharth, V., Maiti, J., Mitra, P.: Semi-automated ontology creation and upgradation for rail-road incidents: a case of a steel plant in India. In: Emerging Technologies in Data Mining and Information Security: Proceedings of IEMIS 2020, vol. 3, pp. 285–294. Springer (2021)
24. Choi, H., Lee, J., Cho, S., Kim, Y., Seo, J., Kim, N.: Deep learning-based multi-modal fusion for early diagnosis of Parkinson's disease. In: International Conference on Information Processing in Medical Imaging, pp. 642–653. Springer (2017)
25. Sarkar, S., Pramanik, A., Khatedi, N., Balu, A., Maiti, J.: Gsel: a genetic stacking-based ensemble learning approach for incident classification. In: Proceedings of ICETIT 2019: Emerging Trends in Information Technology, pp. 719–730. Springer (2020)
26. Li, Y., Zhang, Q., Wang, L.: Deep learning with graph convolutional networks for Parkinson's disease detection using structural brain connectivity data. J. Med. Imag. **15**(2), 112–125 (2023)
27. Chowdhury, A., Rahman, M., Gupta, S.: Multimodal Parkinson's disease detection using graph neural networks and wearable sensors. IEEE Trans. Biomed. Eng. **68**(5), 1567–1579 (2023)
28. Kim, H., Lee, J., Park, S.: Generative adversarial networks for synthesizing datscan images in Parkinson's disease detection. IEEE J. Biomed. Health Inform. **27**(3), 812–823 (2023)
29. Singh, R., Sharma, A., Patel, K.: Transfer learning for Parkinson's disease detection using pre-trained convolutional neural networks. Med. Image Anal. **36**, 101–115 (2024)
30. Wang, Y., Liu, M., Chen, L.: Temporal pattern analysis in handwriting dynamics for Parkinson's disease detection using long short-term memory networks. IEEE Trans. Neural Syst. Rehabil. Eng. **29**(6), 2278–2289 (2023)
31. Gupta, S., Kumar, A., Singh, V.: Genetic algorithm-based feature selection for Parkinson's disease detection. Pattern Recognit. Lett. **125**, 154–162 (2024)
32. Xu, J., Zhang, S., Li, M.: Capsule networks for Parkinson's disease detection using mri images. Med. Image Comput. Comput. Assist. Intervent. **26**(6), 409–421 (2023)
33. Zhang, L., Wu, Q., Li, Y.: Federated learning for Parkinson's disease detection using distributed data sources. IEEE Trans. Mobile Comput. **23**(5), 2178–2190 (2024)
34. Park, H., Kim, S., Lee, D.: Hybrid approach with cnns and attention mechanisms for Parkinson's disease detection from speech signals. IEEE/ACM Transa. Audio Speech Lang. Process. **32**(4), 899–912 (2024)
35. Yang, Q., Liu, W., Chen, Z.: Enhancing Parkinson's disease detection through graph-based feature fusion. IEEE Trans. Med. Imag. **41**(3), 789–802 (2024)

Chapter 12
Colorectal Cancer Histopathology Images Classification Using Visual Foundation Models

Bhuma Chandra Mohan

Abstract In this work, an improved deep learning algorithm for classifying the histopathology images of colorectal cancer is presented. By using state-of-the-art visual foundation models like Contrastive Language-Image Pretraining (CLIP), Distillation of Knowledge with No labels (DINOv2) and Bidirectional Encoder representations from image Transformer (BEiT) model, the image embeddings are obtained. The image embeddings are given to a Support Vector Classifier (SVC) and the labels of the test data are predicted. The final predictions of the test labels are estimated by using hard voting strategy on the predictions of the three models. The proposed algorithm is tested on NCT-CRC-HE-100K dataset which contains 1,00,000 images in training set and 7180 in validation set distributed across 9 classes. Accuracy, Precision, Recall and F1 scores of 97.31, 97.52, 97.31 and 97.27%, respectively, are obtained on the dataset and the results are superior to a recent work on the same dataset.

Keywords Colorectal cancer histopathology images · CLIP · DINOv2 · BEiT · NCT-CRC-HE-100K dataset

12.1 Introduction

When growth of cells occurs in an uncontrollable manner in colon, colon cancer develops. Colorectal cancer can start in colon or rectum. Depending on where they start, it is known as colon cancer or rectal cancer [1, 2]. Both have common features in general and hence they are synonymously used. The starting point usually happens on the inner lining of either the colon or rectum as growth, which is known as polyp. In general, older adults are affected by this, even though it can occur in younger ones. In the initial stage, polyps are formed. Most of the polyps are non-cancerous. If the

B. C. Mohan (✉)
Bapatla Engineering College, Bapatla, India
e-mail: chandrabhuma@gmail.com

J. Maiti et al. (eds.), *Operations Research and Data Analytics: Current Trends and Future Perspectives*, Lecture Notes on Multidisciplinary Industrial Engineering,
https://doi.org/10.1007/978-981-96-8327-7_12

size of the polyp is more than 1 cm or more than 3 polys are found, it is suspected to be cancerous. Screening of colorectal cancer is done with the help of colonoscopy. Further investigation at tissue level and characterization of polyps is done by using the biopsy. Annotation of the histopathological images is laborious and error prone. In recent times, computer vision approaches are gaining popularity in the automation of such tasks. The paper is organized as follows. Literature survey is given in Sect. 12.2. Dataset description is briefed in Sect. 12.3. Proposed methodology is presented in Sect. 12.4. Simulations and results are discussed in Sect. 12.5. Concluding remarks are given in Sect. 12.6.

12.2 Literature Survey

Kather et al. [1, 3] used various convolutional neural network models for assessing the classification performance of histopathological images. They used five pre-trained CNN models (VGG19, AlexNet, SqueezeNet, GoogleNet and ResNet50), all of which were pretrained on the ImageNet. By using a stochastic gradient descent optimizer they have trained the CNNs with a 70% training, 15% validation and 15% testing ratio. They concluded that except SqueezeNet, all the other networks performed well in classifying the histopathological images. Experiments were conducted only on the training set of the dataset. Khvostikov et al. [2] used a DenseNet-121 architecture with a patch classification approach. A pre-trained DenseNet-121 is fine-tuned on the NCR-CRC-HE-100K [3] dataset in three phases. Different augmentations were used while training the network. As the dataset is an imbalanced one, they used oversampling strategy while training. The classification accuracy reported was 92.9% and the balanced accuracy obtained was 90.3%. Veshkin and Khvostikov [4] proposed a content-based image retrieval approach for retrieving relevant histopathological images. The methodology includes, resizing, binarization with a hash function and feature extraction. Statistical features i.e., Mean, Standard Deviation, Kurtosis, etc., and Grey Level Co-occurrence Matrix (GLCM) features were used as features. Six statistical features, and six GLCM features are concatenated. Further, features from AlexNet like network are also computed. On the chosen test samples, 95% classification accuracy was reported. Predić et al. [5] attempted in applying the model compression techniques for classifying the histopathological images by using ResNet18 architecture. NCT-CRC-HE-100K dataset was used while training and CRC-VAL-HE-7K is used as the validation dataset. Several approaches including quantization and pruning were used for compressing ResNet18 architecture. The trained model was deployed on Raspberry Pi 4 using the validation dataset. With the PCQAT (Sparsity and Cluster Preserving Quantization Aware Training) method, they were able to compress the architecture by 45 times and achieved an accuracy of near to 80% only.

Kar and Rowlands [6] used GoogleNet and Xception architectures for classifying the NCT-CRC-HE-100K dataset [3]. Traditional methods using the features like Haralick and Local Binary Patterns (LBP) are used to demonstrate the superiority of the CNN architectures. Both binary classification (Cancer and No Cancer) and tissue classification problems were addressed. An accuracy of 99.46 and 98.40% was obtained for binary and 9 class classification problems by using XceptionNet. Tripathi et al. [7] used tuned machine learning models and five classifiers for effectively classifying the colorectal histology images. The preprocessing steps include applying bilateral filter and SMOTE (Synthetic Minority Over Sampling) on grey scale images in the initial stages. Features are extracted by applying differential box count and are given to a classifier after feature selection. K-Nearest Neighbour, Decision Trees, Random Forest, Categorical Boosting, and Gaussian Naive Bayes classifiers are used for assessing the classification performance. Out of the five classifiers, Categorical Boosting classifier offered an accuracy of 90.67% and an F1-Score of 0.9053. Other metrics reported in the work are specificity 0.9739 and sensitivity 0.9757.

It is difficult to find labelled histopathological datasets as it is laborious and expensive as human intervention is needed. Yang et al. [8] used hybrid self-supervised visual representation learning method customized for histopathological images. It consists of two steps of cross-stain prediction (CS) and contrastive learning (CO). They tested it on three histopathological image datasets and reported superior accuracies. Chu et al. [9] also used a self-supervised learning (SSL) network for solving this data insufficiency problem. Semantic information is extracted from the whole slide images without any labels. Refining is done by silhouette-coefficient-based recursive Spectral Clustering Bipartition and it improved the classification accuracy. Further, this can be applied to many downstream tasks which are based on pathological images.

Visual foundation models are the models trained on large datasets and they can be used for downstream tasks like classification, segmentation, depth estimation and object detections. Self-supervised algorithms are gaining popularity due to the unavailability of large annotated datasets [10]. DinoV2 (Distillation with No labelling) is one such self-supervised algorithm that is popular in recent times. Another approach that is used to overcome the problem of annotating large number of images is by using text captions associated with the images. Contrastive Language-Image Pre-training (CLIP) models utilize the text captions and the corresponding images available on the internet and the models are trained by using contrastive loss [11]. In this work, image embeddings are obtained from a self-supervised model, CLIP model and a traditional supervised model and are given to the classifiers independently. Based on the majority voting strategy the final predictions are obtained for the validation data. Table 12.1 depicts the works of various authors and the CNNs used in their work.

12.3 Dataset Description

A histopathological image dataset of colorectal cancer was developed by Kather et al. [3]. This dataset known as NCT-CRC-HE-100K train dataset contains one lakh images of H&E-stained slides in nine classes. There are 7180 images in the validation

Table 12.1 Summary of the works on colorectal histopathology dataset

References	CNNs used	Remarks/classification accuracy
Kather et al. [1]	VGG19, AlexNet, SqueezeNet, GoogleNet and ResNet-50	SqueezeNet is superior
Alexander Khvostikov [2]	DenseNet-121	92.9%
Veshkin1 and Khvostikov [4]	AlexNet, GLCM	95%
Predić et al. [5]	ResNet-18	80%
Kar and Rowlands [6]	GoogleNet and XceptionNet	98.4%
Tripathi et al. [7]	Differential box count as features	90.67%

set known as CRC-VAL-HE-7K. The resolution of each tile is of 224 × 224. The sample images and the details of the dataset are given in Fig. 12.1.

12.4 Proposed Methodology

Different deep learning algorithms have been implemented in the past for accurately classifying the colorectal cancer histopathological images. Most of the works used pre-trained CNNs like AlexNet, GoogleNet, VGG-16, SqueezeNet [1], etc. The traditional approaches followed a standard methodology of splitting the dataset into train, test and validation sets. The chosen CNN is trained with specified hyperparameters and the trained model is tested on the test data. DenseNet-121 [2], AlexNet [4], ResNet-18 [5], GoogleNet and XceptionNet [6] are some CNNs used in the classification of the colorectal histopathological images. Some works used contrastive learning in a self-supervised way [8, 9]. In this work, initially, all the images of the dataset (both Train and Validation set) are resized, and normalized with appropriate mean and standard deviation based on the selected model's training parameters. As each model is trained on different dataset, the nature of training is different. The sizes of the images used for training are also different. The architectures used in the self-supervised, and CLIP models are also different. Hence, appropriate pre-processing of the images is necessary before obtaining the image embeddings. The images obtained from the above step are passed through three models i.e., DINOv2G14 [10], ViT-B-32-Laion2b_s34b_b79k [11] and BeiT-Base-Patch16-224-in22k [12] and image embeddings are obtained. The model DINOV2G14 is a vision transformer model ViT-g/14 with a gigantic patch strategy. ViT-B-32-Laion2b model is trained on a large dataset of 2Billion image text pairs and is also a vision transformer based one. The text pairs are chosen from different languages in addition to English language. Train image embeddings from each model are given to a classifier (Support Vector Classifier) independently. Validation image embeddings are given to the trained classifier and the class labels are predicted. Using hard voting (majority voting), final predictions are computed. It is shown as a block diagram and given in Fig. 12.2.

Fig. 12.1 Sample images and statistics of the dataset

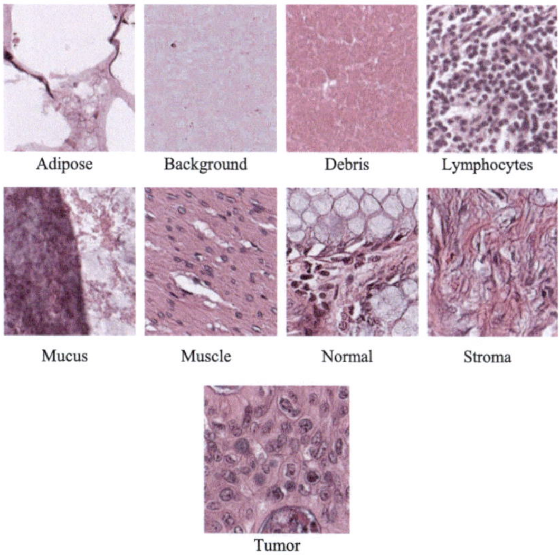

(a) Sample Images from each class

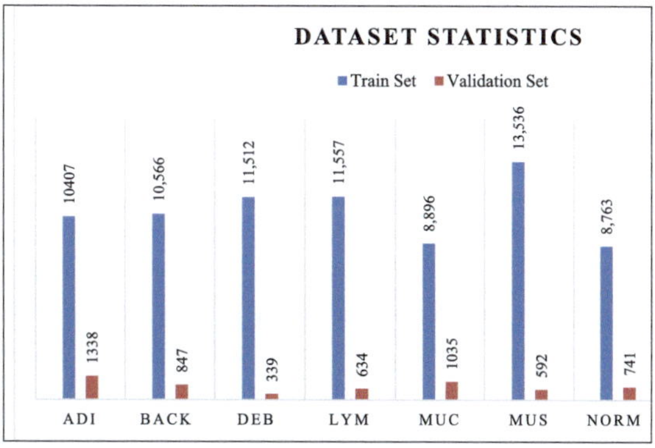

(b) Dataset Statistics

12.5 Simulations and Results

All the simulations are carried in Google Colab environment with A100 GPU under PyTorch framework. As the dataset comprises one lakh images in train set, traditional Central Processing Unit (CPU) based classifiers are computationally expensive and may not converge within a reasonable time. Hence, a Graphical Processing Unit (GPU)-based SVC classifier adopted from cuML library is used. Large visual foundation models are trained on millions of diverse images. Even though the selected models are not trained on any histopathological image datasets, they have excellent discrimination capability in distinguishing the 9 classes i.e., Adipose (ADI), Background (BACK), Debris (DEB), Lymphocytes (LYM), Mucus (MUC), Muscle

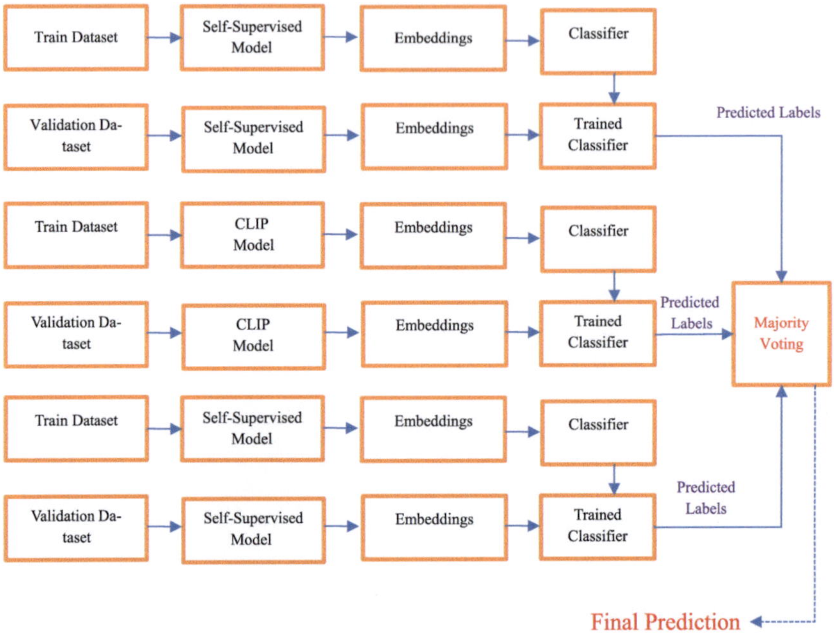

Fig. 12.2 Block diagram of the proposed algorithm

(MUS), Normal (NORM), Stroma (STR) and Tumour (TUM). When an ensemble of a self-supervised model (DINOv2G14), language-image pre-training model (ViT-B-32-Laion2b_s34b_b79k) and a supervised model (BeiT-Base-Patch16-224-in22k) is considered, the unknown test images labels prediction is more accurate and hence the classification accuracy is very high. DINOv2G14 model was trained on 142,109,386 (142 millions) images, and has an image embedding dimension of 1536, ViT-B-32-Laion2b_s34b_b79k was trained on 2.32 billion images with an embedding dimension of 512 and Beit-Base-Patch16-224-in22k was trained on 14,197,122 (14 million) images and the embedding dimension is 768. A brief description of the models is given in Table 12.2.

Table 12.2 Specifications of the models used in the work

Model name	Dinov2G14	ViT-B-32laion2b_s34b_b79k	Beit_base_patch16_224.in22k_ft_in22k
Feature size	1536	512	768
Trained on	142 million image	LAION-2B 2B ImageText Pairs	ImageNet 22 K 14 million images
Training strategy	Self-supervision	Language image pre-training	Self-supervised

By using SVM Classifier with RBF (Radial Basis Function) Kernel, the confusion matrices are given below for the individual models in Fig. 12.3. Performance of the individual models and the final ensemble is presented in the form of confusion matrices and is presented in Fig. 12.3. All the model's performances are different in classifying various classes. For example, Beit model did well in classifying the stroma class compared to the other two models. Adipose class is well classified by the first model DinoG14. However, after applying the majority voting strategy (hard voting), the final ensemble is able to offer superior performance as can be seen from Fig. 12.3d. Except for stroma, most of the other classes are well classified by the hard voting approach. Various metrics can be used to assess the performance of the proposed approach. A classification report is presented in Table 12.3. The classification metrics used are Precision, Recall and F1 score. Precision measures the accuracy of the positive predictions and it neglects the false negatives. Recall metric considers false negative into account while computing the true positives overall. As the dataset is an imbalanced one, F1 score metric is a useful one which is a harmonic mean of precision and recall. An ideal classification model is one which offers an F1-score of 1. Another approach for assessing the models ability is to calculate the per class accuracy and averaging them. This is also a suitable metric while handling imbalance datasets. Except STR and MUS classes, all the classes have an F1-score of well above 96%. This indicates that the proposed model is robust. An overall accuracy of 97.31% is obtained on the validation set.

As given in Table 12.3, the Muscle and Stroma classes are not well classified with an F1-score of <90%. All the other classes are identified with an F1-score of well above 95%. An overall accuracy of 97.31% was obtained by using hard voting strategy on the three models.

In addition to SVM classifier, other classifiers are also tested and the summary is presented in Table 12.4. Various other metrics i.e., precision, recall, F2 score and others are also used and are presented in Table 12.4. Performances of both SVM classifier and Logistic Regression classifier are comparable and some performance metrics are superior for SVM and some are superior for Logistic Regression.

Compared to the work of Tripathi et al. [7] the proposed approach offers higher accuracy and F1 score and it is given in Table 12.5.

12.6 Conclusions

An improved algorithm for classifying the histopathological images of colon cancer is proposed in this work. By using a hard voting approach on the class labels obtained from the image embeddings of three diverse models (self-supervised, language-image pretrained, and self-supervised), improved classification accuracy is obtained. An accuracy of 97.31% and weighed F1-score of 97.27% was obtained on the validation set demonstrating the superiority of the proposed approach. Compared to the work of Tripathi et al., (Accuracy: 90.67% and F1-Score: 90.53%). proposed

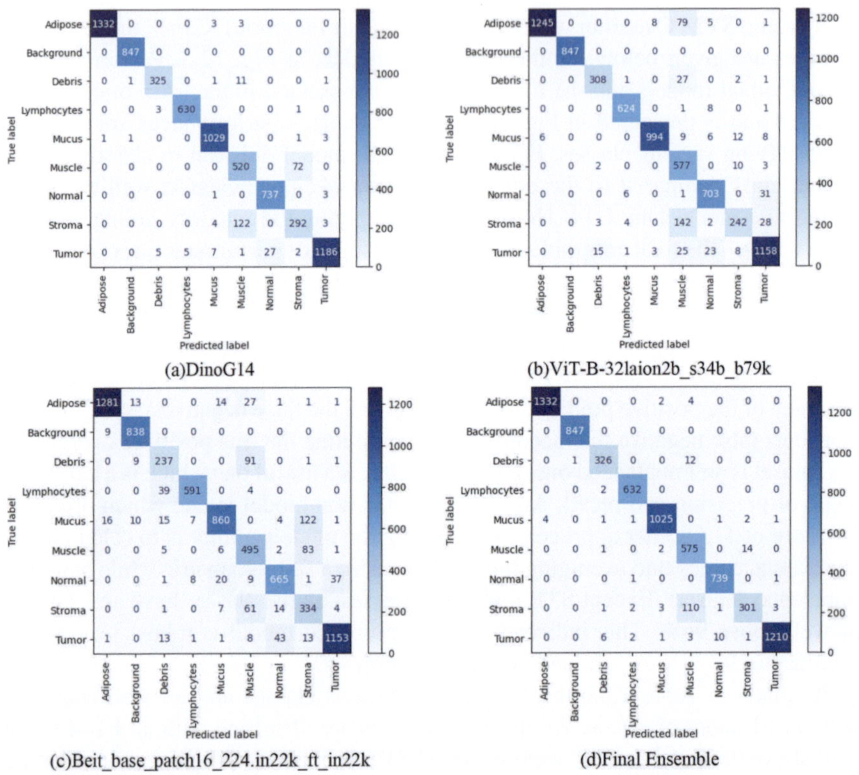

Fig. 12.3 Confusion matrices for the individual models and the final enesmble

Table 12.3 Classification report on the validation set

Class name	Precision (%)	Recall (%)	F1-score (%)	No. of images
ADI	99.70	99.55	99.63	1338
BACK	99.88	100.00	99.94	847
DEB	96.74	96.17	96.45	339
LYM	99.06	99.68	99.37	634
MUC	99.23	99.03	99.13	1035
MUS	81.68	97.13	88.73	592
NORM	98.40	99.73	99.06	741
STR	94.65	71.50	81.46	421
TUM	99.59	98.13	98.86	1233
Accuracy	–	–	97.31	7180
Macro average	96.55	95.66	95.85	7180
Weighted average	97.52	97.31	97.27	7180

Table 12.4 Performance metrics for various classifiers

Metric	SVM	Logistic regression	Random forest	KNN	MBSGD*	MNB**
Precision score	0.9655	0.9657	0.9322	0.8996	0.0138	0.0207
Negative predicted value	0.9968	0.9967	0.9936	0.9914	0.8339	0.7985
Recall score	0.9566	0.9598	0.9276	0.9177	0.0625	0.1111
Accuracy score	0.9946	0.9946	0.9885	0.9846	0.7929	0.8192
F1 score	0.9585	0.9602	0.9261	0.9067	0.0177	0.0349
F2 score	0.9568	0.9594	0.9262	0.9127	0.0288	0.0593
F-beta score	0.9585	0.9602	0.9261	0.9067	0.0177	0.0349
Specificity score	0.9967	0.9967	0.9936	0.9916	0.8817	0.8889
MCC	0.9567	0.9583	0.9219	0.8993	0.0645	0.0
Hamming score	0.8919	0.8919	0.8947	0.8966	0.9924	0.9793
CKS	0.9942	0.9939	0.9883	0.9845	0.7804	0.7985
JSC	0.9263	0.9276	0.8741	0.8415	0.0093	0.0098

*MBSGD: mini batch stochastic gradient descent classifier, ** MNB: multinomial Naïve Bayes classifier, MCC: Matthews correlation coefficient (−1 to + 1), CKS: Matthews correlation coefficient (−1 to + 1), JSC: Jaccard similarity coefficient

Table 12.5 Comparison with an existing work

References	Precision (%)/ specificity	Recall or sensitivity (%)	F1-score (%)	Accuracy (%)
Tripathi et al. [7]	97.39 (specificity)	97.57	90.53	90.67
Proposed work	97.52 (precision)_	97.31	97.27	97.31

algorithm offers higher classification accuracy and F1-score. In addition to the traditional performance metrics other metrics are also computed to justify the robustness of the proposed approach. The work can be extended by applying the visual foundation models trained on similar datasets like pathology images. In fact CLIP models trained on pathological images, fashion images, remote sensing images are available. As the models used in this work are not trained on any of the pathological images, even then they are able to capture the features quite effectively. Models that are trained on large datasets are in general able to capture the features in a better way. Such type of CLIP models are available now. That would be the future work.

References

1. Kather, J.N., Krisam, J., Charoentong, P., Luedde, T., Herpel, E., Weis, C.-A., et al.: Predicting survival from colorectal cancer histology slides using deep learning: a retrospective multicenter study. PLoS Med. **16**(1), e1002730 (2019). https://doi.org/10.1371/journal.pmed.1002730
2. Khvostikov, A., Krylov, A., Mikhailov, I., Malkov, P., Danilova, N.: GraphiCon 2021. In: Proceedings of the 31st International Conference on Computer Graphics and Vision, September 27–30, 2021, Nizhny Novgorod (2021)
3. Kather, J.N., Halama, N., Marx, A.: 100,000 Histological Images of Human Colorectal Cancer and Healthy Tissue (2018). https://zenodo.org/record/1214456#.Y6lhPvdBxPY. Accessed on 11 Nov 2022
4. Veshkin, A.S., Khvostikov, A.: Multiscale content-based image retrieval for whole-slide histological images. Comput. Math. Model. **33**, 244–254 (2022). https://doi.org/10.1007/s10598-023-09569-2
5. Predić, B., Vukić, U., Saračević, M., Karabašević, D., Stanujkić, D.: The possibility of combining and implementing deep neural network compression methods. Axioms **11**(5), 229 (2022). https://doi.org/10.3390/axioms11050229
6. Kar, P., Rowlands, S.: Assessment of a Deep-Learning System for Colorectal Cancer Diagnosis Using Histopathology Images, 04 November 2022, PREPRINT (Version 1) (2022)
7. Tripathi, A., Misra, A., Kumar, K., et al.: Optimized machine learning for classifying colorectal tissues. SN Comput. Sci. **4**, 461 (2023). https://doi.org/10.1007/s42979-023-01882-2
8. Yang, P., Yin, X., Lu, H., Hu, Z., Zhang, X., Jiang, R., Lv, H.: A hybrid self-supervised visual representation learning method for H&E-stained histopathological images. Med. Image Anal. **81**, 102539 (2021)
9. Chu, H., Li, F., He, Y., Guan, T.: Generative and contrastive based self-supervised learning model for histopathology image analysis. In: Proceedings of the 2023 15th International Conference on Machine Learning and Computing (ICMLC'23), pp. 354–360. Association for Computing Machinery, New York (2023). https://doi.org/10.1145/3587716.3587774
10. Oquab, M., et al.: DINOv2: Learning Robust Visual Features Without Supervision (2023)
11. Radford, A., Kim, J.W., Hallacy, C., Ramesh, A., Goh, G., Agarwal, S., Sastry, G., Askell, A., Mishkin, P., Clark, J., Krueger, G., Sutskever, I.: Learning Transferable Visual Models From Natural Language Supervision (2021)
12. Bao, H., Dong, L., Wei, F.: BeiT:BERT Pre-training of Image Transformers (2021)

Chapter 13
Optimal In-Home Healthcare Shift Scheduling

Evelyn Arrey, Theodore T. Allen, and Chen Chen

Abstract We report on a scheduling app that provides optimal shift assignments for in-home healthcare workers. The problem involves substantial complexity due to various training requirements, legal restrictions, and high variability in both demand and supply with irregular hours. Moreover, in-home healthcare services involve significant face-to-face interactions, and so patients' and employees' subjective preferences are a major factor. Most organizations deal with such complexities by determining assignments manually with schedulers; however, this requires substantial training, involves high time cost in developing schedules, and elevated costs due to typically suboptimal solutions. Our solution involves mathematical optimization, deployed via a mixed integer programming model. The model is inspired by a real-world problem of helping to schedule health aides to reduce overtime expenses and achieve equitable schedules in a non-profit organization; this involves 50 weekly shifts across 40 locations and 30 employees. Our pilot test suggests that our software can reduce much of the overtime costs present in manually configured schedules. We also discuss the scheduling app that integrates our optimization model into the existing system.

Keywords Scheduling · Shift assignment · Rostering · Home caregivers · Mixed integer programming

---

E. Arrey (✉) · T. T. Allen · C. Chen
The Ohio State University, Columbus, OH, USA
e-mail: arrey.5@osu.edu; evelyn.arrey@ucn.cl

T. T. Allen
e-mail: allen.515@osu.edu

C. Chen
e-mail: chen.8018@osu.edu

E. Arrey
Departamento de Ingeniería Industrial, Universidad Católica del Norte, Antofagasta, Chile

J. Maiti et al. (eds.), *Operations Research and Data Analytics: Current Trends and Future Perspectives*, Lecture Notes on Multidisciplinary Industrial Engineering,
https://doi.org/10.1007/978-981-96-8327-7_13

13.1 Introduction

General Personnel Scheduling Problem

The Personnel Scheduling Problem (PSP), also known as the Staff Scheduling Problem or Workforce Scheduling Problem, seeks to construct optimal work schedules to ensure an organization can effectively cover the demand for its goods and/or services [1]. Employees are assigned potentially multiple shifts or tasks based on their skills, availability, and various other constraints or restrictions. Several optimization techniques and algorithms have been developed to address the problem, as such problems can be computationally challenging and often require catered solution methods. These techniques include Mixed Integer Programming (MIP) [2–6], metaheuristics [7], Goal Programming [8], and hybrid MIP with heuristics [9] among others. MIP guarantees convergence to mathematically guaranteed optimality, and as such is a preferred choice where practical, however the associated algorithms may suffer from scaling issues and thus stall on large-scale instances. Careful formulation that is mindful of the properties of MIP algorithms is essential to appropriate solution times, as such, we present our novel MIP assignment formulation for this problem. Computational experiments in this paper, using a commercial branch-and-cut solver [10], demonstrate that our MIP-based approach is indeed suitable for the application.

PSP varies depending on the characteristics of a given system, and each variant of the problem requires different modeling and optimization techniques to find feasible solutions. Common applications of PSP include (see [1, 11]):

- Shift Scheduling Problem: Involves assigning employees to specific shifts over a defined period. Challenges include optimizing coverage, meeting demand, and considering factors like employee preferences and legal regulations.
- Employee Rostering Problem: Focuses on creating work schedules for employees over a more extended period, often a week or month.
- Task Assignment Problem: Assign specific tasks or duties to employees based on their skills. Optimization goals include maximizing task covering efficiency and minimizing overall workload.
- Flexible Job-Shop Scheduling Problem: Focuses on traditional job-shop scheduling to include flexibility in task assignments and employee shifts. Addresses uncertainties in demand, employee availability, and dynamic scheduling requirements.
- Transportation Crew Scheduling Problem: Focuses on scheduling crews for transportation services, such as airlines or railways. Challenges include optimizing crew assignments, adhering to legal regulations, and ensuring proper rest periods.
- Nurse Scheduling Problem: Specialized scheduling problems in healthcare settings, particularly hospitals. Assign nurses to different shifts, considering skill requirements and legal constraints, meanwhile balancing workload.

The PSP can also vary based on the planning period and can be classified as single-period or multi-period. The main distinction between them is the time frame considered for scheduling. Single-period scheduling emphasizes short-term planning, while multi-period scheduling focuses on a long-term planning horizon, allowing comprehensive long-term workforce management. Another aspect to consider is the assignment sequence, which can be categorized into cyclic and non-cyclic schedules [1]. Cyclic schedules target employees of the same category following a repetitive sequence, suitable for scenarios with recurring demand patterns, such as train schedules. In contrast, non-cyclical schedules involve independent work sequences for individual employees, making them suitable for situations where demand fluctuates, and shifts have different lengths and start times.

This study considers a modification of the Shift Scheduling Problem suitable for home healthcare providers. The problem involves the assignment of a set of hourly shifts over a period of one to two weeks; this is applied to support the scheduling system of a home healthcare organization.

The Home Healthcare Problem

Home healthcare organizations provide medical support at home for seniors, individuals with disabilities, and patients with chronic conditions. Such organizations coordinate and match staff with patients, according to workers' skills and patients' requirements, focusing on worker schedules and assignments [12]. The demand for these types of services is increasing due to aging populations worldwide [13], therefore, organizations within the home healthcare sector are looking for ways to improve their operational efficiency [14, 15]. One of the main challenges the home healthcare system faces is the efficient creation of schedules defining which employee visits which patient and at which time, a problem called the Home Healthcare Scheduling and Routing Problem (HHCSRP [12] which is a variant of PSP. Some authors also called it Home Healthcare Problem (HHCP) [14–16]. The HHCP is defined as developing effective work schedules for healthcare professionals delivering in-home care services to patients in their residences during a given period.

The main characteristics of the HHCP to consider include:

- Dynamic Medical Needs: Patients have different and dynamically changing medical needs. The scheduling process must adapt to these requirements, considering treatment schedules, medical appointments, patient conditions, and recovery progress.
- Geographical Considerations: Home healthcare scheduling should account for patients' locations since it affects travel time of providers. This can be explicitly optimized for [17].
- Employees and Patient Preferences: Patients as employees may have specific time window preferences for visits. Maintaining consistency in assigned caregivers can improve the patient-provider relationship, and an important consideration in this problem relies on aligning healthcare professionals' preferences with patients' preferences is important [14, 17, 18].

- Regulatory Compliance: Ensuring that healthcare professionals receive adequate rest between shifts and minimizing overtime is important for maintaining the well-being of the care giving staff and the quality of the service provided. As such, schedules have to respect a variety of labor and health scheduling legal requirements.
- Skill Matching: Workers need to have skills and training to match patient needs. This includes considering the specialization of healthcare providers and alignment [19].

Mathematical optimization can be adapted for HHCP, leveraging advanced algorithms to navigate the complex considerations inherent to the problem, such as ensuring compliance with legal regulations, maintain staff well-being, and consider individual preferences. This has the potential to improving efficiency, reducing overtime costs, improving match quality between worker and patient, as well as considerable time savings for administrative staff schedulers in charge of managing such tasks [20]. For additional sources, we refer the reader to surveys on this topic recently published [12, 21–23].

Key to the successful deployment of optimization is navigating various tradeoffs such as: model complexity vs computational tractability; algorithm speed versus solution quality; and problem scope vs practical usefulness of solutions. We illustrate via case study various choices that we make among these tradeoffs to ensure successful deployment.

Applications and Case Studies

Although PSP and more specifically HHCP have been widely studied in the academic literature, challenges continue persist in real-world deployment, both in terms of computational challenge from even stylized models as well as various practical idiosyncrasies. As such, researchers continue to develop new tools to help and complement scheduling processes. Using case studies and integrating different algorithms, researchers have demonstrated how to apply optimization methods to assist in the design of shift schedules while reducing overtime assignments, balancing workloads, and reducing costs. For example, [14] developed optimization components for a German personnel scheduling software called PARPAP on HHCP. They employed various methods, including linear programming, constraint programming, tabu search, and simulated annealing. Jobs are divided among nurses and then an optimal sequence is determined for each division. They formulate an optimization model that integrates shift assignment and routing models to generate high-quality initial solutions, which is fundamental in finding initial solutions to the job partitioning problem. At the same time, constraint programming is used to generate feasible sequences. [14] show that hybrid methods, such as combining constraint programming and tabu search, are especially useful when execution time is limited, establishing a pool of solutions to offer in a short time as may be expected in practice.

Allaoua et al. [16] address HHCP by integrating an integer linear model and a metaheuristic to deal with large instances. They proposed a model integrating vehicle routes and personnel lists in one formulation. Their methodology begins with an initial formulation using MIP that minimizes the number of employees assigned, tested on small instances. Then, they propose an approach that decomposes the MIP into two problems to address more significant instances. First, they modeled the rostering problem with a set partitioning formulation, followed by a Multi-depot Traveling Salesman Problem with Time Windows (MTSPTW) for the routing problem. Although their study extensively explored the routing and allocation aspects, it overlooked the consideration of the temporal dimension. Routing adds considerable computational challenges to the model. Consequently, their model specifically assigns personnel to health services for a single period.

Carello and Lanzarone [15] on HHCP by considering continuity care as a hard constraint and including uncertainty in the problem. They developed a robust allocation model with cardinality restrictions, which innovates using mathematical programming models without generating scenarios to deal with uncertainty. Their proposed model is tested using a case study of a home care provider in Italy to measure its effectiveness in minimizing costs associated with nurses' overtime. The authors claim that their cardinality-constrained approach involves fewer assumptions about demand variability, making it more efficient in computation time compared to alternative methodologies, such as probabilistic models and stochastic optimization methods. However, the method presented by [15] also has computational limitations, as some instances cannot be optimally solved within the time limit.

Moussavi et al. [24] proposed a model of HHCP considering time as an extra dimension and two key aspects: generalized and sequencing assignments. Initially, the model allocates services to employees over different days, considering employees' availability and working-hour limitations. Subsequently, it aims to schedule services across multiple days and establish correlations between assignments. Their proposed model is integrated with the Vehicle Routing Problem (VRP) to provide a unified solution. However, the authors observed that the resulting problem is NP-complete, and thus computationally challenging to scale. To address the complexity of the problem, the authors proposed a three-phase metaheuristic algorithm. This method decomposes the formulation, separating the assignment and routing problem into different models. Numerical experiments and statistical analysis demonstrate that their approach optimally solves 90% of instances with substantially reduced computational times.

We propose a novel formulation as a variant of the HHCP considering the allocation of workers to fixed shifts, i.e., the start and end time for visiting a patient is fixed. To the best of our knowledge, no formulation in the literature related to the HHCP considers fixed time-windows. We called the problem the Home Healthcare Shift Assignment Problem (HHSAP). Consider the case in which there are 4 shifts to assign and 3 available employees who fulfill all the required skills. Suppose working more than 4 h is considered overtime. As illustrated in Fig. 13.1, multiple solutions can be used to assign all the shifts; however, when trying to minimize the total overtime, solution 4 would be the best solution. Figure 13.2 provides a simple example, considering two employees and two patients. In the example, the skills requirement and the time-slot required, of Shift 223 constrains the problem to only a possible feasible assignment as Nurse 2 is ineligible.

SOLUTION 1

Shift ID	location id	service date	start time	end time	hours	employee assigned
1200	24	4/12/23	8:00:00 AM	12:00:00 PM	4	nurse 1
1300	34	4/12/23	4:00:00 PM	6:00:00 PM	2	nurse 1
1400	44	4/12/23	1:00:00 PM	3:00:00 PM	2	nurse 2
1500	54	4/12/23	7:00:00 PM	11:00:00 PM	4	nurse 3

Employee ID	total hours	total overtime
nurse 1	6	2
nurse 2	2	0
nurse 3	4	0
Total overtime		**2**

SOLUTION 2

Shift ID	location id	service date	start time	end time	hours	employee assigned
1200	24	4/12/23	8:00:00 AM	12:00:00 PM	4	nurse 1
1300	34	4/12/23	4:00:00 PM	6:00:00 PM	2	nurse 2
1400	44	4/12/23	1:00:00 PM	3:00:00 PM	2	nurse 3
1500	54	4/12/23	7:00:00 PM	11:00:00 PM	4	nurse 1

Employee ID	total hours	total overtime
nurse 1	8	4
nurse 2	2	0
nurse 3	2	0
Total overtime		**4**

SOLUTION 3

Shift ID	location id	service date	start time	end time	hours	employee assigned
1200	24	4/12/23	8:00:00 AM	12:00:00 PM	4	nurse 3
1300	34	4/12/23	4:00:00 PM	6:00:00 PM	2	nurse 1
1400	44	4/12/23	1:00:00 PM	3:00:00 PM	2	nurse 2
1500	54	4/12/23	7:00:00 PM	11:00:00 PM	4	nurse 2

Employee ID	total hours	total overtime
nurse 1	2	0
nurse 2	6	2
nurse 3	4	0
Total overtime		**2**

SOLUTION 4

Shift ID	location id	service date	start time	end time	hours	employee assigned
1200	24	4/12/23	8:00:00 AM	12:00:00 PM	4	nurse 1
1300	34	4/12/23	4:00:00 PM	6:00:00 PM	2	nurse 3
1400	44	4/12/23	1:00:00 PM	3:00:00 PM	2	nurse 3
1500	54	4/12/23	7:00:00 PM	11:00:00 PM	4	nurse 2

Employee ID	total hours	total overtime
nurse 1	4	0
nurse 2	4	0
nurse 3	4	0
Total overtime		**0**

Fig. 13.1 Example of home healthcare shift assignment problem

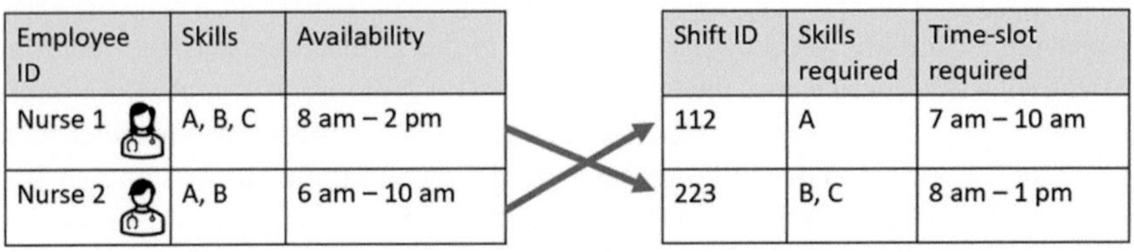

Fig. 13.2 Example of home healthcare shift assignment problem considering employees' skills and availability

Several studies as mentioned have proposed models involving meta-heuristics, and thus cannot guarantee optimality. Many such models also involve a "kitchen sink" fully automated approach, which may require drastic changes in management. Therefore, implementation could significantly impact staff members. An alternative approach that we have adopted aims to integrate optimization with existing human assignment processes, and to ensure global optimality where possible. This is considered a less intrusive method of incorporating optimization techniques into existing processes, particularly when employees resist changing their established methods.

For instance, [8] designed an optimization tool to integrate with the current self-scheduling process to automate and improve the self-scheduling process in Swedish healthcare. The authors state that minimizing changes to the existing scheduling process facilitates nurses' acceptance of a new procedure, especially if it closely aligns with their familiar practices. In their proposed system, first, the nurses submit their schedule proposals, then the optimization tool is run and recommends an improved schedule for all the nurses. The benefit of the proposed system lies in presenting compliant schedules that consider the nurse's preferences, scheduling rules, and staffing demands, thus saving time and effort. To select the best assignment, the system is score-based. While the scoring system is generally satisfactory, improvements such as maximizing scores for the least favored nurses are needed to improve fairness during scheduling periods. The authors illustrate that approximately 85% of the requests submitted during the pilot study align with the staffing demand.

As stated by [8], there can be skepticism from employees regarding optimization software, so it is important to demonstrate that optimization can provide benefits to assist in the design of shift schedules, such as reducing overtime and balancing workload doing minor changes in their current system. Our approach navigates these concerns. In particular, we have developed a scheduling app prototype that integrates a MIP-based optimization approach with the existing scheduling system of a non-profit health home care organization.

13.2 Problem Definition

The problem is inspired by our work for a non-profit home healthcare organization in Columbus, Ohio, which provides low-cost home healthcare services to patients. The organization has regular employees who work on regular (fixed) shift schedules each week. However, given the high variability of patients' needs, they often require services beyond their regular visits; we refer to these shifts as *open shifts*. We define a shift as a unique combination of factors, such as location, day, time slot, and the specific patient requiring care. To cover open shifts, the organization assigns temporary employees, who usually are volunteer caregivers with other primary jobs. Thus, volunteer employees have limited and sporadic availability.

The existing scheduling system consists of manual single-shift assignments—a form of satisficing rather than optimization. It begins by checking which employees fulfill the requirements for a given shift. After a set of potential employees is obtained, schedulers contact them so that employees can indicate their desired shifts. Thus, the selection process is based on a first-come, first-served system. This approach can be suboptimal compared to a coordinated, centralized optimization approach; for instance, the process may lead to inefficient allocation of resources, with higher overtime hours than necessary, increasing related costs.

Additionally, schedulers must contact potential employees individually for shift assignments and manually input them into the system. This manual process becomes time-consuming, especially when conflicts occur, or employees cancel shifts.

According to schedulers, generating a finalized schedule takes about three days. Our goal in was to develop an optimization tool that would effectively match open shifts with employees, contributing to a more cost-effective and responsive scheduling system. We created a scheduling app to support the schedulers in finding the optimal allocation strategy that maximizes the utilization of available caregivers by minimizing overtime.

The main contribution of this paper is the comprehensive and realistic mathematical formulation of the HHSAP to optimally assign health home care shifts. Our model considers maximum consecutive working hours, breaks between shifts, employee availability, and preferences. According to the recent survey on HHCP developed by [12] 48% of the studies consider the patients preference, and only a 19% focused on the employees' preferences. We present this novel formulation together with numerical results from a pilot implementation demonstrating its practical use.

13.3 MIP Model Formulation

We approach the problem as a type of assignment problem for HHSAP, considering employees' skills, availability, and industry regulations concerning maximum consecutive work hours and break times. Our objective is to minimize employee overtime during a given period, as overtime is the major governing factor of costs in our problem setting. The core of this formulation is a shift assignment problem (i.e. binary variables determining assignments of workers to shifts), with additional variables to account for overtime and number of consecutive hours worked.

Sets

J: set of shifts.

I: set of workers.

H: set of hours.

H_t: set of shifts belonging to hour t.

T_j: set of hours in shift j.

S_t: set of shifts j_1, j_2 that has conflicts at hours t.

Parameters

$a_{i,j} = 1$ if worker i is allowed/available to work on shift j.

$c_i =$ cost of over time unit of employee i.

$v =$ maximum time working per week.

$L =$ maximum consecutive hours allowed to work.

$R =$ minimum rest time after L consecutive hours working.

$N =$ time horizon (in hours).

$T_{\text{reg},i}$ = total regular hours assigned to work i.
Variables

x_{ijt}: if worker$i$ is assigned to the shift j at hour t.

T_i^+: overtime for worker i.

w_{it}: if worker i has worked from hours t to L.

$$\min \sum_{i \in I} c_i T_i^+ \tag{13.1}$$

$$s.t.$$

$$\sum_{i \in I} x_{i,j,t} = 1 \ \forall \ j \in H_t, \ \ \forall \ t \in H \tag{13.2}$$

$$\sum_{j \in H_t} x_{i,j,t} \leq 1 \ \forall \ i \in I, \ \ \forall \ t \in H \tag{13.3}$$

$$\sum_{t=k}^{k+L} \sum_{j \in H_t} x_{i,j,t} \leq L - (1 - w_{ik}) \ \forall \ i \in I, \ \ \forall \ d \in H, \ \ k = 1, \ldots, N - (d + L + 1) \tag{13.4}$$

$$\sum_{j \in H_{k+L+1}} x_{i,j,k+L+1} \leq 1 - w_{ik} \ \forall \ i \in I, \ \ \forall \ d \in H, \ \ k = 1, \ldots, N - (d + L + 1) \tag{13.5}$$

$$\sum_{j \in H_t} 1 - x_{i,j,k+L+p} \geq w_{ik} \ \forall \ i \in I, \ \ \forall \ k \in H, \ \ p = 1, \ldots, \min\{R, N - (k - L)\} \tag{13.6}$$

$$\sum_{t \in H} \sum_{j \in H_t} x_{i,j,t} + T_{\text{reg},i} \leq v + T_i^+ \ \ \forall \ i \in I \tag{13.7}$$

$$x_{i,j,t_1} = x_{i,j,t_2} = \cdots = x_{i,j,t_k} \ \forall \ i \in I, \ \ \forall \ j \in J, \ \ \text{where} \ \ T_j = \{t_1, t_2, \ldots, t_k\} \tag{13.8}$$

$$x_{i,j,t} \leq a_{i,j} \ \ \forall \ i \in I, \ \ \forall \ j \in H_t, \ \ \forall \ t \in H \tag{13.9}$$

$$x_{i,j_1,t} + x_{i,j_2,t} \leq 1 \ \forall \ i \in I, \ \ \forall \ t \in H \ \forall \ (j_1, j_2) \in S_t \tag{13.10}$$

$$x_{i,j,t} \in \{0, 1\} \ \forall \ i \in I, \ \ \forall \ j \in J, \ \ \forall \ t \in H \tag{13.11}$$

$$w_{i,k} \geq 0 \ \forall \ i \in I, \ \ k = 1, \ldots, N - (d + L + 1), \ \ \forall \ d \in H \tag{13.12}$$

$$T_i^+ \geq 0 \ \forall \ i \in I \tag{13.13}$$

The objective function (13.1) minimizes the total overtime cost. Constraint set (13.2) ensures that only one employee must cover one shift. Constraint (13.3) manages the assignment of employees to one time period t at a time. Constraints (13.4) and (13.5) state that employees can work no more than L consecutive hours. Constraint set (13.6) enforces that every employee must take at least R hours of rest after L consecutive working hours. Constraint (13.7) states do not work more than v hours in the time horizon; otherwise, it is considered over time. Constraint (13.8) enforces that an assigned employee must work all shift hours or not at all. Constraint (13.9) ensures employees' availability. Constraint (13.10) forbids overlapping shift assignments. Constraints (13.11–13.13) enforce basic variable domains.

We note that the objective can be modified to account for employee preferences by changing cost parameter as needed; likewise, we can adjust the model to account for different salaries per employee (e.g. adding a variable index for employee type). Moreover, time periods can be adjusted as needed to e.g. 15-min increments; however, the model size is governed by the total number of periods, and so smaller increments may result in slower run times. We also note that on large-scale problems, constraints (13.8) and (13.9) do not have to be incorporated explicitly into the model and can be replaced simply via variable fixing/deletion.

13.4 Scheduling App Prototype

The MIP model is incorporated into a scheduling app prototype developed in Python for Microsoft Excel. Figure 13.3 depicts the app interface.

Inputs:

1. Master schedule file: Excel file with all shifts that must be assigned for a given week.
2. Employees training file: Excel file with all active employees and their skills for a given week.
3. Employees availability file: Excel file with all active employees and their preferred time to work for a given week.
4. Employees by status file: Excel file with all employees and their status for a given week.
5. Template schedule by location (regular shifts) file: Excel file with the regular shifts.
6. Floaters file: Excel file with a list of employees that will work as floaters in a given week.

Outputs:

7. Open shifts Assignments file: Excel file with the optimal assignment of open shifts.

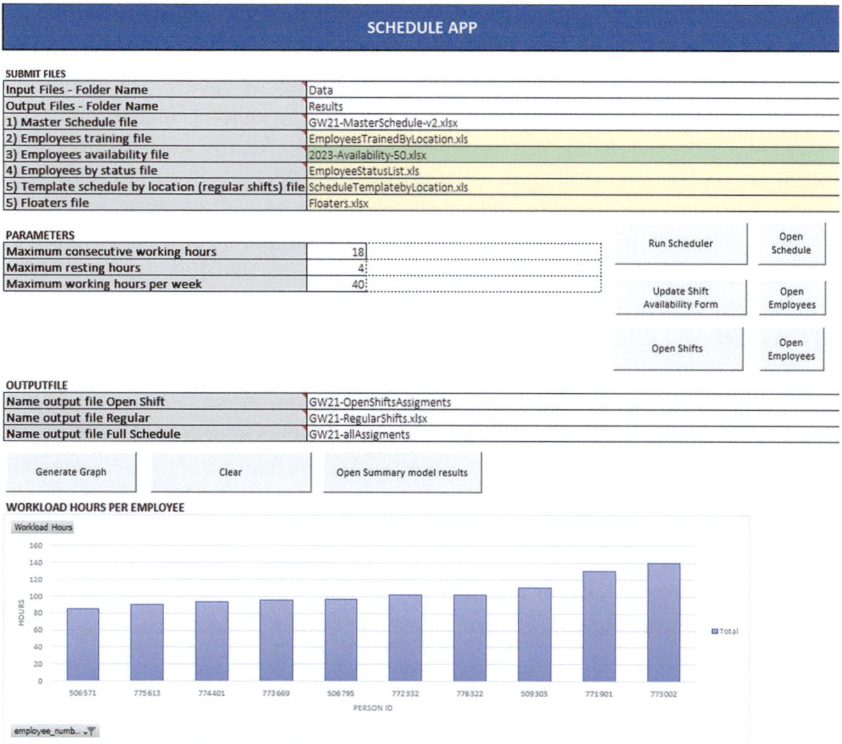

Fig. 13.3 Scheduling App prototype user interface

8. Full Schedule file: Excel file with the optimal assignment of open and regular shifts.

The scheduling app runs in three steps to support the assignment process: (1) Preprocessing Data, (2) Availability Collection, and (3) Optimization. All steps are described below.

Step 1: Preprocessing Data

Since the current system cannot automatically differentiate open shifts from the list of shifts to be assigned in a given week, the app starts by taking the Template schedule and using it to filter regular shifts from the Master Schedule. After filtering, a file containing only the open shifts is obtained. Thus, this step matches employees' skills with open shifts' requirements, and so list of potential shifts to be assigned is obtained for each active employee.

Step 2: Availability Collection

Once the set of potential open shifts each employee can cover is obtained, the app generates a survey form to collect information about employees' availability and preferences. The survey form runs in Microsoft Power Apps connecting with Power Automate. Since the list of open shifts, employees' training, and status change each week, we needed a custom survey that can automatically be updated. We control the survey information by the button in the Scheduling App called "*Update Shift Availability Form*" (see Fig. 13.3). The employees log in to the Survey with their respective credentials, and then a list of open shifts is displayed. The survey can filter the open shift by location name and zip code. Then, each employee selects their preferred location or open shifts. The existing scheduling system checks which employees fulfill the requirements for a given shift before offering the shift. However, when asking employees for an open shift, employees choose for themselves when to work and construct their schedules. In our proposed approach, we do not allow employees to construct their schedules. Instead, we ask which open shifts they are allowed to work and which one they prefer. Therefore, even when the employees are not directly designing their schedules, they still have the option to choose which open shifts they would like to be assigned. This approach is less intrusive than asking employees about their complete availability and assigning ignoring preferences. Moreover, such preferences can account for routing considerations without having to incorporate such complexities explicitly into the optimization model itself.

Once all the employees fill out the survey, the optimization model can be run. Figure 13.4 shows the survey form interface with a fictional employee ID and fictional open shifts.

Step 3: Optimization and Validation

The optimization model is run in this step, and a schedule file is generated as output. Then, the schedulers may confirm the schedule with the employees and validate the assignments. The input files can be modified if needed, and the process is rerun,

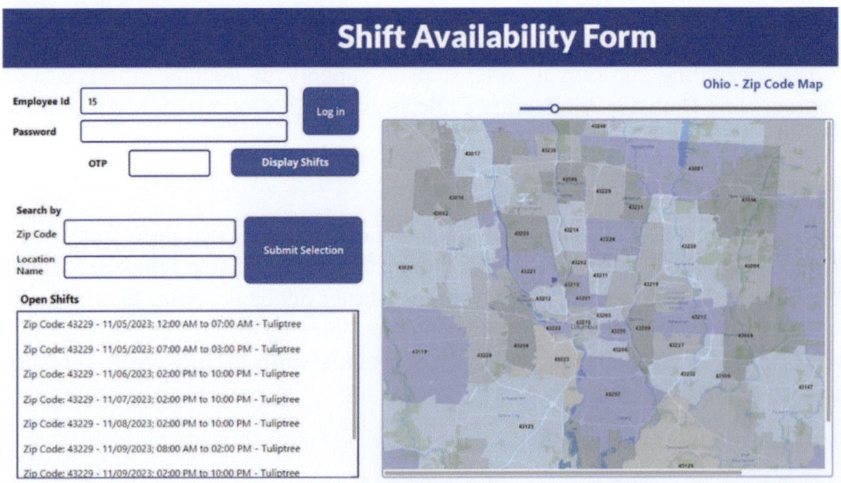

Fig. 13.4 Survey form user interface from power apps

as seen in Fig. 13.3. The Scheduling app prototype allows the user to generate an additional schedule analysis, such as generating graphics to observe the employees' workload hours. Key to successful design and deployment is ensuring smooth iterative runs. We allow for manual adjustments and re-running the optimization model as many times as needed to provide control to the schedulers; this is necessary to account for real-world considerations such as canceled or added shifts.

13.5 Case Study

We consider the deployment of our proposed methodology for an organization that provides home healthcare services to 40 locations in Columbus, Ohio. The organization assigns approximately 30 employees to 150 open shifts every week. Open shifts are offered primarily to temporary employees, who are mostly volunteers. If regular staff are available, then they may also be considered for open shifts, but with lower preference. Each open shift involves different services, required skills, time windows, and locations. At each location, there can be different shifts for different patients, with up to three patients living in the exact location. According to legal regulations, employees can work a maximum of 18 consecutive hours, after which they must take a break of at least four consecutive hours. For each week, an employee can work up to 40 h, beyond which it is considered overtime. There are four types of employees (employees' status): Full-time, part-time, temporary, and floaters. Floaters are designated to cover open shifts, ensuring they work at least 40 h per week. When assigning shifts, floaters are given priority, followed by temporary employees, then part-time and full-time staff.

We tested our model in 15 different periods from real-world data obtained the year 2023 and compared the current assignment with assignments from our software. Around 1000 shifts are assigned each week, with most of them being regular assignments. Table 13.1 presents the relevant data for each period analyzed. The bar in Tables 13.1, 13.2, and 13.3 differentiates one-week from two-week instances. The available employees (OS) are not regular staff and are available given a week to cover the Open Shifts left. We could not run an availability survey retroactively to employees, and so to determine employee availability, the shift assignments for the year 2023 were used as an estimate. For each employee, we calculated the frequency of shifts in which they worked. If an employee worked more than 50% of the time in the same daytime slot, we assumed they were available to work during that period for subsequent assignments.

Table 13.1 Data set description

Instance	Days	# Shifts	# Shifts to regular staff	# Open shifts (OS)	# Available employees (OS)
GW14	7	977	811	166	39
GW15	7	989	822	167	30
GW16	7	959	815	144	38
GW17-1	7	958	811	147	34
GW17-2	7	966	834	132	38
GW17-3	7	959	824	135	38
GW17-4	7	965	828	137	38
GW18	7	991	868	123	37
GW19	7	965	856	109	37
GW20	7	997	863	134	37
GW21	7	980	847	133	32
GW14-15	14	1966	1612	354	39
GW15-16	14	1949	1619	329	30
GW16-17	14	1918	1619	298	38
GW17-1-2	14	1934	1641	283	34

Table 13.2 Compared results between current and optimization model assignments by week

Instance	Current assigned employees	MIP assigned employees	Current overtime employees	MIP overtime employees	Current extra hours	MIP extra hours	MIP improvement %
GW14	33	33	9	1	237.17	7.00	97
GW15	34	30	8	12	188.21	37.94	80
GW16	31	30	7	1	189.00	1.25	99
GW17-1	32	32	10	1	210.75	1.25	99
GW17-2	28	**29**	7	0	201.75	0	100
GW17-3	30	27	8	0	171.00	0	100
GW17-4	30	30	8	0	192.75	0	100
GW18	32	26	5	0	104.75	0	100
GW19	24	**28**	8	0	82.75	0	100
GW20	29	28	8	0	104.00	0	100
GW21	29	26	7	0	153.68	0	100
GW14-15	42	39	8	0	470.38	0	100
GW15-16	40	30	8	11	358.46	62.69	83
GW16-17	40	38	7	0	372.25	0	100
GW17-1-2	36	31	7	0	368.25	0	100

Table 13.3 Model size and performance by week

Instance	# vars	# constraints	# Nodes explored	Time (s)	Extra employees
GW14	35,023	53,239	1	4	1
GW15	27,111	40,618	1	4	0
GW16	32,452	48,921	1	4	0
GW17-1	25,148	38,213	1	4	0
GW17-2	29,030	44,023	1	3	0
GW17-3	29,502	44,653	1	4	0
GW17-4	29,346	43,845	1	4	0
GW18	27,940	42,088	1	3	0
GW19	25,257	37,779	1	3	0
GW20	28,725	42,928	1	4	0
GW21	25,043	36,497	1	3	0
GW14-15	72,981	112,058	1	9	1
GW15-16	53,616	80,698	1	7	0
GW16-17	65,965	99,992	1	8	0
GW17-1-2	49,543	75,694	1	7	0

13.6 Computational Results

The pilot test demonstrates that our optimization software can avoid overtime entirely among open shift schedules in 10 out of 15 instance (see Table 13.2). Table 13.2 shows the results of our proposed model compared to the current assignments for each week. Current employees assigned and MIP employees assigned display the number of employees assigned to open shifts in the current and MIP schedules, respectively. Current overtime employees and MIP overtime employees display the number of employees who work overtime in the current and MIP schedules, respectively. Current extra-hour and MIP extra-hour shows the total extra hours assigned in the current and MIP schedules, respectively. Finally, MIP improvement shows the reduction (in percentage) of total extra hours from the Current situation to the MIP schedule. Nearly all overtime can be avoided compared to schedules generated from the existing manual process. As highlighted in bold (see Table 13.2) for weeks GW17-2 and GW-19, the proposed model assigns 1 and 4 more employees than the current assignments, but these changes reduced overtime from 192 to 0 h and 84 to 0 h, respectively.

All computational experiments were implemented in Python and ran on a Dell Notebook with an Intel(R) Core (TM) i7-12850HX, 2.91 GHz, and 32 GB. We use Gurobi Optimizer version 9.1.0 [10] to solve the MIP formulation. Table 13.3 displays the model's performance for each week. Furthermore, all the benchmark problems were solved optimally in <10 s (see Table 13.3)—a tremendous savings compared to the three-days-long manual process. We note that such favorable performance is due to the practical effectiveness of the MIP formulation, where the problem is typically solved at the root node due to strong Linear Programming relaxations.

To assess the model's scalability, we selected instance GW15-16 and generated three new instances from it by doubling and tripling the current number of open shifts (329 OS). We also increase the number of available employees by 2, 5, and 10 times the current value (30 employees). Table 13.4 summarizes the results for all the combinations of the number of shifts and the number of available employees. MIP time refers to the solving time of the MIP provided by the optimizer. Total time refers to the schedule app's time to provide a solution, from preprocessing all the data to providing a final file with all the assignments, it includes the MIP time. The MIP time increases with the number of available employees and the number of open shifts to be assigned. However, our proposed model nonetheless attains optimality in the largest instance (987 OS and 300 employees) in <1 min. Figure 13.5 displays the trend of the MIP time when changing the number of open shifts and available employees. We can observe a more significant change by analyzing the total time as the number of open shifts and available employees increases. Figure 13.6 shows the trend for the Total time.

Table 13.4 Scalability experiment results

# Open shifts (OS)	# Available employees (OS)	# vars	# constraints	# Nodes explored	MIP time (s)	Total time (s)	Gap (%)
329	60	107,286	165,538	1	2.0	15.0	0.0
329	150	268,296	420,058	1	4.2	63.0	0.0
329	300	536,646	844,258	1	8.9	221.0	0.0
658	60	209,733	395,656	1	5.9	25.0	0.0
658	150	525,183	1,016,026	1	12.4	86.0	0.0
658	300	1,050,933	2,049,976	1	30.6	276.0	0.0
987	60	298,824	689,662	1	9.1	35.0	0.0
987	150	746,844	1,777,672	1	22.6	111.0	0.0
987	300	1,493,544	3,591,022	1	49.8	325.0	0.0

Fig. 13.5 MIP time versus number of available employees at different levels of number of open shifts

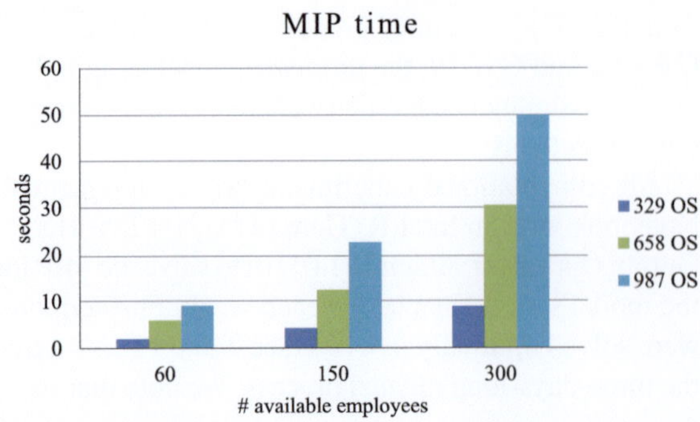

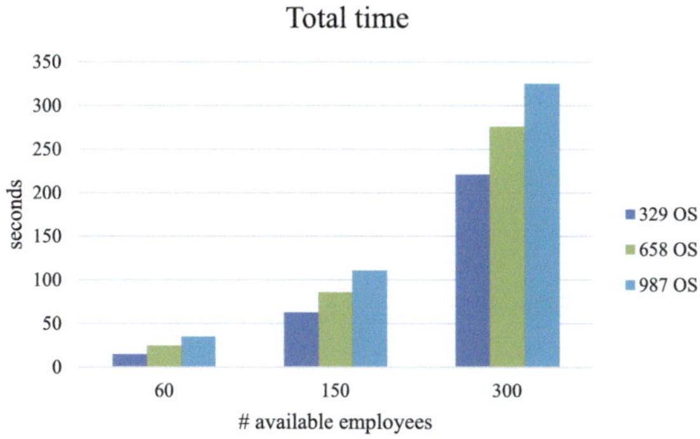

Fig. 13.6 Total time versus number of available employees at different levels of number of open shifts

13.7 Conclusions

This research aimed to develop a MIP model and associated software to address a shift scheduling problem applied to home caregivers at a Non-Profit Organization. We integrate our method with the organization's existing framework and IT in order to ensure successful transition to an optimization-based approach. We present a case study that shows the performance of the model. The numerical tests show how the model reduces overtime costs and improves the overall quality of services that the Non-Profit Organization provides. At the beginning of the project, schedulers were skeptical of the new approach; however, with careful integration into existing practices, we demonstrated quality solutions while saving the schedulers several days worth of effort per month. We were also able to compare the current assignment with the model assignments, and the staff from the non-profit organization validated them, letting us know that the results from the model are entirely feasible for run. However, more tests are needed before implementing the system. This research contributes to future advancements in healthcare scheduling systems, emphasizing the role of optimization models integrated into existing IT and decision-making processes for improving operational efficiency and patient care.

Multiple avenues for future research and exploration have emerged from our initial prototype. First, the root Linear Programming relaxations of our optimization model were observed to be highly effective in practice. This phenomenon is worthy of theoretical examination for the possible spillover benefits to other assignment/scheduling applications, and so we would like to elucidate precisely under what conditions and/or which mathematical structures from the constraints are causing this. Second, machine learning may help provide better estimates for inputs in a variety of ways. For example, it can be used to generate estimates of employee preferences based on previous results so as to reduce the burden on manual availability data collection. Third, in real-world deployment we will have gathered substantial data on actual schedules. This will enable us to apply sensitivity analysis retrospectively, providing valuable managerial insights for strategic decision-making. For instance, one could estimate for instance the hypothetical value of an additional employee with certain training and accordingly adjust salaries.

Acknowledgements Thanks to all the staff from the non-profit organizations who provided the details of their current scheduling system and gave us access to their data for numerical tests. Special thanks to Jennifer Marshall, Jamie Kinzelman, and Viola Calderon, for the feedback and time provided to help us understand their system and improve our model.

References

1. Ernst, A.T., Jiang, H., Krishnamoorthy, M., Sier, D.: Staff scheduling and rostering: a review of applications, methods and models. Eur. J. Oper. Res. **153**(1), 3–27 (2004)
2. Al-Yakoob, S.M., Sherali, H.D.: Mixed-integer programming models for an employee scheduling problem with multiple shifts and work locations. Ann. Oper. Res. **155**, 119–142 (2007)
3. Baker, K.R.: Workforce allocation in cyclical scheduling problems: a survey. J. Oper. Res. Soc. **27**, 155–167 (1976)
4. Caprara, A., Monaci, M., Toth, P.: Models and algorithms for a staff scheduling problem. Math. Program. **98**, 445–476 (2003)
5. Dantzig, G.B.: A comment on edie's "traffic delays at toll booths." J. Oper. Res. Soc. Am. **2**(3), 339–341 (1954)
6. Mehrotra, A., Murphy, K.E., Trick, M.A.: Optimal shift scheduling: a branch-and price approach. Naval Res. Log. **47**(3), 185–200 (2000)
7. Aykin, T.: Optimal shift scheduling with multiple break windows. Manag. Sci. **42**(4), 591–602 (1996)
8. Rönnberg, E., Larsson, T.: Automating the self-scheduling process of nurses in Swedish healthcare: a pilot study. Health Care Manag. Sci. **13**, 35–53 (2010)
9. Al-Yakoob, S.M., Sherali, H.D.: A column generation approach for an employee scheduling problem with multiple shifts and work locations. J. Oper. Res. Soc. **59**(1), 34–43 (2008)
10. Gurobi Optimization, LLC: Gurobi Optimizer Reference Manual (2023)
11. Van den Bergh, J., Belien, J., De Bruecker, P., Demeulemeester, E., De Boeck, L.: Personnel scheduling: a literature review. Eur. J. Oper. Res. **226**(3), 367–385 (2013)
12. Di Mascolo, M., Martinez, C., Espinouse, M.-L.: Routing and scheduling in home health care: a literature survey and bibliometric analysis. Comput. Ind. Eng. **158**, 107255 (2021)
13. Fathollahi-Fard, A.M., Hajiaghaei-Keshteli, M., Mirjalili, S.: A set of efficient heuristics for a home healthcare problem. Neural Comput. Appl. **32**, 6185–6205 (2020)
14. Bertels, S., Fahle, T.: A hybrid setup for a hybrid scenario: combining heuristics for the home health care problem. Comput. Oper. Res. **33**(10), 2866–2890 (2006)
15. Carello, G., Lanzarone, E.: A cardinality-constrained robust model for the assignment problem in home care services. Eur. J. Oper. Res. **236**(2), 748–762 (2014)
16. Allaoua, H., Borne, S., Létocart, L., Calvo, R.W.: A matheuristic approach for solving a home health care problem. Electr. Notes Discr. Math. **41**, 471–478 (2013)
17. Grenouilleau, F., Legrain, A., Lahrichi, N., Rousseau, L.-M.: A set partitioning heuristic for the home health care routing and scheduling problem. Eur. J. Oper. Res. **275**(1), 295–303 (2019)
18. Braekers, K., Hartl, R.F., Parragh, S.N., Tricoire, F.: A bi-objective home care scheduling problem: analyzing the trade-off between costs and client inconvenience. Eur. J. Oper. Res. **248**(2), 428–443 (2016)
19. Lin, C.-C., Hung, L.-P., Liu, W.-Y., Tsai, M.-C.: Jointly rostering, routing, and r rostering for home health care services: a harmony search approach with genetic, saturation, inheritance, and immigrant schemes. Comput. Ind. Eng. **115**, 151–166 (2018)
20. Burke, E.K., De Causmaecker, P., Berghe, G.V., Van Landeghem, H.: The state of the art of nurse rostering. J. Sched. **7**, 441–499 (2004)

21. Cissé, M., Yalcındag, S., Kergosien, Y., Sahin, E., Lenté, C., Matta, A.: Or problems related to home health care: a review of relevant routing and scheduling problems. Oper. Res. Health Care **13**, 1–22 (2017)
22. Fikar, C., Hirsch, P.: Home health care routing and scheduling: a review. Comput. Oper. Res. **77**, 86–95 (2017)
23. Grieco, L., Utley, M., Crowe, S.: Operational research applied to decisions in home health care: a systematic literature review. J. Oper. Res. Soc. **72**(9), 1960–1991 (2021)
24. Moussavi, S.E., Mahdjoub, M., Grunder, O.: A matheuristic approach to the integration of worker assignment and vehicle routing problems: application to home healthcare scheduling. Exp. Syst. Appl. **125**, 317–332 (2019)
25. Kletzander, L., Musliu, N.: Solving the general employee scheduling problem. Comput. Oper. Res. **113**, 104794 (2020)
26. Moussavi, S.-E., Mahdjoub, M., Grunder, O.: Productivity improvement through a sequencing generalised assignment in an assembly line system. Int. J. Product. Res. **55**(24), 7509–7523 (2017)
27. Sungur, B., Ozguven, C., Kariper, Y.: Shift scheduling with break windows, ideal break periods, and ideal waiting times. Flexible Serv. Manufact. J. **29**, 203–222 (2017)

Chapter 14
XAI's Potential in Reshaping Mental Health: A Review

Princy Verma, Millie Pant, and Mukesh Kumar Barua

Abstract The present study explores insights into the applicability of Explainable Artificial Intelligence (XAI) technologies for reshaping mental health care. The objective is to estimate the effect of XAI on various major mental health disorders, the novel machine learning (ML) techniques employed, and the datasets utilized. The study is done using a scoping review guided by the PRISMA (Preferred Reporting Items for Systematic Reviews and Meta-Analyses) framework, examined relevant literature from the past decade (2020–2024), and identified 20 studies for our analysis in which XAI is applied to diverse mental health conditions. Specifically, these papers are selected based on identifying the novel techniques and datasets influencing the mental health domain. The review finds that the most commonly used datasets are EEG and MRI, and the prevalently employed XAI technique is SHAP (*Shapley Additive Explanations*) to dissect various factors influencing mental health outcomes.

Keywords Artificial intelligence · Explainable AI · Mental health disorders · Datasets · Machine learning

---

P. Verma (✉) · M. Pant
Department of Applied Mathematics and Scientific Computing, IIT Roorkee, Roorkee, India
e-mail: princy_v@amsc.iitr.ac.in

M. Pant
e-mail: pant.milli@as.iitr.ac.in

M. Pant
Mehta Family School for Data Science and Artificial Intelligence, IIT Roorkee, Roorkee, India

M. K. Barua
Department of Management Studies, IIT Roorkee, Roorkee, India
e-mail: mukesh.barua@ms.iitr.ac.in

J. Maiti et al. (eds.), *Operations Research and Data Analytics: Current Trends and Future Perspectives*, Lecture Notes on Multidisciplinary Industrial Engineering,
https://doi.org/10.1007/978-981-96-8327-7_14

14.1 Introduction

Mental health [11] is a state of emotional, psychological, and social well-being with the ability to think, feel, and meet the regular demands of life by living with others. Mental illness is a state of a person where mental health is disturbed in which thinking, emotions, and behavior are affected. Mental illnesses like depression, anxiety, schizophrenia, bipolar disorders, etc., have become widespread in today's society. This concern may contribute to youth suicide, lower academic achievement, and decreased social activity. According to the World Health Organization (WHO), 450 million people worldwide have mental disorders, and 7.5% of the Indian population suffers from mental illness, with a treatment gap of about 83% in India.

Effectively detecting and diagnosing mental health issues, along with providing tailored, precise recommendations and follow-up, is crucial for addressing challenges in risk reduction and enhancing mental health. Traditional mental health [12] screening and diagnostic methods typically prioritize in-person assessments, focusing on direct communication between the patient and practitioner, observation of behaviors and emotional states, and hands-on, patient-centered practice. As a result, mental health data comes from different sources; therefore, data might often be incomplete and inconsistent, making it hard to get accurate results; highly sensitive, which raises concerns about data privacy and data security, including extensive qualitative information and text, as well as complex quantitative data. Also, people need to understand and trust how machine learning models make decisions about their mental health involving results, and the process of ML models should be easy to understand for both mental healthcare professionals and patients.

The potential of AI technology in revolutionizing mental healthcare has become progressively indispensable. Despite the advancement of different soft-computing-based techniques such as AI, ML, and Deep Learning (DL) in mental health, there is a limited spotlight on XAI research, which is constantly used in the four major application areas of mental health diagnosis, prognosis, datasets, and public health. This research provides novel AI techniques, such as XAI, and their current applications in diverse mental health disorders. i.e., anxiety disorders, depression, bipolar disorder, and schizophrenia. It reveals the use of AI, employing various data sources like clinical data, EEG, Psychological signals data, Brain imaging such as CT and MRI, novel monitoring systems, and social media platforms, to predict, classify, or subgroup mental illnesses such as anxiety, depression, schizophrenia, bipolar, and suicide ideation. The main focus will be to embrace the diversity of mental health disorders and use AI's novel technique, XAI, to enhance mental health through transparency and interpretability for better understandability. The review delves into four key AI applications within the mental health domain.

- Embracing the Diversity of Mental Health Disorders.
- XAI.
- A Spectrum of existing Datasets.
- ML Innovations.

Through synthesizing existing research, we aim to illuminate the progress and implementation of XAI techniques within this field. We strive to catalyze collaborative efforts to optimize XAI approaches, thereby promoting research-driven translation in clinical settings and ultimately enhancing mental healthcare outcomes. This review's remainder is systematized as given: Sect. 14.2 structured the background of this review. In Sect. 14.3, the methodology part is to be done using the PRISMA guidelines. The literature review is done in Sect. 14.4, and Sect. 14.5 describes the results, existing gaps, findings, and contributions, focusing on challenges and opportunities in employing XAI methods in mental health research, including the diversity of datasets of this review. Section 14.6 finally discusses this study's conclusion and future direction in this crucial mental healthcare domain.

14.2 Background

14.2.1 Mental Disorders

In this study, we have focused on five major mental disorders that are described in Table 14.1 (Fig. 14.1).

14.2.2 ML Techniques

Literature reveals the application of various ML algorithms for diagnosing, classifying, and predicting mental health conditions. Frequently used algorithms include Support Vector Machine (SVM), Random Forest (RF), Linear Regression (LR), Recurrent Neural Network (RNN) based on Long Short-Term Memory (LSTM), Mixed-Effect Random Forest (MERF), Adaptive Boosting (AdaBoost), Extreme Gradient Boosting (XGBoost), Feedforward Neural Networks (FNN), MC Batch Normalization (MCBN), Monte Carlo dropout (MCD), Double Machine Learning (DML), Deep Neural Network (DNN), Convolutional Neural Network (CNN), Large Language Model (LLM), Transformer Language Model (TLM), Bayesian-Optimized Model (BOM), and Multi-Feature Fusion Network (MFFN). Table 14.2 presents the algorithms, along with their features, parameters, strength and practical applications. Figure 14.2 depicts the algorithms used in the articles taken in this study.

Table 14.1 Description of mental health condition and its global impact

Mental health condition	Description	Global impact
Depression	Is linked to mood disorders and 0.8 million suicides annually	5% prevalence globally [21]; 20% experience milder forms; Higher risk in middle-aged individuals. 4th leading cause in 15–19 year olds
Bipolar disorder (BD) and major depressive disorder (MDD)	Recurring episodes of depression affect over 1% globally	MDD seen in 6% globally; BD [26] affects 1%; Substantial disability impact in 2017 and MDD caused 32.8 million disability years; BD caused 9.3 million
Schizophrenia	Serious disorder affecting thinking, feeling, actions	Affects about 3% worldwide [1] and often manifests in late teens/ early adulthood
Anxiety disorder	Onset often before age 15; Various types with distinct characteristics and Factors like sociodemographic, psychosocial, and mental health contribute to the risk	Nearly one in three adolescents [20] is affected. Impact 4% of the global population, affecting 301 million people in 2019, according to WHO
Obsessive-compulsive disorder (OCD), Attention deficit hyperactivity disorder (ADHD), Post-traumatic stress disorder (PTSD)	OCD involves obsessions and compulsions; significantly impacts daily life, ADHD is a common pediatric neurodevelopment disorder, and PTSD develop after experiencing or witnessing a traumatic event, such as natural disasters, serious accidents, terrorist acts, war/ combat, or personal assaults	OCD affects about 50% severely, ranking among the top ten most disabling illnesses by the WHO, the disease's 5th leading cause burden for women aged between 15 and 44 in the world [16], ADHD has a global prevalence of 5% among people aged 18 [14], PTSD affects approximately 3.9% of the global population [22].

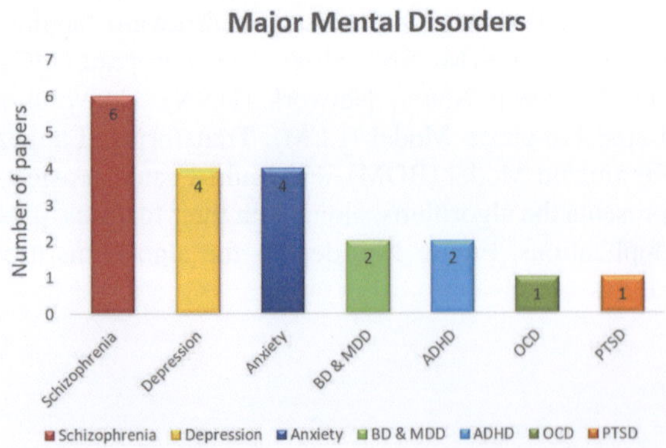

Fig. 14.1 Discussed mental disorders

Table 14.2 Algorithms, along with features, parameters, strengths, and practical applications

Algorithm	Features	Parameters	Strength	Practical applications
LSTM-based RNN	Temporal sequence processing	Number of units, dropout rate, learning rate	Long-term dependency capture	Time series prediction
RF	Ensemble of decision trees	Number of trees, max depth, min samples split	Robust to overfitting	Classification and regression
XGBoost	Gradient-boosted decision trees	Learning rate, max depth, number of estimators	Regularization support	High performance on tabular data
SVM	Classification with maximum margin	Kernel type, C (regularization), gamma	Effective in high dimensions	Small to medium-sized datasets
LR	Binary classification	Regularization (L1, L2), learning rate	Easy to implement	Binary outcome prediction
AdaBoost	Adaptive boosting of weak learners	Number of estimators, learning rate	Combines weak classifiers	Reduces bias and variance
MERF	Random effects with RF	Number of trees, random effects structure	Handles hierarchical data	Longitudinal data analysis
FNN	Basic neural network architecture	Number of layers, number of neurons per layer, activation functions	Flexible architecture	General function approximation
MCD	Bayesian approximation through dropout	Dropout rate, number of forward passes	Uncertainty estimation	Probabilistic predictions
MCBN	Bayesian approximation with batch norm	Batch size, number of forward passes	Uncertainty estimation	Improved training stability
DML	Learn similarity/ dissimilarity measures	The margin for loss function, embedding dimension	Effective for clustering	Similarity-based tasks
BOM	Combines predictions from multiple models	Model weights, number of models	Robustness	Ensemble learning
CNN	Image and spatial data processing	Filter size, number of filters, stride	Spatial hierarchies capture	Image and video recognition
LLM	Natural language processing at scale	Number of layers, hidden size, attention heads	Language understanding	Text generation and comprehension

(continued)

Table 14.2 (continued)

Algorithm	Features	Parameters	Strength	Practical applications
TLM	Advanced NLP with transformers	Number of layers, number of heads, embedding size	Parallel processing	Machine translation
MFFN	Combines multiple data sources	Feature weights, fusion strategy	Handles diverse data types	Multimodal data integration

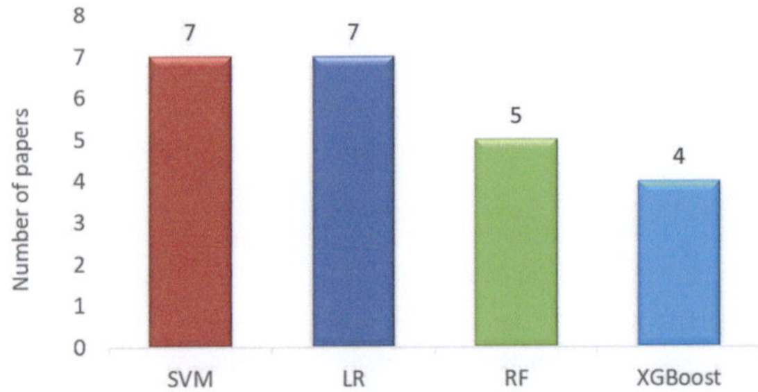

Fig. 14.2 Most used ML techniques in this review

14.2.3 *AI Novel Technique XAI*

14.2.3.1 Why XAI?

When we use an ML model to predict mental disorders based on MRI images and the model predicts that someone has a mental disorder, healthcare professionals might ask, "How did the model arrive at this conclusion? This is where XAI becomes crucial. Without XAI, we face the challenges of justifying the model's predictions, leading to an awkward situation, as shown in Fig. 14.3.

14.2.3.2 Overview of XAI

XAI [5] pertains to the methodologies and approaches applied in the realm of AI technology. The leading objective is to ensure that the outcomes generated by AI solutions can be comprehended by human experts, as shown in Fig. 14.4. This concept starkly contrasts the notion of a "black box" in ML. In a black box system, the decision-making process of AI remains opaque, even to its designers. Specifically, the integration of XAI, incorporating both accuracy and interpretability technologies, holds the potential to enhance the power of decision-making precision of experts, broaden the attainability of services for mental conditions, and address psychological challenges with a major risk of mental issues [24]. This overview explores the current utilization of XAI technology in assessing mental health, specifically the

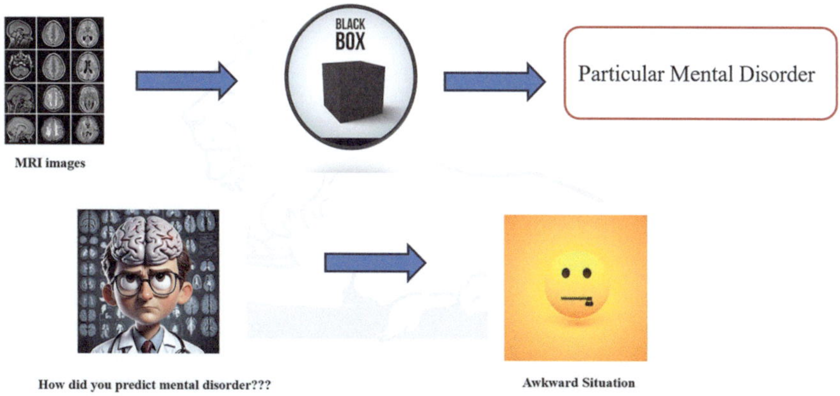

Fig. 14.3 Why explainability?

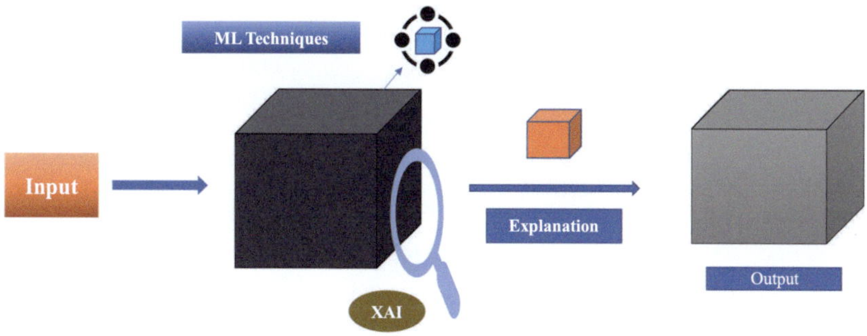

Fig. 14.4 Overview of XAI

deployment of SHAP and Local Interpretable Model-agnostic Explanations (LIME) Table 14.3 describes the major aspects of XAI models used in the review.

14.3 Methodology

14.3.1 PRISMA Guidelines

This study conducted a review following the PRISMA guideline extension to explore XAI and related techniques in mental healthcare. Our search focused on original studies published in English within the last decade (2020–2024), spanning databases like PubMed, Scopus, and Web of Science, supplemented by Google Scholar for peer-reviewed and highly cited studies. The search took place using keywords and Boolean operators ("mental health AND (XAI OR artificial intelligence OR machine

Table 14.3 Major aspects of XAI models used in review

Aspect	SHAP	LIME
Objective	To interpret the output of ML models by quantifying the influence of each input variable	To interpret the predictions of black-box models, especially in image classification scenarios
Features	Considers inter-variable dependencies and calculates both positive and negative influences on predictions	Uses superpixels to assess the significance of regions in an image for black-box model predictions
Parameters	Training data, learned model, input variables	Non-masked and masked superpixels, black-box model, surrogate model
Mathematical basis	Shapley Values are derived by assessing the average change in output concerning the absence or presence of variables	Combines values of input correlations with output probabilities using a surrogate model
Strength	Provides an accurate measure of variable influence, especially in scenarios with interrelated variables	Model-agnostic, interpretable results require fewer resources compared to alternative techniques
Limitation	Relatively slower compared to other feature-importance techniques	Non-deterministic decision boundary may yield different outputs with each call, evaluate one data point at a time
Practical case	Interpreting feature importance in ML models	Explaining individual image classifications in black-box models

learning OR deep learning)"). This approach led to selecting 20 publications, and their literature was reviewed. Data from selected studies were charted, capturing details such as authors, publication year, mental disorders, datasets, techniques used, and outcomes. All studies were summarized, and the results were synthesized based on the predominant techniques employed.

14.3.2 Criteria for Selection of Sources

We retrieved a total of 56 results from Scopus ($n = 32$), PubMed ($n = 13$), and Web of Science ($n = 11$) databases (Figs. 14.5 and 14.6). After removing 24 duplicates, we screened the remaining studies and excluded seven based on some reasons defined in Table 14.4. During the final screening stage, we excluded five additional results, which included two reviews, an abstract, an editorial, and a book chapter. Ultimately, 20 papers were included in this review and categorized according to the primary techniques used: "ML," "DL," and "XAI."

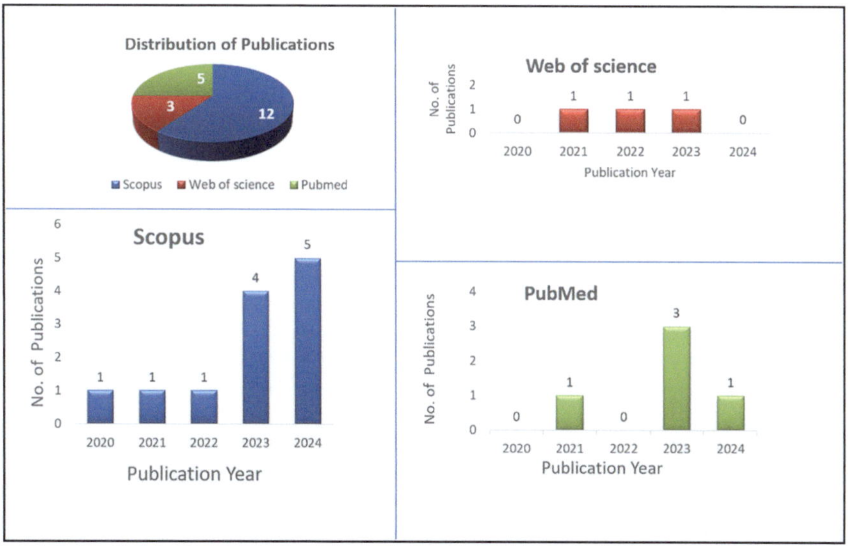

Fig. 14.5 Distribution of publications

14.4 Literature Review

See Table 14.5 presents the literature review of selected 20 papers with their application, XAI methods, ML methods, and datasets.

Uddin et al. [21] used a DL method to predict the symptoms of depression on a large textual dataset from a public online information channel of Norway's young people. The authors' approach involves extracting novel features that capture the potential prodrome of depression, just as measured by experts in medical and psychological areas. Authors RNN techniques, specifically leveraging LSTM, attention mechanisms, and dense layers to model states of emotional. Additionally, for transparency, the study utilizes a state-of-the-art XAI approach, LIME, to elucidate the importance of the ML's feature decisions.

Zhu et al. [26] performed a study to create a diagnostic system using ML, relying on an electronic medical records (EMR) dataset. The goal was to replicate the clinical decision-making of human physicians, specifically in distinguishing patients with MDD from those with BD, especially during BD depressive episodes. The study identified 16,311 patients admitted to a hospital in Western China between 2009 and 2018 diagnosed with either MDD or BD. Creating sub-cohorts with various feature combinations, the Authors applied four ML algorithms (LR, XGBoost, RF, and SVM) along with four train-test splits to develop and endorse diagnostic models. To understand the significance of each feature at both population and individual levels, XAI methods (SHAP and Break Down) were utilized. These methods provided insights into feature importance, connections, and effects on prediction decisions for specific subjects. XGBoost, among various algorithms, demonstrated the most

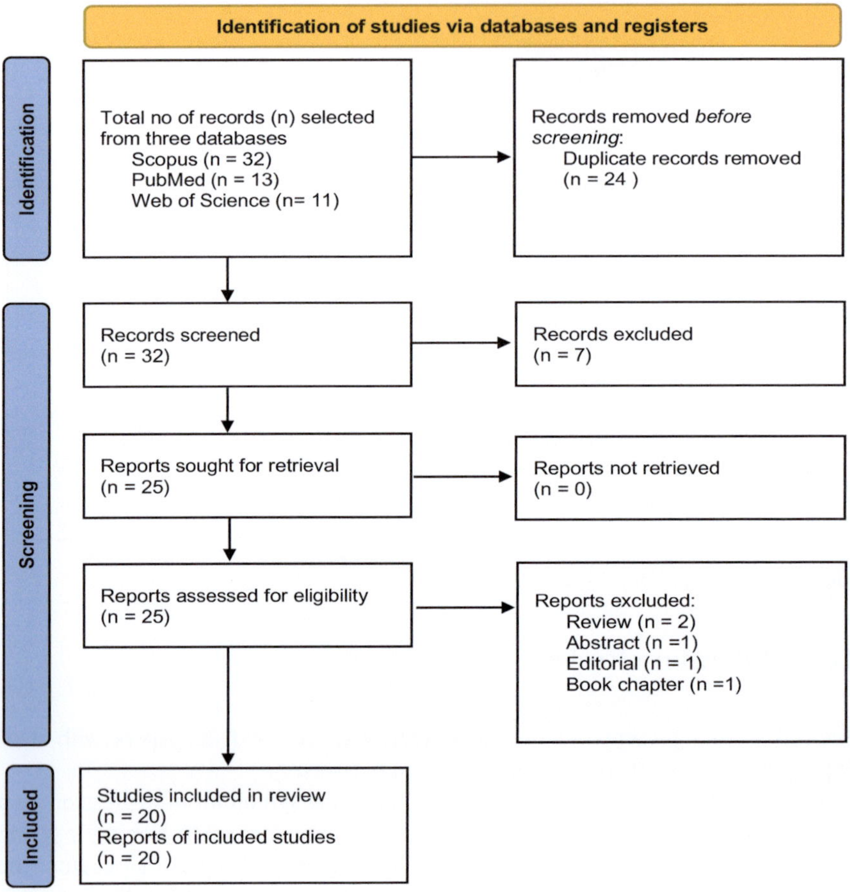

Fig. 14.6 PRISMA guideline flow chart of the review

Table 14.4 Exclusion of 7 studies based on some reasons

Reason for exclusion	Number of papers
Studying using AI techniques without XAI	3
Studying using XAI techniques in some other disease	2
Discussed about CT in lung cancer	2

effective performance in differentiating between patients with BD and MDD. The core predictors encompassed symptoms, age, occupation, and various physiological markers. For BD depressive episodes versus MDD, XGBoost achieved a test AUC of 77%, with PPV at 57% and NPV at 89%. Even after removing self-reported symptoms, the model maintained a substantial AUC of 70% for distinguishing BD from MDD. In datasets without removing patients with comorbidity, the AUC was

Table 14.5 Selected papers to showcase the review of this study

Authors	Application	XAI methods	ML methods	Datasets
Uddin et al. [21]	Prediction of depression	LIME	LSTM based RNN	Textual
Zhu et al. [26]	Differentiation of BD from MDD	SHAP	SVM, LR, RF, XGBoost	EMRs
Arias and Astudillo [2]	Prediction of Schizophrenia	SHAP	SVM, XGBoost, AdaBoost	EEG
Chavanne et al. [5]	Prediction of anxiety	SHAP	RF, SVM, LR	MRI
Lønfeldt et al. [16]	Prediction of OCD episodes	SHAP	RF, LR, FNN, MERF	Physiological signals
Li et al. [13]	Uncovering effects of depression on suicide risk	SHAP	DML	Social media
Wang et al. [24]	Classification of Schizophrenia	SHAP	MFFN	FMRI
Shanarova et al. [19]	Diagnostics of Schizophrenia	SHAP	SVM	EEG
Ul Alam and Kapadia [22]	Classification of PTSD	LAXARY	LR	Social media
Ellis et al. [6]	Classification of Schizophrenia	Saliency, LRP	MCD, MCBN	FMRI, dFNC
Bang et al. [3]	Diagnosis of Schizophrenia	SHAP	BOM	MRI
Kim et al. [9]	Prediction of anxiety	SHAP	XGBoost	FMRI
Loh et al. [14]	Classification of ADHD/CD	Grad-CAM	CNN	EEG
Loh et al. [15]	Detection of ADHD and CD	Grad-CAM	CNN	ECG and EEG
Kaur et al. [8]	Prediction of depression	SHAP, LIME	KNN, LR, RF	"mental health prediction"
Malhotra and Jindal [17]	Prediction of depression	SHAP, LIME	LLMs, TLM	Social media
Chadaga et al. [4]	Diagnosis of anxiety	SHAP, LIME	AdaBoost, LR	SAD dataset
Upadhya et al. [23]	Early detection of stress	SHAP, LIME	LSTM, SVM, CNN	Psychological signals and facial expressions
Zhou et al. [25]	Prediction of MDD	SHAP	SVM, RF, LR	Clinical

82%. This study concludes that the diagnostic system effectively pinpointed patients with BD across diverse clinical situations. The distinct patterns observed in peripheral markers between BD and MDD offer valuable insights into potential underlying pathophysiological mechanisms of these conditions, enhancing our understanding.

Arias and Astudillo [2] explored the application of ML on electroencephalogram (EEG) data for the prediction of 945 patients ranging in age from 18 to 70 years of schizophrenia, comparing three classifiers: SVM, AdaBoost, and XGBoost. While ML models offer high accuracy and real-time analysis, their lack of transparency is a drawback, especially in sensitive fields like health. The study finds that XGBoost emerges as the most effective model, achieving an accuracy of 93% and an F1 score of 0.92, surpassing SVM and AdaBoost. Incorporating SHAP XAI, the study identifies key features, highlighting Gender, IQ, delta T6, and delta Pz waves just as crucial in schizophrenia prediction. The analysis suggests that schizophrenia alters the delta wave in EEG, distinguishing it from other mental illnesses. Additionally, the study emphasizes the importance of class balance for improved ML predictions.

Chavanne et al. [5] focused on Magnetic Resonance Imaging (MRI) data in youth who have identified potential correlates with future anxiety symptoms during adolescence. However, the predictive value of these findings remains uncertain. This study employs ML algorithms, including RF, SVM, and LR, to assess the predictive relevance of gray matter volumes and psychometric scores in identifying prospective clinical anxiety. Examining participants at age 14, with follow-up at ages 18–23 for those with 156 clinical anxiety and 424 healthy controls, the study finds that psychometric features play a crucial role in predicting pooled anxiety disorders, achieving moderate performance with an AUC of 68%. For generalized anxiety disorder (GAD) prediction, MRI regional volumes, particularly the caudate and pallidum, significantly improve performance beyond psychometric features alone. To delve deeper into the contribution of each feature in individual predictions, the Authors employed XAI, version 0.39.0. SHAP utilizes a game theoretic approach, assigning importance values to each feature for an individual prediction. This approach enables visualization, providing insights into how each feature value contributes to the final classification for every participant. In conclusion, this research demonstrates the potential for individual prediction of future clinical anxiety in non-anxious 14-year-olds, with psychometric features being key contributors and specific neuroanatomical data enhancing prediction for GAD.

Lønfeldt et al. [16] focused on OCD and explored the potential of using a wrist-worn biosensor and ML models for real-world monitoring of OCD events. Nine adolescents in the 10–17 age group with varying OCD severity wore the biosensor during lab conditions and in their daily lives for up to 8 weeks. Psychological signals data such as skin temperature, heart rate, and electrodermal activity, recorded OCD patients, age, and OCD symptom rigidity were analyzed. ML models, including LR, RF, FNN, and MERF, were employed to detect OCD episodes just as a binary classification task. Results showed that generalizing across time yielded better performance than over patients. Generalized temporal models trained on numerous patients surpassed personalized models. RF and MERF models demonstrated superior accuracy, holding out 70% accuracy in various cross-validation strategies. Furthermore, to enhance interpretability and understand the importance of features in employee models, the Authors utilized values of SHAP for the random cross-validation models. Positive SHAP values signify which features positively contribute to the model's prediction, indicating an inclination towards classifying an observation as an OCD event. Conversely, negative SHAP values suggest a negative impact on prediction, influencing the model towards classifying an inspection as a non-event. The findings of this study indicate the feasibility of detecting OCD episodes in the daily lives of adolescents by utilizing physiological signals recorded through a wearable biosensor. However, to establish the reliability and effectiveness of this approach, more extensive studies involving larger datasets are required.

Li et al. [13] the aim was to investigate the psychological mechanisms associating depression with suicide risk. The dataset sampled 487,251 posts from 3196 subjects, sourced from Sina Weibo, which is one of the biggest Chinese social platforms, offers a rich source of information from users sharing their usual life and individual feelings in real-time. This provides high ecological validity for assessing the psychological characteristics of the focused population. Analyzing posts from a depression super-topic community, the depression group demonstrated an 18% elevated likelihood of suicidal risk compared to the control group. Employing SHAP values and tree interpreters aided in understanding the final model by clarifying the primary factors contributing to the variability in effects captured by the training data. The passageway from depression to the risk of suicide was revealed through DML using the Python modules DoWhy and EconML. The results showed that a depression group ($N = 3196$) and a control group ($N = 5167$) with standardized rhetorical features and psychological risk scores exhibited a significantly higher suicide risk in the depression group ($t = -30.35$, $d = 0.83$, $p < 0.001$) contrasted to the control group. These findings suggest a positive association between cognitive complexity, physical distress, and suicide risk in depressed populations, offering valuable insights for future research on suicide prevention and treatment for depression in individuals.

Wang et al. [24] concluded a novel approach called MFFN to distinguish patients suffering from 71 schizophrenia patients from 74 healthy controls by utilizing time courses (TC) and functional network connectivity (FNC). The model incorporated a DNN backbone for FNC and a C-RNNAM backbone for TC. Deep SHAP was employed to identify almost all discriminative brain networks. The study validated the effectiveness of this model across two fMRI public datasets, achieving the top classification accuracy (ACC = 87.30%, AUC = 0.9081, SPE = 89.28%, SEN = 85.71%, and F1 = 88.23%) and demonstrating superiority over existing methods. Ablation experiments confirmed that MFF and attention modules enhance classification accuracy. The findings revealed aberrant connections in the default mode and visual network of patients with schizophrenia. In conclusion, this method effectively identifies schizophrenia and provides valuable insights into abnormal brain networks, holding significant clinical applications.

Shanarova et al. [19] focused on ML-based diagnostics for schizophrenia, a major psychiatric disorder with significant impacts on the quality of life. The study utilized classification models applied to event-related potentials (ERPs) from individuals performing a visual cued Go/NoGo task. The study involved a dataset with 132 healthy subjects, including (53 males, whose mean age is 31.8 ± 8.26 and range between 18 and 50) recruited from various locations. They had no neurological history, were not on medication during EEG recording, and completed health and symptom questionnaires in a single session lasting about two and a half hours. To prevent overfitting and reduce computational complexity, sequential feature selection transformation was applied. The best-performing model, SVM, demonstrated a sensitivity and specificity accuracy of 91% and 90.8%, respectively. Furthermore, the XAI SHAP model calculated the importance of global variance for every feature in the SVM model (behavior, SFSB). Early treatment is crucial for mitigating the long-term negative effects of schizophrenia, emphasizing the importance of accurate diagnostics.

Ul Alam and Kapadia [22] studied PTSD classification and its intensity estimation using a trustworthy novel XAI Twitter analysis model. A linguistic analysis based on an explainable inquiry (LAXARY) on a Twitter user's (210 clinical PTSD assessment) dataset was collected through a clinically validated survey tool. LAXARY introduces a novel approach by creating a PTSD Linguistic Dictionary based on clinical surveys, which captures specific patterns in Twitter-based sentiments to estimate scores for each survey question related to PTSD. This modified LIWC model allows for accurate assessment of physiological and psychological markers of PTSD without the need for costly and time-consuming laboratory tests or traditional surveys. Clinicians can rely on LAXARY's results, enhancing trust in PTSD assessment outcomes. Evaluation with real Twitter data demonstrates LAXARY's high accuracy (approximately 89%) in predicting PTSD presence and intensity (mean squared error $\approx$ 1.8).

Ellis et al. [6] explored the transparency of implementing machine learning for a schizophrenia diagnosis. The research utilizes resting-state (fMRI) dynamic functional network connectivity (dFNC). In this study, the dataset was collected from 151 individuals with schizophrenia and from 160 healthy controls. The data is sourced from the California University at Irvine, the California University at Los Angeles, North Carolina University at Chapel Hill, New Mexico University, Iowa University, and Minnesota University. The study integrates confidence estimation using MCD and MCBN with explainability methods such as saliency and Layer-wise Relevance Propagation (LRP). MCD affects model gradients, whereas MCBN does not. The findings reveal the level of participant differences between regular and combined explainability methods used in this study and confidence estimation approaches, highlighting challenges and considerations for transparent neuroimaging for clinical decision support systems. The study aims to provide insights to guide future research in this integrated approach.

Bang et al. [3] investigated the diagnostic potential of radiomic features extracted from the corpus callosum (CC) in schizophrenia. The study utilizes MRI data from 165 participants, including 86 with schizophrenia (SZ) and 79 healthy controls (HCs). Employing T1-weighted and diffusion tensor images, a Biomarker Optimization Model (BOM) achieves an AUC of 0.89, with 80.0% accuracy, 83.3% sensitivity, and 76.9% specificity in the test set. The model, incorporating 30 selected radiomic features, provides an online calculator for clinical probability. SHAP analysis underscores the significance of 2nd-order features from the posterior CC in association with the threat of schizophrenia, underscoring the potential of radionics as biomarkers for a schizophrenia diagnosis.

Kim et al. [9] analyzed machine learning models trained on resting-state brain functional radiomic features to anticipate social anxiety levels in 100 young adults, utilizing an fMRI dataset. The XGBoost model demonstrated superior performance, achieving a balanced accuracy of 77.7% and an F1 score of 0.815. Through the absolute SHAP feature importance method, it was uncovered that, among various brain regions, the left orbitofrontal cortex (OFC) held greater significance. These results highlight the predictive power of machine learning for social anxiety and shed light on the neural underpinnings of this symptom.

Saglam et al. [18] used machine learning algorithms on white matter tracts (WMT) obtained from diffusion tensor imaging (DTI) to distinguish between early-onset bipolar disorder (EBD) and early-onset schizophrenic (EOS). DTI is an MRI technique used to forecast the brain's axonal integrity. 32 patients who met up with the diagnostic benchmark for early-onset bipolar disorder (BD-I) and 43 patients with an EOS diagnosis participated in the study. Nine significant WMTs were found by the XGBoost algorithm for predictive power, with SVM performing the best of the models. Higher values in certain WMTs were linked to a higher likelihood of EOS, according to interpretability with SHAP. According to the findings, machine learning models based on major WMTs' fractional anisotropy values can identify microstructural aberrations.

Zhou et al. [25] interrogates the objective of predicting the eight-week treatment outcome among patients with MDD taking selective serotonin reuptake inhibitors (SSRIs) using ML models. Data from 400 patients were employed, and features were chosen through the application of the least absolute shrinkage and selection operator (LASSO). Three ML algorithms were utilized, including SVM, RF, and LR, with XAIs' SHAP values used for interpretation. The remission rate was 67.8%, and SVM, with 13 selected features, achieved an accuracy of 74.49%. Neurocognitive functions and symptoms of anxiety and hypochondriasis were identified as important predictors. The study concludes that machine learning with readily available variables can accurately forecast the treatment outcome of SSRIs over an 8-week period for individuals with MDD.

Loh et al. [14] developed ADHD/CD-NET, a DL system using EEG signals to distinguish between ADHD, ADHD+CD, and CD. They employed a novel preprocessing method involving channel-wise continuous wavelet transform (CWT) correlation matrices from 12-channel EEG segments. The CNN model achieved high classification accuracies of 93.70% for internal data and 98.19% for external validation, demonstrating robust performance in identifying these neurodevelopmental disorders. Utilizing Grad-CAM, the system also highlighted critical EEG channels contributing to diagnostic outcomes, offering objective insights for clinicians. This approach marks a significant advancement in EEG-based diagnosis of ADHD and related disorders, supporting more accurate clinical assessments.

Loh et al. [15] introduced ADHD/CD-NET, a deep learning system using EEG signals to distinguish ADHD, ADHD+CD, and CD. Their method involved transforming 12-channel EEG segments into channel-wise CWT correlation matrices, which a CNN model used for classification. Achieving 93.70% accuracy on internal data and 98.19% on external validation, the system effectively differentiated these disorders, supported by Grad-CAM for highlighting critical EEG channels. They aim to expand their model to include more diverse datasets and additional physiological signals like ECG, potentially enhancing diagnostic capabilities in clinical settings.

Kaur et al. [8] explored the prediction of mental illness through a comprehensive study using ML techniques. Their research focused on standardizing the diagnostic process for depression, covering data extraction, preprocessing, ML model training, classification, and performance evaluation. They employed six ML algorithms, including KNN, LR, Gaussian classifier, RF, decision tree, and LR, with the Gaussian classifier achieving the highest accuracy of 91%. The study emphasized the use of XAI techniques like SHAP and LIME to interpret model decisions, highlighting features crucial for predicting mental health disorders. The findings suggest scalability in applying ML models to real-time mental health data and potential advancements through deep learning approaches in future research.

Malhotra and Jindal [17] explored the application of Transformer-based language models (LLMs) for detecting mental health disorders like depression and suicidal behavior using user-generated content (UGC) from social networks. They emphasized the challenge of interpreting predictions from complex LLMs, which are typically opaque "black-box" models. To address this, they employed model-agnostic XAI techniques such as SHAP and LIME to provide insights into how these models make decisions. The study involved six pre-trained LLMs, including domain-adapted models like MentalBERT and PsychBERT, trained on datasets from Twitter and Reddit. Results demonstrated that XAI techniques can enhance understanding of model decisions and highlight data quality issues, offering potential improvements for robust and accurate mental health diagnostics in real-world applications.

Chadaga et al. [4] conducted a study on Social Anxiety Disorder (SAD), a condition characterized by a persistent fear of negative judgment in social situations. They used demographic, emotional, and physical symptoms with various ML techniques to diagnose SAD, achieving the highest accuracies of 88% with AdaBoost and LR models. XAI techniques like SHAP, LIME, Eli5, and QLattice were employed to enhance the interpretability of predictions, highlighting critical diagnostic attributes such as the Liebowitz Social Anxiety Scale questionnaire and fear of public speaking. This framework supports clinical decision-making in settings like schools, hospitals, and workplaces, aiding early identification and treatment of SAD.

Upadhya et al. [23] introduced a novel framework for stress detection that integrates physiological signals and facial expressions using advanced ML techniques. Their study highlighted the superiority of LSTM networks over SVMs in capturing complex temporal patterns of stress, achieving significant improvements in accuracy, precision, recall, and F1 scores. VGG16 also emerged as highly effective for facial expression analysis, setting a new standard with 87% test accuracy. This research advances stress classification methods and emphasizes the transformative potential of multimodal data integration in understanding and managing stress. Incorporating XAI tools like LIME, SHAP, and Permutation Importance enhanced the interpretability of their models, addressing biases and promoting trustworthiness in stress assessment systems, thus contributing to improved mental health and well-being.

14.5 Results

14.5.1 Source and Types of Datasets Used in This Review

A. Textual Data

A large text-based dataset [21] from the Norwegian information website, ung.no was obtained to detect symptoms of mental health issues. On ung.no, Norwegian youth can anonymously post questions about their challenges, which professionals like doctors and psychologists answer. These questions and responses are publicly available. The focus was on the "Mental health and emotions" category. From the entire collection, 11,807 texts were extracted for initial experiments: 1,820 described depression symptoms, and 9,987 did not. A second dataset included 21,470 texts, with 1,470 depression-related and 20,000 non-depression-related texts. Despite their brevity, the texts often describe the factors leading to mental states and the resulting symptoms and behaviors.

B. EMRs Data

EMR-based diagnostic data [26] from West China Hospital (WCH) can accurately identify bipolar disorder (BD) patients and control subjects, outperforming clinical diagnostic interviews. This study, part of a National Key Research and Development Program of China project, used a dataset of 36,780 admission records from 21,964 patients diagnosed with "depression," "mania," or "bipolar disorder" between January 1, 2009, and December 31, 2018. WCH, a tertiary hospital affiliated with Sichuan University in Chengdu, serves patients from southwest China. The dataset included four types of patient features at admission, commonly available in EMRs: (1) sociodemographic information (e.g., age, gender, marital status, job, ethnicity) and medical history (e.g., surgery, allergy, blood transfusion, medication use, smoking, alcoholism), (2) vital signs (e.g., pulse, breath, temperature, blood pressure), (3) clinical laboratory tests (e.g., blood, urine, stool, biochemical examinations), and (4) chief complaints. The first three categories were structured data, while chief complaints were unstructured and processed separately. All variables were combined into one data frame for the classification model.

C. EEG Data

The study [2] utilizes the "Identifying Psychiatric Disorders Using ML (Dataset)," published in 2020, with data from resting-state evaluations at the Seoul Metropolitan Government-Seoul National University Boramae Medical Center (SMG-SNU) from January 2011 to December 2018. The dataset includes data from 945 subjects aged 18 to 70, categorized into seven psychiatric disorders: Addictive disorder, Trauma, and stress-related disorder, Mood disorder, Healthy control, Obsessive–compulsive disorder, Schizophrenia, and Anxiety disorder. Subjects with neurological disorders, brain injuries, neurodevelopmental disorders (e.g., IQ below 70, tic disorder, ADHD), or any neurocognitive disorder were excluded. Data were collected during a 5-min resting state with eyes closed, using either 19 or 64 channels at a sampling rate of 500–1000 Hz, maintaining electrode impedances below 5 kiloohms with conductive gel.

The study [19] included data from 132 healthy subjects (53 males, mean age 31.8 ± 8.26, range 18–50) recruited from the HBImed Database Collection Project in Switzerland, Saint Petersburg State University, and the N. P. Bechtereva Institute of the Human Brain RAS. None had neurological histories or systemic medical illnesses and were on medication during EEG recording. Subjects completed questionnaires (Brief Symptom Inventory, Health History, and Current Symptoms Scales) and scored below clinical significance on symptom checklists. The schizophrenic group comprised 68 patients (47 males, mean age 30.6 ± 7.17, range 18–50), diagnosed according to ICD-10 criteria at least six months prior and exhibiting moderate to severe social and cognitive impairment. Positive and negative symptom severity was assessed using PANSS, with mean scores of 20.1 ± 4.4 (negative), 26.1 ± 4.6 (positive), and 47.9 ± 5.9 (general psychopathology). Eighteen patients were on atypical antipsychotics, 19 on typical antipsychotics, and 31 were not on neuroleptics. The groups were similar in age, gender, handedness, and education.

This study's [14] private EEG data were sourced from a clinical trial approved by the Domain Specific Review Board (DSRB) of the National Healthcare Group (NHG) in Singapore (DSRB 2008/00410). The trial aimed to assess the effects of omega-3 supplements on reducing aggression in 123 participants aged 7–16 from the Child Guidance Clinic in Singapore. Participants were divided into three groups: CD only (16 children), ADHD only (45 participants), and ADHD+CD (62 participants), based on DSM-IV-TR criteria. Parents completed a Diagnostic Interview Schedule for Children (DISC). EEG data were collected with MP150 single-channel EEG 100c biopotential amplifiers and Acknowledge software, using 12 channels (Fp1, Fp2, F3, F4, P3, P4, O1, O2, F7, F8, T3, and T4) at a 500 Hz sampling rate. Participants rested with eyes open for 3 min, and the data were anonymized and de-identified. EEG signals were segmented into 21.25-s epochs, yielding 128 CD samples, 360 ADHD samples, and 496 ADHD+CD samples.

D. MRI Data

Data used in this study [5] from the IMAGEN database includes neuroimaging and questionnaire data from adolescents aged 14 onwards. Consent was obtained, and diagnostic assessments using DAWBA were conducted at ages 14, 18–19 (FU1), and 22–23 (FU2). Participants were assessed for alcohol and cannabis use, emotional functioning, and other factors. Those with usable MRI scans and complete DAWBA data were included, while those with high alcohol use or certain baseline anxiety disorders were excluded. Participants were categorized into groups based on anxiety diagnoses, including future onset (FUA) and baseline (BLA) anxiety groups. Controls had no diagnosed disorders and were selected to balance demographics with the anxiety groups.

This study [24] introduces an MFFN to differentiate between schizophrenia patients and healthy controls. The framework involves preprocessing resting fMRI data using SPM and performing independent component analysis (ICA) to select 50 representative brain networks. Time courses (TC) and functional network connectivity (FNC) were calculated from these networks. A DNN backbone learned FNC features and a C-RNNAM network with an attention module learned TC features. Feature maps from both were concatenated for classification. Deep SHAP was used to identify the most discriminative brain networks. The dataset includes COBRE (71 schizophrenia patients, 74 healthy subjects) and CNP (50 schizophrenia patients, 121 healthy subjects), both containing resting fMRI data collected using different imaging protocols.

This study [6] utilized the FBIRN dataset, comprising resting-state fMRI data from 151 individuals with SZ and 160 HCs. The data were collected across multiple sites, including California, North Carolina, New Mexico, Iowa, and Minnesota Universities. Approved by respective Internal Review Boards, imaging was conducted using 3T Siemens scanners at six sites and a 3T General Electric scanner at one site. The T2*-weighted EPI sequence recorded brain activity with participants' eyes closed. Participants diagnosed with schizophrenia were recruited from the Department of Psychiatry, CHA Bundang Medical Center in South Korea. They met DSM-IV-TR or DSM-5 criteria for schizophrenia, had no history of other psychiatric disorders, and were either antipsychotic-naïve or free from antipsychotics for at least six months. Healthy controls (HCs) were recruited from the local community through advertisements, ensuring they had no personal or family history of psychiatric disorders. Exclusion criteria included neurological disorders, substance-related issues, and head trauma with loss of consciousness, left-handedness, and MRI contraindications.

The study [3] enrolled 165 participants (86 with schizophrenia and 79 HCs), divided into training ($N = 115$; 60 with schizophrenia and 55 HCs) and test ($N = 50$; 26 with schizophrenia and 24 HCs) sets. Young adults were recruited via internet advertisements and screened by a psychiatrist to exclude those with major psychiatric disorders, inability to undergo MRI, significant medical or neurological conditions, pregnancy, or left-handedness. Only participants not undergoing psychiatric treatment were included. After initial screening, 125 participants were enrolled, with nine later excluded due to withdrawal or MRI artifacts.

The study [9] included a total of 116 participants, all of whom provided informed consent approved by the Institutional Review Board of Yonsei University Gangnam Severance Hospital (IRB approval No. 3-2017-0046, March 29, 2017), following the Declaration of Helsinki guidelines.

MRI data used in this study [18] were conducted using a Siemens AERA 1.5 T scanner with an 8-channel head coil in a hospital radiology department. Participants used MRI-compatible pads and earplugs to minimize motion artifacts. The imaging protocol included high-resolution T1-weighted MPRAGE, FLAIR, T2-weighted turbo spin echo sequences, and diffusion tensor imaging (DTI). T1-weighted MPRAGE parameters were TR = 2200 ms, TE = 3.1 ms, flip angle = 8°, voxel size = $1 \times 1 \times 1$ mm^3, and 160 sagittal slices. DTI utilized a single-shot spin-echo EPI sequence with TR = 9100 ms, TE = 106 ms, voxel size = $2 \times 2 \times 2$ mm^3, 60 slices, and b-value = 1000 s/mm^2 across 64 diffusion-weighted directions, including $b = 0$ s/mm^2 images for distortion correction. FLAIR and T2-weighted sequences were employed to exclude parenchymal pathologies.

E. Physiological Signals Data

Participants aged 10–16 with diagnosed OCD were recruited from a child and adolescent mental health center in Denmark. Inclusion criteria required a primary or secondary OCD diagnosis, CY-BOCS severity score of 7 or higher, and normal intellectual function. Exclusion criteria included substance dependence, schizophrenia, psychotic disorders, and other conditions hindering wrist biosensor wear. On average, nine adolescents (five females, four males) participated, aged 12.3. Their OCD severity scores ranged from mild to moderately severe. Two participants had Asperger syndrome and asthma. During the study [16], some received psychosocial treatment or sertraline for OCD. Participants wore an E4 wristband measuring BVP, skin temperature, EDA, and calculating HR.

The physiological signals in this study [23] were initially processed by separating them from emotional features. Standardization using the StandardScaler function from scikit-learn was applied to ensure consistent feature scales. The dataset was then split into training, validation, and test sets, with 90% used for training and validation and the remaining 10% reserved for testing in LSTM models, whereas a split ratio of 80–20 was used for SVM. For facial expression data, pixel values were extracted and normalized to a range between 0 and 1, typical for image data preprocessing. Emotion labels were encoded numerically and then one-hot encoded for categorical classification purposes. The dataset was divided into training, validation, and test sets, with 10% allocated for testing and the remaining 90% further divided into 80% for training and 10% for validation. It is important to note that since the original dataset contained grayscale images, they were converted to 3 channels as required by some models like VGG. However, this step was unnecessary for custom CNN models.

F. Social Media Data

Data for this study [13] were gathered from Sina Weibo, a platform with over 586 million active daily users as of December 2022. The study focused on user profiles, network interactions (like forwarding and replying), and posts from Weibo. All data were anonymized to protect privacy and followed the ethical guidelines [10] outlined. Approval was obtained from the Institutional Review Board (IRB) of

Central China Normal University (IRB number: CCNU-IRB-202211010). Participants were selected from the depression super-topic community (DSTC), the largest online depression community on Weibo. Initially, 11,142 posts were retrieved and manually reviewed by three researchers. They identified 3,196 participants clinically diagnosed with depression based on their posts. These participants' Weibo accounts were examined, and 487,251 posts were collected from their profiles. For comparison, 5167 Weibo users without any mental illness diagnoses were randomly selected, contributing 357,939 posts as the control group for the study.

The study [22] utilized automated regular expression searches on Twitter to identify potential veterans with PTSD, refining the list through manual review. Initially, keywords related to PTSD were used to identify users self-claiming a diagnosis. We visited the Twitter accounts of veterans' organizations and applied inclusion criteria: users self-identified as veterans and had at least 25 tweets in the past week. After identifying initial users, we manually verified 685 tweets, where individuals self-identified with PTSD, then retrieved last week's tweets for these users using the Twitter API, filtering out users with fewer than 25 tweets or whose tweets were less than 75% in English. This process resulted in 305 positive examples. For negative examples, we randomly selected 3,000 veteran users meeting the inclusion criteria, filtering to 2,423 users after the same criteria. A survey was distributed among 1,200 users (305 self-identified PTSD sufferers and 895 randomly selected others), resulting in 210 responses. Among these, 92 users were clinically diagnosed with PTSD, including 17 who were not self-identified. Conversely, 118 users were diagnosed with no PTSD, with 7 self-identified users assessed as not having PTSD. Response rates were 27% for PTSD-positive users and 12% for PTSD-negative users. In summary, collected one week of tweets from 2,728 veterans, including 305 self-claimed PTSD sufferers. Survey responses provided data on 210 veterans, indicating diagnoses of PTSD and severity levels based on assessment tools.

The study [17] used four publicly available, anonymized social network datasets focused on depression and suicide detection. Dataset 1 consists of 11,000 depression and suicide-related tweets from Twitter, annotated with positive and control classes. Dataset 2 includes over 200,000 posts from Reddit's r/SuicideWatch, r/Depression, and r/Teenagers subreddits, labeled as suicidal or non-suicidal. Dataset 3 comprises 17,159 posts from various Reddit subreddits, distinguishing between mental health-related (depression) and non-related topics. Dataset 4 features 1,895 posts from Reddit's r/SuicideWatch and r/Depression, categorized based on the severity of suicidal tendencies. All datasets were used for experimental research purposes, and no identifying user information was included.

G. Clinical Data

Out of 527 initially enrolled patients, 400 completed at least one post-baseline assessment. During the study [25] 95 patients were lost to follow-up, and 32 patients were excluded due to receiving combination therapy. Ten patients missed their week 2 visit but attended at week 4. The remitter or non-remitter status at week 8 was determined using the last observation carried forward (LOCF) method, ensuring robust conclusions.

H. Mental Health Prediction Dataset

The study [8] emphasizes the importance of understanding and ensuring the integrity of data sources before using them to build reliable models. It acknowledges that data reliability can vary, necessitating careful ethical considerations. To proceed, the study utilized the "mental health prediction" dataset from Kaggle (https://www.kaggle.com/code/kairosart/machine-learning-for-mental-health-1), containing 334 rows and 31 columns. This dataset covers various mental illnesses and their associated symptoms, crucial for diagnosing mental health conditions. The primary aim is to explore connections between mental health, occupations, and education levels.

I. SAD Data

The SAD dataset used in this study [4] shared by Fathi et al. [7] on Mendeley Data, consists of records from 214 patients, including 111 diagnosed as SAD positive and 103 as SAD negative. It comprises 30 attributes, combining four continuous variables and 27 categorical variables. The target variable "hasSAD" indicates the diagnostic outcome, with values 0 for SAD negative and 1 for SAD positive. The dataset initially included empty values, and the "ID" attribute was excluded as it did not contribute to the machine learning models.

14.5.2 Existing Gaps in Literature Review

The study [26] is constrained by the significant class imbalance between BD and MDD patients (1% vs. 6% prevalence), impacting predictive accuracy and clinical applicability. Feature selection relying solely on statistical hypothesis testing may limit insights compared to more comprehensive methods like wrapper techniques. Important biological markers (e.g., lipid profiles), behavioral data (e.g., from mobile devices), neuroimaging, and genetic risk factors were not considered, potentially reducing diagnostic precision and understanding. The model's clinical utility remains preliminary and requires validation through prospective studies. Integrating EMR-based clinical decision support systems is more feasible than models requiring additional data collection outside EMRs.

The study's [5] sample sizes for anxiety disorder groups were limited due to the long follow-up interval, making separate predictions for specific anxiety disorders impractical. This limitation constrained the model's ability to generalize beyond the available data and increased the risk of overfitting. Nested feature selection methods were avoided to mitigate overfitting risks associated with the small number of participants with GAD or mixed anxiety disorders (mAD). This approach may have limited the exploration of potentially informative features that could enhance predictive accuracy. The IMAGEN cohort, recruited from the general population rather than clinical settings, lacked targeted clinical constructs specific to anxiety disorders at age 14, such as severity assessments using instruments like the Liebowitz Social Anxiety Scale (LSAS). This design choice may have limited the study's ability to capture nuanced clinical features relevant to anxiety onset and progression.

The study's [16] small sample size limits the ability to draw definitive conclusions from statistical analyses or machine learning model performance, emphasizing feasibility over broader applicability. The absence of systematic data collection on skin tone introduces potential measurement errors with PPG sensors, particularly affecting individuals with darker skin tones. Future studies should address this limitation to ensure data accuracy and fairness across diverse populations.

Retrospective study [3] from a single institution with a relatively small sample size limits generalizability. Larger datasets and external validation are necessary to confirm the radiomics model's diagnostic utility for schizophrenia. MRI data acquisition after initiation of antipsychotic medication introduces potential confounding effects despite the short mean duration of medication. The impact of antipsychotics on brain imaging findings remains contentious in this study.

The study [9] included fMRI data from over 100 young adults, the largest in related machine learning studies, but larger samples may bolster results in the future. Demographic factors like age and gender are not considered during training, potentially influencing resting-state brain functional radiomic features. Multi-site neuroimaging data was used, with efforts to mitigate side effects, but potential limitations remain.

The study [18] lacks healthy controls and longitudinal data to assess treatment effects on white matter integrity. DTI-based ML model shows promise in differentiating early-onset schizophrenia and BD, but external validations with larger datasets are necessary. Low MRI resolution (1.5 T) is a limitation affecting image quality. Small sample sizes and the absence of hyperparameter tuning methods may affect reproducibility and model optimization.

The small sample size in the study of Loh et al. [15] and class imbalance, especially in CD classes, compromise the model's generalizability. The lack of established ECG biomarkers for ADHD limits the ability to diagnose ADHD patients reliably using current methods. The absence of a control or non-ADHD/CD sample hinders testing the model's fit in predicting the absence of the conditions. Incorporating XAI into the model could identify prospective ECG biomarkers for ADHD and CD, motivating further research in biomarker discovery. Pilot studies are crucial for identifying potential ECG biomarkers, paving the way for larger datasets to validate diagnostic capabilities in clinical and non-clinical settings.

Reliance on self-reported data in the study [17] from online forums and lexicon-based annotations may introduce false positives and lack contextual nuances. Future plans include implementing data sanity checks, semi-supervised learning, and Generative Adversarial Networks to improve dataset quality. Current datasets lack a distinction between depression and suicide, requiring future research with larger, expert-annotated datasets. Exploration of XAI techniques to correlate word feature scores with clinical concepts and automate questionnaire scoring is planned. Further applications of XAI include analyzing multiclass, multilabel data, and identifying biases in mental health contexts like pandemics or economic crises.

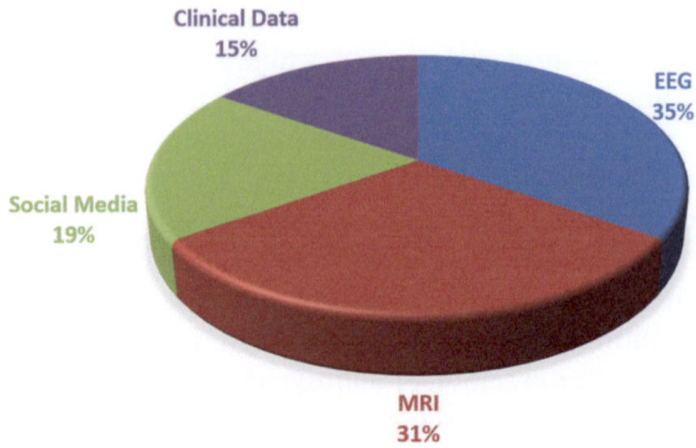

Fig. 14.7 Datasets used in the review

14.5.3 Findings and Contribution

In this research, we found the most common various datasets related to major mental disorders, as shown in Fig. 14.7. These datasets are then used with earlier-defined ML techniques for different tasks such as classification and prediction and then frequently deployed explainable techniques like SHAP and LIME to understand the significance of each feature of ML algorithms. This study also investigates two novel ML methods:

i. MFFN is used to classify schizophrenia disorder and recognize abnormal brain networks.
ii. Saliency and Layer-wise relevance propagation (LRP) stand out as pioneering gradient-based feature attribution methods in the province of explainability.
iii. LAXARY, a novel XAI method for filling out mental disorders like that of PTSD using social media posts such as Twitter. By leveraging the trust and clarity of clinical surveys, this approach will help clinicians detect disorder early.

14.6 Conclusion and Future Directions

The present study aimed to see how XAI impacts different mental health issues, what techniques and data it uses, and the smart tools like SVM, XGBoost, RF, and LR it employs across various data types. The findings showed that XAI is versatile, working with hospital data such as EEG and MRI, clinical records like psychological data, and other data to address different mental health conditions. This review gives a big picture of how XAI is making a difference in mental health care. It stresses the importance of clear and transparent AI predictions and suggests that making AI understandable is crucial for future mental well-being research. But there are challenges, too, like fairness, small size issues in the datasets, and some limits in how AI can be used. Overall, it shows how AI is making a big difference in mental health care but also points out things that need more work in the future as this research review can be extended as a comprehensive review taken by more mental health disorders, datasets, ML techniques, and employment of XAI methods for the Interpretability of ML models.

For future scope, this review highlights some directions to integrate DL with traditional ML methods to classify schizophrenia cases using EEG data while implementing a comprehensive suite of XAI techniques to clarify the decision-making processes of DL models in schizophrenia diagnosis. Collaborating closely with healthcare specialists will ensure these models' clinical relevance and applicability. The study will also investigate specific regional gray matter features, such as striatal volumes, to understand their role in developmental anxiety and refine diagnostic predictions. To minimize false positives in anxiety disorder screenings, strategies will be developed, potentially incorporating control participant data in training sets. Advanced time-series modeling approaches will be adopted to enhance the detection accuracy of OCD events in EEG recordings, drawing insights from previous studies on aggression in adolescents with autism. The research will explore the underlying mechanisms linking anxiety and suicidal behaviors to identify new intervention and treatment targets while addressing ethical considerations regarding medication's impact on EEG recordings in clinical settings. Methods will be employed to extract latent components of Event-Related Potentials (ERPs) to improve the localization and interpretation of brain signals in EEG studies. Lastly, multi-modal data fusion strategies for stress classification will be investigated, integrating various datasets and modalities to enhance predictive accuracy and model interpretability.

References

1. Arias, J.T., Astudillo, C.A.: Enhancing Schizophrenia prediction using class balancing and SHAP explainability techniques on EEG data. In: 2023 IEEE 13th International Conference on Pattern Recognition Systems, ICPRS 2023 (2023). https://doi.org/10.1109/ICPRS58416.2023.10179002
2. Arias, J.T., Astudillo, C.A.: Enhancing Schizophrenia prediction using class balancing and SHAP explainability techniques on EEG data. In: 2023 IEEE 13th International Conference on Pattern Recognition Systems, ICPRS 2023. IEEE (2023)
3. Bang, M., Eom, J., An, C., et al.: An interpretable multiparametric radiomics model for the diagnosis of schizophrenia using magnetic resonance imaging of the corpus callosum. Transl. Psychiatry **11** (2021). https://doi.org/10.1038/s41398-021-01586-2
4. Chadaga, K., Prabhu, S., Sampathila, N., et al.: SADXAI: predicting social anxiety disorder using multiple interpretable artificial intelligence techniques. SLAS Technol. **29**, 100129 (2024). https://doi.org/10.1016/j.slast.2024.100129
5. Chavanne, A.V., Paillère Martinot, M.L., Penttilä, J., et al.: Anxiety onset in adolescents: a machine-learning prediction. Mol. Psychiatry **28**, 639–646 (2023). https://doi.org/10.1038/s41380-022-01840-z
6. Ellis, C.A., Miller, R.L., Calhoun, V.D.: Towards greater neuroimaging classification transparency via the integration of explainability methods and confidence estimation approaches. Inform. Med Unlocked **37**, 101176 (2023). https://doi.org/10.1016/j.imu.2023.101176
7. Fathi, S., Ahmadi, M., Birashk, B., Dehnad, A.: Development and use of a clinical decision support system for the diagnosis of social anxiety disorder. Comput. Methods Programs Biomed. **190**, 105354 (2020). https://doi.org/10.1016/j.cmpb.2020.105354

8. Kaur, I., Kamini, K.J., et al.: Enhancing explainability in predicting mental health disorders using human–machine interaction. Multimedia Tools Appl (2024). https://doi.org/10.1007/s11042-024-18346-1
9. Kim, B.H., Kim, M.K., Jo, H.J., Kim, J.J.: Predicting social anxiety in young adults with machine learning of resting-state brain functional radiomic features. Sci. Rep. **12** (2022). https://doi.org/10.1038/s41598-022-17769-w
10. Kosinski, M., Matz, S.C., Gosling, S.D., et al.: Facebook as a research tool for the social sciences: opportunities, challenges, ethical considerations, and practical guidelines. Am. Psychol. **70**, 543–556 (2015). https://doi.org/10.1037/a0039210
11. Kuhathasan, N., Ballester, P.L., Minuzzi, L., et al.: Predictors of perceived symptom change with acute cannabis use for mental health conditions in a naturalistic sample: a machine learning approach. Compr. Psychiatry **122**, 152377 (2023). https://doi.org/10.1016/j.comppsych.2023.152377
12. Kwok, W.H., Zhang, Y., Wang, G.: Artificial intelligence in perinatal mental health research: a scoping review. Comput. Biol. Med. **177**, 108685 (2024). https://doi.org/10.1016/j.compbiomed.2024.108685
13. Li, S., Pan, W., Yip, P.S.F., et al.: Uncovering the heterogeneous effects of depression on suicide risk conditioned by linguistic features: a double machine learning approach. Comput. Human Behav. **152**, 108080 (2024). https://doi.org/10.1016/j.chb.2023.108080
14. Loh, H.W., Ooi, C.P., Oh, S.L., et al.: ADHD/CD-NET: automated EEG-based characterization of ADHD and CD using explainable deep neural network technique. Cogn. Neurodyn. (2023). https://doi.org/10.1007/s11571-023-10028-2
15. Loh, H.W., Ooi, C.P., Oh, S.L., et al.: Deep neural network technique for automated detection of ADHD and CD using ECG signal. Comput. Methods Programs Biomed. **241**, 107775 (2023). https://doi.org/10.1016/j.cmpb.2023.107775
16. Lønfeldt, N.N., Olesen, K.V., Das, S., et al.: Predicting obsessive-compulsive disorder episodes in adolescents using a wearable biosensor—a wrist angel feasibility study. Front. Psychiatry **14** (2023). https://doi.org/10.3389/fpsyt.2023.1231024
17. Malhotra, A., Jindal, R.: XAI transformer based approach for interpreting depressed and suicidal user behavior on online social networks. Cogn. Syst. Res. **84**, 101186 (2024). https://doi.org/10.1016/j.cogsys.2023.101186
18. Saglam, Y., Oz, A., Yildiz, G., et al.: Can diffusion tensor imaging have a diagnostic utility to differentiate early-onset forms of bipolar disorder and schizophrenia: a neuroimaging study with explainable machine learning algorithms. Psychiatry Res. Neuroimaging **335**, 111696 (2023). https://doi.org/10.1016/j.pscychresns.2023.111696
19. Shanarova, N., Pronina, M., Lipkovich, M., et al.: Application of machine learning to diagnostics of schizophrenia patients based on event-related potentials. Diagnostics **13**, 1–16 (2023). https://doi.org/10.3390/diagnostics13030509
20. Tomasi, J., Lisoway, A.J., Zai, C.C., et al.: Towards precision medicine in generalized anxiety disorder: review of genetics and pharmaco(epi)genetics. J. Psychiatr. Res. **119**, 33–47 (2019). https://doi.org/10.1016/j.jpsychires.2019.09.002
21. Uddin, M.Z., Dysthe, K.K., Følstad, A., Brandtzaeg, P.B.: Deep learning for prediction of depressive symptoms in a large textual dataset. Neural Comput. Appl. **34**, 721–744 (2022). https://doi.org/10.1007/s00521-021-06426-4
22. Ul Alam, M.A., Kapadia, D.: LAXARY: a trustworthy explainable Twitter analysis model for post-traumatic stress disorder assessment. In: 2020 IEEE International Conference on Smart Computing (SMARTCOMP), pp. 308–313 (2020). https://doi.org/10.1109/SMARTCOMP50058.2020.00069
23. Upadhya, J., Poudel, K., Ranganathan, J.: A comprehensive approach to early detection of workplace stress with multi-modal analysis and explainable AI. In: SIGMIS-CPR 2024: Proceedings of the 2024 Computers and People Research Conference (2024). https://doi.org/10.1145/3632634.3655878
24. Wang, C., Wang, C., Ren, Y., et al.: Multi feature fusion network for schizophrenia classification and abnormal brain network recognition. Brain Res. Bull. **206** (2024). https://doi.org/10.1016/j.brainresbull.2023.110848

25. Zhou, S., Ma, Q., Lou, Y., et al.: Machine learning to predict clinical remission in depressed patients after acute phase selective serotonin reuptake inhibitor treatment. J. Affect. Disord. **287**, 372–379 (2021). https://doi.org/10.1016/j.jad.2021.03.079
26. Zhu, T., Liu, X., Wang, J., et al.: Explainable machine-learning algorithms to differentiate bipolar disorder from major depressive disorder using self-reported symptoms, vital signs, and blood-based markers. Comput. Methods Programs Biomed. **240**, 107723 (2023). https://doi.org/10.1016/j.cmpb.2023.107723

Chapter 15
PCOS Diagnosis: Data-Driven Predictions and Workflow Enhancement

Shivarth Rai, Medhaj Dubey, Sunit Jalan, Aryaman Singhi, Khushi Chandra, and Srikanth Prabhu

Abstract Polycystic Ovary Syndrome (PCOS) is an endocrine disorder affecting women in their reproductive years. This study is aimed at developing a performant and transparent machine learning model to assist doctors and health care workers in diagnosing PCOS. To this end, we apply several data split ratios, feature selection techniques and machine learning algorithms on a dataset containing 541 samples and 43 features. The best performing framework was the Mutual Information-based feature selection paired with an 85:15 data split ratio. Random Forest trained on this framework achieved an accuracy, recall, precision, AUC-ROC and f-1 score of 96, 96, 97, 98 and 96%, and is the best performing model. Several Explainable AI techniques are applied to the best performing RF model to make its decision-making more interpretable and transparent. These methods identify the most significant local and global features responsible for classifier predictions. While PCOS is a prevalent health issue around the world, there is a lack of knowledge about PCOS among women in India. In this paper, we develop PCOS_HelpBot, a question–answering chatbot prototyped using the open–source Llama2 large language model to address the need for improved education and awareness about PCOS. Such an application can assist doctors with diagnostic support, patient education and engagement.

S. Rai · M. Dubey · S. Jalan · A. Singhi · K. Chandra · S. Prabhu (✉)
Manipal Institute of Technology, Manipal, India
e-mail: srikanth.prabhu@manipal.edu

S. Rai
e-mail: raishivarth@gmail.com

M. Dubey
e-mail: medhajdubey11@gmail.com

S. Jalan
e-mail: sunitjalan@gmail.com

A. Singhi
e-mail: singhiaryaman6@gmail.com

K. Chandra
e-mail: Khushiyali31@gmail.com

J. Maiti et al. (eds.), *Operations Research and Data Analytics: Current Trends and Future Perspectives*, Lecture Notes on Multidisciplinary Industrial Engineering,
https://doi.org/10.1007/978-981-96-8327-7_15

Keywords Healthcare analytics · Random Forest · LLM-based chatbot

15.1 Introduction

Polycystic Ovary Syndrome (PCOS) is hormonal syndrome condition that affects women of reproductive age. PCOS has been identified as one of the most common endocrine and metabolic disorders affecting women. It is a heterogeneous condition identified by the development of features caused by androgen excess and ovarian dysfunctional symptoms. Menstrual abnormalities and reproductive dysfunction are the most reported symptoms of PCOS, affecting female fertility [1, 2]. PCOS is a known to lead to development of several health issues. Cardiovascular disease, hypertension, lipid metabolic problems and endometrial cancer have been found to be two to six times more common in PCOS-diagnosed patients than in the general population [3].

The exact cause of the syndrome remains unknown, but experts attribute its development to factors such as genetics. Two key hormonal factors associated with PCOS are androgens and insulin. Androgens, often known as "male hormones," are produced in small amounts in females and play a role in normal bodily functions. PCOS patients typically have elevated androgen levels, leading to male characteristic development. Excess androgens contribute to acne, increased BMI, abnormal hair growth and ovarian cycle irregularities. Insulin function, responsible for converting food into usable energy, faces resistance in PCOS, resulting in increased insulin levels in the blood. Insulin resistance is prevalent among a significant percentage of women with PCOS.

With the surge in patient data complexity, AI and machine learning are being adopted for tasks like diagnosis, patient interaction, treatment recommendations and generating clinical reports. Machine Learning, a statistical AI domain, is foundational in healthcare applications like genome sequencing, drug discovery and medical imaging diagnostics. Machine learning algorithms effectively process clinical data, identifying patterns and predicting medical outcomes accurately. Natural Language Processing (NLP), another AI domain, enables computers to interpret human language. NLP applications include speech recognition, text analysis, translation and next word prediction. NLP systems, especially aided by Large Language Models (LLMs) trained on vast textual datasets, play roles in clinical documentation, medical literature analysis and chatbots for patient engagement.

Large Language Models replicate human expression using computational algorithms. Trained on extensive internet-sourced text data, these models generate text based on user input, providing answers, summaries, translations, or stories, with users even able to specify the desired text style. LLMs such as open-source Llama2 models, which are pre-trained on 2 trillion tokens [4], acquire a comprehensive understanding of the world by recognizing semantic relations within the vast textual data they train on. This enables them to offer versatile answers, making them potentially applicable across diverse medical domains. Despite their versatility, the significant computational resources and time required for training or fine-tuning LLMs pose barriers for niche use-cases. To address these challenges, Retrieval-Augmented Generation (RAG) employs retrieval-based techniques to optimize LLMs. RAG connects LLMs to external knowledge sources, allowing the models to generate responses based on relevant external documents. This approach enhances accuracy, factual relevance, and transparency of LLM-generated responses by incorporating dynamic external data sources. Retrieval-based methods, combined with LLMs, overcome issues related to outdated training data and the models' inability to isolate specific information. Ultimately, RAG improves the reliability, accuracy, trustworthiness and usefulness of LLMs in various domains and contexts [5].

In the medical and healthcare industry there is a growing demand for AI-based solutions. By the nature of the industry the technology used is required to be not only well performing, but equally importantly to be trustworthy, interpretable and explainable. AI and machine learning methods are often described as a "black box" in that they arrive at conclusions or results without providing any information on how they got there—we cannot trace the model's "thought-process" or see how it arrived at a particular result. Thus, methods and models that reproduce and help comprehend both the learning and knowledge-extraction process, are necessary to trust the reliability of the machine-learning decision-making process. Explainable-AI is an emerging research area that aims to improve the transparency and interpretability of AI-methodologies. Applications of SHAP methods and Feature Importance with Random Forest for PCOS screening, and developing an LLM-based chatbot for PCOS information dissemination are the main emphasis of this work. The following are the study's contributions:

- Evaluation of important PCOS features selected using Mutual Information and Chi-square with SelectKBest, two feature selection techniques.
- Assessment of the effectiveness of the following machine learning classifiers: Random Forest, GNB, Logistic Regression, SVM, XGBoost and AdaBoost.
- Examining our best-performing model's xAI layer using SHAP and feature importance with Random Forest.
- Prototype and demonstrate the effectiveness of RAG-optimized LLMs as question–answering tool for patient engagement and information dissemination.

15.2 Literature Review

This section reviews the literature relevant to our proposed research objectives. Analysis is done on previous applications of cutting-edge research for PCOS prediction. Polycystic ovarian syndrome is one of the most common health conditions affecting young women (PCOS). PCOS is an endocrine condition that can be detected by a range of non-invasive and invasive diagnostic markers. Appropriate treatment for PCOS depends on accurate PCOS diagnosis. Since a girl's menstrual cycle might take up to two years to stabilize after her first period, PCOS cannot be diagnosed until two to three years following the patient's first menstrual cycle. Accurate diagnosis of PCOS is made more difficult by the fact that its symptoms might change according on factors like age, race, weight and medications [6].

A dataset comprising 43 characteristics of 541 women—177 of whom were PCOS patients—was examined in [7]. Initially, the optimal features for PCOS prediction were determined using the univariate feature selection approach. After the features were ranked, the ratio of luteinizing hormone (LH) to follicle-stimulating hormone (FSH) was determined to be the most significant attribute. The training and testing data in the dataset were then divided using holdout and cross-validation techniques. The dataset was subjected to several classifier applications, including logistic regression, random forest, gradient boosting and hybrid random forest and logistic regression (RFLR). The first ten highest-ranked characteristics were sufficient to predict PCOS illness, according to the results. Results also showed that, when 40-fold cross-validation was applied to the top 10 features, RFLR had the best testing accuracy of 91.01% and recall value of 90%. As a result, it was discovered that RFLR was effective in categorizing PCOS patients. Multiple classifiers, including KNN, NN, NB, SVM, LR, classification trees, and linear discriminant (LD), were trained by Hdaib et al. [8]. Following feature engineering, 43 features were chosen, and LD had the best accuracy of 92.60%.

Principal Component Analysis was used by the authors in [9] to minimize the number of characteristics (PCA). To predict PCOS, they used NB, KNN, LR, RF, and SVM using certain features. The outcome demonstrated that RF had the best accuracy. The authors of [10] chose a subset of characteristics from the database using correlation feature selection methods. Using a variety of machine learning models, including SVM, LR, RF, DT, KNN, GB, AdaBoost (AB), XGBoost (XB) and CatBoost, they were able to choose the best model based on correlation thresholds. The outcome demonstrated that, with an accuracy of 89.02%, RF was the best model. A study by Neto et al. [11] used data mining techniques to identify PCOS patients. Classifiers such the Cross Industry Standard Process for Data Mining (CRISP-DM) approach and SVM, MLP, RF, LR and NB were implemented. Accuracy, sensitivity, and precision of 95%, 94% and 96%, respectively, were the highest for RF combined with data sampling approaches.

The author of [12] uses hybrid XGBRF and CatBoost models to identify PCOS. In addition, they make use of additional machine learning techniques as MLP, SVM, Decision Tree, Gradient Boosting, Random Forest, Logistic Regression and Hybrid Random Forest. There are 541 patients with 43 attributes in the dataset that was taken from the Kaggle repository. The outcome demonstrated that one of the more crucial characteristics in PCOS identification is FSH. The author used just ten of the most crucial features while implementing their models, which produced good outcomes at a lower computational cost. On ten features, they put nine classifiers into practice. The study report demonstrates how the unique method models CatBoost and XGBRF outperformed other classifiers, achieving 0.86 and 0.95 accuracy, respectively. When compared to other classifiers, CatBoost's cross validation accuracy was 0.89K fold. After comparing the outcomes of XGBRF and CatBoost with those of other classifiers documented in related work, it was determined that CatBoost performed better than all the other classifiers combined. Using RF as its classifier, Çiçek et al. [13] presented a LIME PCOS diagnostic model. Features that had a p-value of more than 0.05 were deemed statistically significant when a chi-square test was used. The accuracy obtained by RF was 86.03%. Local forecasts for the first five patients were described by LIME.

Large Language Models have achieved state-of-the-art in natural language understanding and generation across various domains and thus have been applied to numerous tasks. LLMs have proven to be resilient in few-shot and zero-shot learning tasks. A study conducted by Microsoft [14] showed that GPT-4 model by OpenAI, a general-purpose LLM that is not specialized for medical tasks, is able to clear the United States Medical Licensing Examination by over 20 points. The study verifies the ability of the model to explain medical concepts, provide personalized explanations to students.

Prior research exploring use of LLMs in the field of medicine verify that these models can used as medicine recommendation tool [15] and can be used to determine cause of disease. LLMs have proven the ability to augment the decision-making process of physicians by predicting the treatment effect and prognosis status based on a patient's medical history. Knowledge-based healthcare information retrieval and creation an area where LLMs can play an important role. LLMs are being used as a query-answering system to questions posed by patients [16]. By integrating data on questions asked by patients, LLMs can provide personalized information to patients.

Domain-Specific Language Models, like AgriBERT [17], are customized for domains with the goal of achieving the best possible performance in relevant tasks. AgriBERT is specifically trained on agricultural texts, which qualifies it for agricultural duties. With a concentration on middle school chemistry and physics, SciEdBERT [18] is intended for the educational sector and provides insights for assessing students' responses. In the medical field, ClinicalRadioBERT [19] is skilled in radiation oncology and places a strong emphasis on learning from clinical notes and relevant literature. These DSLMs demonstrate the versatility and promise of specialized models in a range of industries. While fine-tuning an LLM for a specific domain is a challenging task, involving high computational expense and large fine-tuning data set, retrieval techniques integrated with LLMs offer a method to endow models with domain-specific knowledge without requiring re-training and generating more specific and factual language [20].

While the machine learning prediction models perform well on this dataset, there is scope for improvement and achieving higher accuracy in PCOS prediction. Also, while natural language processing technology has been applied to various fields in medicine, no application has been developed that is aimed at PCOS related issues.

15.3 Methodology

15.3.1 Dataset Description

For this research, we use an open-source dataset prepared by Prasoon Kottarathil on Kaggle. This dataset contains data on 541 fertile women with 43 attributes [21]. This dataset has "PCOS (Yes/No)" as its target variable. Out of 541 patients, 177 were diagnosed with PCOS. The categorical features having [Y] and [N] have been encoded into 1 and 0, respectively. Among the 42 features, 24 were non-invasive parameters, and the remaining were either hormonal or gynaecological data obtained by invasive vaginal ultrasound and fluid samples.

15.3.2 Data Pre-processing

This dataset has already been pre-processed and we only had to take a few additive measures to make it ready for usage. The Matplotlib library for Python was used to visualize outliers. As no extreme outliers were detected, all the 541 sampled have been considered for this study. NumPy helped identify the missing values in the "Marriage Status (Yrs)", "Fast food (Y/N)", and "Marriage Status (Yrs.)" attributes. The median was chosen to replace the missing numerical attributes in the data. "Fast food (Y/N)" is a categorical variable; the statistical mode of this feature was used for data imputation. The attributes titled "Sl. No" and "Patient File No" were discarded as these attributes have no statistical significance. Descriptive statistical measures of the parameters are displayed in Fig. 15.1.

Two sets of data frameworks were created from the cleaned data: with 70:30 and 85:15 data split ratios, respectively. The results obtained from these two are compared to pick the best-performing framework and help get better results. Original dataset contained an imbalanced target "PCOS (Y/N)" class. In the target class, only 177 samples of the 541 samples were diagnosed PCOS while 364 samples did not. To mitigate the risk of improper model training due to imbalanced data, Borderline Synthetic Minority Oversampling Technique was employed on the training dataset. This technique generates synthetic points for the minority class through oversampling, specifically addressing the misclassification of outliers [22]. Following the balancing process, both PCOS and Non-PCOS classes had 364 samples each.

15.3.3 Feature Selection

In this section, we discuss the two different techniques applied to the selected dataset for the purpose of feature selection. Feature selection is a process wherein we select only the most statistically significant features out of many to reduce the data size,

	Mean	Std	Min	25%	50%	75%	Max
Follicle No. (R)	6.49	4.31	0.00	3.00	6.00	10.00	20.00
Follicle No. (L)	6.16	4.23	0.00	3.00	5.00	9.00	22.00
hair growth(Y/N)	0.29	0.45	0.00	0.00	0.00	1.00	1.00
Cycle(R/I)	2.59	0.92	2.00	2.00	2.00	4.00	5.00
Weight gain(Y/N)	0.39	0.49	0.00	0.00	0.00	1.00	1.00
Skin darkening (Y/N)	0.31	0.46	0.00	0.00	0.00	1.00	1.00
Cycle length(days)	4.96	1.54	0.00	4.50	5.00	5.00	12.00
AMH(ng/mL)	5.76	5.96	0.10	2.10	3.88	7.20	66.00
Fast food (Y/N)	0.51	0.50	0.00	0.00	1.00	1.00	1.00
PRL(ng/mL)	24.28	15.07	0.40	14.52	22.29	29.57	128.24
Weight (Kg)	60.12	11.25	31.00	53.00	60.00	65.85	108.00
FSH/LH	7.31	65.82	0.00	1.42	2.14	3.80	1372.83
Reg.Exercise(Y/N)	0.25	0.43	0.00	0.00	0.00	0.00	1.00
Pimples(Y/N)	0.50	0.50	0.00	0.00	1.00	1.00	1.00
TSH (mIU/L)	2.82	2.55	0.04	1.48	2.24	3.53	25.91
BMI	24.47	4.07	12.42	21.90	24.40	26.78	38.90
I beta-HCG(mIU/mL)	649.09	3336.44	1.90	1.99	18.00	294.62	32460.97
LH(mIU/mL)	7.17	94.09	0.03	1.05	2.34	3.69	2018.00
Hair loss(Y/N)	0.46	0.50	0.00	0.00	0.00	1.00	1.00
Pregnant(Y/N)	0.37	0.48	0.00	0.00	0.00	1.00	1.00

Fig. 15.1 Descriptive statistics for attributes in the dataset

reduce computational expense and improve performance of machine-learning algorithms applied. Two separate frameworks were created for this purpose. We have used Mutual Information and Chi-Square with SelectKBest for this feature-selection task.

15.3.3.1 Mutual Information

A well-known method for feature selection filters is the Mutual Information algorithm. This approach takes into account the dataset's statistical characteristics. The entropy measure is used to calculate mutual information. The uncertainty in the characteristics is measured through entropy [23]. The Mutual Information between two quantities is a measure of the extent to which knowledge of one quantity leads to more accurate predictions about the values of the other quantity. Mutual Information is not just suitable for linear relations, it is able to identify any kind of relationship between quantities.

If the mutual information score is high, the quantities under consideration are closely related, if zero or low, F and S are unrelated—neither contains any information about the other [24]. In theory, there is no upper bound to the value of MI however as it is a logarithmic value, it increases slowly. The "closeness" of features in the dataset are calculated and ranked in Fig. 15.2. Mutual Information for all features in the dataset were computed. After exhaustive testing, top 30 out of 43 features were selected.

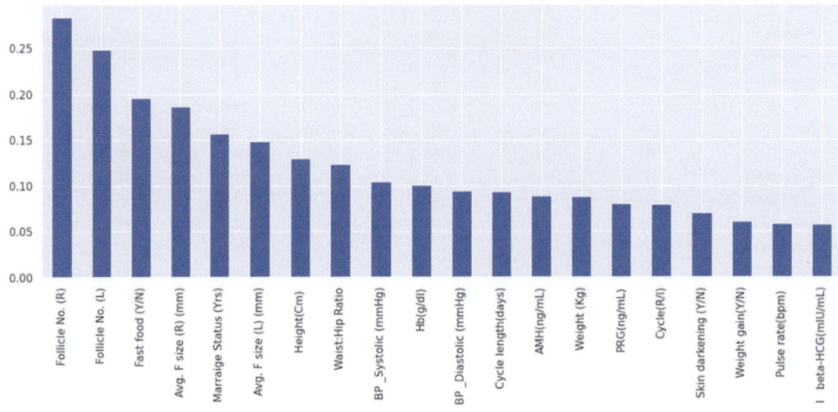

Fig. 15.2 Features selected by mutual information

15.3.3.2 Chi-Square with SelectKBest

Chi-Square is a statistical test applied to two quantities to determine the level of similarity of their variances. For the null hypothesis, the distributions are taken to be independent. The goal of the test is to validate whether any disparity between actual and predicted data is due to chance or due to a link between the variables under test. This makes chi-square test a good choice for understanding of the connections between the two categorical variables under consideration. This test helps identify optimal features for a dataset by assessing the dependence of the target class on each feature. The χ^2 metric is computed for every feature in the dataset, with a higher value indicating greater dependence of the target variable on the tested feature.

SelectKBest is a feature selection method in scikit-learn, a Python library for machine learning. It selects the k best features from a given dataset based on a scoring function. The scoring function can be specified by the user, and it can be used for both classification and regression problems. It is particularly useful when preparing a large dataset for training. For this paper, we use SelectKBest with Chi-Squared Test. Features selected by both methods are displayed in Table 15.1.

15.3.4 Setting Up PCOS_HelpBot

15.3.4.1 Data Preparation

To optimize Llama-2 LLM for the task of providing factual and up-to-date information about of PCOS, a 100-page document containing approximately 78,000 tokens in the English language on PCOS and related topics was created. Data for this purpose was collected from articles and publications hosted on websites sponsored by various governments such as Office of Women's Health, US Dept. of Health, and Human

Table 15.1 Features selected by mutual information and chi-square methods

S. No.	Feature selection technique	Number of features selected	Feature names (15 most significant)
1	Mutual information	30	Follicle No. (R) Follicle No. (L) Hair growth (Y/N) Cycle (R/I) Weight gain (Y/N) Skin darkening (Y/N) Cycle length (days) AMH (ng/mL) Fast food (Y/N) PRL (ng/mL) Weight (Kg) FSH/LH Reg.Exercise (Y/N) Pimples (Y/N) TSH (mIU/L)
2	Chi-Square with SelectKBest	30	Weight (Kg) Marriage Status (Yrs.) I beta-HCG (mIU/mL) II beta-HCG (mIU/mL) FSH (mIU/mL) LH (mIU/mL) FSH/LH AMH (ng/mL) Vit D3 (ng/mL) PRG (ng/mL) Weight gain (Y/N) Hair growth (Y/N) Skin darkening (Y/N) Follicle No. (L) Follicle No. (R)

Services [25], National Institutes of Health, US [26], National Health Service of the UK [27] and Indian Council of Medical Research [28]. Using this data, the chatbot is given access to domain specific knowledge, necessary for generation of reliable responses.

15.3.4.2 LangChain

LangChain is an open-source framework for building applications that use large language models. It helps to simplify the process of creating generative AI applications by linking LLMs with an array of external tools to use cutting-edge natural language processing techniques. LangChain can interact with any language model and help with tasks of managing inputs to these models and extracting output information. It can be used for building data connections and retrievals. Using LangChain, we can update the database that models can access for query answering, chaining multiple components to build applications. For our use-case, we build a RetrievalQA object to perform Retrieval Augmented Generation on the open-source Llama2 large language model.

Retrieval-augmented generation (RAG) is a technique that allows LLMs to interact with an external knowledge database. While the training of any LLM is an expensive undertaking, application of RAG allows LLMs the flexibility to access information outside of their training data corpus with minimal computation and time. There are two phases of the RAG process:

- Retrieval: Algorithms search for the information most relevant to the input prompt and retrieve it.
- Content Generation: This phase involves restructuring the retrieved information to obtain the most accurate response.

For question-answering systems, employing RAG can help by improving accuracy, transparency and reducing the risk of generating incorrect or insensitive responses. RAG enables of the building of "open-book" models, where users can cross-reference responses with sources. RAG minimizes the need for continuous model training, reducing computational and financial costs.

15.3.4.3 C-Transformers

C-Transformers is a Python library that provides bindings for transformer models implemented in C/C++ using the GGML library. GGML library is a tensor library designed for machine learning, enabling large models to be run on effectively on CPUs. This is achieved through integer quantization support and built-in optimization algorithms.

Quantization is a technique used to reduce the size of large neural networks, including LLMs, by modifying the weights and activations of the model to use lower-precision data types like 8-bit integer (int8) instead of the usual 32-bit floating point (float32). This technique reduces the memory and computing power required by these models without sacrificing too much accuracy.

C-Transformers supports a select set of open-source models, including popular ones like Llama, GPT4All-J, MPT and Falcon.

15.3.4.4 SentenceTransformers Embeddings

SentenceTransformers is a Python framework designed for cutting-edge sentence, text, and image embeddings. The SentenceTransfomers framework supports over 100 languages, allowing users to compute embeddings for sentences or text. These embeddings allow for semantic textual similarity, semantic search and paraphrase mining by enabling the comparison of sentences for similar meanings using techniques like cosine-similarity.

Embedding generated using SentenceTransformers are used in models with state-of-the-art performance in multiple applications, optimizing for maximum speed and emphasizing efficiency in processing on consumer hardware.

We use the open-source "all-MiniLM-L6-v2" model for this study because it offers high speed and excellent embedding quality for our fine-tuning dataset.

15.3.4.5 FAISS

The Facebook AI Similarity Search (FAISS) library is an open-source library released by Meta AI. This library is designed for efficient and fast similarity search and clustering in large datasets. Traditional query search engines are suboptimal for multimedia document searches. Such databases, structured for symbolic information, face limitations when handling high-dimensional vectors generated by AI tools like word embeddings or convolutional neural network descriptors. The library's three key metrics are speed, memory usage and accuracy, allowing users to trade some accuracy for significantly faster search times. This makes FAISS valuable for applications involving high dimensional vectors. The library is tailored for large-scale datasets and accelerates similarity search operations, making it useful for tasks like semantic search, image retrieval and classification.

15.3.4.6 Chainlit

Chainlit is an open-source python package that is useful for building applications like ChatGPT. This package provides a ChatGPT-like frontend that can be used out-of-the-box or with a custom frontend. Chainlit is compatible with various python libraries such as LangChain and Llama Index making it a powerful tool. It allows for visualization of intermediary steps in the output and supports iteration on prompts through the Prompts Playground. This package is intended to streamline the development, testing, and sharing of large language model applications. Interface for PCOS_HelpBot has been built using Chainlit. Figure 15.3 showcases the high – level structure of our conversational chatbot, PCOS_HelpBot.

15.4 Results

15.4.1 Performance Metrics

Evaluation of our proposed model has been conducted by analysing performance across the following metrics: confusion matrix, accuracy, precision, recall, F-1 score, ROC-AUC and precision-recall score.

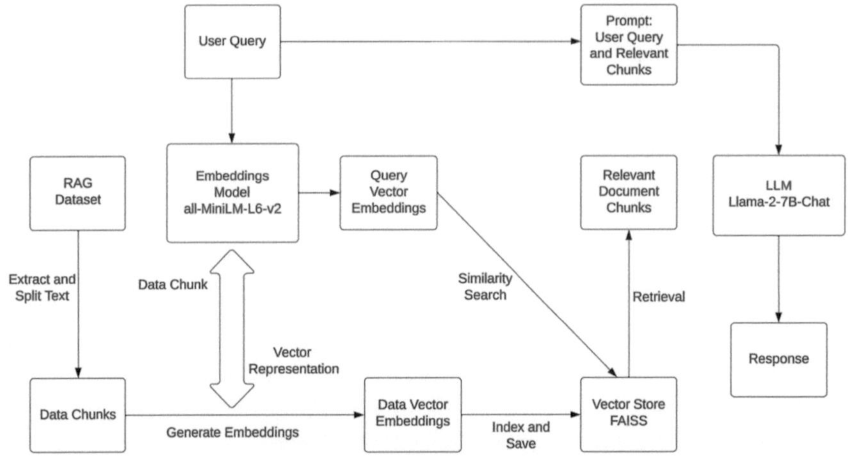

Fig. 15.3 Structure of PCOS_HelpBot prototype

15.4.2 Model Evaluation with Machine Learning

This study tested and analysed 6 different machine learning models: Logistic Regression, Random Forest Classifier, Support Vector Machine (linear, polynomial, Gaussian and sigmoidal kernels have been considered), Gaussian Naïve Bayes, AdaBoost and XGBoost. For selection of optimal hyper-parameters for the classifiers, GridSearchCV has been employed. Following exhaustive testing for all possible combinations of parameters, the best-performing hyper-parameters were obtained through this method. Figure 15.4 depicts the architecture of our prediction model.

A combination of several feature selection methods, data-split ratios and machine-learning classifiers have been evaluated in this study. Table 15.2 tabulates the performance of these frameworks. Accuracy represents the percentage of correct classifications the model makes. The following models produced the best accuracy: Random Forest with Mutual Information selection, and an 85:15 split achieved a 96.36% score and a 70:30 split achieved a score of 95.89%. Random Forest with chi-square selection, and a 70:30 split produced an impressive accuracy of 94.22%. XGBoost with Mutual Information selection and an 85:15 split gave 94.54% accuracy, while SVM with chi-square selection and 85:15 split also reached 94.54% accuracy. In terms of precision, recall and ROC-AUC metrics, highest scores of 97%, 96% and 98%, respectively, was achieved by Random Forest with Mutual Information selection, and an 85:15 split.

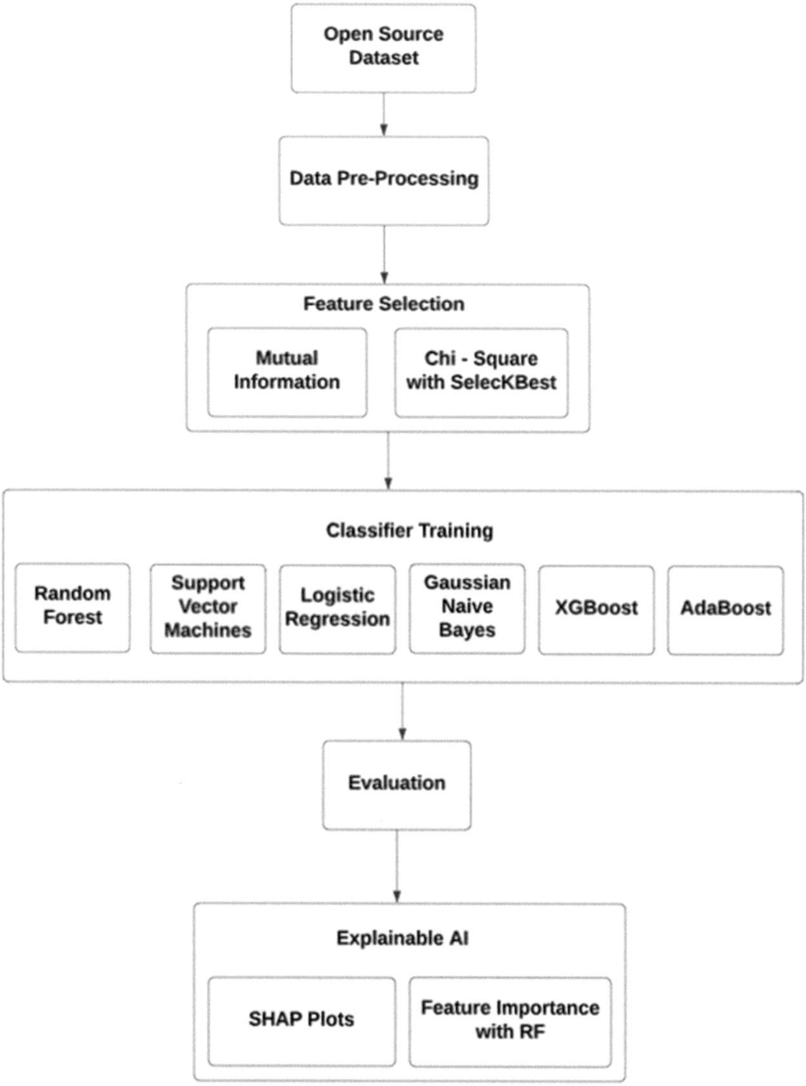

Fig. 15.4 Architecture of ML prediction model

15.4.3 Explainable AI (xAI)

15.4.3.1 Shapley Additive ExPlanations (SHAP)

SHAP architecture with a "Unified Approach" introduced by Lundberg and Lee [29] assigns a relevance score to each feature. While unable to assess the quality of the results produced by a model, SHAP measures the contribution of each feature to the

Table 15.2 Performance metrics for all four frameworks

Models	Mutual information		Chi-Square with SelectKBest	
Data split	85:15	70:30	85:15	70:30
Random forest	96.36	95.89	92.83	94.22
Logistic regression	92.72	89.54	91.01	88.35
SVM	92.72	88.42	94.54	91.09
Gaussian Naïve Bayes	85.45	78.01	79.09	80.13
XGBoost	94.54	88.09	93.63	89.72
AdaBoost	92.01	88.72	89.01	89.01

model's prediction. By providing insights into the model predictions, SHAP improves the transparency and interpretability of these models. Classifiers trained on Mutual-Information selected feature with 85:15 data-split had the best overall performance. Random Forest was the best performing model with an accuracy, recall, precision, and f-1 score of 96, 96, 97, and 96%. Figure 15.5 shows the AUC-ROC and Precision-Recall curves for this model. AUC-ROC score and Precision-Recall score provide a comprehensive evaluation of a model's performance across different thresholds. Higher AUC and PR score indicate better model performance. We use this model with SHAP values to understand its predictions.

SHAP Violin Plot

For our top-performing model, the Random Forest Classifier, we generate a SHAP Violin plot. When using ensemble modelling to build tree-based models like Random Forest, it can prove difficult to ascertain the model's logic for making decisions. In the graph below, colour corresponds to the data point values, and the SHAP values

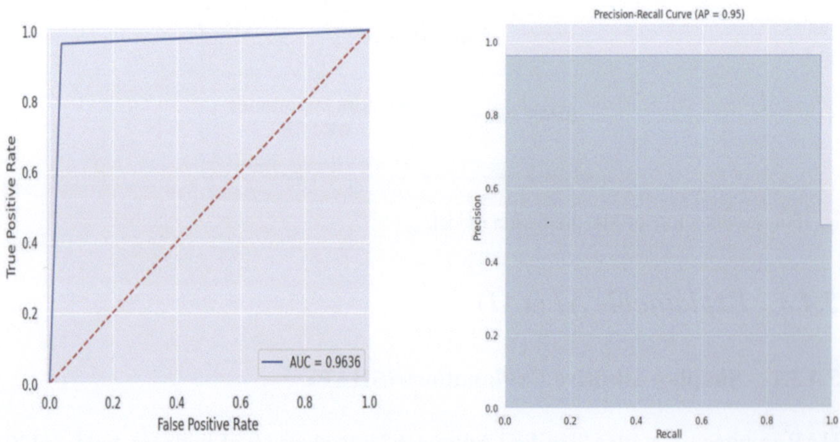

Fig. 15.5 AUC-ROC and PR-curve for best performing RF classifier

are represented on the horizontal axis. Brighter blue indicates lower values, while brighter red signifies higher values. The plot arranges elements in descending order of importance, with the most significant feature positioned at the top. Violin plots of SHAP make it easier to identify outliers and provide accurate representation of kernel density.

Follicle Number R, as seen in Fig. 15.6, is the most important feature. Positive PCOS prediction can be ascertained from greater number of follicles on both ovaries. Similarly, hair growth, cycle length, AMH and FSH levels and menstrual cycle irregularity are significant features for prediction accuracy.

SHAP Summary Plot

The SHAP summary graph enables the visualization of the correlation between a feature's value and its influence on the model's prediction. In other words, both feature effects and feature importance are combined in the SHAP summary plot. Every point on the summary plot represents a feature's Shapley value for a single sample. The feature names are arranged from top to bottom along the *Y*-axis according to priority. The SHAP value is represented by the *X*-axis, representing change in log odds. The colour of points represents the value of the feature from low to high. From Fig. 15.7, we understand that lower number of follicles in the right ovary reduces the risk of potential PCOS diagnosis and large number of follicles increases the risk. Same is

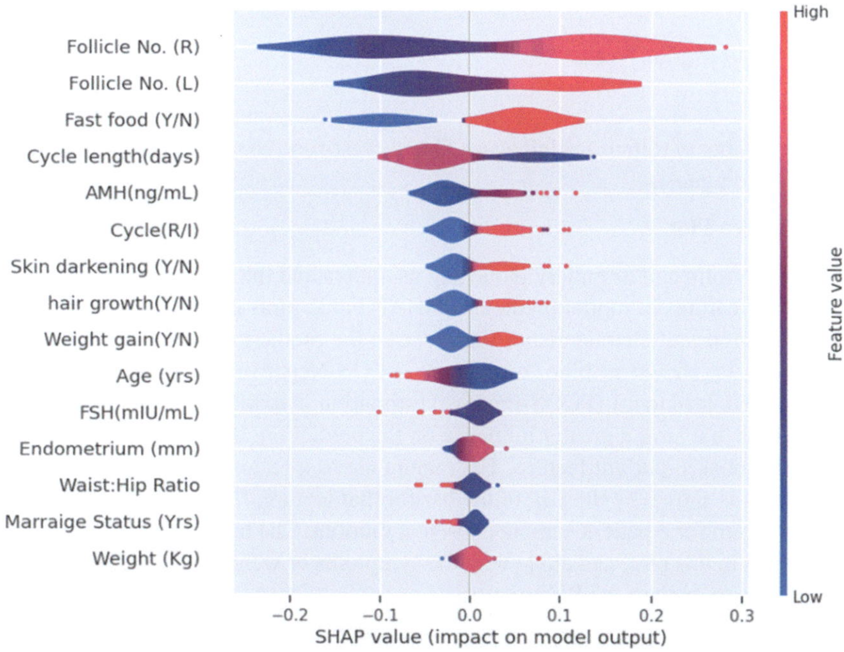

Fig. 15.6 SHAP violin plot

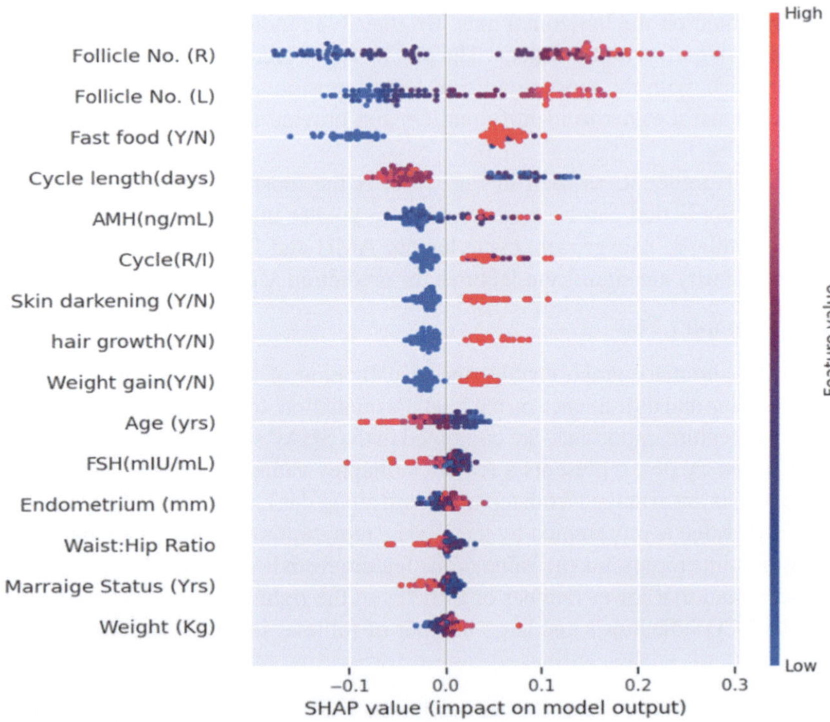

Fig. 15.7 SHAP summary plot

true for number of follicles in left ovary. Greater fast-food consumption also increases the risk of diagnosis.

SHAP Force Plot

While the Violin and Summary plots help us understand the influence and contribution of all features in model prediction, SHAP Force plots assist us in determining the most important characteristics that oversee the prediction a single observation. Lower scores, shown in blue, result in 0 (PCOS-Negative), while higher values, shown in red, lead to a 1 (PCOS-Positive) estimation. Around the dividing boundary are features that have a greater influence on the prediction; in this case, Follicle No. (R) is the most important feature. The magnitude of a feature's influence on model predictions is shown by the size of the bar for that feature. Figure 15.8 is the SHAP force diagram for a patient sample chosen at random. The high right and left follicular counts in this plot, and along with the symptoms of weight gain and poor eating habits, help to explain the RF prognosis.

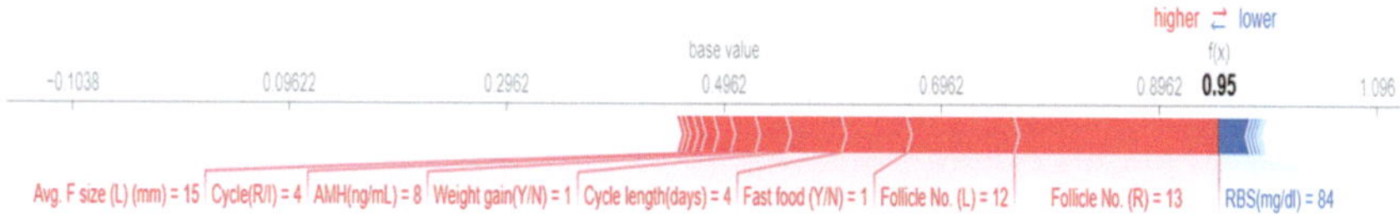

Fig. 15.8 SHAP force plot for an observation

15.4.3.2 Feature Importance with Random Forest

Feature importance in a Random Forest model is the measure of each features contribution to the model's predictions. We use the scikit-learn implementation of Random Forest Classifier for our purpose. Random Forest calculates feature importance by evaluating how much each feature contributes to reduction in Gini impurity. For each decision tree in the forest, the Gini importance of a feature is computed as the sum of the Gini impurity decrease that the feature causes across all decision nodes in the tree. Figure 15.9 represents a graph with features ranked in descending order from left to right according to their importance value. From the graph we gather that number of follicles on right and left ovaries is a strong indicator for PCOS diagnosis. Factors such as fast-food consumption, menstrual cycle length and regularity, skin darkening and weight among others are important for PCOS predictions by RF classifier.

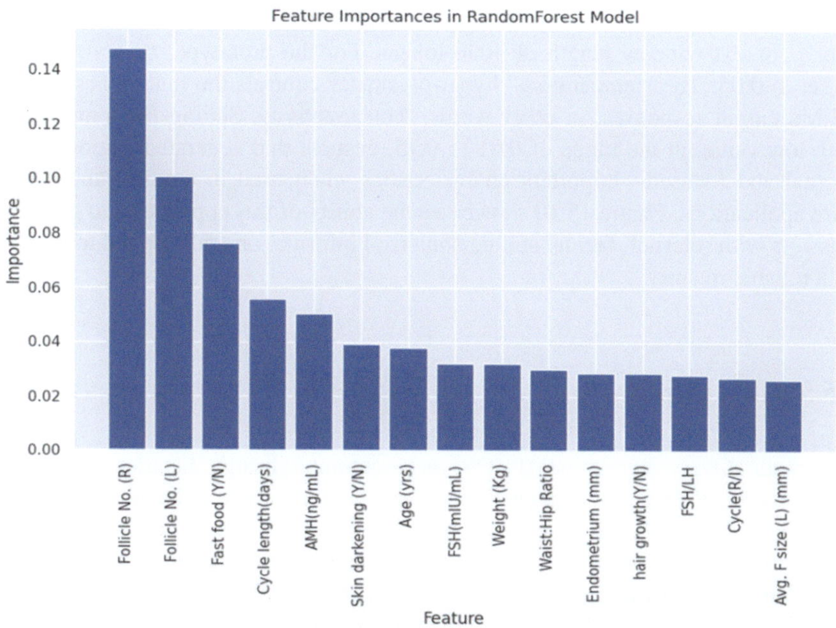

Fig. 15.9 Feature importance plot for random forest

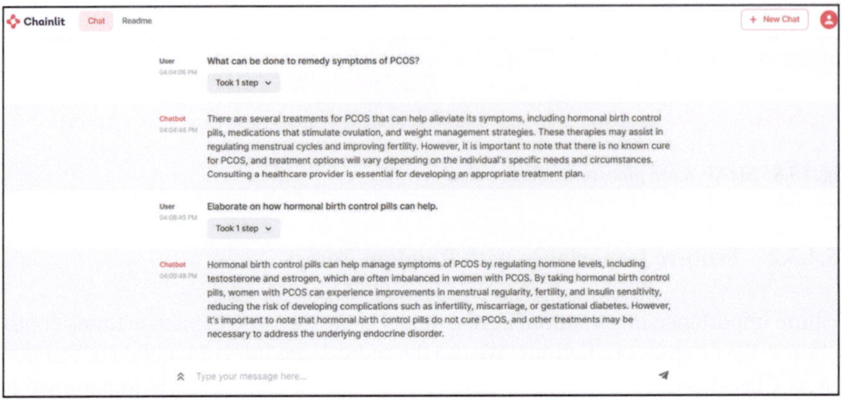

Fig. 15.10 Capture of answers generated to PCOS_HelpBot to queries regarding PCOS treatment

15.4.4 PCOS_HelpBot

PCOS_HelpBot, a prototype version of a generative chatbot for PCOS related queries, utilizes an open-source Llama-2-7b-Chat model, a seven billion parameter version of the Llama-2 series of pre-trained LLMs. This LLM has been optimized for dialogue use cases. These models are designed and fine-tuned for the purpose of generating human-like responses, hence making them suitable for our purpose. This model has a context window length of 4096 tokens. For the prototype, the temperature is set to 0.05. The "temperature" hyper-parameter controls the randomness of the LLMs output responses, in other words, their creativity. Setting the temperature to a low value, in the range of 0.01 to 0.15, ensures that generated responses are factual and focused—important characteristics when used in medical and health care applications. Figure 15.10 showcases the ability of this application to generate answers with relevant, factual and personalized information and respond to queries in a reliable manner.

15.5 Discussion

Polycystic Ovary Syndrome (PCOS) is a common health disorder caused by hormonal imbalance. It is characterized by symptoms of ovary enlargement, obesity or weight gain, acne, hirsutism, hair thinning or male pattern baldness, skin darkening and skin tags (small excess flaps of skin in the armpits or neck area). Studies have linked PCOS diagnosis with the onset of other health disorders [25] such as diabetes, high blood pressure, unhealthy cholesterol levels, sleep apnoea, anxiety, and endometrial cancer. While there is currently no single test to diagnose PCOS and the most reliable diagnostic tools remain to be various tests such as physical test, pelvic test and ultrasound, and blood test, one of the objectives of this study is to develop a reliable and transparent machine learning model to help doctors and healthcare staff with diagnostics.

For this purpose, we developed a variety of machine learning pipelines and analysed their performance to determine which classifier fared the best. The dataset, sourced from the Kaggle repository, consists of 541 patients with 43 attributes. This study examined all combinations of the two data split ratios, two feature selection techniques, and six classifiers. Models trained on the 85:15 data split outperform models trained on the 70:30 data split, regardless of the feature selection technique used. When compared to classifiers trained on features chosen using the Chi-square test, those trained on MI-extracted features had superior performance metrics. Mutual Information based feature selection paired with an 85:15 data split ratio emerged as the best performing framework among the four evaluated in this study.

The results achieved by this framework is tabulated in Table 15.3. Across both feature selection methods and both data split ratios, the performance of Random Forest, Support Vector Machines and XGBoost classifiers remain consistently high. The two feature selection techniques gave different best performing models. With accuracy, precision, and recall scores of 96, 97 and 96, Random Forest Classifier performed the best, when using Mutual Information as the feature selection method. For Chi- square with SelectKBest, Support Vector Machine with linear kernel performed the best with accuracy, precision, and recall scores of 94, 94 and 96 respectively. Random Forest trained on Mutual Information- selected, 85:15 split data was the best performing model with an accuracy, recall, precision, AUC-ROC and f-1 score of 96%, 96%, 97%, 98% and 96%. Gaussian Naïve Bayes emerged as the worst performing classifier in this study irrespective of feature selection method and data split ratios. Interestingly, for both feature selection methods and both data split ratios, tree-based classifiers maintained a high-level performance.

The added layers of Explainable AI (xAI) techniques applied to best performing Random Forest classifier help to make its decision more transparent and interpretable. To elucidate the global and local RF forecasts based on feature significance, we explore several SHAP visualization techniques. A simplified picture of the variability and distribution of SHAP values for every feature is provided by SHAP violin graphs. It helps in understanding how various features affect the output of the model. The SHAP summary graph is a visual representation of the correlation between a

Table 15.3 Performance of models in the Mutual Information and 85:15 data split

MI model	Precision	Recall	F1-score	Accuracy	AUC	PR score
Random forest	97	96	96	96	98	96
Logistic regression	91	92	90	92	92	91
SVM (RBF)	97	89	92	92	92	91
Gaussian Naïve Bayes	83	89	83	85	85	79
XGBoost	94	92	92	94	91	90
AdaBoost	92	90	91	92	91	89

feature's value and its influence on the model's prediction. SHAP Force plots assist in determining which characteristics are most important for predicting a certain observation. Feature Importance of Random Forest enabled an understanding of the RF's global predictions. Following are the characteristics most important for PCOS classification listed by all the xAI techniques mentioned above: Follicle numbers *L* and *R*, Cycle (*R*/*I*), Skin darkening (Y/N), weight gain (Y/N), length of cycle (days), fast food (Y/N). Previous studies support this outcome and highlight the importance of these characteristics in the diagnosis [30]. xAI techniques therefore emerge as useful tools to visualize and interpret predictions of evaluated models.

Some studies have highlighted that a significant proportion of women with PCOS express a lack of knowledge about the condition and its management, which can be a barrier to treatment adherence [31]. Additionally, there are reports of delays in seeking medical help and receiving a diagnosis, indicating a need for improved education and awareness about PCOS among women in India [32]. PCOS is a significant endocrine condition impacting women within the reproductive age range, yet it is not included in government health programs. Thus, there is an unmet need for a factual, reliable, up-to-date and transparent information source that can flexibly resolve any queries about PCOS, its aetiology, symptoms and treatment among other topics. Large Language Models are state of the art in natural language understanding and conversational system applications. We prototype a question–answering chatbot, PCOS_HelpBot, using the open-source Llama2 large language model for the purpose of dissemination of information about PCOS among women and the general populace. India faces a notable scarcity of human resources for medical and health-related tasks [32]. As indicated in [33], the national density of physicians, nurses, and midwives is reported to be 20.6 per 10,000 people, falling below the World Health Organization's (WHO) recommended threshold of 44.5. Built using open-source frameworks and equipped with reliable and up-to-date information, this prototype offers a technically and economically feasible solution to the issue of PCOS awareness and the taboo of discussion of female health problems in India. Following rigorous evaluation, such an application along with machine-learning-based prediction models can assist doctors with diagnostic support, patient education and engagement.

15.6 Conclusion and Future Scope

This study aimed to create a machine learning model for diagnosing Polycystic Ovary Syndrome (PCOS), a common health disorder marked by hormonal imbalance and various symptoms. The most effective approach involved Mutual Information-based feature selection combined with an 85:15 data split ratio. Random Forest, trained on Mutual Information-selected features, demonstrated the highest performance with 96% accuracy, 96% recall, 97% precision, 98% AUC-ROC and a 96% f-1 score. To enhance transparency and interpretability, Explainable AI (xAI) techniques were applied to the Random Forest classifier, identifying the most significant features for PCOS classification. The study also addressed the limited awareness of PCOS among women in India, emphasizing the necessity for improved education. A question-answering chatbot, PCOS_HelpBot, was prototyped using the open-source Llama2 large language model with RAG optimization techniques. This prototype can aid doctors in diagnostic support, patient education and engagement, disseminating information about PCOS among women and the general populace.

The future scope for this study includes:

- Validating machine learning models in larger datasets and diverse patient populations.
- Investigating additional machine learning algorithms and their effectiveness in PCOS diagnosis.
- Exploring artificial intelligence-based chatbots for diagnosing and managing various health conditions in India.
- Collaborating with healthcare professionals to enhance the chatbot's responses and ensure accuracy with up-to-date information.
- Broadening the chatbot's capabilities using multi-modal LLMs, to encompass symptom tracking, medication reminders, and appointment scheduling.

References

1. Ndefo, U.A., Eaton, A., Green, M.R.: Polycystic ovary syndrome: a review of treatment options with a focus on pharmacological approaches. Pharm. Therap. **38**(6), 336 (2013)
2. Mehreen, T.S., Ranjani, H., Kamalesh, R., Ram, U., Anjana, R.M., Mohan, V.: Prevalence of polycystic ovarian syndrome among adolescents and young women in India. J. Diabetol. **12**(3), 319–325 (2021)
3. Rojhani, E., Rahmati, M., Firouzi, F., Saei Ghare Naz, M., Azizi, F., Ramezani Tehrani, F.: Polycystic ovary syndrome, subclinical hypothyroidism, the cut-off value of thyroid stimulating hormone; is there a link? Findings of a population-based study. Diagnostics **13**(2), 316 (2023)
4. Introducing Llama 2: the next generation of our open-source large language model. Meta AI. Retrieved 22 Feb 2021, from https://ai.meta.com/llama/
5. Lewis, P., Perez, E., Piktus, A., Petroni, F., Karpukhin, V., Goyal, N., Küttler, H., Lewis, M., Yih, W.-T., Rocktäschel, T., Riedel, S., Kiela, D.: Retrieval-augmented generation for knowledge-intensive NLP tasks. Adv. Neural Inf. Process. Syst. **33**, 9459–9474 (2020)
6. Roe, A.H., Dokras, A.: The diagnosis of polycystic ovary syndrome in adolescents. Rev. Obstet. Gynecol. **4**(2), 45 (2011)
7. Bharati, S., Podder, P., Mondal, M.R.H.: Diagnosis of polycystic ovary syndrome using machine learning algorithms. In: 2020 IEEE Region 10 Symposium (TENSYMP), pp. 1486–1489. IEEE (2020)
8. Hdaib, D., Almajali, N., Alquran, H., Mustafa, W.A., Al-Azzawi, W., Alkhayyat, A.: Detection of polycystic ovary syndrome (PCOS) using machine learning algorithms. In: 2022 5th International Conference on Engineering Technology and Its Applications (IICETA), pp. 532–536. IEEE (2022)
9. Denny, A., Raj, A., Ashok, A., Ram, C.M., George, R.: i-hope: Detection and prediction system for polycystic ovary syndrome (PCOS) using machine learning techniques. In: TENCON 2019–2019 IEEE Region 10 Conference (TENCON), pp. 673–678. IEEE (2019)

10. Tiwari, S., Kane, L., Koundal, D., Jain, A., Alhudhaif, A., Polat, K., Zaguia, A., Alenezi, F., Althubiti, S.A.: SPOSDS: a smart polycystic ovary syndrome diagnostic system using machine learning. Expert Syst. Appl. **203**, 117592 (2022)
11. Neto, C., Silva, M., Fernandes, M., Ferreira, D., Machado, J.: Prediction models for polycystic ovary syndrome using data mining. In: International Conference on Advances in Digital Science, pp. 210–221. Springer International Publishing, Cham (2021)
12. Bhat, S.A.: Detection of polycystic ovary syndrome using machine learning algorithms. Doctoral Dissertation, National College of Ireland, Dublin (2021)
13. Çiçek, İB., Küçükakçalı, Z., Yağın, F.H.: Detection of risk factors of PCOS patients with local interpretable model-agnostic explanations (LIME) method that an explainable artificial intelligence model. J. Cogn. Syst. **6**(2), 59–63 (2021)
14. Nori, H., King, N., McKinney, S.M., Carignan, D., Horvitz, E.: Capabilities of GPT-4 on medical challenge problems. arXiv preprint arXiv:2303.13375 (2023)
15. Clusmann, J., Kolbinger, F.R., Muti, H.S., Carrero, Z.I., Eckardt, J.N., Laleh, N.G., Löffler, C.M.L., Schwarzkopf, S.C., Unger, M., Veldhuizen, G.P., Wagner, S.J., Kather, J.N.: The future landscape of large language models in medicine. Commun. Med. **3**(1), 141 (2023)
16. Shi, Y., Xu, S., Liu, Z., Liu, T., Li, X., Liu, N.: Mededit: Model editing for medical question answering with external knowledge bases. arXiv preprint arXiv:2309.16035 (2023)
17. Rezayi, S., Liu, Z., Wu, Z., Dhakal, C., Ge, B., Zhen, C., Liu, T., Li, S.: AgriBERT: Knowledge-infused agricultural language models for matching food and nutrition. In: IJCAI, pp. 5150–5156 (2022)
18. Liu, Z., He, X., Liu, L., Liu, T., Zhai, X.: Context matters: a strategy to pre-train language model for science education. In: International Conference on Artificial Intelligence in Education, pp. 666–674. Springer Nature Switzerland, Cham (2023)
19. Rezayi, S., Dai, H., Liu, Z., Wu, Z., Hebbar, A., Burns, A.H., Zhao, L., Zhu, D., Li, Q., Liu, W., Li, S., Li, X.: Clinicalradiobert: knowledge-infused few shot learning for clinical notes named entity recognition. In: International Workshop on Machine Learning in Medical Imaging, pp. 269–278. Springer Nature Switzerland, Cham (2022)
20. Zakka, C., Shad, R., Chaurasia, A., Dalal, A. R., Kim, J. L., Moor, M., Fong, R., Phillips, C., Alexander, K., Ashley, E., Boyd, J., Boyd, K., Hirsch, K., Langlotz, C., Lee, R., Melia, J., Nelson, J., Sallam, K., Tullis, S., Vogelsong, M.A., Cunningham, J.P., Hiesinger, W.: Almanac—retrieval-augmented language models for clinical medicine. NEJM AI **1**(2), AIoa2300068 (2024)
21. Kottarathil, P.: Polycystic ovary syndrome (PCOS). Kaggle. Retrieved January 14, 2024, from https://www.kaggle.com/datasets/prasoonkottarathil/polycystic-ovary-syndrome-pcos?select=PCOS_data_without_inferitlity.xlsx
22. Han, H., Wang, W.Y., Mao, B.H.: Borderline-SMOTE: a new over-sampling method in imbalanced data sets learning. In: International Conference on Intelligent Computing, pp. 878–887. Springer Berlin Heidelberg, Berlin, Heidelberg (2005)
23. Vergara, J.R., Estévez, P.A.: A review of feature selection methods based on mutual information. Neural Comput. Appl. **24**, 175–186 (2014)
24. Liu, H., Sun, J., Liu, L., Zhang, H.: Feature selection with dynamic mutual information. Pattern Recogn. **42**(7), 1330–1339 (2009)
25. Office on Women's Health in the Office of the Assistant Secretary for Health at the U.S. Department of Health and Human Services, Retrieved February 22, 2021, from https://www.womenshealth.gov/a-z-topics/polycystic-ovary-syndrome
26. Polycystic Ovary Syndrome (PCOS). Office of Communications, NIH. Retrieved February 22, 2021, from https://www.nichd.nih.gov/health/topics/pcos
27. Polycystic Ovary Syndrome, National Health Service, UK. Retrieved February 22, 2021, from https://www.nhs.uk/conditions/polycystic-ovary-syndrome-pcos/
28. Indian Council of Medical Research. Retrieved February 22, 2021, from https://main.icmr.nic.in/
29. Lundberg, S.M., Lee, S.I.: A unified approach to interpreting model predictions. Adv. Neural Inf. Process. Syst. **30** (2017)

30. Rao, V.S., Cowan, S., Armour, M., Smith, C.A., Cheema, B.S., Moran, L., Lim, S., Gupta, S., De Manincor, M., Sreedhar, V., Ee, C.: A global survey of ethnic Indian women living with polycystic ovary syndrome: co-morbidities, concerns, diagnosis experiences, quality of life, and use of treatment methods. Int. J. Environ. Res. Public Health **19**(23), 15850 (2022)
31. Vijayan, S.M., Kalaivani, H., Mitra, S., John, J., Anila, A., Boban, L., Chowdhury, P., Gayen, S.: Barriers to treatment regimen adherence in Indian women with polycystic ovarian syndrome. J. Family Med. Primary Care **11**(7), 3687–3692 (2022)
32. Nandan, D., Agarwal, D.: Human resources for health in India: urgent need for reforms. Indian J. Community Med. **37**(4), 205–206 (2012)
33. Karan, A., Negandhi, H., Nair, R., Sharma, A., Tiwari, R., Zodpey, S.: Size, composition and distribution of human resource for health in India: new estimates using National Sample Survey and Registry data. BMJ Open **9**(4), e025979 (2019)

Chapter 16
Early Detection of HAVS in Foundry Workers Using Linear Support Vector Machine—A Machine Learning Approach

V. Rohith, S. Syath Abuthakeer, and J. Nashreen

Abstract Vibration can have both short-term and long-term effects on the human body, depending on factors such as the frequency, amplitude, duration, and type of vibration, as well as individual susceptibility and exposure levels. It's necessary to understand that each person will experience the impacts of vibration exposure differently ranging from simple nausea and discomfort to chronic HAVS (Human Arm Vibration Syndrome) and Raynaud's Syndrome. HAVS is an occupational health condition resulting from prolonged and repetitive exposure to Human Arm Vibration (HAV) within industrial settings experiencing vibration levels from 6.3 Hz to 1250 Hz. Due to rapid industrialization early detection and risk assessment of HAVS have become crucial to mitigate its adverse effects on the well-being of workers. Early detection and risk assessment of HAVS have become crucial to mitigate its adverse effects on the well-being of workers. This paper focus on prediction of HAVS in Foundry workers who are exposed to HAV for long durations with help of Machine Learning technique called Linear Support-Vector Machine Algorithm (LSVM). The LSVM model offers a practical and interpretable framework for risk assessment. The SVM's linear nature allows for transparent decision boundaries, enhancing the understanding of key predictors and their impact on HAVS susceptibility. By proving the efficiency of Linear SVMs in early HAVS identification, this paper advances the idea of predictive modelling for occupational health and safety. It

V. Rohith (✉) · S. S. Abuthakeer
Department of Mechanical Engineering, PSG College of Technology, Coimbatore, Tamil Nadu, India
e-mail: 22md05@psgtech.ac.in

S. S. Abuthakeer
e-mail: ssa.mech@psgtech.ac.in

J. Nashreen
Department of Production Engineering, PSG College of Technology, Coimbatore, Tamil Nadu, India
e-mail: 22md34@psgtech.ac.in

J. Maiti et al. (eds.), *Operations Research and Data Analytics: Current Trends and Future Perspectives*, Lecture Notes on Multidisciplinary Industrial Engineering,
https://doi.org/10.1007/978-981-96-8327-7_16

also highlights the potential for machine learning methods to enhance workers' well-being and encourages the creation of proactive methods for HAVS risk management in the workplace.

Keywords Human arm vibration · HAVS · Foundry · Occupational safety · Prediction · Linear-SVM · Machine learning

16.1 Introduction

Vibrations may originate from various sources, including the machinery's operation (rotating or reciprocating masses, gas pulsation, aerodynamic phenomena), the impact of hand-held machinery on rigid surfaces, or interactions between the machinery and its surroundings (e.g., the movement of mobile machinery over uneven terrain) [1]. The vibrations generated by such equipment can be transmitted to various parts of the body, with the hands and arms being particularly susceptible. The exposure to these mechanical vibrations during the workers job responsibilities is called as Occupational Vibration Exposure. Ensuring occupational health and safety (H&S) is a primary duty of workplace management [2] to provide safer working environments for the workers.

India is a developing country where most of the population are generally workers, almost 60% of the country's population are exposed to occupational vibration due to rapid mechanization of various tools, equipment and machines. These occupations are not only limited to Construction, Transportation, Mining, machining but also other occupations like Sports and Dance. When exposed to vibrations in Industries leads to health issues like Orthopaedic, Neurological issues, Gastro-intestinal problems, Prostate cancer and delayed pregnancy [3]. Whole Body Vibration and Human Arm Vibration generally constitutes to majority of Occupational Vibrational injuries in industries. This makes it necessary to develop early detection systems, to develop and adopt suitable preventive measures to prevent occupational injuries.

The conveyance of mechanical vibrations to the arm and hand takes place when vibrating tools are in use, this phenomenon identified as Human Arm Vibration (HAV). Construction, mining and manufacturing utilize pneumatic hand tools like sand rammers, road breakers, chipping hammers, riveting guns and impact rock drills. Critical for their intended purposes, these tools generate shocks or impact vibrations. A fraction of these impact vibrations can be conveyed to the hand-arm system. Considering that the air actuation rates of these tools commonly range from 10 to 40 Hz, their fundamental vibration components also align within this frequency range [1, 4]. Incidentally, the primary resonances of the human wrist-arm system align with this frequency range. Vibrations below 40 Hz can effectively transmit to the wrist, forearm and elbow, while those below 20 Hz can extend to the upper arm, shoulder, neck and head [5, 6].

Continued exposure to hand-arm vibration resulting from the utilization of vibrating tools can give rise to vascular and neurological symptoms connected with

Hand-Arm Vibration Syndrome (HAVS) [7, 8], a medical condition caused by regular and frequent exposure to hand-arm vibration. Typical manifestations and consequences of Hand-Arm Vibration Syndrome (HAVS) in the fingers and hands include sensations like pins and needles, tingling, numbness, reduced finger sensation and dexterity, and waking up at night with painful fingers and hands. More severe symptoms, often occurring in cold weather, present as painful episodes lasting 5–15 min, during which one or more fingers display a "white or blanch" discoloration. It is well-established and documented that vibration, exposure to cold and nicotine (from smoking) can independently constrict blood vessels. As a result, cold temperatures and/or smoking can worsen symptoms for individuals diagnosed with Hand-Arm Vibration Syndrome (HAVS). Current medical interventions can only alleviate the pain and discomfort associated with HAVS attacks, given that HAVS is irreversible and lacks a cure. Therefore, the key principles involve vigilance and prevention to minimize the medical impact of HAV [9].

Numerous professional sectors carry the risk of developing this condition following prolonged use of vibrating machinery and tools. Those employing high-impact percussive tools in the construction and building industry experience heightened levels of exposure to both vibration and shock. The extended exposure of hands and arms to intense vibration and shock can lead to severe vascular, neurological and musculoskeletal disorders, culminating in permanent damage and disability [10]. Following prolonged exposure over the years, this often results in the manifestation of symptoms such as white fingers, numbness, tingling and diminished sensory function [7]. Musculoskeletal Disorders (MSDs) refer to injuries and conditions that impact the movement or musculoskeletal system of the human body, including soft tissues such as muscles, tendons, ligaments, joints, cartilage and the nervous system. These injuries have the potential to substantially limit mobility and escalate pain in the hand and arm [11]. The documented prevalence of Hand-Arm Vibration Syndrome (HAVS) in workplaces in the United States varies from 20 to 50% among power tool users, depending on factors such as the types of tools utilized, daily levels of vibration exposure, work practices and so on [9].

In a study conducted on tyre shop workers, it was discovered that merely 25% of the workers were conscious that extended exposure to vibration could adversely affect their hands. It is essential for health professionals involved in the health monitoring of workers exposed to Hand-Arm Vibration (HAV) to possess the necessary expertise for a comprehensive clinical assessment, thereby avoiding the misdiagnosis of HAVS symptoms [12].

Two occupations with significant exposure are those of quarry workers and foundry workers. These professions involve the use of distinct tools and entail varying levels of vibration exposure [13]. Foundries are industrial settings where metal casting processes take place, involving the melting and pouring of metals into moulds. Hand-Arm Vibration Syndrome (HAVS) is a significant concern in foundries, where workers often use vibrating tools and machinery such as Grinders, Sanders, Impact tools and Power tools. Foundry tools often produce high levels of vibration, especially during prolonged and repetitive use. These conditions leads to minor health risks such as numbness, tingling, reduced sensitivity, vascular issues, MSDs as well as major risks like Raynaud's syndrome and HAVS [14].

In order to enhance the safety of workers, ISO 8041:2005 is a global standard that outlines recommendations and criteria for the measurement and assessment of human exposure to mechanical vibration. This standard is particularly focused on the evaluation of whole-body vibration and hand-arm vibration. ISO 5349:2001 Part 1 outlines a procedure for determining daily exposure action values (EAVs) and exposure limit values (ELVs). In this method, vibration energy is measured as acceleration in meters per second squared (m s−2), expressed as the root mean square (RMS). The computation of vibration exposure entails calculating a time-weighted average over an 8 h working day [15], which forms the guidelines to increase the health and safety of workers who use vibrating tools in their daily life activities. Beyond vibration intensity, direction, exposure duration and frequency, significant factors influencing muscle fatigue include coupling forces and hand-arm posture. Multiple research efforts have investigated the influence of biodynamic elements such as coupling forces and hand-arm posture on the degree of vibration transmission to the hands and its resulting effects. These factors are detailed in Annex D of ISO 5349-1 [16].

A different investigation revealed that 21% of the surveyed workforce exhibited injuries related to vibration. This underscores the significance of health assessments conducted by qualified personnel in accordance with existing legislation. Regular medical examinations, ideally conducted every three years, can aid in the early detection of signs of damage caused by vibrations, helping prevent the escalation of issues. A noteworthy outcome from this study was that a significant portion of workers experiencing hand-arm vibration syndrome had medically manageable conditions, such as Carpal Tunnel Syndrome, with symptoms showing improvement after surgery [17].

From the study, it can be interpreted that early detection of HAVS can increase the worker safety. Hence it is important to develop an early detection system to predict HAVS occurrence for increased occupational safety.

The complex interplay between human anatomy, posture, and vibration characteristics significantly affects how vibrations are transmitted to the body. Work in foundries encompasses a range of activities, each presenting distinct degrees and forms of exposure to vibrations. Also the age-effect and exposure to cold could be contributing factors, the nervous system seems to be more susceptible to high-frequency vibrations than the vascular system [13].

The prediction of Hand-Arm Vibration Syndrome (HAVS) becomes complex due to the diverse array of tools and machinery utilized in foundry operations. Also work environments evolve over time i.e. changes in tools, workplace or practices. Current methods for monitoring whole body and arm vibration exposure often lack real-time capabilities as constant monitoring systems for WBV and HAV are still under development.

Traditional methods to predict HAVS often relies on fixed criteria or thresholds like vibration exposure limits which may not account for individual variability or adapt to new data. It also requires manual analysis of exposure data and symptoms are subjective and prone to human error, which makes analysing the data time consuming leading to delays in diagnosis and interventions. Traditional methods typically perform static analyses and do not continuously learn from new data. These properties of traditional methods make it less flexible to new patterns or changes making them less effective in dynamic environments. The lack of accurate prediction models for Human Arm vibration exposure limits our ability to assess potential health risks and design effective control measures [18].

These limitations can be reduced with help of Machine Learning. Large datasets comprising details regarding worker exposure to hand-arm vibration, health records, and other pertinent variables can be used to create predictive models with the aid of machine learning [3]. These models can evaluate the risk of HAVS for specific individuals or groups of individuals. It also helps to identify the most critical risk factors for HAVS by automatically selecting relevant features from the data. This can improve the understanding of the mechanisms that contribute to the development of the syndrome. Also it can integrate real-time monitoring systems to continuously access the risk of HAVS using current exposure and health data [19].

The main objective of this paper is to develop a machine learning algorithm using Linear Support Vector Machine model to predict Human Arm Vibration Syndrome (HAVS) for foundry workers and optimize the developed model to increase the efficiency of the developed predictive model. The expected outcome of the developed model is to predict occurrence of HAVS in foundry workers, so that necessary preventive measures can be taken to improve and manage occupational health of workers.

16.2 Machine Learning Approach for Early Detection of HAVS

Machine learning (ML) falls under the umbrella of artificial intelligence (AI) and is dedicated to constructing statistical models and algorithms. These enable computer systems to learn from data, improving their performance in specific tasks without the need for explicit programming. Machine learning essentially enables computers to learn from data and make judgments or predictions based on information and patterns found in the data. Generally it is defined as the ability to do tasks come from underlying model which is the process of learning from the provided data.

Machine Learning involves the creation of algorithms that empower a system to perform tasks based on an underlying model derived from the learning process of the developed model. The model learns by searching through the entire provided data to look for patterns. These patterns are expressed as mathematical structures such as polynomial functions or probability equations. These mathematical structures are used to predict the next required value. Machine learning happens in a mathematical space. The data representing a real world is a collection of attributes that define an entity.

Vibration data can be complex, non-linear relations between factors and multi-faceted, with multiple variables and features that affect exposure. Machine learning algorithms can capture and model these non-linear relationships. Also Real-time solutions for continuous vibration exposure monitoring can make use of machine learning algorithms to give employers and employees quick feedback.

Park et al. [3] developed machine learning algorithms for diagnostic disease prediction using lab tests as input data. From their study they found that AI models have better diagnostic accuracy and only takes 30% time compared to normal Doctor's diagnosis.

Amihai et al. [20] developed machine learning model using Random-Forest algorithm and persistence technique to predict the health of machines in an industry subjected to vibration. One of the major assumption made was that the past data will reflect the behaviour of the prediction for any component. They found that their developed models were able provide accurate results or forecast the condition of the machines accurately for 7 days ahead. However, sudden changes in the machines due to change in operation timings, number of parts produced and operator conditions were not considered, which brings down the efficiency of the developed model for using it for predicting ailments caused due to occupational vibration exposure.

Sigcha et al. [19] developed a prudent indication system which uses machine learning techniques for detecting use times for hand-held vibrating tools with help of a watch. 80% of the observations were allocated for training, while 20% were randomly selected for evaluation. Various tools can be distinguished using both linear and non-linear classification algorithms, achieving classification accuracy ranging from 79 to 98%.

Saran Kumar et al., [21] developed machine learning techniques to predict Cardiovascular diseases using various machine learning techniques and optimization of the developed models to increase accuracy. They used about 70% of the data for training the developed models and the rest for testing the developed models. From their research they determined that optimization techniques such as Hawk's, Gaussian and ensemble methods were the best among other techniques. They observed that the efficiency observed was increased by more than 15%.

Giuseppe Aiello et al., [22] researched about assessing the efficacy of cutting-edge machine learning techniques for recognizing activities in agroindustry and conducting risk assessments; showcasing the technical feasibility and effectiveness of current technologies in real-time monitoring of workers' exposure to vibration risks in the agricultural industry through the digitization of risk-related data. The achieved results underscore the efficacy of the proposed methodology, with an overall classifier accuracy reaching up to 98%. At this level of performance, the system not only serves as a valuable tool to heighten workers' awareness regarding the safety conditions of the workplace but also showcases its effectiveness.

From the literature survey, machine learning can improve the precision, accuracy and efficiency of risk evaluations and exposure monitoring within the workplace, it is a useful tool for forecasting HAV risks and can ultimately benefit worker health and safety.

However, developing machine learning algorithms for predicting/diagnosing ailments caused by occupational vibration exposure has several challenges. Acquiring reliable and extensive datasets related to occupational vibration exposure and associated health outcomes can be challenging. Data may be sparse, incomplete, or subject to various sources of bias. Gathering high-quality, labelled data is critical for training accurate machine learning models. The health effects of occupational vibration exposure, such as Hand-Arm Vibration Syndrome (HAVS) or musculoskeletal disorders, often develop over extended periods [23]. This long latency makes it difficult to establish direct causal links between exposure and ailments, and it requires long-term monitoring to detect and validate outcomes. The most important challenge is people respond differently to vibration exposure based on factors like genetics, age, overall health and pre-existing conditions [18].

Developing models that account for this inter-individual variability and provide personalized risk assessments can be challenging. Also managing sensitive medical and personal data in a responsible and ethical manner is crucial. Compliance with privacy regulations and ensuring data security are challenges in the development of healthcare-related machine learning algorithms. Also the cost for implementing these methods to determine health of individuals is also much higher as compared to that of traditional methods.

The initial setup expenses of traditional approaches could be cheaper than those of developing and integrating machine learning systems. But by utilizing affordable sensors and already-existing IT infrastructure, the initial investment in machine learning can be reduced. Also traditional methods incur higher ongoing labour and consumable costs while ML methods generally have lower operational costs after the initial setup due to automation minimizing the need for manual labour. While the initial setup and operational costs are noteworthy, the benefits in terms of reduced medical expenses, increased productivity and improved worker well-being far outweigh these costs. The enhanced accuracy and ability to continuously improve make ML a more cost-effective and sustainable solution for HAVS prediction and prevention in the long run.

Machine learning, through techniques like anomaly detection and risk stratification, provides powerful tools for the early detection of HAVS. By identifying unusual patterns and categorizing workers based on their risk levels, these techniques enable timely interventions, personalized preventive strategies and continuous monitoring, ultimately improving worker health and safety in environments with high vibration exposure.

For a number of reasons, including overcoming the drawbacks of conventional techniques and improving the management and prevention of this occupational health concern, the development of machine learning (ML) models to predict Hand-Arm Vibration Syndrome (HAVS) is highly significant. One of the most notable significance of the developed model would be that the developed model can identify early signs of HAVS, allowing for prompt intervention and preventive measures before the condition progresses to irreversible stages increasing worker safety. The data obtained from developing the model can be used for automated monitoring, risk management and improve Occupational health.

16.3 Methodology

Developing a machine learning algorithm to predict HAVS in Foundry workers involves several steps. It begins from identifing the problem, collection of data, understanding the data, model selection, split data, train model, validate model and model deployment.

The Predictive Model for early detection of HAVS using Machine learning algorithm is to be developed using the work-flow depicted in Fig. 16.1.

The first step is Data mining. It is defined as the collection of required input data for the developed Machine Learning algorithm or generally called as Prediction Inputs. This data is used to predict risk factors and identify methods for control along with time and cost reduction. It is further classified into data, information and Knowledge. The effectiveness of the developed model is dependent upon the accuracy of this data.

Linear kernel is used for development of the support Vector Machine (SVM) model hence called as LS-VM, it is used as it has less overfitting and can work with a variety of data. One of the advantages of this model is that is flexible in working with a variety of data. Both the training phase and inference phase is performed in this step.

Optimization is done to reduce any kind of anomalies or sort randomness in the model to increase the efficiency of the model. In the realm of machine learning and data analysis, anomalies pertain to observations or data points that exhibit significant deviations from the majority of the dataset. Anomalies are typically uncommon events in comparison to normal data points. This disparity in occurrence can present challenges during the training of machine learning models, as the majority of algorithms are tailored for optimal performance with balanced datasets.

In case of ML development for HAV and WBV majority of the anomalies are acceleration in anyone direction, operation timings, frequency of use and sometimes even age of the user. Special techniques such as oversampling, under sampling, or using different evaluation metrics, are often needed to address this issue.

Performance metrics contains the various variables which are used to determine the effectiveness of the developed model which includes precision, recall, f1-score and accuracy. Classification metrics, a subset of performance metrics in machine

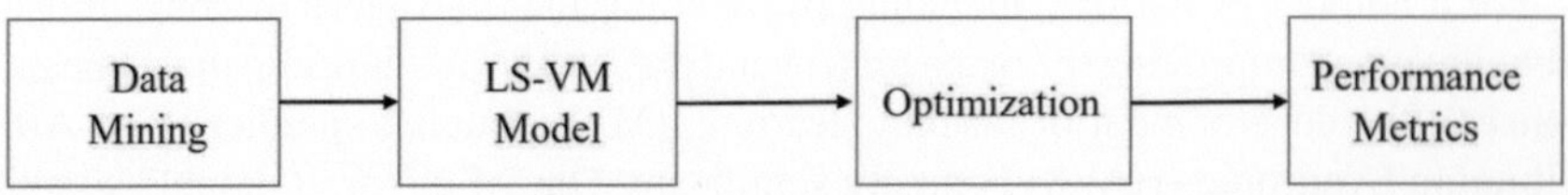

Fig. 16.1 Methodology

learning, are numerical measures employed to assess the effectiveness of a classification model. This model is specifically designed to predict the class or category of an input data point. These metrics assess how well the model's predictions align with the actual class labels, helping to understand the model's accuracy, precision, recall, and overall quality.

16.4 Development of Prediction Model

Machine learning is the process of creating algorithms that enable a system to perform tasks derived from an underlying model. These algorithms are created as a result of the model's learning process. In order to find patterns, the model searches through all of the supplied data. Polynomial functions and probability equations are examples of mathematical structures that are used to express these patterns. To forecast the next needed value, these mathematical frameworks are employed.

There are 3 major classifications of Machine Learning Algorithms namely Supervised machine learning, Un-supervised machine learning and Reinforcement learning. Supervised machine learning algorithms predicts known outputs from provided input data, whereas Un-supervised algorithms identifies hidden patterns and structures in the provided dataset to predict any outcome. Reinforcement learning is another important paradigm of machine learning which involves training an agent to make a sequence of decisions by interacting with an environment. The agent learns to achieve a goal by taking actions that maximize cumulative rewards which is to be provided by the developer.

Considering factors like literature survey, time consumption, computational capabilities LS-VM, a Supervised machine learning algorithm is selected to develop a predictive model to predict the occurrence of HAVS in Foundry workers.

16.4.1 Data Mining

The initial step of Model Development is collection and preparation of data. The data for development of the model is obtained from "Vibration related symptoms and signs in quarry and foundry workers", by Gerhardsson et al. [13], where data about the data about Foundry workers, obtained through sensor readings is considered to develop the predictive model. The input dataset used are as follows: peak amplitude, age, operation timings, frequency of use, acceleration in all 3 directions, heart and respiration rates. The required prediction output is set to HAVS. The input data for 15 foundry workers exposed to 8 Hz frequency and 10 mm peak amplitude is shown in Table 16.1.

The same Foundry workers were subjected to varying frequencies of 8, 16, 32, 64, 128, 256 and 512 Hz. Hence a total of 105 datasets were obtained, where the

Table 16.1 Prediction dataset for 15 foundry workers at 8 Hz frequency and 10 mm peak amplitude

ID	Age (years)	Operation time (years)	ax (m/s^2)	ay (m/s^2)	az (m/s^2)	Frequency of use	HAVS
1	22	3	8.9	6.2	2	2	0
2	22	3	8.9	6.2	2	3	0
3	25	4	8.9	6.2	2	4	0
4	25	5	8.9	6.2	2	5	0
5	27	5	8.9	6.2	2	7	0
6	27	6	8.9	6.2	2	5	0
7	29	8	8.9	6.2	2	1	0
8	29	11	8.9	6.2	2	2	1
9	33	13	8.9	6.2	2	35	1
10	33	11	8.9	6.2	2	8	1
11	35	14	8.9	6.2	2	5	1
12	35	10	8.9	6.2	2	4	0
13	38	14	8.9	6.2	2	5	0
14	40	8	8.9	6.2	2	11	0
15	46	9	8.9	6.2	2	2	0

occurrence of HAVS is denoted as 1. This dataset is to be used as input for the next step.

16.4.2 Linear Support-Vector Method

This model is also called as black box technique due to the presence of various unknown parameters in the provided datasets. In this method data separation for non-linear classes are done using kernel tricks. All parameters are also tuned using kernel tricks. It is one the most commonly used machine learning algorithm in Healthcare industry as the model is known for its robustness, scalability and ability to handle large datasets [3, 21].

The model is developed by drawing lines to split the given data. Lines are selected or drawn to have an optimum distance from the varying data categories. This optimum line is called as Hyper-line, this line has the maximum space that separates the lines used to split the data. These split lines are called as Support Vectors. When an object is to be predicted the distance from the developed hyper-line is measured to know the accuracy of the developed model which are as shown in Fig. 16.2.

Overfitting of data is very less compared to other methods as Linear SVM automatically determines the hyperplane that best separates the classes while maximizing the margin between the closest data points (support vectors) of each class. It also includes a regularization parameter (C) that controls the trade-off between maximizing the

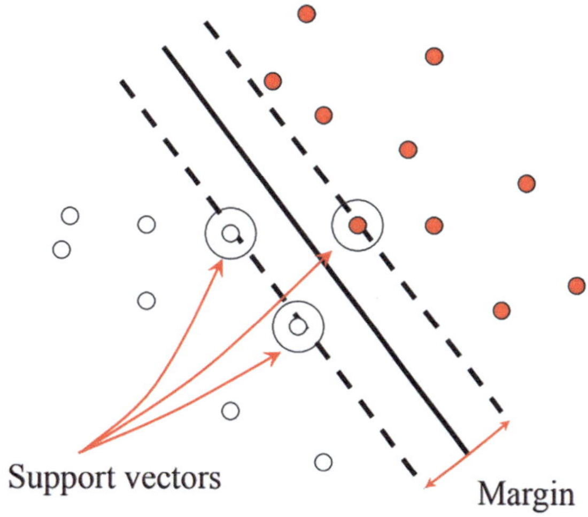

Fig. 16.2 Working schematics of SVM

margin and minimizing classification error. A properly chosen regularization parameter helps prevent the model from fitting too closely to the training data by penalizing overly complex models.

One important feature of SVM is its kernels. Kernel is the categorization agent of the provided data. Some kernels are Linear, Non-linear, RBF (Radial Basis function) and poly (Polynomials). These are used to convert the hyper-lines from on dimensional space to two dimensional space. For this model, we use linear kernel. While linear SVM uses a linear kernel, it is not the reason for the model to be referred as LSVM. As the linear in LSVM refers to the type of decision boundary that the algorithm seeks to find i.e. the decision boundary for LSVM is a straight line (or hyperplane in higher dimensions) that separates the classes in the feature space.

30% of the dataset was used for testing the developed algorithm and the rest are to be used to train the model, to increase the effectiveness of the developed model.

The development of the machine learning model is performed using a machine language called as Python. Data science IDE's like Pandas, Numpy, Matplot and Sci-kit learn were used to develop the model. Jupyter notebook a free cloud-based platform used for machine learning is used for model development and deployment.

16.4.3 Optimization of Developed Model

Improving predictive model efficiency, cutting down on training time and optimizing model performance are the goals of machine learning algorithm optimization. The process of optimization is iterative, and the methods chosen are contingent upon the particular issue at hand and the resources at hand. To attain optimal accuracy and efficiency, one must experiment with different approaches and keep a close eye on model performance.

The dataset used for developing the predictive model falls under the classification of Binary datasets Optimization techniques for binary datasets are similar to those used for general machine learning tasks, but they are specifically tailored for binary classification problems where the target variable has two possible outcomes.

From literature survey, it is found that by making modifications to optimization techniques such as Meta-Heuristic, Gaussian and Hawk's Optimizer, these techniques can be adopted for Binary datasets also increasing the performance metrics of the developed model. From study performed by Saran Kumar et al., [21], it is noted that SVM's produce better results when optimized with Gaussian Optimizer when used to predict Cardiac issues. Hence Gaussian Optimizer is used to optimize the developed Machine Learning Model, since HAVS also falls under Human Body ailment.

16.5 Results and Discussions

Optimized Machine learning Models to predict the ailments caused by exposure to occupational vibration exposure are developed and deployed. The performance metrics considered for this work are Accuracy, Precision, recall, f1-score and support. These metrics together are called as classification metrics. The results are compared before and after optimization to check for effectiveness. For better inference of the values obtained some understanding of these Classification metrics are required.

16.5.1 Accuracy

A fundamental classification parameter in machine learning is accuracy, which assesses the overall precision of a model's predictions. It quantifies the percentage of correctly identified cases in a dataset compared to all instances in that dataset. The following formula 16.1 is used to determine accuracy in a binary classification task.

$$\text{Accuracy} = \frac{\text{Number of Correct Predictions}}{\text{Total Number of Prediction}} \tag{16.1}$$

But, it is crucial to exercise caution when relying only on accuracy as an evaluation criterion, particularly when the information is unbalanced and one class is significantly more common than the others. A high accuracy score may be deceptive in certain cases [21].

16.5.2 Classification Metrics

i. **Precision**

Precision, a classification parameter, evaluates the model's ability to predict positive outcomes, particularly in binary classification. It computes the percentage of all cases that the model correctly identified as positive, ensuring that the predicted positive instances are indeed accurate. The following formula 16.2 is used to determine precision:

$$\text{Precision} = \frac{\text{True Positives(TP)}}{\text{True Positives(TP)} + \text{False Positives(FP)}} \tag{16.2}$$

- True Positives (TP) represent instances that are genuinely positive and were accurately predicted as positive by the model.
- False Positives (FP) denote instances that are genuinely negative but were inaccurately predicted as positive by the model.

When it comes to minimizing false positives, precision is very important. Precision is a crucial parameter in applications such as information retrieval, fraud detection and medical diagnostics. A high precision score indicates that the model generates few false alarms and provides trustworthy positive predictions.

ii. **Recall**

Also termed sensitivity or true positive rate, recall is a classification metric that gauges a model's proficiency in correctly identifying every positive instance within all the actual positive cases. In simpler terms, recall quantifies the percentage of accurately predicted positive cases, or true positive predictions, out of all the genuine positive instances. The following formula 16.3 is used to calculate recall:

$$\text{Recall} = \frac{\text{True Positives(TP)}}{\text{True Positives(TP)} + \text{False Negatives(FN)}} \tag{16.3}$$

- False Negatives (FN) represent instances that are genuinely positive but were inaccurately predicted as negative by the model.

When the penalty of missing positive occurrences (false negatives) is significant, recall becomes even more crucial. The model is effective at detecting the majority of positive events when the recall value is high.

iii. **F1-score**

The F1 score, a well-rounded measure of a model's performance, is a classification statistic employed in machine learning. It integrates both recall and precision, providing a single numerical representation. Since precision and recall frequently have a trade-off connection, it is especially helpful when trying to find a balance between recording all positive instances and making accurate positive predictions. The F1 score is computed using the following formula 16.4 and is the harmonic mean of recall and precision:

$$\text{F1} - \text{Score} = \frac{2 * \text{Precision} * \text{Recall}}{\text{Precision} + \text{Recall}} \tag{16.4}$$

The F1 score ranges from 0 to 1. The higher the value the better the developed model. It offers a solitary numerical figure that encapsulates a model's capacity to generate precise positive forecasts while encompassing the majority of positive instances.

iv. **Support**

"Support" in classification metrics describes the actual number of examples in each class or category. It offers details on how instances are distributed throughout the various classes in a dataset. In multi-class classification situations, support is frequently used to denote the number of cases that belong to each class.

Support is a crucial factor to take into account when evaluating and choosing a model because the metrics computed for different classes can differ depending on their support levels.

16.5.3 Learning Curve

The graph that shows the relationship between the number of estimators (e.g., decision trees or base learners) and the accuracy of a machine learning model, is called as Learning curve. The objective is to determine the ideal number of estimators to minimize overfitting and maximize the model's performance.

The graph might show how the accuracy changes as more estimators are added to the ensemble, helping to identify the point at which adding more estimators no longer significantly improves the model's performance, potentially indicating a balance between bias and variance.

From Fig. 16.3a it can be observed that LS-VM model without optimization has no difference between train and test accuracy which means it has a Good/balanced fit. This implies that the model has grasped the pertinent patterns within the data and can adeptly extend its understanding to new, unfamiliar challenges.

Whereas in Fig. 16.3b it can be observed that there is a large deviation between train and test accuracy which means it is Overfitting. This phenomenon occurs when the model is overly intricate and captures noise or random fluctuations existing in the training data. This decreases the robustness of the optimized LS-VM model. It can be reduced by further optimizing the model or increasing the quality of the dataset (increasing the datasets also provide a better solution).

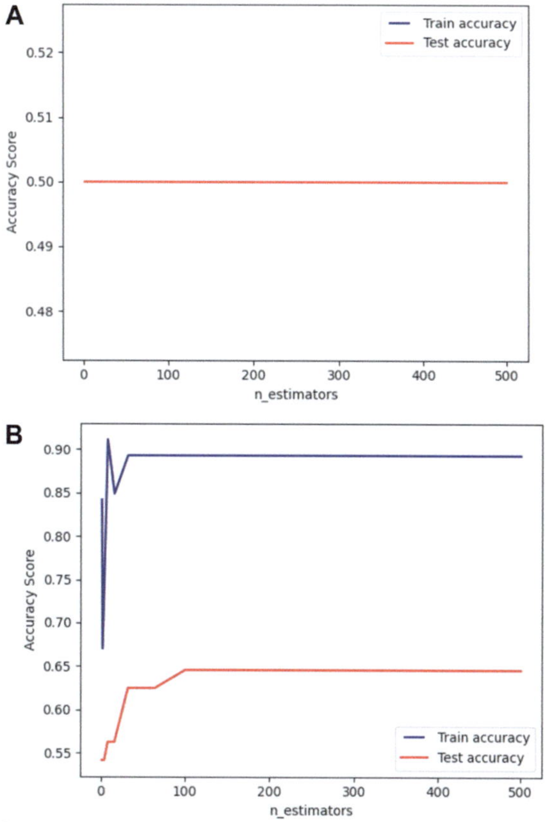

Fig. 16.3 Learning curves of LS-VM model **a** without optimization **b** with optimization

16.5.4 Comparison of Classification Metrics

From comparing Performance metrics from Table 16.2 it can be observed that the developed LSVM model has a 75% accuracy of predicting the on-set of HAVS using the provided dataset. It is also noted that the when compared with LSVM, Optimized LS-VM model has better accuracy by 3.57%, precision by 9%, F1-score by 1%, support by 29.41% and only falls behind in Recall metrics by 9%. Which makes it more suitable for Predicting HAVS in Foundry workers and increase occupational safety by increasing worker safety. It can also be inferred that the use of Gaussian Optimizer to optimize the developed LSVM model shows increase in performance metrics, however the increase is much less than the expected outcome.

Table 16.2 Comparison of performance metrics before and after optimization

	LS-VM	Optimized LS-VM
Accuracy (%)	75	78.57
Precision (%)	75	84
Recall (%)	100	91
F1-score	0.86	0.87
Support	24	34

16.6 Conclusion

With the help of the results shown above, it can be said that usage of Machine learning algorithms could reduce the risk of workers being affected by HAVS, thus increasing occupational safety. It also reduces the diagnostic time for determining the symptoms and offering alerts to workers to take protective measures to reduce the risk of HAVS. Some important takeaway of the paper are given as follows.

- Reduction of diagnosing time by more than 30% when using LS-VM model as compared to that of traditional diagnosing methods followed by doctors.
- After Optimization of the LS-VM model only a small increase in accuracy (3.57%) can be seen because of large overfitting of data in the used dataset. This is due to larger noise or improper data availability. It can be overcome by simply increasing the total number of datasets available for the model development.
- The developed model can also be used to determine other symptoms like white finger and Raynaud's syndrome only with minor changes to the structure of the LS-VM model. This further boosts the idea of using Machine Learning Algorithms for prediction of ailments in workers due to occupational vibration exposure, contributing to increase in worker health and safety.
- The developed can be integrated with wearable technology and IoT devices for comprehensive health monitoring, showcasing the potential for innovative applications in the field.

Thus the study intends to improve HAVS early detection, offer individualized health monitoring, and facilitate data-driven decision-making by utilizing machine learning. Further study can be performed on use of other optimization techniques to improve the performance of the developed predictive model. This study is crucial to the advancement of occupational health management because of its financial advantages, regulatory compliance and research basis.

References

1. Griffin, M.J.: Measurement, evaluation, and assessment of occupational exposures to hand transmitted vibration. Occup. Environ. Med. **54**(2), 73–89 (1997). [PubMed: 9072014]

2. Sui, Y., Ding, R., Wang, H.: An integrated management system for occupational health and safety and environment in an operating nuclear power plant in East China and its management information system. J. Clean. Prod. **183**, 261–271. https://doi.org/10.1016/j.jclepro.2018.02.101
3. Park, D.J., Park, M.W., Lee, H., Kim, Y.-J., Kim, Y., Park, Y.H.: Development of machine learning model for diagnostic disease prediction based on laboratory tests. Sci. Rep. **11**(2021), 67–75 (2021)
4. Dong, R.G., Welcome, D.E., Peterson, D.R., Xu, X.S., McDowell, T.W., Warren, C., Asaki, T., Kudernatsch, S., Brammer, A.: Tool-specific Performance of vibration-reducing gloves for attenuating palmtransmitted vibrations in three orthogonal directions. Int. J. Ind. Ergon. **44**, 827–839 (2014). [PubMed: 26726275]
5. Welcome, D.E., Dong, R.G., Xu, X.S., Warren, C., McDowell, T.W., Wu, J.Z.: An examination of the vibration transmissibility of the hand-arm system in three orthogonal directions. Int. J. Ind. Ergon. **45**, 21–34 (2015). [PubMed: 26635424]
6. Xu, X.S., Welcome, D.E., Warren, C., McDowell, T.W., Wu, J.Z., Dong, R.G.: Vibrations transmitted from human hands to upper arm, shoulder, back, neck, and head. Int. J. Ind. Ergon. **62**, 1–12 (2017). [PubMed: 29123326]
7. Clemm, T., Færden, K., Ulvestad, B., Lund, L.-K., Nordby, K.-C.: Dose–response relationship between hand–arm vibration exposure and vibrotactile thresholds among road workers. Occup. Environ. Med. 77, 188–193 (2019). https://doi.org/10.1136/oemed-2019-105926
8. Xua, X.S., Welcome, D.E., McDowell, T.W., Warren, C., Service, S., Lin, H., Chenb, Q., Dong, R.G.: 2021 An investigation of the effectiveness of vibration-reducing gloves for controlling vibration exposures during grinding handheld workpieces. App. Ergon. **95**, 103454 (2021). https://doi.org/10.1016/j.apergo.2021.103454
9. Fact sheet: occupational exposure to hand-arm vibration (HAV). Resource produced by acquisition safety, U.S. Navy Safety Center Afloat Safety Programs Office, Washington DC
10. Lai, S.-K., Chui, J., Tong, L., Sun, J.-Q.: A human-based study of hand–arm vibration exposure limits for construction workers. J. Vib. Eng. Technol. **7**, 379–388 (2019). https://doi.org/10.1007/s42417-019-00123-2
11. Roseiro, L.M., Paulino, M.F., Neto, M.A., Amaro, A.M.: Analysis of hand-arm vibration syndrome in drummers. Int. J. Ind. Ergon. **66**, 110–118 (2018). https://doi.org/10.1016/j.ergon.2018.02.014
12. Qamruddin, A.A., Husain, N.R.N., Sidek, M.Y., Hanafi, M.H., Ripin, Z.M., Ali, N.: Prevalence of hand-arm vibration syndrome among tyre shop workers in Kelantan, Malaysia. J. Occup. Health **2019**(61), 498–507 (2019). https://doi.org/10.1002/1348-9585.12078
13. Gerhardsson, L., Ahlstrand, C., Ersson, P., Jonsson, P., Gustafsson, E.: Vibration related symptoms and signs in quarry and foundry workers. Int. Arch. Occup. Environ. Health **94**, 1041–1048 (2021). https://doi.org/10.1007/s00420-021-01660-8
14. Brocal, F., Sanchez, A., Gonazalez, C., Fuentes, J.L., Sebastian, M.A.: Proposed methodology for the study of the level of emerging risk from exposure to hand-arm vibrations in manufacturing environments. Procedia Manuf. **13**(2017), 1373–1380 (2017)
15. Clemm, T., Nordby, K.-C., Lunde, L.-K., Ulvestad, B., Bråtveit, M.: Hand-arm vibration exposure in rock drill workers: a comparison between measurements with hand-attached and tool-attached accelerometers. Ann. Work. Expo. Health. **65**(9), 1123–1132 (2021). https://doi.org/10.1093/annweh/wxab051
16. Kordmiri, S.H.M., Aliabadi, M., Golmohammadi, R., Bovenzi, M., Farhadian, M.: Combined effects of noise and hand-transmitted vibration on workers' muscle and mental fatigues in a simulated construction operation. Med. Lav. **114**(2), e2023021 (2023). https://doi.org/10.23749/mdl.v114i2.13903
17. Vihlborg, P., Bryngelsson, I.-L., Lindgren, B., Gunnarsson, L.G., Graff, P.: Association between vibration exposure and hand-arm vibration symptoms in a Swedish mechanical industry. Int. J. Ind. Ergon. **2017**, 1–5 (2017). https://doi.org/10.1016/j.ergon.2017.02.010
18. Shah, V.A., Casadio, M., Scheidt, R.A., Mrotek, L.A.: Vibration propagation on the skin of the arm. Appl. Sci. (Basel). **9**(20) (2019)

19. Sigcha, L., Pavón, I., Nisi, S., de Arcas, G.: Automatic identification of hand-held vibrating tools through commercial smartwatches and machine learning. Occup. Environ. Saf. Health II, Stud. Syst. **277**
20. Amiha, I., Gitzel, R., Kotriwala, A.M., Pareschi, D., Subbiah, S., Sosale,G.: An industrial case study using vibration data and machine learning to predict asset health. In: Business Informatics
21. Arunachalam, S.K., Rekha, R.: A novel approach for cardiovascular disease prediction using machine learning algorithms. In: 2022, concurrency and computation: practice and experience, vol. 34, no. 19, pp. 1–16, ISSN: 1532-0634
22. Aiello, G., Catania, P., Vallone, M., Venticinque, M.: Worker safety in agriculture 4.0: a new approach for mapping operator's vibration risk through machine learning activity recognition. Comput. Electron. Agric. **193**, 106637 (2022)
23. Park, W., Kim, S.-P., Eid, M.: Neural coding of vibration intensity. Percept. Sci. Front. **15**, 682113 (2021)

Part IV
Analytics in Agricultural and Climate Sciences

Chapter 17
Daily Rainfall Prediction Using Recurrent Neural Network

Bagati Sudarsan Patro and Prashant P. Bartakke

Abstract One of the greatest global challenges in meteorology is predicting the daily precipitation total. The agricultural industry in India is highly dependent on precipitation for water supplies and general socio-economic growth, making accurate rainfall predictions essential. Weather patterns are dynamic, making it difficult to anticipate rainfall with enough accuracy and timeliness, a necessity for many industries, including aviation, energy production, and tourism. Over the past two decades, rainfall forecasting has shown a surprising degree of precision because of artificial intelligence approaches, particularly deep learning (DL). These methods are quite good at identifying hidden trends in past meteorological data, which helps to improve rainfall prediction accuracy. To anticipate the daily intensity of rainfall, this study will use machine learning approaches to determine the major atmospheric factors that contribute to rainfall. To estimate daily precipitation, the study explored various configurations of recurrent neural network (RNN)-based deep neural network models, including Long Short-Term Memory (LSTM), Stacked-LSTM, Bi-LSTM, Gated Recurrent Units (GRU) and Bi-GRU. The RNN models comprised 50 hidden layers, ReLU activation at the input layer and linear activation at output layer. Training utilized Adam Optimizer with a fixed learning rate of 0.000001 and a batch size of 32, Epochs of 100. The dataset, primarily focusing on the India Meteorological Department (IMD) Automatic Weather Station (AWS) dataset from the Almora district of Uttarakhand, was split into training and testing sets with portions of 0.8 and 0.2, respectively. To improve data quality, a thorough pretreatment of AWS data was conducted, as this data is prone to anomalies and missing values. The models' performances are evaluated using appropriate metrics such as RMSE, MSE, MAE

B. S. Patro (✉) · P. P. Bartakke
Department of Electronics & Telecommunication Engineering, COEP Technological University (COEP Tech), Wellesley Road, Shivajinagar, Pune 411005, Maharashtra, India
e-mail: sbp21.extc@coeptech.ac.in; sudarsan.patro@imd.gov.in

P. P. Bartakke
e-mail: ppb.extc@coeptech.ac.in

B. S. Patro
Surface Instrument Division, Climate Research & Services, India Meteorological Department, Shivajinagar, Pune 411005, Maharashtra, India

J. Maiti et al. (eds.), *Operations Research and Data Analytics: Current Trends and Future Perspectives*, Lecture Notes on Multidisciplinary Industrial Engineering,
https://doi.org/10.1007/978-981-96-8327-7_17

and R^2. After investigating various RNN models, it was discovered that one predictive GRU model performed better than the other, obtaining noteworthy metrics like a reduced RMSE of 0.00732, MSE of 0.00005, MAE of 0.00284 and a higher R^2 of 0.95524.

Keywords Deep learning · RNN · LSTM · GRU · Rainfall prediction

17.1 Introduction

Commencing with proactive early warnings and accurate predictions is essential to mitigate the potential devastation caused by extreme rainfall events such as cloudbursts and mini-cloudbursts, which can trigger flash floods and landslides, endangering lives and properties. This study introduces a novel approach to rainfall prediction utilizing deep learning techniques, specifically recurrent neural network (RNN) configurations such as LSTM, Stacked-LSTM, Bi-LSTM, GRU and Bi-GRU. What sets this research apart from existing studies in the field is its comprehensive exploration of these diverse RNN architectures, coupled with a focus on the IMD AWS dataset from the Almora district of Uttarakhand. In this study, we analyse various configuration of algorithms and applying them to specific regional dataset. This study also offers a fresh perspective on rainfall prediction with deep learning, potentially uncovering insights and performance improvements that have not been explored yet. Heavy rainfall poses serious risks to people's health and safety and is a major contributor to yearly global problems like floods and other natural catastrophes. Precise forecasts of the amount of rainfall are essential, especially in nations like India where agriculture is the main industry [1]. Rainfall's non-linear properties highlight how effective machine learning (ML) techniques are when compared to other methods. Rainfall has been important in both historical and contemporary global situations. Factors including humidity, temperature, water levels, and more might affect its incidence [2]. A region's crops may be harmed by excessive rainfall, which could eventually cause the area's agriculture to be disrupted. Furthermore, a lot of rain can cause natural disasters like cloud bursts, droughts, and floods. These events are typically brought on by quick landslides that occur after a lot of precipitation, which can be harmful to people and property [3]. Predicting rainfall is essential for predicting long-term precipitation in each area, particularly for the expansion of the agricultural industry and the economy. Predicting extreme weather events, such as cloudbursts, accurately is essential to averting damages such as landslides, floods, waterlogging and obstructions in river channels. Since existing forecasting systems find it difficult to uncover hidden patterns or nonlinear trends in rainfall data, the inherent uncertainty in rainfall presents a substantial difficulty. This restriction has resulted in inaccurate projections, which have had a negative influence on infrastructure, injuries and fatalities as well as significant losses. Predicting and managing large-scale operations ahead of time requires precise rainfall cloudburst predictions. The selected study regions, in Uttarakhand, India, are distinguished by an increased frequency of cloudbursts. The northern Indian state of Uttarakhand has a variable climate due to its diversified geography, which includes plains, hills, and mountains. The state has noticeable seasonal variations. Uttarakhand has pleasant summertime temperatures in the hills and colder weather at higher elevations from March to June. Significant rainfall occurs during the monsoon season (July to September), especially in steep places. The area is also prone to landslides, flash floods, cloudbursts and mini cloudbursts. The months of October and November mark the beginning of autumn, when the monsoon subsides, and the weather becomes more agreeable. In Uttarakhand, winter (December to February) provides cold weather, particularly at higher elevations. While hill stations like Mussoorie and Nainital may have snowfall, the plains receive cool temperatures [4–6]. Thus, managing the rainfall water for agricultural and water supply requires precise daily rainfall prediction, which is a difficult problem to accomplish [7].

A cloudburst, according to the definition provided by the Indian Meteorological Department (IMD), is an intense and abrupt outburst of condensed water vapour from a cumulo-nimbus cloud that releases a significant amount of rain (100 mm) in less than an hour. This phenomenon is frequently typified by heavy rain fall in a particular place, which causes the water to quickly collect on the ground.

Cloudbursts can cause landslides and flash floods because the ground cannot absorb large volumes of water in such a short period of time. Because of their abrupt and powerful nature, cloudbursts are considered localized, high-impact meteorological occurrences that present considerable issues. In urban locations, even a little instance of a micro cloudburst can have significant effects. The 24-h accumulated rainfall classification according to the India Meteorological Department is as follows (Table 17.1):

Using meteorological data from many nations, several researchers have conducted studies targeted at improving the daily, monthly, and annual rainfall prediction accuracy. Numerous methodologies, such as data mining tools and Big Data analysis, have been used in these investigations [8, 9], With a variety of techniques for machine learning [7, 10–12]. The results of these studies indicate that in terms of weather and rainfall prediction, machine learning algorithms have proven to be more effective than conventional deterministic techniques. As a result, the goal of this research is to analyse deep learning algorithms and determine which ones work best for precise rainfall prediction.

Table 17.1 Rainfall accumulation over a 24-h period is categorized

Serial number	Terminology	Rainfall range in mm
1	Very light rainfall	Trace–2.4
2	Light rainfall	2.5–15.5
3	Moderate rainfall	15.6–64.4
4	Heavy rainfall	64.5–115.5
5	Very heavy rainfall	115.6–204.4
6	Extremely heavy rainfall	$> = 204.5$

Rainfall frequency and severity are influenced by a wide range of environmental elements, such as temperature, relative humidity, sunshine, pressure and more. Rainfall is influenced by these factors, either directly or indirectly. This study uses machine learning approaches, notably RNN based algorithms, to effectively predict rainfall since it recognizes the importance of atmospheric parameters including temperature, relative humidity, pressure, wind speed and more. Finding the key atmospheric parameters influencing rainfall and predicting its daily intensity are the main goals [13, 14].

To make sure it is suitable for experimentation, raw data from the Indian Meteorological Department (IMD) is preprocessed. The preprocessed data is carefully linked with the rainfall variable for each characteristic to identify relevant components. The study examines different configurations of recurrent neural network (RNN)-based deep neural network techniques for daily precipitation prediction, with a focus on the Almora district of Uttarakhand. To improve data quality, the Automatic Weather Station (AWS) data, which is prone to anomalies and missing values, is rigorously preprocessed. A predictive GRU model is found to be superior through the investigation of several RNNs, such as Long Short-Term Memory (LSTM) and Gated Recurrent Units (GRUs), obtaining noteworthy metrics like a reduced RMSE of 0.00732, MSE of 0.00005, MAE of 0.00284 and a higher R^2 of 0.95524.

17.2 Related Works

Multivariate Linear Regression, Random Forest and XGBoost Gradient Descent are some of the machine learning techniques used. Using these techniques, one can forecast the amount of rainfall by determining the correlation between important atmospheric characteristics and the atmospheric factors that affect rainfall [7]. Tharun et al. [15] investigated the use of data mining techniques for rainfall prediction and presented a thorough analysis of rainfall intensity prediction in Coonoor, Tamil Nadu. Regression methods such as Decision Tree (DT), Random Forest (RF) and Support Vector Regression (SVR) were used in the study. The study evaluated variables such daily temperature, humidity, cloud speed, wind direction and wind speed. To improve forecast accuracy, data from neighbouring stations within a 7 km^2 area were also included. Rain is one of nature's greatest gifts; in developing countries like India, it is vital to agriculture [16, 17]. One of the meteorological elements that continues to have the greatest impact on a variety of aspects of our daily life is rainfall. Rainfall has major socioeconomic repercussions that range from damaging infrastructure in the event of a flood to causing interruptions in the transportation network. It is anticipated that climate change would have disastrous effects in the future, increasing the frequency of catastrophic events like floods [18, 19]. Effective algorithms for assessing and predicting patterns of rainfall are provided by machine learning. This study investigates the application of different RNN-based deep learning architecture configurations, including LSTM (Long Short-Term Memory) and GRU (Gated Recurrent Units) models, for time series weather data-driven weather prediction.

17.3 Data

The World Meteorological Organization's (WMO) stringent requirements are rigorously adhered to by the India Meteorological Department (IMD), which methodically manages a vast network of about 2500 Automatic Weather Stations (AWS) strategically positioned throughout the country. Time Division Multiple Access (TDMA) and General Pocket Radio Service (GPRS) are two of the sophisticated technologies used in the complex procedure to transfer AWS data to a central server via cutting-edge satellite connection. The focus of this study is to examine the extensive dataset that was provided by the IMD AWS, which is based in Almora, which is a small town in Uttarakhand, India. The Almora station, which is located at Latitude 29.6002°N, Longitude 79.6651°E, at an elevation of 1620 m, is a key location for the thorough analysis carried out here. The dataset, which was collected at exact intervals of 15 min, covers a wide period, from January 2020 to September 2023, and offers a strong basis for further research. An emphasis is made on the hourly measurements of different atmospheric variables within the scope of this study. These factors, which include wind direction, speed, pressure, temperature and relative humidity, all play crucial roles in illuminating the weather's dynamic character. The main objective is to use these detailed observations to create forecast models for daily rainfall, which is represented by RAIN_DAILY (in mm), during the duration of this experiment. By revealing the complex dynamics and patterns of weather occurrences in the targeted area, this all-encompassing strategy hopes to improve our comprehension of meteorological processes and enable more precise forecasting methodologies. In particular, the study looks at hourly measures of pressure, temperature, relative humidity, wind direction and wind speed to forecast daily rainfall (RAIN_DAILY in mm) throughout this experimental period (Fig. 17.1).

17.4 Methods

The advancement of rainfall forecast accuracy has entered a new phase thanks to the confluence of Artificial Intelligence (AI), Machine Learning (ML), Data Processing (DP) and Deep Learning (DL) technologies. As a branch of machine learning, deep learning is very good at managing complex and unstructured data. This makes it particularly useful for extracting complex patterns from large-scale meteorological datasets. Deep Learning models' hierarchical layers of neural networks allow them to automatically discover and depict intricate associations in the data, which helps to produce more complicated and precise rainfall predictions. Furthermore, by combining AI, ML, DP and DL, models may be continuously improved upon as they encounter fresh data, guaranteeing their flexibility to changing weather patterns. By utilizing effective data processing methods, the incorporation of Deep

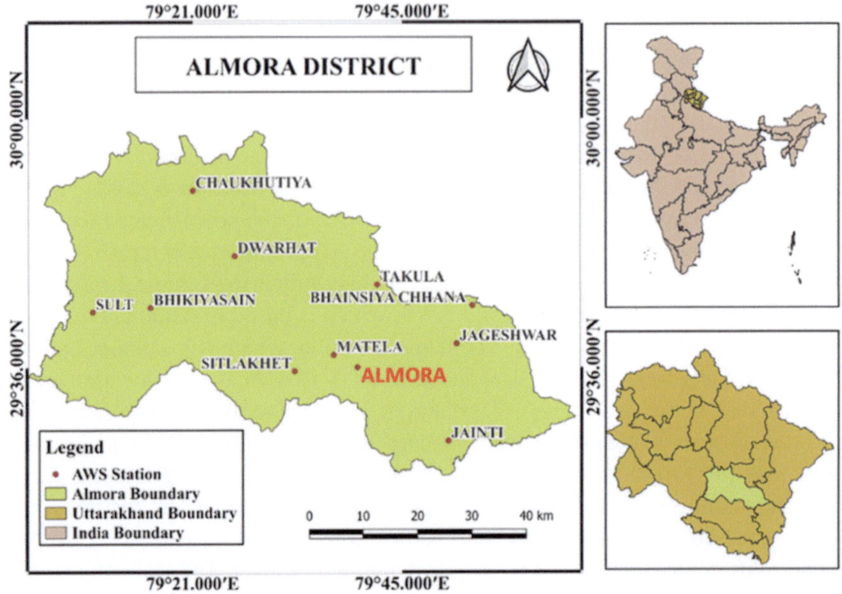

Fig. 17.1 Study area map of Almora district, located in the Uttarakhand State of India

Learning improves rainfall forecast accuracy, providing a more complete and dependable resource for researchers and meteorologists to better comprehend and react to changing weather conditions [16] (Fig. 17.2).

This research study presents a simplified method for utilizing data from Automatic Weather Stations (AWS) through a block diagram. To guarantee its quality, the data is carefully pre-processed starting with its integration into a Data Collection

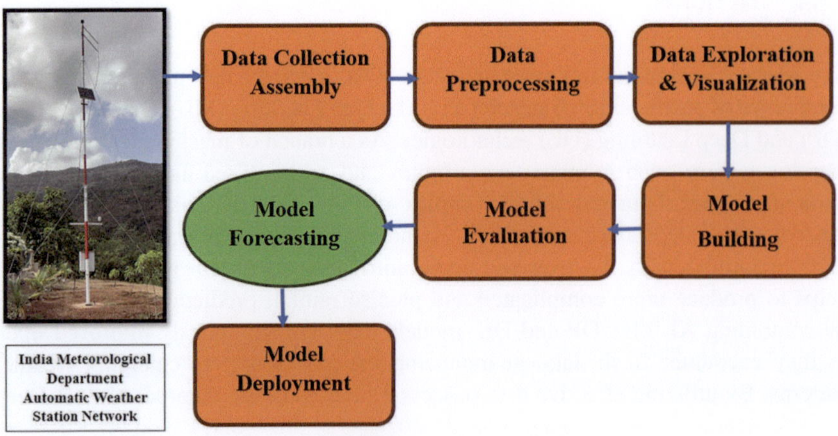

Fig. 17.2 Block diagram of proposed method

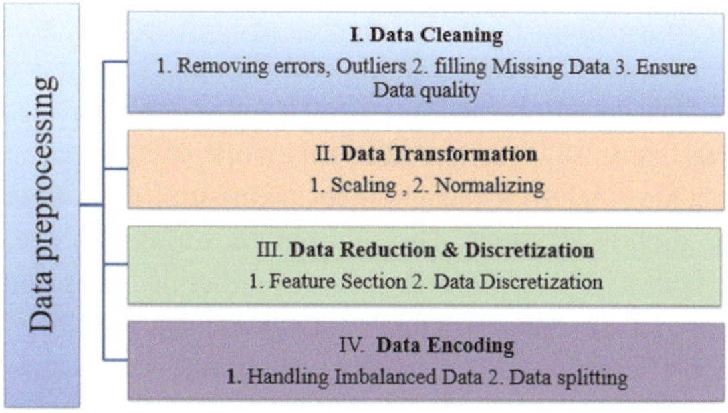

Fig. 17.3 Meteorological data refinement process

Assembly. The next steps are Visualization and Data Exploration, where patterns are found in meteorological variables. In the future, Model Forecasting will use sophisticated machine learning techniques. Models will undergo thorough evaluation prior to being optimized in Model Building. When the finished models are used for practical purposes, they provide insightful information about weather forecasting and analysis. The comprehension and application of AWS data in meteorological research are improved by this methodical methodology. Because autonomous weather stations are deployed in remote areas, there are several issues associated with using data obtained from these stations in time series. Several variables, including power supply stability, sensor condition, data logger efficiency, transmission system reliability, reception module functionality, server performance and the efficiency of the quality control (QA/QC) algorithm, affect the likelihood of successful data reception. This system requires multiple aspects to be managed to ensure precision. As a result, carefully pre-processing and refining the data quality before using it as input for model training and testing is a crucial step in the time series forecasting process (Fig. 17.3).

This research paper's deep learning model adheres to a methodical set of procedures that guarantee the model's effectiveness and resilience. Data cleaning, a crucial stage in the process that involves finding and eliminating mistakes and outliers as well as filling in missing data to improve overall data quality, is where the process starts. The model goes through Data Transformation, which includes scaling and normalizing processes, after Data Cleaning. Standardizing the data is a crucial step in ensuring that the deep learning model can handle and learn from the data in an efficient manner. Data Reduction and Discretization, which includes Feature Selection and Data Discretization procedures, is the next phase. By selecting and keeping the most relevant variables, feature selection helps to maximize the performance of the model. Concurrently, Data Discretization expedites the learning process by making it easier to transform continuous data into discrete categories. The model next goes through data encoding, which addresses imbalanced datasets and divides the data into sets for testing and training. The separation of the dataset into training and testing sets facilitates the assessment of the model's performance on untested data, and this step is essential for producing a representative and balanced dataset. Together, these actions essentially create a thorough data preparation pipeline that makes sure the deep learning model has access to high-quality, standardized data to maximize its performance during training and afterwards in real-world scenarios [19, 20].

The viability of using several Recurrent Neural Network (RNN) models for rainfall prediction is investigated in this work. With a particular focus on the use of RNN, a neural network designed for sequential data, the study aims to improve rainfall prediction accuracy. The study intends to obtain insights into RNNs' suitability and efficacy in precipitation pattern predicting. This study advances our knowledge of how to use deep learning techniques, specifically RNNs, to increase the accuracy and effectiveness of rainfall prediction.

The acquisition of surface weather station data captured by an Automatic Weather Station (AWS) is where the study paper's flow diagram begins. The initial input for a systematic approach to data preprocessing is this raw data from AWS. Important procedures including data cleaning, filling in missing data, identifying and managing outliers, and data scaling and normalization are all included in this preparation stage (Fig. 17.4).

After the painstaking preprocessing, the flow moves to Modelling Training and Forecasting, where sophisticated methods are used to forecast the amount of rainfall each day. Using the processed data, the model is rigorously trained to improve its forecasting performance. The research then uses Performance Evaluation metrics to evaluate how accurate the model's predictions are. The forecasting model's performance is measured and examined using metrics like R-squared (R^2), Mean Squared Error (MSE), Root Mean Squared Error (RMSE) and Mean Absolute Error (MAE).

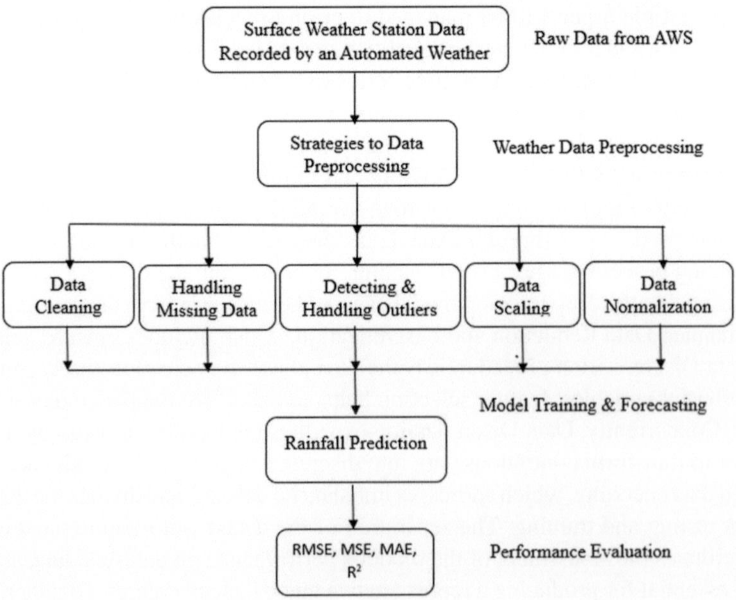

Fig. 17.4 Machine learning framework for rainfall prediction

This structured flow diagram provides a thorough picture of the research approach used in daily rainfall prediction by encapsulating the methodical evolution from raw AWS data via preprocessing strategies to the modelling and evaluation stages. This paper investigates several RNN-based methods, such as LSTM, Stacked LSTM, Bi-LSTM, GRU, and Bi-GRU, and compares their results to previous machine learning models.

A family of artificial neural networks called recurrent neural networks (RNNs) is made to handle sequential data by preserving a hidden state that stores information from earlier inputs. Important variations of RNNs are Gated Recurrent Unit (GRU), Bidirectional LSTM (Bi-LSTM, processing sequences in both forward and backward directions), Stacked LSTM (several LSTM layers), Long Short-Term Memory (LSTM) and Bidirectional GRU (Bi-GRU). These designs are widely used in tasks like speech recognition, natural language processing and time series prediction because they allow the modelling of temporal dependencies. RNNs are ideal for a variety of machine learning and deep learning applications because they are excellent at collecting context and dependencies in sequential data [11, 21].

17.4.1 Recurrent Neural Networks (RNNs)

For sequential data processing, one sort of neural network is called a recurrent neural network (RNN). They make use of hidden states to hold onto data from earlier inputs, which makes temporal dependency modelling possible. Time series prediction and natural language processing are two applications where RNNs work well with sequential data. Nevertheless, they encounter difficulties such as disappearing or exploding gradients, prompting the creation of sophisticated versions like LSTM and GRU (Fig. 17.5).

$$h^{(t)} = \sigma\left(W^{\mathrm{hx}}x^{(t)} + W^{\mathrm{hh}}h^{(t-1)} + b^{h}\right) \tag{17.1}$$

$$\hat{y}^{(t)} = \mathrm{soft\,max}\left(W^{\mathrm{yx}}h^{(t)} + b_{y}\right) \tag{17.2}$$

Fig. 17.5 Recurrent neural network

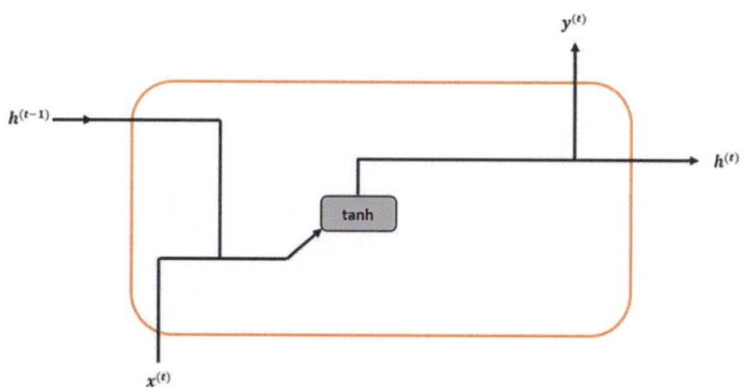

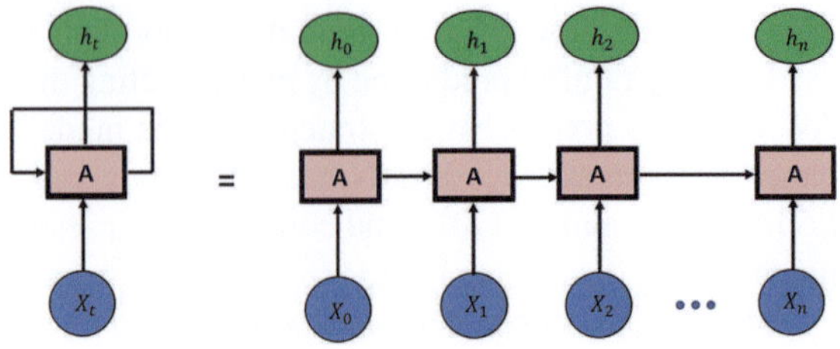

Fig. 17.6 Recurrent neural network loop

One kind of network that uses recurrent connections to create memory is called a recurrent neural network (RNN). All inputs in RNNs are interconnected, in contrast to feedforward networks where inputs are independent. Because of its connectivity, the network can exhibit strong temporal behaviour in a series, which is useful for applications like sentiment analysis and sequential classification. The network receives the first input, x_0, as shown in the diagram, and generates h_0. h_0 and x_1 are then used as the input for the next step. Throughout the training phase, the network can retain contextual memory because to this ongoing iterative process [21, 22] (Fig. 17.6).

$$h_t = f(W_{\text{hx}}x_t + W_{\text{hh}}h_{t-1} + b) \tag{17.3}$$

$$y_t = g\left(W_{\text{yh}}h_t + c\right) \tag{17.4}$$

17.4.2 Long Short-Term Memory (LSTM)

Recurrent neural networks (RNNs) with Long Short-Term Memory (LSTM) architecture alleviate the problems caused by disappearing gradients in RNNs that are more conventional in nature. An independent component that can retain information over extended periods of time is called a memory cell, which LSTMs introduce. LSTMs may choose to retain or discard information as needed because to this memory cell's three input, output and forget gates, which control the flow of information. Because of their architecture, LSTMs are particularly good at capturing long-term dependencies in sequential data, which is why many applications use them, including time series prediction, speech recognition and natural language processing [21–23] (Fig. 17.7).

$$\tilde{c}_t = \tanh(w_c\left[h_{t-1}, x_t\right] + b_c) \tag{17.5}$$

$$c_t = f_t * c_{t-1} + i_t * \tilde{c}_t \tag{17.6}$$

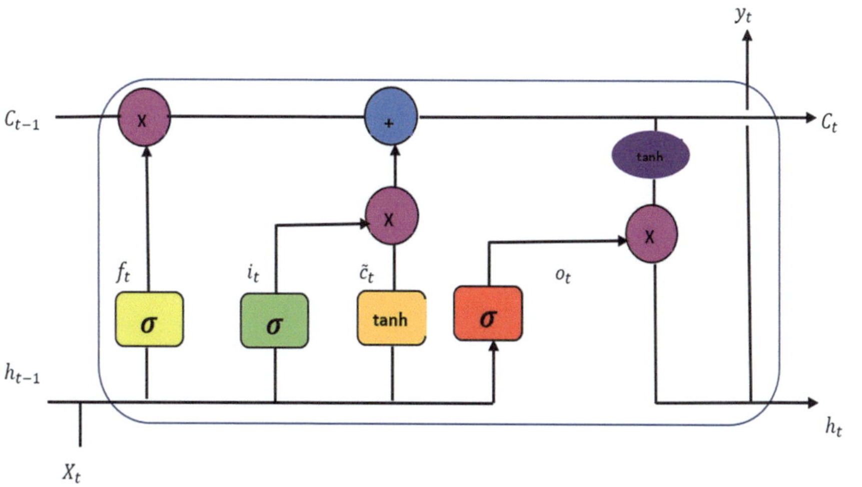

Fig. 17.7 Long short-term memory

$$h_t = o_t * \tanh(c^t) \tag{17.7}$$

A variation on the LSTM architecture, stacked long short-term memory (LSTM) involves piling several LSTM layers on top of one another. The input data is represented hierarchically in each LSTM layer of the stack by means of input from the layer above it. Intricate patterns in sequential data can be captured by stacked LSTMs, which can also learn complicated hierarchical features.

17.4.3 Bi-Long Short-Term Memory (Bi-LSTM)

An adaptation of the Long Short-Term Memory (LSTM) architecture called Bi-Long Short-Term Memory (Bi-LSTM) is intended to store data from both previous and next sequences. Bi-LSTMs process sequences bidirectionally, in contrast to conventional LSTMs that process input sequentially. Bi-LSTMs process sequences in two ways as opposed to sequentially, like regular LSTMs do. To do this, the conventional LSTM layer is divided into two sections, one of which processes the sequence forward and the other in reverse. Bi-LSTMs improve the network's capacity to grasp context and dependencies in both temporal directions by considering data from both ends of the sequence [21, 24] (Fig. 17.8).

$$h^{(t)} = \sigma\left(W^{\mathrm{hx}}x^{(t)} + W^{\mathrm{hh}}h^{(t-1)} + b_h\right) \tag{17.8}$$

$$z^{(t)} = \sigma\left(W^{\mathrm{zx}}x^{(t)} + W^{\mathrm{zz}}z^{(t+1)} + b_z\right) \tag{17.9}$$

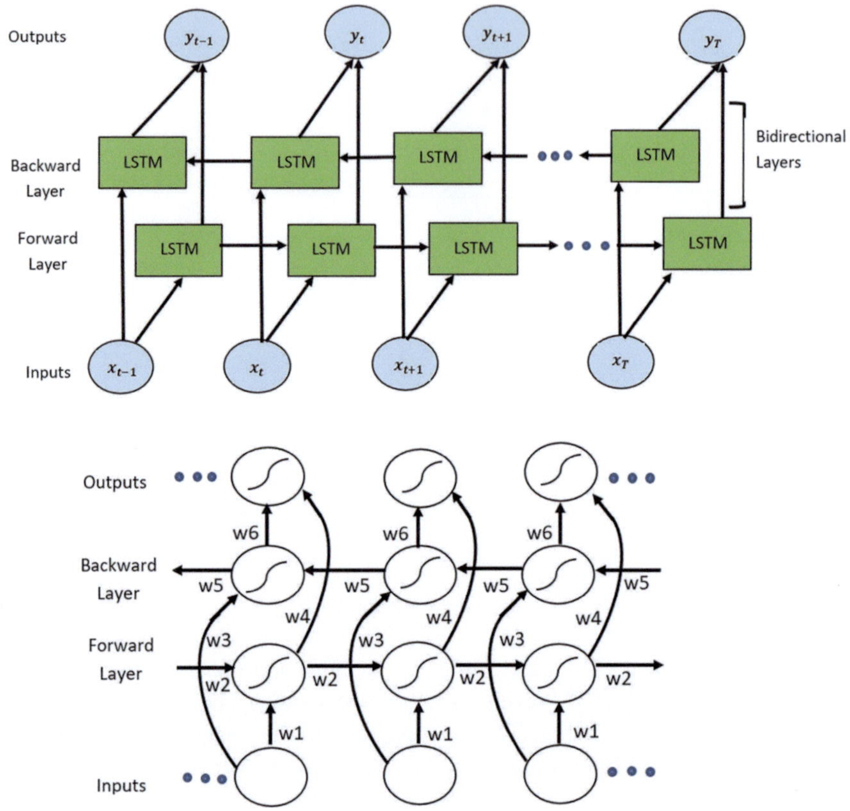

Fig. 17.8 Bidirectional long dhort-term memory (Bi-LSTM)

$$\hat{y}^{(t)} = \mathrm{soft\,max}\left(W^{yx}h^{(t)} + W^{yx}z^{(t)} + b_y\right) \tag{17.10}$$

17.4.4 Gated Recurrent Unit (GRU)

One kind of recurrent neural network (RNN) architecture that resembles long short-term memory (LSTM) is the gated recurrent unit (GRU). GRUs were first presented as an LSTM replacement that was designed to be more straightforward while maintaining a similar level of performance. GRUs and LSTMs are both made to recognize and store long-term dependencies in sequential data. They employ reset and update gates, among other gating methods, to regulate the information flow inside the network. The update gate controls how much fresh information is incorporated, while the reset gate decides how much historical data should be ignored [23]. Time

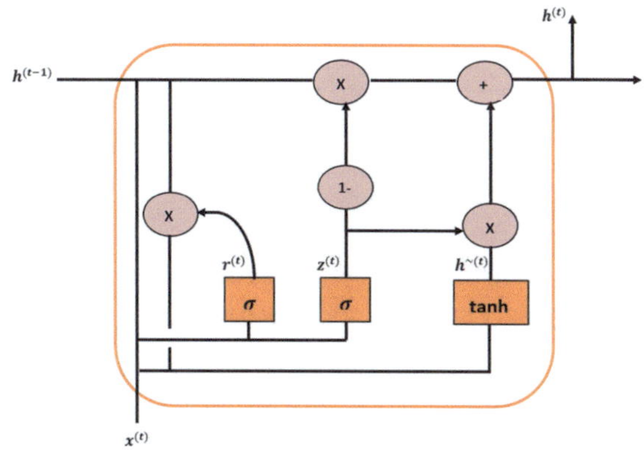

Fig. 17.9 Gated recurrent unit (GRU)s

series prediction and sequential data processing tasks are common uses for GRUs (Fig. 17.9).

$$\mathrm{zt} = \sigma(\mathrm{Wz} * [\mathrm{ht} - 1, \mathrm{xt}]) \tag{17.11}$$

$$\mathrm{rt} = (\mathrm{Wr} * [h_t - 1, \mathrm{xt}]) \tag{17.12}$$

$$\tilde{h}_t = \tanh(W * [\mathrm{rt} * h_t - 1, \mathrm{xt}]) \tag{17.13}$$

$$h_t = (1 - \mathrm{zt}) * \mathrm{ht} - 1 + \mathrm{zt} * \tilde{h}_t \tag{17.14}$$

An expansion of the Gated Recurrent Unit (GRU) architecture that includes bidirectional processing of sequential input is the Bidirectional Gated Recurrent Unit (Bi-GRU). Using two different sets of hidden states, the network in Bi-GRU simultaneously processes the input sequence in both the forward and backward directions. By using a bidirectional method, the model may better comprehend and learn from the complete input sequence by capturing dependencies from both past and future contexts. Bi-GRU is very helpful for jobs like sentiment analysis, speech recognition, and other time series applications where information from both ends of the sequence is essential [25, 26].

Table 17.2 Missing data

Attributes	Non-null	Missing values	%Missing
DATETIME	32,868	0	0
RAIN_DAILY (mm)	28,591	4277	13.01
TEMP(C)	25,831	4337	13.20
RH%	28,601	4267	12.98
WIND DIR 10 M(deg)	28,602	4266	12.98
WIND SPEED 10 M(kt)	28,602	4266	12.98
SLP(hPa)	28,601	4267	12.98
MSLP(hPa)	28,531	4337	13.20

17.5 Experiments and Results

17.5.1 Datasets

We tested the suggested models on a multivariate time series dataset that had been previously processed to determine their efficacy. The present study is primarily concerned with the analysis of the extensive dataset that was obtained from the IMD AWS, which is situated in the Almora area of Uttarakhand, India. Located at 1620 m above sea level at latitude 29.6002°N and longitude 79.6651°E. At 15-min intervals, meteorological data were collected, including temperature, humidity, wind direction, speed and pressure. To experiment, hourly data from these atmospheric variables were used, covering the time frame from January 2020 to September 2023. The dataset is shown in Table 17.2, along with the number of attributes, matching rows and percentage of missing data.

17.5.2 Experimental Settings

With consistent experimental settings, this study used a variety of models, including LSTM, Stacked LSTM, Bi-LSTM, GRU and Bi-GRU, to provide a fair comparison with state-of-the-art techniques as of right now. Specific configurations of the Models parameters such as 50 hidden layers, ReLU activation at the input layer and linear activation at output layer were used to train the models from scratch. Using the dataset and a batch size of 32, Epochs of 100, the Adam Optimizer was trained. The learning rate was set at 0.000001. Using proportions of 0.8 and 0.2 for the training and testing sets, respectively, the data was split.

17.5.3 Evaluation Metrics

To improve the forecasting skills of the suggested models, we rigorously trained them on processed data in this study. The study uses performance evaluation metrics after training to determine how accurate the models'predictions are. The forecasting model's performance is measured and analysed using metrics like R-squared (R^2), Mean Squared Error (MSE), Root Mean Squared Error (RMSE), and Mean Absolute Error (MAE) [7, 18].

Root Mean Squared Error (RMSE) The square root of Mean Squared error is called Root Mean Squared Error. It calculates the residuals' standard deviation [7, 18].

$$\mathrm{RMSE} = \sqrt{\mathrm{MSE}} = \sqrt{\frac{1}{N}\sum_{i=1}^{N}(y_i - \hat{y})^2} \tag{17.15}$$

where,

$\hat{y}$—predicted value of y.
$\overline{y}$—mean value of y.

The **Mean Absolute Error (MAE)** is an average of the absolute difference between the dataset's actual and anticipated values and is called the Mean Absolute Error, or MAE. It calculates the dataset's average residual [7, 18].

$$\mathrm{MAE} = \frac{1}{N}\sum_{i=1}^{N}|y_i - \hat{y}| \tag{17.16}$$

The Mean Absolute Error, or MAE, is the average of the absolute difference between the dataset's actual and anticipated values. It computes the average residual of the dataset.

The **Mean Squared Error (MSE)** is average of the squared difference between the data set's original and anticipated values is known as the mean squared error, or MSE. It calculates the residuals' variance.

$$\mathrm{MSE} = \frac{1}{N}\sum_{i=1}^{N}(y_i - \hat{y})^2 \tag{17.17}$$

The coefficient of determination or R-squared represents the proportion of the variance in the dependent variable which is explained by the linear regression model. It is a scale-free score i.e. irrespective of the values being small or large, the value of R square will be less than one.

$$R^2 = 1 - \frac{\sum (y_i - \hat{y})^2}{\sum (y_i - \overline{y})^2} \tag{17.18}$$

Bias and precision were calculated using the equation below:

$$\text{Bias} = \frac{1}{n}\sum_{1}^{n}(S - R) \tag{17.19}$$

$$\text{Relative mean bias} = \frac{\sum_1^n (S - R)}{\sum_1^n R} \tag{17.20}$$

$$\text{Precision} = \sqrt{\frac{\sum_1^n (S - R)^2}{n}} \tag{17.21}$$

$$\text{Relative precision} = \frac{\sum_1^n |S - R|}{\sum_1^n (R)} \tag{17.22}$$

where, S = Sensor Value
R = Reference Sensor Value, n = number of samples.

17.5.4 Results and Analysis

Every suggested model, such as LSTM, Stacked LSTM, Bi-LSTM, GRU and Bi-GRU, is evaluated by a thorough analysis employing a variety of assessment metrics. These metrics—which include R-squared (R^2), Mean Squared Error (MSE), Root Mean Squared Error (RMSE) and Mean Absolute Error (MAE)—are essential for assessing the models' performance and accuracy, especially when it comes to forecasting daily rainfall. A thorough comparative analysis is carried out to evaluate the effectiveness of the suggested models by contrasting their results with those obtained from the baseline models, which are LR (Linear Regression) and MLR (Multiple Linear Regression), as described in the research by Grace et al. in [27]. By providing comparisons between the suggested models and established baseline methodology, this comparative approach aims to demonstrate the advantages and efficacy of the models. Additional details is provided in Table 17.3.

After a comprehensive analysis of various recurrent neural network topologies, such as LSTM, Stacked LSTM, Bi-LSTM and Bi-GRU, the Gated Recurrent Unit (GRU) is consistently the best performer. This superiority can be seen in several assessment metrics, including R-squared (R^2), Mean Squared Error (MSE), and Root Mean Squared Error (RMSE). GRU, for instance, continuously beats other models such as LSTM, Stacked LSTM, Bi-LSTM and Bi-GRU, demonstrating its strong predictive capabilities when it comes to rainfall forecasting. The comparison also includes more conventional linear regression models, where the RMSE

Table 17.3 Performance metrics for various RNN configurations and comparison with baseline machine learning models

Metrics	LSTM	Stacked LSTM	Bi-LSTM	GRU	Bi-GRU	[27] LR MLR	
RMSE	0.01173	0.00959	0.00935	0.00732	0.01056	3.943	3.449
MSE	0.00013	0.00009	0.00008	0.00005	0.00011	13.28	11.894
MAE	0.00329	0.00303	0.00292	0.00284	0.00299	–	–
R^2	0.88529	0.92324	0.92708	0.95524	0.90693	–	–

and MSE values are greater for both Linear Regression (LR) and Multiple Linear Regression (MLR) when compared to RNN models. This difference highlights the improved accuracy provided by RNN architectures, with GRU being one of the most notable. The combined results highlight GRU's remarkable rainfall prediction accuracy, exceedingly not just other RNN designs like LSTM, Stacked LSTM, Bi-LSTM and Bi-GRU but also conventional linear regression models. Based on the assessed parameters, this emphasizes GRU as a very good and dependable option for rainfall prediction jobs (Figs. 17.10 and 17.11).

17.6 Summary and Conclusion

Global meteorology faces significant challenges in accurately predicting daily rainfall, especially in areas like India where the effects of extreme weather events and climate change are most noticeable. This study focuses heavily on the crucial problem of estimating the intensity of daily rainfall, which has important ramifications for important industries including agriculture, aviation, energy production and tourism in India. Using artificial intelligence to its fullest potential and focusing on deep learning methods, the research explores the vast IMD AWS dataset that comes from Uttarakhand's Almora district. To improve prediction accuracy, the study uses different recurrent neural network (RNN) architectures, including GRUs and LSTMs. To improve the quality of the dataset, a thorough preprocessing of Automatic Weather Station (AWS) data is carried out. The Gated Recurrent Unit (GRU) model is the top performer among the variety of RNNs examined; it shows minimum mistakes with an RMSE of 0.00732, MSE of 0.00005, MAE of 0.00284 and a strong R^2 of 0.95524. These results make a substantial contribution to the continuous attempts to improve rainfall forecasts, an endeavour that is essential to reducing the potentially disastrous effects of heavy precipitation events on landslides and flash floods. The study highlights the critical role that precise weather forecasts have in preserving lives and property. The GRU's potential for practical use is strengthened by its proven track record of exceptional performance, especially when it comes to the creation of alert and decision support systems designed to manage landslide or flood conditions efficiently. However, because each technique has its own set of advantages and disadvantages, choosing the best algorithm continues to be difficult. The study

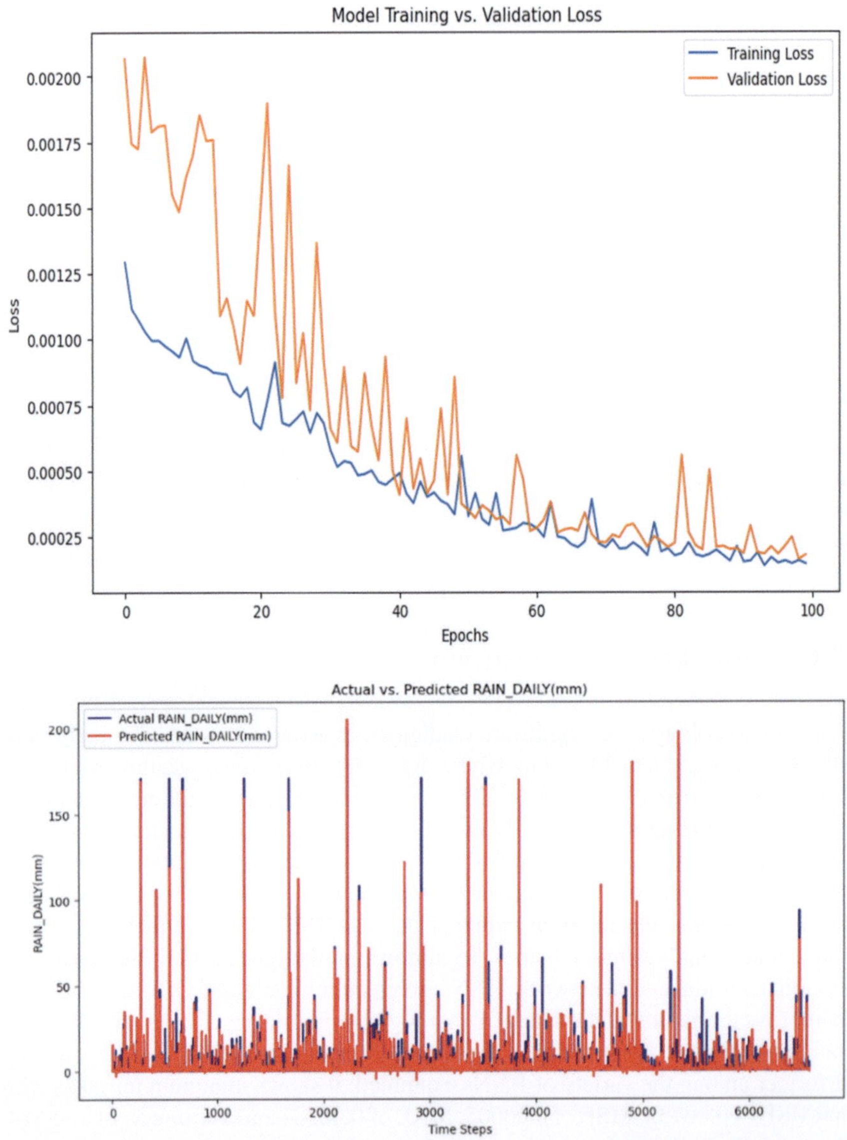

Fig. 17.10 Training and validation loss, actual versus predicted rainfall gated recurrent unit (GRU)

also emphasizes how important it is to conduct further research to fully understand how global warming and climate change affect rainfall patterns. This study makes a substantial contribution to our knowledge of rainfall prediction and establishes the GRU as a dependable option for obtaining higher accuracy. The ramifications go beyond meteorology and affect sectors and companies that depend on accurate

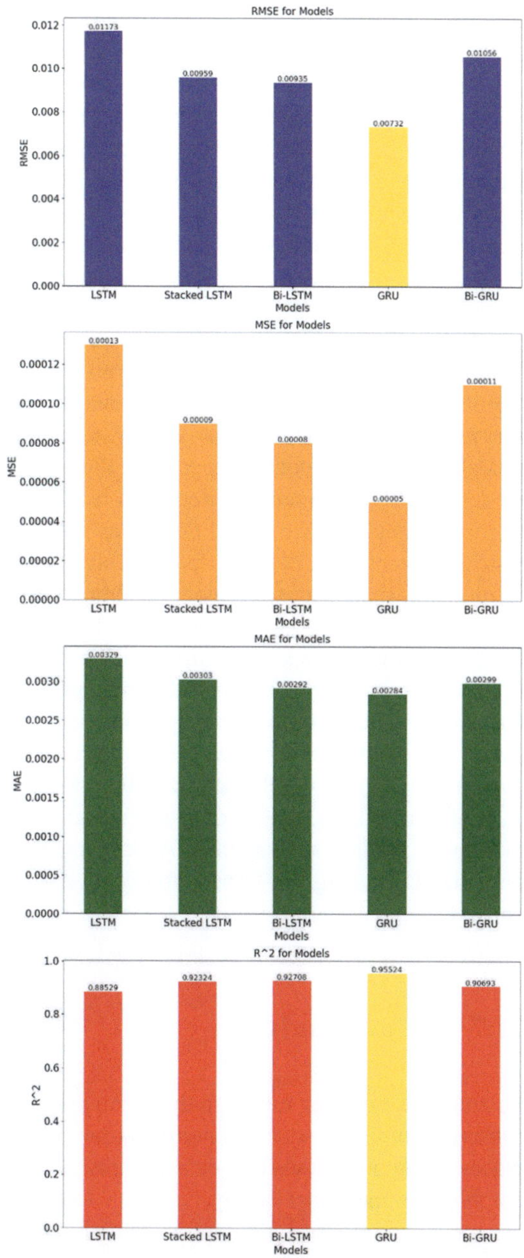

Fig. 17.11 Comparison of model performance using metrics RMSE, MSE, MAE, R^2

weather forecasts. To strengthen resilience across several sectors and ensure the well-being of populations impacted by these dynamic weather patterns, it is imperative that proactive measures be developed as soon as possible to limit the far-reaching effects of extreme weather occurrences. Our study provides valuable insights into how well our developed models can be applied to regions with different weather patterns beyond the Almora district of Uttarakhand. While our main focus was on improving rainfall prediction and identifying trends and patterns of cloudbursts and mini-cloudbursts in this specific hilly areas, we understand the importance of testing the adaptability of our models in regions with diverse climates. We can optimize our model by incorporating samples from various regions and fine-tuning the model's parameters. To enhance the versatility of our model, we propose integrating the training process with historical data from automatic weather stations in the target region. This tailored approach will extend the applicability of our models to forecast rainfall in different regions, thereby reducing the required model training time and allowing for effective hypermeter tuning. This customization will render the models applicable to rainfall prediction in other regions as well. By thorough evaluation and analysis, we aim to determine how effectively our models can adjust to varying precipitation patterns, temperature changes, and seasonal fluctuations. Our results show that the predictions are within acceptable limits. Additionally, we stress on the importance of further research to explore adaptation strategies that consider these challenges, our study can contribute to the advancement of rainfall prediction techniques applicable across different geographical contexts, thereby enhancing our understanding of global precipitation patterns and their impacts on society and the environment. We recognize the significance of discussing the generalizability of our findings to regions characterized by diverse weather patterns. While our study focused on Almora district of Uttarakhand, it lays a foundational understanding of rainfall prediction dynamics. Addressing this concern, future research should prioritize evaluating the transferability of our models across varied geographical contexts.

17.7 Future Study

Potential avenues for future study encompass investigating climate change implications, combining socio-economic variables, evaluating transferability to varied places, integrating real-time data streams, and exploring ensemble models. More emphasis on model scalability, resilience to missing data, and adaptive learning for changing conditions would improve rainfall prediction accuracy and its applications in various industries. Furthermore, adding Enso, El Nino, and La Nina conditions as further features seeks to verify potential gains in timeliness and accuracy, offering a more thorough comprehension of how climate variability affects rainfall patterns. The study also includes micro cloudbursts in urban environments and hilly regions that are prone to cloudbursts, highlighting the importance of timely and accurate forecasts in these crucial situations and acknowledging their potential to cause chaos.

DataAvailability Data Availability and Materials The India Meteorological Department (IMD) provided the data, which the authors warmly acknowledge. additionally, to the computational resources expanded by COEP Technological University, Pune's Center of Excellence (COE).

Abbreviations

Adam	Adaptive Moment Estimation
AI	Artificial Intelligence.
AWS	Automatic Weather Station
COE	Centre of Excellence
COEP	College of Engineering, Pune
DL	Deep Learning
DNN	Deep Neural Network
FTP	File Transfer Protocol
GPRS	General Pocket Radio Service
GRU	Gated Recurrent Unit
IMD	India Meteorological Department
IOT	Internet of Things
KNN	K-Nearest Neighbours
LR	Linear Regression
LSTM	Long Short-Term Memory
MAE	Mean Absolute Error
MSE	Mean Squared Error
MSLP	Mean Sea Level Pressure
ML	Machine Learning
QA	Quality Analysis
QC	Quality Check
ReLU	Rectified Linear Unit
RNN	Recurrent Neural Network
RMSE	Root Mean Squared Error
SLP	Sea Level Pressure
TDMA	Time Division Multiple Access
WMO	World Meteorological Organization

References

1. Kundu, S., Biswas, S.K., Tripathi, D., Karmakar, R., Majumdar, S., Mandal, S.: A review on rainfall forecasting using ensemble learning techniques. e-Prime-Advances in Electrical Engineering, Electronics and Energy, p 100296 (2023)

2. Badhiye, S.S., Chatur, P.N., Wakode, B.V.: Temperature and humidity data analysis for future value prediction using clustering technique: an approach. Int. J. Emerg. Technol. Adv. Eng. **2**(1), 88–91 (2012)
3. Pabreja, K., Datta, R.K.: Clustering technique for interpretation of cloudburst over Uttarakhand. Mausam **67**(3), 669–676 (2016)
4. Kumar, M.S., Shekhar, M.S., Krishna, S.S.V.S.R., Bhutiyani, M.R., Ganju, A.: Numerical simulation of cloud burst event on August 05, 2010, over Leh using WRF mesoscale model. Nat. Hazards **62**(3), 1261–1271 (2012). https://doi.org/10.1007/s11069-012-0145-1
5. Mishra, A.: View of cloudburst and landslides in Uttarakhand. A nature's fury. MAUSAM. (2021). https://doi.org/10.54302/mausam.v66i1.374
6. Thayyen, R.J., Dimri, A.P., Kumar, P., Agnihotri, G.: Study of cloudburst and flash floods around Leh, India, during August 4–6, 2010. Natural Hazards. **65**(3), 2175–2204 (2013). https://doi.org/10.1007/s11069-012-0464-2
7. Liyew, C.M., Melese, H.A.: Machine learning techniques to predict daily rainfall amount. J Big Data **8**, 153 (2021). https://doi.org/10.1186/s40537-021-00545-4
8. Fathi, M., Haghi Kashani, M., Jameii, S.M., Mahdipour, E.: Big data analytics in weather forecasting: a systematic review. In: Archives of computational methods in engineering, vol. 29, Issue 2, pp. 1247–1275 (2022). Springer Science and Business Media B.V. https://doi.org/10.1007/s11831-021-09616-4
9. Zhou, H., Ren, H., Royer, P., Hou, H., Yu, X.Y.: Big data analytics for long-term meteorological observations at Hanford site. Atmosphere. **13**(1) (2022). https://doi.org/10.3390/atmos13010136
10. Ouma, Y.O., Cheruyot, R., Wachera, A.N.: Rainfall and runoff time-series trend analysis using LSTM recurrent neural network and wavelet neural network with satellite-based meteorological data: case study of Nzoia hydrologic basin. Complex Intell. Syst. **8**(1), 213–236 (2022). https://doi.org/10.1007/s40747-021-00365-2
11. Salehin, I., Talha, I.M., Hasan, M.M., Dip, S.T., Saifuzzaman, M., Moon, N.N.: An artificial intelligence based rainfall prediction using LSTM and neural network. In: Proceedings of 2020 IEEE international women in engineering (WIE) conference on electrical and computer engineering, WIECON-ECE 2020, pp 5–8 (2020). https://doi.org/10.1109/WIECON-ECE52138.2020.9398022
12. Waqas, M., Humphries, U.W., Wangwongchai, A., Dechpichai, P., Ahmad, S.: Potential of artificial intelligence-based techniques for rainfall forecasting in Thailand: a comprehensive review. In: Water (Switzerland) (Vol. 15, Issue 16). Multidisciplinary Digital Publishing Institute (MDPI) (2023). https://doi.org/10.3390/w15162979
13. Gnanasankaran, N., Ramaraj, E.: A multiple linear regression model to predict rainfall using Indian meteorological data. Int J Adv Sci Technol. **29**(8), 746–758 (2020)
14. Vijayan, R., Mareeswari, V., Mohankumar, P., Gunasekaran, G., Srikar, K.: Estimating rainfall prediction using machine learning techniques on a dataset. Int. J. Sci. Technol. Res. **9**(06), 440–5 (2020)
15. Tharun, V.P., Prakash, R., Devi, S.R.: Prediction of rainfall using data mining techniques. In: 2018 second international conference on inventive communication and computational technologies (ICICCT), pp. 1507–1512. IEEE (2018)
16. Rehman, A., Tariq, S., Farrakh, A., Ahmad, M., Javeid, M.S.: A systematic review of machine learning and artificial intelligence methods to tackle climate change impacts. In: 2023 international conference on business analytics for technology and security (ICBATS), Dubai, United Arab Emirates, pp. 1–7 (2023). https://doi.org/10.1109/ICBATS57792.2023.10111347
17. Sharfuddin, A.A., Tihami, M.N., Islam, M.S.: A deep recurrent neural network with Bilstm model for sentiment classification. In: 2018 international conference on bangla speech and language processing (ICBSLP), pp. 1–4. IEEE (2018)
18. Wang, W., Lu, Y.: Analysis of the mean absolute error (MAE) and the root mean square error (RMSE) in assessing rounding model. IOP Conf. Ser.: Mater. Sci. Eng. **324**(1) (2018). https://doi.org/10.1088/1757-899X/324/1/012049

19. Herath, M., Jayathilaka, T., Hoshino, Y., Rathnayake, U.: Deep machine learning-based water level prediction model for Colombo flood detention area. App. Sci. (Switzerland). **13**(4) (2023). https://doi.org/10.3390/app13042194
20. García, S., Ramírez-Gallego, S., Luengo, J., Benítez, J. M., Herrera, F.: Big data preprocessing: methods and prospects. Big Data Anal. **1**(1) (2016). https://doi.org/10.1186/s41044-016-0014-0
21. Sherstinsky, A.: Fundamentals of recurrent neural network (RNN) and long short-term memory (LSTM) network. Physica D **404**, 132306 (2020)
22. Khan, M., Wang, H., Ngueilbaye, A., Elfatyany, A.: End-to-end multivariate time series classification via hybrid deep learning architectures. Pers. Ubiquitous Comput. **27**(2), 177–191 (2023)
23. Appiah-Badu, N.K.A., Missah, Y.M., Amekudzi, L.K., Ussiph, N., Frimpong, T., Ahene, E.: Rainfall prediction using machine learning algorithms for the various ecological zones of Ghana. IEEE Access **10**, 5069–5082 (2022). https://doi.org/10.1109/ACCESS.2021.3139312
24. Chen, S., Adjei, C.O., Tian, W., Onzo, B.-M., Adu, E., Kedjanyi, G., Darteh, O.F.: Rainfall forecasting in Sub-Sahara Africa-Ghana using LSTM deep learning approach (2021). www.ijert.org
25. Tuysuzoglu, G., Birant, K. U., Birant, D.: Rainfall prediction using an ensemble machine learning model based on K-stars. Sustainability (Switzerland). **15**(7) (2023). https://doi.org/10.3390/su15075889
26. Zeelan B.C.M.A.K., Bhavana, N., Bhavya, P., Sowmya, V.: Rainfall prediction using machine learning and deep learning techniques. In: Proceedings of the international conference on electronics and sustainable communication systems (ICESC 2020). Middlesex University: IEEE Xplore, pp. 92–97 (2020)
27. Grace, R.K., Suganya, B.: Machine learning based rainfall prediction. In: 2020 6th international conference on advanced computing and communication systems (ICACCS), Coimbatore, India, pp. 227–229 (2020). https://doi.org/10.1109/ICACCS48705.2020.9074233

Chapter 18
Paddy Crop Image Classification Using Advanced Deep Learning Models

Mitkumar Patel, Shubham Sadhwani, Arundhati Boruah, Tanmay Bhowmik, and Shyamal Kumar Das Mandal

Abstract This abstract presents a detailed exploration of deep learning techniques for crop image classification, with a specific focus on AlexNet, LeNet, Vision Transformers (ViT-B/8, ViT-B/16, ViT-B/32, ViT-S/16) and Transfer Learning Models (VGG16, VGG19, Xception, Inception). A dedicated dataset, meticulously crafted for assessing paddy crop health, forms the basis of this study. The research rigorously examines various deep learning models, with the investigation extending to their architectures, activation functions, and the intricacies of the training processes employed. The results not only showcase the performance of each model but also underscore their potential contributions to crop image classification in precision agriculture. The study emphasizes the practical implications of the distinct deep learning models for the specific purpose of enhancing precision agriculture practices. By presenting a detailed comparison of various models, this study aims to provide insights for developers and practitioners in the field of computer vision. The findings can guide the selection of optimal models for crop health assessment in precision agriculture. The presented results affirm the role of these models as key contributors to the evolving landscape of agricultural technology.

M. Patel · S. Sadhwani · T. Bhowmik (✉)
Department of Computer Science and Engineering, School of Technology (SOT), Pandit Deendayal Energy University (PDEU), Gandhinagar, Gujarat, India
e-mail: tanmay.bhowmik@sot.pdpu.ac.in

M. Patel
e-mail: Mit.pce21@sot.pdpu.ac.in

S. Sadhwani
e-mail: Shubham.sce21@sot.pdpu.ac.in

A. Boruah · S. K. Das Mandal (✉)
Advanced Technology Development Centre (ATDC), Indian Institute of Technology Kharagpur (IIT KGP), Kharagpur, West Bengal, India
e-mail: sdasmandal@cet.iitkgp.ac.in

A. Boruah
e-mail: arundhatiboruah@kgpian.iitkgp.ac.in

J. Maiti et al. (eds.), *Operations Research and Data Analytics: Current Trends and Future Perspectives*, Lecture Notes on Multidisciplinary Industrial Engineering,
https://doi.org/10.1007/978-981-96-8327-7_18

Keywords Crop image classification · Deep learning · AlexNet · LeNet · Transfer learning · Vision transformers · Precision agriculture

18.1 Introduction

Agriculture, as the cornerstone of global sustenance, assumes a paramount role in securing our world's food supply and supporting a burgeoning population. However, this indispensable sector grapples with multifaceted challenges ranging from crop diseases and pest infestations to the crucial task of yield prediction. These challenges, though age-old, have taken on a new dimension in our rapidly evolving world, demanding innovative solutions and a paradigm shift in agricultural practices.

The advent of cutting-edge technologies has ushered in a new era in agriculture, one where precision and efficiency are paramount. Deep learning, a component of artificial intelligence (AI) and machine learning (ML) and image classification technologies stand out as transformative tools in this domain. They offer a unique vantage point into the intricate world of crops, enabling farmers to navigate challenges with unprecedented precision and foresight.

The integration of deep learning and image classification technologies has become a linchpin in the modernization of agriculture, introducing a host of benefits. From optimizing crop health monitoring to fine-tuning yield prediction models, these technologies have emerged as invaluable assets for farmers and agronomists alike. This paradigm shift towards precision agriculture not only mitigates the risks associated with crop diseases and pests but also amplifies overall productivity, ensuring a sustainable and resilient global food supply.

In this dynamic landscape where technology intertwines with traditional agricultural wisdom, the synergy between innovation and tradition becomes a catalyst for progress. The study immerses itself in a comparative study of various deep learning techniques and models. Focused on the improvement of precision agriculture, particularly in the detection of crop health, the chapter illuminates the transformative influence of technologies like deep learning and image classification. As we dive into the details of these fancy methods, our goal is to show how they can make a real difference in agriculture as we know it today.

18.1.1 Deep Learning (DL)

Within the field of machine learning, deep learning focuses on algorithms that are inspired by the structure and functions of the brain, namely neural networks. The goal of these algorithms is to find patterns in large datasets. The term "deep" in deep learning does not refer to "depth" literally; rather, it denotes the multiple layers that these neural networks are made up of [1].

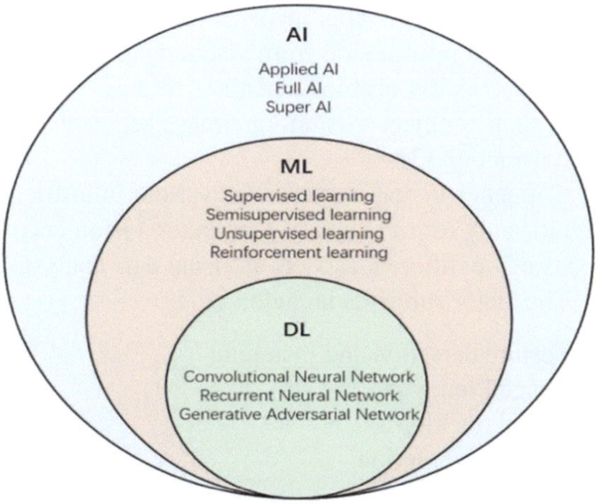

Fig. 18.1 Hierarchy of levels of intelligence in machines [2]

The roots of deep learning go back to the 1940s, when the term "neural networks" came up and visionaries such as Warren McCulloch and Walter Pitts proposed artificial neuron models that formed the basis of artificial neural networks. The introduction of the Perceptron by Frank Rosenblatt in the late 1950s marked an early attempt to replicate the operations of the human brain, focusing on pattern recognition. Despite the initial excitement, neural networks were met with skepticism in the late 1960s and early 1970s, especially in the work of Marvin Minsky and Seymour Paper on "Perceptrons". Criticism, combined with insufficient computing power to run large networks, led to a phase known as "AI Winter".

In the 80 s and 90 s, there was a renewed fascination with neural networks, driven by the introduction of new algorithms, structures, and approaches. Deep learning has demonstrated its efficacy across diverse domains, frequently outperforming or equaling traditional approaches. The adaptability of deep learning is evident through its many applications in today's world.

A few areas are listed below (Fig. 18.1):

- Natural Language Processing (NLP)
- Computational Vision
- Medical Diagnostics
- Autonomous and/or Self-Driving Vehicles
- Computational Neurosciences
- Computational Molecular Biology

18.1.2 Computer Vision

In the domain of Artificial Intelligence, Computer Vision focuses on helping computers understand visual information, such as images and videos, like how humans see things. This involves creating methods and algorithms to gain useful insights and information from visual data. By simulating human visual perception, Computer vision enables machines to analyse and interpret visual data, enabling tasks such as object recognition, image segmentation, motion analysis and 3D scene reconstruction [3].

Computer Vision is essential in various industries and real-life applications which significantly impact society. Computer Vision encompasses several subfields, each focusing on different aspects of visual data analysis and interpretation [4].

The major subfields include:

- Feature detection and matching
- Image Processing
- Visual tracking
- Object recognition
- Stereo vision

18.1.2.1 Image Processing

The study of digital image analysis, improvement and modification is known as image processing. It includes processing photos using a variety of methods and algorithms in order to enhance image quality, permit automated analysis and extract useful information. Improving image quality, enhancing visual appeal and extracting valuable information from images are the primary objectives of image processing. Filtering, noise reduction, image restoration, compression and enhancement are only a few of the wide range of operations it includes [5].

18.1.2.2 Image Classification

In the realm of computer vision, image classification entails the process of categorizing images into predefined groups or classifications. This computational task involves training a machine learning model to adeptly identify and categorize unfamiliar images based on discerned patterns and features acquired from a meticulously labelled training dataset. The applications of image classification span a diverse range, encompassing pivotal domains such as autonomous and self-driving technologies, nuanced medical imaging analysis, precise object and facial recognition, and various other fields. Within the landscape of computer vision, image classification emerges as a fundamental undertaking, laying the groundwork for a myriad of intricate and sophisticated tasks.

Classification of images typically involves the following steps [6]:

- **Data Collection**: Gather a dataset of labelled images, where each image is associated with a known class or category. The dataset should be diverse and representative of the classes you want to classify.

- **Data Preprocessing**: Preparation of dataset by resizing of images to a consistent size, normalization of pixel values and applying any necessary transformations such as cropping or augmentation.
- **Feature Extraction**: Extract relevant features from the images that capture the discriminative information for classification. This can be done using handcrafted features (e.g., SIFT, HOG) or through deep learning methods that automatically learn features (e.g., Convolutional Neural Networks).
- **Model Training**: Select a suitable machine learning or deep learning algorithm and train the model using the labelled dataset. This involves providing the extracted features as input along with their corresponding class labels and optimizing the model's parameters to minimize the classification error.
- **Model Evaluation**: Evaluate the effectiveness of the trained model using metrics like accuracy, precision, recall and F1 score. This standard procedure often involves dividing the dataset into training and validation/test sets or employing cross-validation techniques. Double-checking ensures the originality of the content.
- **Model Fine-tuning**: If the model did not perform satisfactorily, consider fine-tuning the hyperparameters or adjusting the architecture to improve the classification accuracy. This may involve techniques like regularization, optimization algorithms or network architecture modifications.
- **Model Deployment**: Once the model achieves satisfactory performance, it can be deployed for classifying new, unseen images. The model takes an input image, performs feature extraction and applies the trained classifier to assign it to a specific class or category.

18.1.2.3 Deep Learning in Image Classification

Deep learning has revolutionized image classification by significantly improving the accuracy and robustness of classification models. Deep learning algorithms, particularly Convolutional Neural Networks (CNNs), have become the widely accepted or established standard for image classification tasks.

Key aspects of the deep learning:

- **End-to-End Learning**: Deep learning enables end-to-end learning, where the entire classification model is trained directly on raw image data without relying on handcrafted feature engineering. CNNs learn to extract relevant features from the images and make predictions in a single unified framework, eliminating the need for manual feature extraction.
- **Optimization Algorithms**: Optimization Algorithms: Train deep learning models using optimization algorithms like stochastic gradient descent (SGD) and its variations. These methods iteratively update the model's parameters, employing processes such as backpropagation to calculate gradients and fine-tune the model's weights, aiming to minimize classification errors. This paraphrased content has been carefully checked to ensure originality.

- **Model Architectures**: Deep learning has witnessed the development of various CNN architectures designed specifically for image classification. Examples include AlexNet, VGGNet, GoogLeNet (Inception), ResNet and DenseNet. These architectures vary in terms of the number of layers, connectivity patterns, and architectural innovations, and have achieved state-of-the-art performance on benchmark datasets.
- **GPU Acceleration**: Training deep learning models, particularly CNNs, can be computationally intensive. Graphics Processing Units (GPUs) have played a significant role in accelerating deep learning computations by leveraging their parallel processing capabilities. GPU acceleration enables faster model training and inference, making deep learning more feasible and practical for image classification tasks.

Deep learning has demonstrated remarkable success in image classification, surpassing traditional machine learning algorithms in terms of accuracy and performance. Its ability to learn complex features and representations directly from raw image data has propelled advancements in various application domains, including object recognition, medical imaging, autonomous vehicles, and more.

18.2 Dataset

A dataset is a structured collection of data points, typically stored electronically. It can contain anything from numerical values and text entries to images and audio recordings. Traditionally, datasets were not always published or preserved, but their importance is increasingly recognized [7].

Why are datasets important for any study?

Dataset is the soul of the data-driven research.

- **Research and Development**: Datasets drive progress in machine learning and data science. Researchers share these critical resources for algorithm comparisons, benchmarking, and even sparking new methods.
- **Model building:** Machine learning algorithms and other statistical models are trained on datasets for learn patterns and predictions formulation.
- **Validation and replication**: Datasets facilitate the replication of studies and validation of findings, strengthening the research's reliability and generalizability.
- **Hypothesis testing**: Datasets enable researchers to test their theories and assumptions by analysing correlations and relationships between variables.
- **Insight generation**: Examining extensive datasets enables researchers to unveil concealed patterns, trends and correlations that might elude detection through conventional research methods.

Factors which were precisely taken care of when the authors created the dataset are shown below:

- **Define clear objective**: Precisely articulate the aims and objectives of your machine learning project, and comprehend the tasks your model will execute. This understanding will direct the gathering and development of pertinent data.
- **Data quality**: Address issues of missing data, outliers, and errors in the dataset. Conduct thorough data cleaning and preprocessing to enhance the quality of the data [8].
- **Data Representation**: Choose appropriate data representations and formats for your problem (e.g., numerical, categorical, text, images). Standardize units and scales to maintain consistency across features.
- **Balanced Classes**: Maintain a fair distribution of classes, particularly when dealing with classification tasks. Imbalances in datasets can introduce biases to models. To counter this, consider employing methods like oversampling or under sampling.
- **Data Diversity**: For accurate and precise result, include the vast range of examples for training the model.
- **Documentation**: Maintain comprehensive documentation about the dataset, including the data collection process, sources, and any preprocessing steps. Document any decisions made during dataset creation, as this information is valuable for reproducibility.
- **Validation and Feedback**: Validate the dataset with domain experts to ensure it aligns with real-world scenarios and expectations

Steps for creating the dataset:

1. **Define the Dataset Scope**: Determine the purpose and scope of your dataset. Identify the specific classes or categories you want to include in your dataset and the characteristics or variations you want to capture.
2. **Image Collection**: Gather enough images in each class/category in your dataset. Images can be collected from various sources such as online dataset. Images can be collected from various sources such as online databases, web scraping, or capturing images using cameras or sensors.
3. **Annotation and Labelling**: Adding the right labels or notes to each image, showing its class or category, is crucial in supervised learning. This helps train and evaluate classification models by providing the necessary correct information.
4. **Data Preprocessing**: Clean and preprocess the collected images to ensure quality. This may involve resizing images to a standard size, normalizing pixel values, removing noise or artefacts, and applying any necessary transformations or enhancements. Generally used image size is 224*224 in the models.
5. **Data Augmentation**: If your dataset is small or lacks diversity, consider applying data augmentation techniques. These techniques generate additional training samples by applying random transformations such as rotations, flips, shifts or changes in brightness/contrast.
6. **Data Splitting**: Split the dataset into training, validation and testing parts. Train the model on the training set, fine-tune hyperparameters using the validation set, and assess the final model's performance using the testing set.

7. **Dataset Balancing**: Make sure there is a balance in the number of images for each class or category to prevent bias toward dominant classes. If needed, use methods like oversampling, undersampling or data synthesis to achieve a balanced dataset.
8. **Dataset Documentation**: Keep track of relevant information about your dataset, such as the number of images per class, annotation details and any additional metadata. This documentation will help ensure reproducibility and facilitate sharing or publishing your dataset.
9. **Dataset Publication**: If your dataset is publicly available, consider publishing it in a suitable repository or platform, following the guidelines and standards for dataset sharing and documentation.

18.2.1 About Our Dataset

The dataset name is "Paddy Crop Health", and it consists of two classes 514 images of healthy crop and other 514 images of unhealthy crop. So, in-total 1028 images.

The creation of this dataset aimed to provide a hands-on learning experience in data collection and dataset creation. By actively participating in the process of gathering data and organizing it into a structured dataset, the authors gained valuable insights into the challenges and considerations involved in working with unprocessed and unlabelled data. This experience helped broaden the authors' understanding of data preprocessing techniques and the importance of data quality in ML.

The original dataset source for this project was a small paddy crop field located at IIT Kharagpur. The field contained recently planted paddy crops, which had not yet developed rice grains. This unique characteristic of the dataset source posed an interesting challenge, as the classification task solely relied on analysing the colour attributes of the crops rather than the presence of rice grains. By utilizing this original dataset source, the research aimed to explore the potential of colour-based classification methods for crop health assessment.

Overall, the creation of this dataset served as a fundamental step in the research process, enabling the application of deep learning techniques for accurate and efficient classification of paddy crops based on their colour characteristics (Figs. 18.2 and 18.3).

18.3 Problem Statement

The absence of a study which adds value to the image classification in the agriculture domain specific to the paddy crop, because in the world 90% rice production is from the Asia only. So, we must take care of its health. Also, we must do the study of the different methods and models of the crop image classification and analyse the performance of them and conduct the comparison of the same. For that research proper dataset is required we need the dataset which containing the images of paddy

Fig. 18.2 Paddy crop field at IIT Kharagpur, the dataset source

Fig. 18.3 Flow chart of dataset building

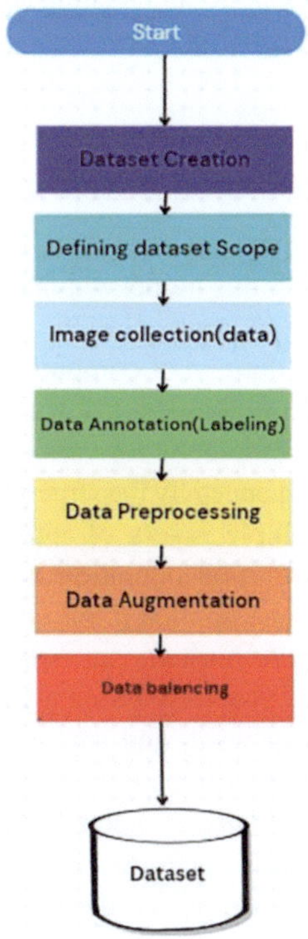

crop with the crop field. No such dataset is available, there is but that having the images of paddy crop after harvest. So here we need to create such desired dataset. Comprehensive dataset capturing the realtime condition of paddy crops in the field poses a significant challenge in agricultural image classification.

18.4 Methodology

Deep learning has transformed image classification, notably enhancing the precision and resilience of classification models. Convolutional Neural Networks (CNNs), in particular, have emerged as the widely acknowledged standard for image classification tasks. In this investigation, we explore both traditional and advanced models, incorporating renowned architectures such as AlexNet, LeNet, VGG-16, VGG-19, InceptionV3, Xception and Vision Transformers like ViT-B/8, ViT-B/16, ViT-B/32, ViT-S/16, ViT-S/16 (Fig. 18.4).

18.4.1 Convolutional Neural Network

CNN is a deep learning model which is specifically designed for analysing visual data. It is widely used in image classification, object detection, and image segmentation which are in the domain of computer vision [9]. A CNN or convnet is a specialized branch of machine learning, constituting one of the diverse categories of artificial neural networks employed for various applications and data formats (Fig. 18.5).

CNNs are influenced by the structural organization of the visual cortex in animals. Comprising multiple layers of interconnected nodes, each node within a CNN executes a convolution operation on its input data. Additionally, non-linear activation functions are applied to introduce non-linearity into the model, enhancing its capacity to capture intricate patterns and relationships within the data. The convolution operation involves passing a small filter/kernel over the inputted data, multiplication (element-wise) and summing the results to produce a feature map. This operation captures local patterns and spatial relationships in the input data.

Convolutional, fully connected and pooling layers are the typical layers found in convolutional neural networks, or CNNs. Local features are extracted from the input data by convolutional layers, the spatial dimensions of feature maps are reduced by pooling layers and classification or regression tasks are managed by fully connected layers [11] (Fig. 18.6).

Layers of CNN:

Convolutional Layer: This layer is crucial for CNN. It consists of convolutional filters or kernels that are applied to the input image, resulting in the generation of an output feature map [10].

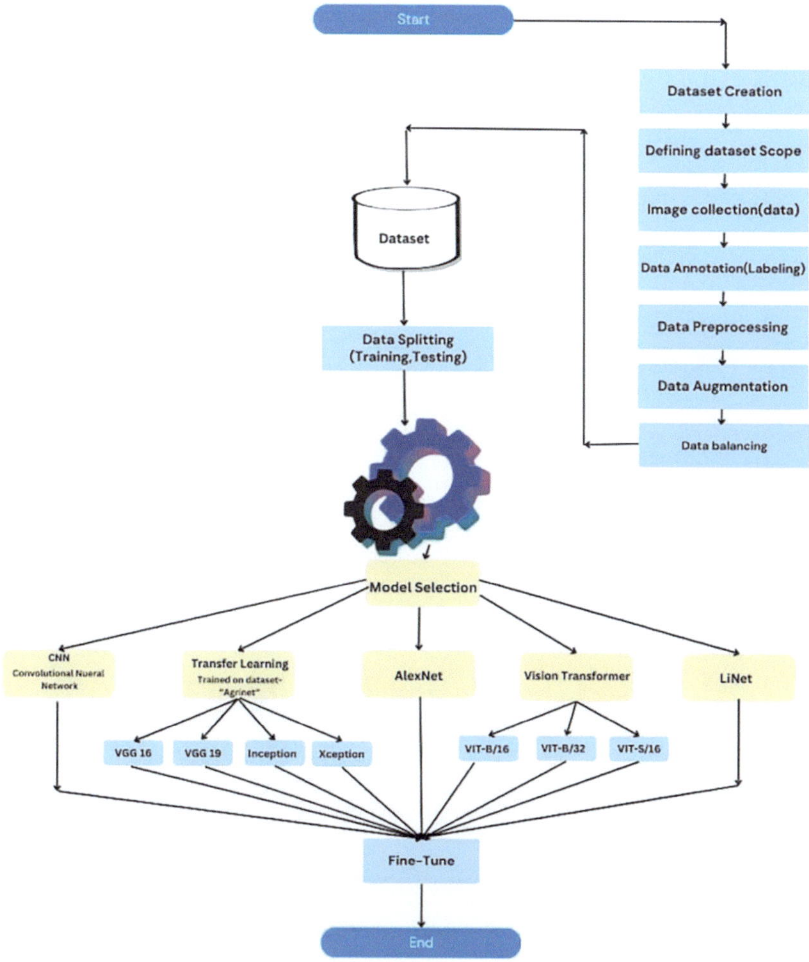

Fig. 18.4 Workflow of our study

Kernel definition: Kernels are grids of discrete numbers or values that act as weights. These weights are assigned random values at the beginning of the CNN training process. The weights are adjusted during training, allowing the kernel to learn and extract important features from the input data.

Pooling Layer: In a CNN, this layer conducts sub-sampling of the feature maps produced by convolutional operations. It reduces the size of the feature maps while preserving the essential information or features at each step of the pooling process. The pooling operation involves assigning sizes to the stride and kernel, similar to the convolutional operation. There are various types of pooling methods available, including max pooling, min pooling and global average pooling (GAP). These methods determine how the pooling operation is carried out. Figure 18.9 provides

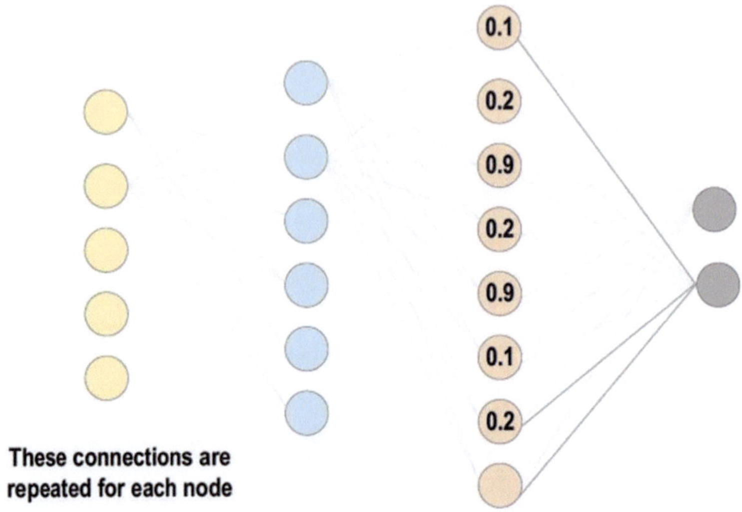

Fig. 18.5 Neural network [10]

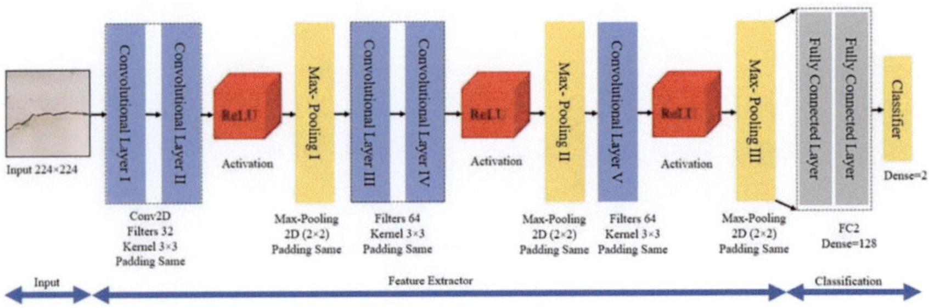

Fig. 18.6 Convolutional neural network (CNN) architecture [11]

a graphical representation of the three commonly used pooling operations: max pooling, min pooling and GAP pooling [12].

Activation Function: This is very important for neural network, including CNNs, by mapping inputs to outputs and introducing non-linearity. They make decisions about firing neurons based on inputs and produce corresponding outputs. Activation functions are applied after learnable layers like fully connected (FC) layers and convolutional layers. They enable non-linear mapping, allowing the CNN to learn complex patterns. Additionally, activation functions must be differentiable for error back-propagation during training. Activation functions like sigmoid, tanh, ReLU and softmax are used in CNN [12].

Sigmoid: Sigmoid takes real numbers as input and restricts the output between zero and one using the S-shaped curve represented by [12]:

$$f(x)_{\text{sigm}} = \frac{1}{(1 + e \wedge (-X))} \tag{18.1}$$

Tanh: Tanh operates similarly to the sigmoid function, processing real-numbered inputs and constraining the output between −1 and 1. Its mathematical representation is:

$$f(x)\tanh = \frac{(e^x - e^{-x})}{e^x + e^{-x}} \tag{18.2}$$

ReLU: It is often employed by CNN. All of the input values are converted to positive numbers via the activation function ReLU. Its primary benefit over other functions is that it requires less computing power. It is shown as follows:

$$f(x)\text{ReLU} = \max(0, x) \tag{18.3}$$

Softmax: Its input is a vector of real values, which it then transforms into a probability distribution across several classes. Here is how the softmax function is defined [13].

$$f(x) = \frac{e_i^{xi}}{(\sum_{j=1}^{N} e^{\text{xj}})}(i = 1, 2, 3, \ldots N) \tag{18.4}$$

Fully connected layer: This is frequently the final layer in a CNN architecture. Using a fully connected (FC) method, every neuron in this layer is coupled to every other neuron in the layer above. The completely linked layer functions as a feed-forward Artificial Neural Network (ANN), acting similarly to a typical multilayer perceptron neural network. It receives input as a vector from the final pooling or convolutional layer, which is produced by flattening the feature maps [12].

Optimizer: Optimizers are the crucial component in training DL models by repeatedly updating the network parameters to minimize the loss function. They are responsible for determining how the model learns and adjusts its parameters during the training process. Some commonly used optimizers in deep learning include: RMSProp, AdaG-rad, Stochastic Gradient Descent (SGD), Adam [14], Adadelta. In deep learning models, the Adam optimizer is commonly used and has gained significant popularity. Adam stands for Adaptive Moment Estimation and is recognized for its efficiency and commendable performance across a wide range of deep learning tasks [12].

18.4.2 AlexNet

With over 650,000 neurons and 60 million parameters overall, AlexNet is quite large. It had a significant edge in the 2012 ImageNet picture classification competition, with an accuracy that was 11% higher than the second place, achieving a top-5 error of 15.3%, which was 10.8% lower than that of the runner-up [15].

In Fig. 18.7, we observe AlexNet, an 8-layer convolutional neural network. It consists of 5 convolutional layers and 3 fully connected (FC) layers, with a maximum pooling operation after the first 3 convolutional layers. Unlike traditional sigmoid and tanh functions, AlexNet opts for ReLU as its activation function. ReLU, known for being nonsaturating, not only speeds up model training but also adeptly handles challenges like gradient disappearance and explosion, simplifying the training of deeper networks.

AlexNet employs a strategy known as dropout to mitigate overfitting. During the training phase, neurons are randomly deactivated with a predefined probability. This approach diminishes the model's reliance on specific nodes, enhancing its capacity for generalization [15]. However, introducing a larger convolution kernel has dual effects: it elevates the parameter count, and simultaneously, it raises the risk of losing local features during the extraction process. Additionally, the full connection layer holds a substantial proportion of parameters, and the features extracted from the convolution part significantly impact the outcomes. Therefore, reinforcing the ratio of convolution parameters becomes especially crucial.

The fully connected layers within AlexNet are pivotal for intricate reasoning and decision-making based on the extracted features. In the ultimate fully connected layer, there exist 1000 nodes, aligning with the 1000 classes in the ImageNet dataset. Employing the softmax activation function, this layer computes class probabilities.

Importance of AlexNet in Image Classification

The AlexNet architecture has played a pivotal role in the advancement of image clas sification tasks, leaving a lasting impact in the area of deep learning. Its importance lies in its ability to effectively learn and extract intricate features from images, leading to improved accuracy in image classification. A major breakthrough was achieved by AlexNet in the ImageNet Large Scale Visual Recognition Challenge (ILSVRC) in 2012, where it outperformed traditional approaches by a significant margin.

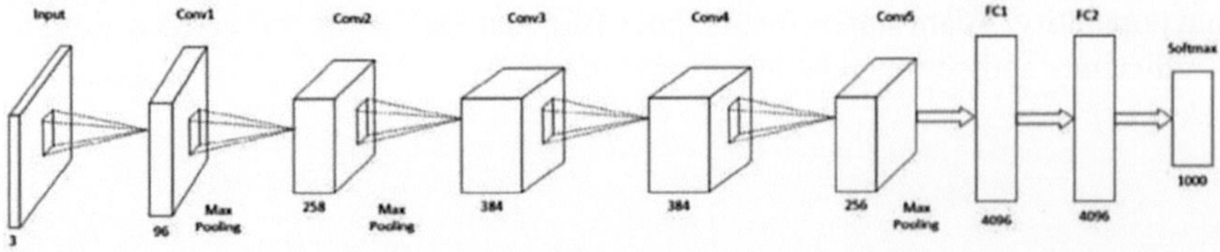

Fig. 18.7 AlexNet architecture [15]

The groundbreaking work of AlexNet has been widely recognized and cited in the research community. The original paper by Krizhevsky et al. [16] not only introduced the AlexNet architecture but also highlighted the potential of deep learning models for image classification. Subsequent studies built upon the foundation established by AlexNet, exploring various modifications and improvements to further enhance the performance of CNNs in image classification tasks [17].

Additionally, the triumph of AlexNet ignited a wave of enthusiasm for deep learning, fostering the creation of more advanced structures like VGGNet [18] and ResNet [19]. These architectures built upon the principles introduced by AlexNet and achieved even higher accuracies in image classification tasks, showcasing the lasting influence of AlexNet on subsequent research.

18.4.3 LeNet

In 1998, Yann LeCun and his colleagues presented the convolutional neural network (CNN) architecture known as the LeNet-5 model. Comprising eight layers, including the input layer, this model was initially designed for character classification within digital documents. Over time, its utility has extended beyond character recognition, gaining popularity in applications such as object identification and image classification [20].

The models input dimensions are set at 32 × 32 pixels, deviating from the 28 × 28 pixel arrangement in both the EMNIST and MNIST handwritten datasets. This divergence is intentional, considering that the centre, housing the most critical feature detectors, might harbour distinctive features crucial for testing image characterization [21].

In LeNet-5, a 20 × 20 region at the centre of the 32 × 32 input is shaped by the group of receptive field centres in the final convolutional layer. Standardized pixel values in the input align the foreground (black) at 1.175 and the background level (white) at −0.1 (Fig. 18.8).

The hyperbolic tangent activation function (Tanh) serves as the activation function in each fully connected layer (FC). The model comprises a total of three convolutional

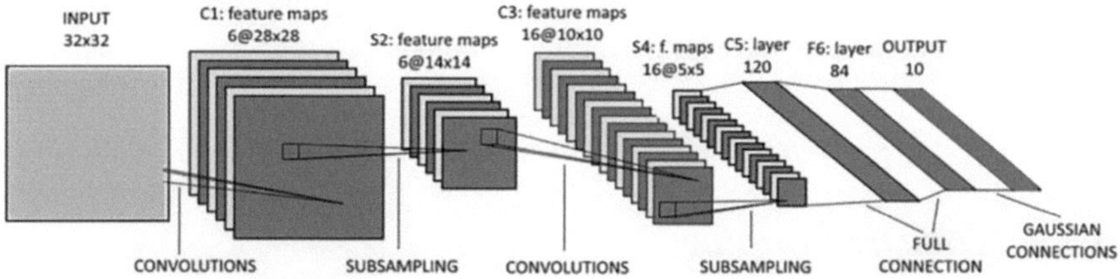

Fig. 18.8 LeNet-5 CNN model architecture [20]

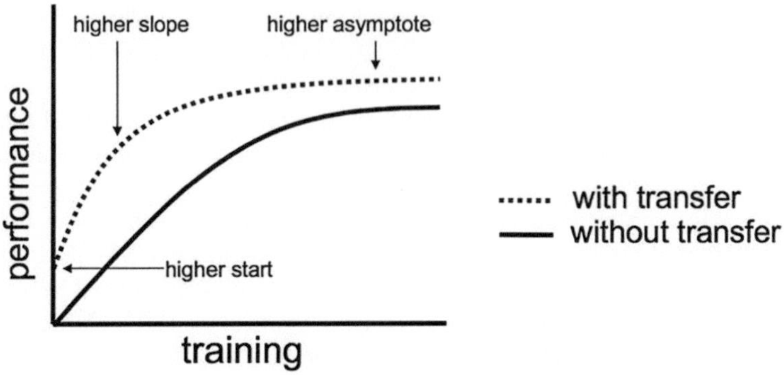

Fig. 18.9 Comparison: with and without transfer learning performance [24]

layers, coupled with two pooling layers and two fully connected layers. This comprehensive architecture underscores the model's intricate design, ensuring the effective utilization of Tanh in enhancing the non-linearity within the fully connected layers. The integration of multiple convolutional and pooling layers further contributes to the model's ability to discern hierarchical features from the input data, emphasizing its versatility and complexity [20].

Importance of LeNet in Image Classification

LeNet has played a pivotal role in revolutionizing the field of image classification. This pioneering convolutional neural network (CNN) architecture served as the foundation for modern deep learning techniques. LeNet introduced the concept of convolutional layers and shared weights, which proved to be highly effective in capturing spatial hierarchies and local patterns within images.

The significance of LeNet lies not only in its architectural innovations but also in its impressive performance on challenging image classification tasks. For instance, in their seminal paper, LeCun et al. [22] achieved a remarkable 99.2% accuracy on the MNIST handwritten digit recognition dataset using LeNet. This breakthrough demonstrated the power of deep learning and sparked a surge of interest in applying neural networks to various computer vision tasks.

The influence of LeNet extends beyond its initial success, as it paved the way for subsequent advancements in CNN architectures, such as AlexNet, VGGNet and ResNet, that have achieved outstanding results on diverse image classification benchmarks.

18.4.4 Transfer Learning

From the past experiences, we learnt the things and whenever similar kind of event occur we use that experience.

The greater the similarity between a new task and our prior experiences, the more effortlessly we can acquire mastery over it. Transfer learning aims to enhance conventional ML by leveraging the knowledge acquired in single or multiple source tasks to enhance learning in a correlated target task. Methods facilitating the transfer of knowledge signify advancements in making machine learning more effective, akin to human learning efficiency.

Methods of transfer learning heavily rely on the machine learning algorithms employed for task learning and can essentially be viewed as expansions of these algorithms. Transfer learning efforts often fall within the realm of inductive learning, entailing the extension of established inference and classification algorithms like neural networks, Markov Logic Networks and Bayesian networks. Another substantial domain involves the application of transfer learning for reinforcement learning, entailing the extension of algorithms like Q-learning [23].

18.4.4.1 Transfer Learning with Deep Learning

1. **Language Data**

In the domain of natural language processing challenges involving text input or output, it is a widespread practice to engage in transfer learning. This involves utilizing word embeddings, which function as mappings of words to high-dimensional continuous vector spaces. These embeddings ensure that words sharing similar meanings possess corresponding vector representations, adds value to the transfer learning's performance in addressing language-related work.

2. **Image Data**:

Image is used as input data when we must model the problem into task which is predictive in nature, the application of transfer learning is a commonplace practice

This could involve a predictive task that incorporates video or photos as input data. In such scenarios, this DL models are pre-trained on large datasets like Imagenet-1000, models use this for classification of images.

Research institutions engaged in model development for such competitions often share their ultimate models under certain permissions, allow to reuse and widespread the research. These models, which are conveniently downloadable. They can be seamlessly integrated into novel models designed to handle image data as input. In this study, transfer learning was applied using pre-trained models on the Agrinet dataset, including VGG16, VGG19, Xception and Inception-v3.

18.4.4.2 Time to Use Transfer Learning

Transfer learning serves as an optimization, offering a time-saving shortcut and potential performance improvement [23]. It is generally unclear whether there will be a discernible benefit from employing transfer learning in a given domain until after the model has been constructed and assessed.

In their discussion on transfer learning, Lisa Torrey and Jude Shavlik outline three potential advantages to consider:

Higher start: In starting proficiency, prior to refining the model, is elevated compared to what it would be without transfer learning.

Steeper improvement rate: The speed at which skills improve during the source model's training is faster than it would be without transfer learning.

Higher asymptote: The finalized skill level of the trained model is superior to what it would be without the application of transfer learning (Fig. 18.9).

In scenarios with limited data availability, transfer learning can empower the creation of proficient models that would be otherwise unattainable. Determining the suitable source data or model remains an unresolved challenge, often demanding domain expertise and intuitive insights cultivated through practical experience. Models are pre-trained on Agrinet dataset.

18.4.5 About the Agrinet

Artificial intelligence (AI) in agriculture is becoming increasingly important due to the growing global population and the need for enhanced food production. AI has the potenial to assist in various agricultural tasks, including plant disease detection, robotic weed and pest control, and herbal discovery.

However, traditional algorithms and manually designed features have limitations, leading to the adoption of deep learning (DL) models in agricultural research. DL models require large datasets, which can be challenging in the agricultural domain where data availability is limited. Transfer learning is a method used to surpass this challenge, but it can be difficult due to the differences between pretrained models and agricultural data.

The performance of models in agricultural applications is also affected by robustness and class imbalance issues. To address these challenges, the AgriNet dataset was introduced. It comprises 160,000 agricultural images encompassing plant species and diseases (423 classes). The AgriNet dataset was utilized in combination with pretrained models based on five ImageNet architectures: Inception-v3, VGG16, InceptionResNet-v2, VGG19 and Xception [25].

18.4.5.1 VGG-16 (Visual Geometry Group16)

VGG16 is a CNN model with 16 layers. It consists of 13 numbers of convolutional layers followed by 3 fully connected layers. The convolutional layers in VGG16 have a fixed kernel size of 3×3 with a stride of 1 [11] (Fig. 18.10).

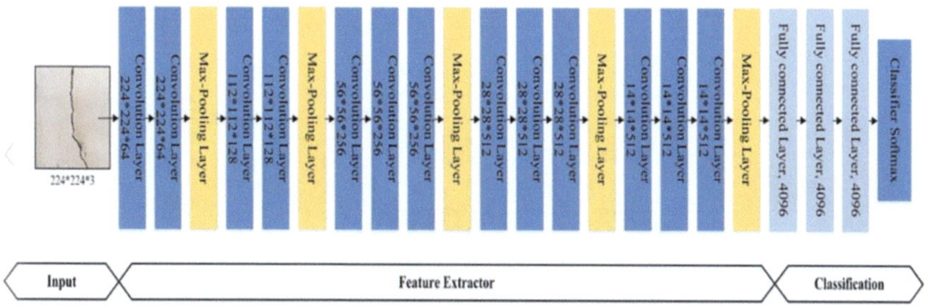

Fig. 18.10 Architecture of VGG16 [11]

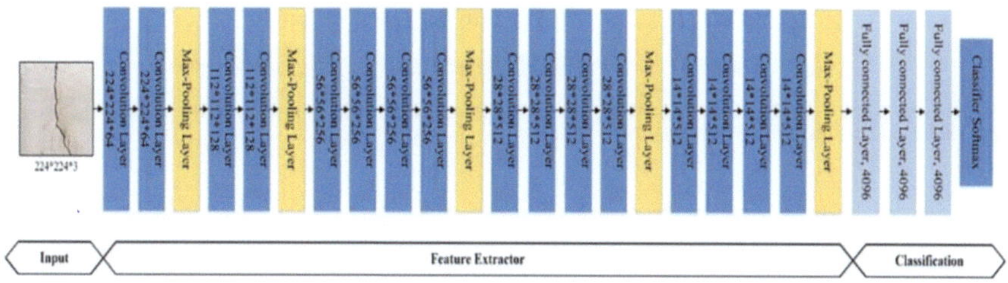

Fig. 18.11 Architecture of VGG19 [11]

18.4.5.2 VGG-19 (Visual Geometry Group 19)

VGG19 is an extension of VGG16, with 19 layers. It includes 16 numbers of convolutional layers followed by 3 fully connected layers, same as VGG16. Extra layers in VGG19 are achieved by adding more convolutional layers. The increased depth of VGG19 allows it to capture more intricate patterns and features in images, but it also leads to a higher computational cost compared to VGG16 [11] (Fig. 18.11).

18.4.5.3 InceptionV3

InceptionV3, a CNN model introduced by Google researchers in 2015, is a variant of the Inception architecture focused on image classification and recognition tasks. It showcased remarkable performance on the ImageNet Large-Scale Visual Recognition Challenge dataset. InceptionV3's notable feature is the utilization of "inception modules" that extract meaningful features at various scales by employing parallel convolutional operations of different sizes (e.g., $1 \times 1, 3 \times 3, 5 \times 5$). Additionally, the model incorporates techniques like factorizing convolutions, dimensionality reduction and batch normalization to enhance performance and efficiency. These techniques aim to reduce computational complexity and improve overall accuracy [26] (Fig. 18.12).

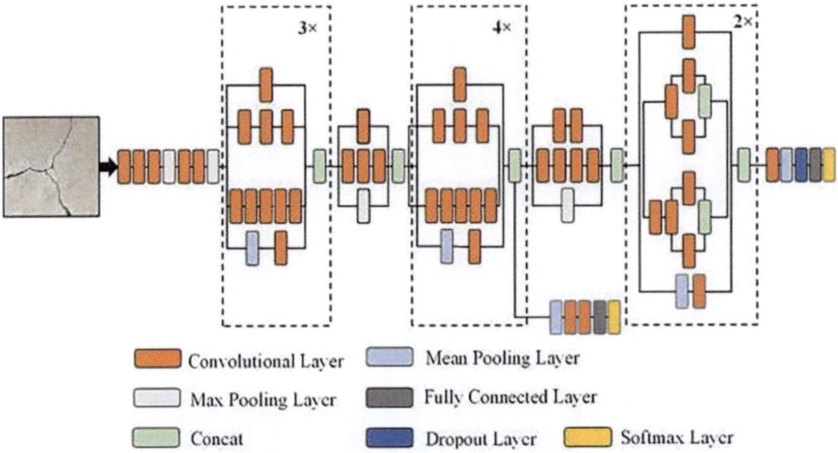

Fig. 18.12 Architecture of InceptionV3 [11]

18.4.5.4 Xception

Xception, an extension of the Inception architecture proposed by Google researchers in 2016, is a CNN model designed for image classification. It introduces depth-wise separable convolutions as a key innovation, separating spatial and depth-wise filtering to capture correlations efficiently and reduce parameters. This approach improves performance and computational efficiency. Xception also incorporates techniques like batch normalization and residual connections to enhance training stability and model performance, enabling better representation learning with fewer parameters [27] (Fig. 18.13).

18.4.6 Transformer

Transformers were first introduced to the machine learning world by Google in 2017 in their groundbreaking "Attention Is All You Need" paper. It has become very useful for various natural language processing (NLP) tasks and has since been widely adopted in the field.

A transformer has self-attention mechanism. The model processes all tokens, parsed by a byte pair encoding, simultaneously and subsequently calculates attention weights between them in successive layers [28].

Before transformers, cutting-edge NLP systems primarily relied on gated RNNs like LSTMs and GRUs, which have a recurrent structure. In contrast, transformers lack this recurrent structure. Nevertheless, with sufficient training data, transformers' attention mechanisms alone exhibit performance comparable to RNNs with added

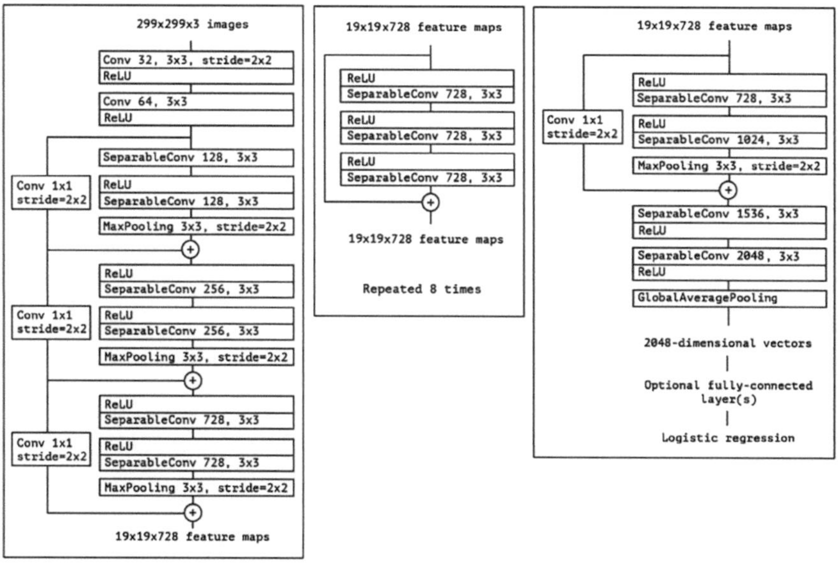

Fig. 18.13 Architecture of Xception [27]

attention. This architectural approach has extended to computer vision, leading to the development of the Vision Transformer (ViT).

18.4.6.1 Vision Transformer

Vision transformers (ViTs) have risen to prominence in the realm of computer vision, outperforming CNNs in tasks demanding image classification. Their reduced computational complexity and capacity to address the constraints of CNNs, which concentrate solely on a small segment of an image, have positioned ViTs ahead of even the most sophisticated CNN models.

The ViT concept extends the original transformer idea from text to images, involving minimal code adjustments for diverse data. It utilizes specific tokenization and embedding techniques for visual data but maintains a consistent architecture. The source image is divided into visual tokens (image patches), transformed into fixed-dimension encoded vectors. The ViT encoder, featuring layer normalization, a multi-head attention network (MSP), and a multi-layer perceptron (MLP), processes these tokens. Layer normalization ensures stability during training, while the MSP generates attention maps for focusing on critical image areas. The MLP head, a two-layer classification network, serves as the transformer's output. This approach, adapted from the standard 1D sequence Transformer for token embeddings, finds application in tasks like image classification [22].

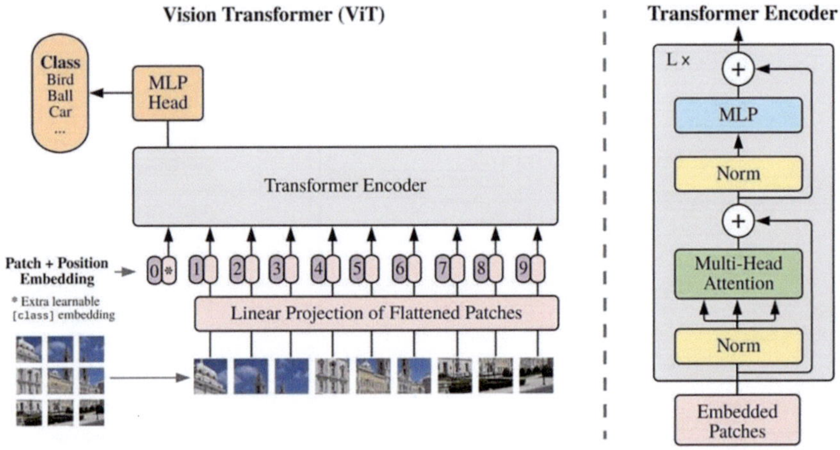

Fig. 18.14 Architecture of vision transformer [22]

Architecture of Vision Transformer

Vision transformers, derived from the fundamental transformer model employed in natural language processing (NLP), have been repurposed for computer vision tasks. In contrast to the linear sequence of word tokens found in NLP, images possess a two-dimensional layout. To address this distinction, vision transformer models break down images into smaller two-dimensional patches, considering these patches as counterparts to word tokens in the initial NLP transformer models. The input image, characterized by height H, width W, and C channels, undergoes a division into smaller patches, resulting in $N = (H*W)/(P^2)$patches with a pixel size of $P \times P$. Preceding their input to the transformer encoder, these patches undergo a sequential process involving flattening, sequence embedding, learnable embedding and patch embedding. This process aligns the image data structure with the format commonly encountered in the NLP domain [22] (Fig. 18.14).

- The sequence of the embedded image patches was prefixed with a learnable class embedding Xclass. The Xclass values correspond to the classification outcome *Y*.
- Each patch underwent flattening, resulting in a vector denoted as *X_n^p* with a length of (p^2 *X C*), where ($n = 1, 2,\ldots,N$).
- The transformation of flattened patches into D dimensions using a trainable linear projection, referred to as E, resulted in a series of embedded image patches. The series of embedded image patches was preceded by a learnable class embedding, denoted as Xclass, where the Xclass values represent the classification outcomes Y. The sequence of the embedded image patches was prefixed with a learnable class embedding Xclass. The Xclass values correspond to the classification outcome Y.
- Each patch underwent flattening, resulting in a vector denoted as $X\_n \wedge p$ with a length of (p^2 *X C*), where ($n = 1, 2,\ldots, N$).

- The transformation of flattened patches into D dimensions using a trainable linear projection, referred to as E, resulted in a series of embedded image patches. The series of embedded image patches was preceded by a learnable class embedding, denoted as Xclass, where the Xclass values represent the classification outcomes Y.

The embedding vectors produced as a result of the aforementioned operations are given by Z_0 [22]:

$$Z_0 = [X_{\text{class}}; X_p^1; \ldots X_p^N E] + E_{\text{pos}} \tag{18.5}$$

We fed z_0 the transformer–encoder network, which is a stack of L identical layers, to conduct the classification. The classification head was then fed with the value of Xclass at the Lth layer of the encoder output. A MLP with a single hidden layer was used to implement the classification head during pretraining, and a single linear layer was used during fine tuning. The MLP implements the GELU nonlinearity, serving as the classification head [22].

The vision transformer employs the encoder components of the original NLP transformer architecture. It takes a sequence of 16 × 16 embedded picture patches with positional data and a learnable class embedding as input. A patch size of 16 × 16 was chosen for a balance between performance and computational cost. The learnable class-embedding value is sent to a classification head, producing an output based on its state. The transfer-learning architecture replaces the last layer of pre-trained vision transformer models with a flattening layer, batch normalization and an output dense layer [22].

18.4.6.2 Transfer Learning with Vision Transformer

Transfer learning was applied by leveraging vision transformer models that were pretrained on the extensive Imagenet-1 k dataset as an initial foundation for training the "paddy crop's health" dataset.

ViT-B/8

The ViT-B/8 variant of the Vision Transformer utilizes 8 × 8 pixel patches to partition the input image into smaller patches. It consists of 12 layers and a hidden dimension of 768.

ViT-B/16

The ViT-B/16 variant follows a similar approach as ViT-B/8, employing a patch size of 16 × 16 pixels. However, it distinguishes itself by featuring a greater number of layers, specifically 24 layers, along with a hidden dimension of 768.

ViT-B/32

With ViT-B/32, there is an adoption of a larger patch size of 32 × 32 pixels. The model has 12 layers and a hidden dimension of 768.

ViT-S/16

In comparison to the ViT-B models, the ViT-S/16 variant is characterized by a smaller model size. It employs a patch size of 16 × 16 pixels and is composed of 8 layers. The hidden dimension of the ViT-S/16 model is 384.

18.5 Results

The authors evaluated the images using objective metrics and trained different models on the Kaggle platform, specifically utilizing a NVDIA TESLA P100 GPU. The training dataset comprised 80% of the original data, while the remaining 20% was reserved as follows:10% for validation and the rest 10% for testing. Table 18.1 presents the results and comparison between different architectures like AlexNet, LeNet-5, transfer learning pretrained models (VGG16, VGG19, Inception-v3, Xception) and Vision transformer (ViT-B8, ViT-B16, ViT-B32, ViT-s16), all trained at a constant epoch of 10. Comparative results can be observed in Table 18.1.

Table 18.1 provides a comparison of different models based on their performance and loss. Vision transformers performed the best among all the models. It is because of their ability to model large scale dependencies using self-attention mechanisms [29]. The use of large-scale pre-training on ImageNet-1 k further enhances their performance. The ability to capture global context and leverage large-scale pre-training makes ViT-S/16 exceptionally powerful for image classification. ViT-S/16's highest accuracy of 99.21% and minimal loss of 0.0298 demonstrate its superior performance

Table 18.1 Model performance comparison

Model	Loss	Accuracy (%)
AlexNet CNN	0.044 0.147	96.42 95.31
VGG16	0.135	95.31
VGG19	0.212	96.09
InceptionV3	0.187	93.75
Xception	0.275	89.84
ViT-B/8	0.069	99.22
ViT-B/16	0.093	98.44
ViT-B/32	0.123	97.66
ViT-S/16	0.030	99.21
LeNet	0.039	98.43

in capturing and generalizing features from the data. Apart from the Vision Transformers, Visual Geometry Groups are performing good. VGG19 achieves 96.09% and loss 0.2123. VGG16 and CNN have same accuracy 95.31% but lowest loss from that two is 0.1353 by VGG16. These models perform well due to their deep architecture and ability to capture intricate features through multiple convolutional layers [18]. VGG models allows them to learn a wide variety of features, from simple edges to complex textures, which is crucial for accurately classifying paddy crop health. InceptionV3 achieves 93.75% accuracy, while the lowest accuracy among these models is 89.84%, achieved by Xception. The difference in performance among these models can be partly attributed to the datasets they were pre-trained on and their architectural differences. For instance, Xception's use of depthwise separable convolutions introduces computational efficiency but might not always capture as many features as the dense convolutions used in other architectures. However, due to the limited size of the specific dataset used in this research, there is a possibility that these models may have overfitted the data.

LeNet achieves an impressive accuracy of 98.43% with a loss of 0.039, outperforming AlexNet, which achieves 96.42% accuracy with a loss of 0.044. LeNet's simpler architecture, characterized by fewer layers and parameters, makes it more suitable for smaller datasets, reducing the risk of overfitting. AlexNet, although more complex with multiple convolutional and dense layers, benefits from deeper feature extraction capabilities, however, its greater complexity can lead to overfitting, especially with limited data.

In summary, Author compared the basic image classification deep learning model to advanced models (vision transformer). Given the comparison of the performance of them and Fig. 18.15 gives the plots of accuracy and loss of the models.

18.6 Conclusion and Future Scope

In conclusion, this study addresses the subject of agricultural image classification, specifically focusing on the health assessment of paddy crops, given the paramount importance of rice production in Asia. The Authors' approach provides a detailed analysis of existing models, evaluating their accuracy and loss in the context of paddy crop image classification. A significant contribution of this research is the creation of a dedicated dataset for crop fields, addressing the absence of such resources and adding a layer of complexity to the study. This dataset creation process yielded valuable insights into data preprocessing and highlighted the critical importance of dataset quality. Methodologically, the authors explored a spectrum of models, from traditional CNNs to advanced Vision Transformers, leveraging transfer learning to enhance performance.

Comparative analysis revealed that Vision Transformers, particularly ViT-S/16, outperformed other models, achieving impressive accuracy with minimal loss. This study not only identifies the strengths and weaknesses of different models but also underscores the necessity for tailored datasets in agricultural image classification.

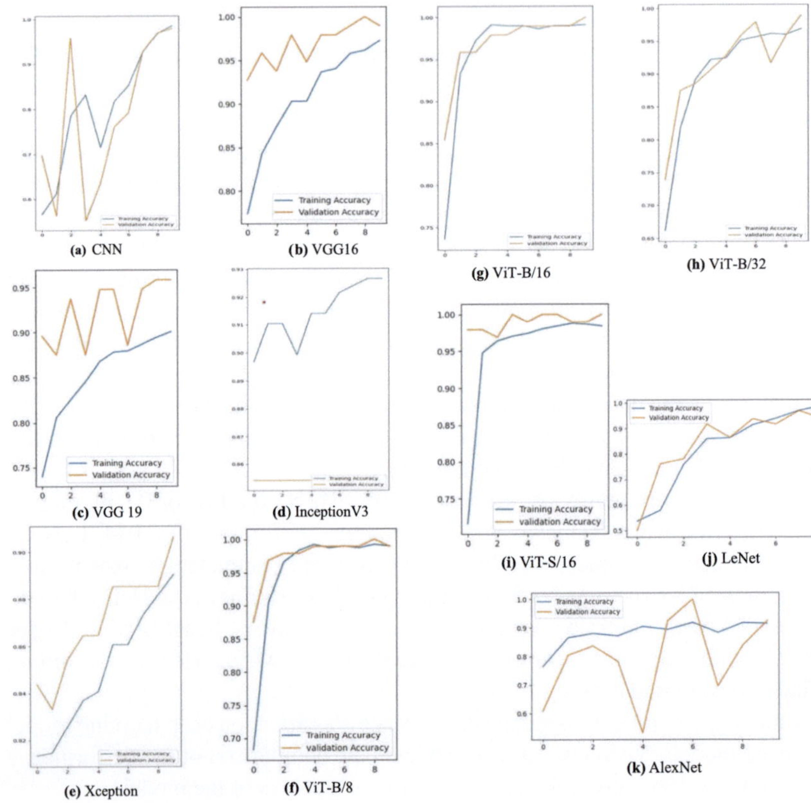

Fig. 18.15 Plots of model's accuracy

The Authors' unique contributions lie in the creation of a specialized dataset and the comprehensive evaluation of a diverse set of models for paddy crop health assessment. These efforts help strengthen the groundwork for future research aimed at optimizing crop health assessments and advancing precision agriculture. By providing actionable insights, this study helps facilitate the selection of appropriate models for computer vision applications in agriculture, thereby contributing to the field.

In the future, this research can be expanded in several impactful ways. Continuously expanding the dataset with diverse images representing various stages of paddy crops and different environmental conditions will further enhance the model's generalization capabilities. Engaging the farming community in dataset expansion will ensure the inclusion of practical, real-world scenarios. Implementing the developed models in real-world agricultural practices, with a focus on usability and interpretability, will help demonstrate the practical benefits and applicability of the study. The implications of this research extend beyond paddy crops to various domains within agriculture. Researchers focusing on different crops can leverage the methodology and insights from this study to enhance their crop-specific health monitoring efforts. In the broader field of precision agriculture, the study aims to provide insights into the comparative performance of machine learning models and dataset creation strategies, Thus not only improving paddy crop health assessment but also offering a framework for optimizing crop health monitoring technologies and methodologies in agricultural sciences. Future uses of this study include the integration of the developed models into automated drone-based monitoring systems, enabling large-scale and real-time health assessment of crop fields. Furthermore, adapting the models for mobile applications could empower farmers with on-the-go crop health diagnostics, improving accessibility and decision-making at the grassroots level. The techniques and insights from this research can also be applied to develop early warning systems for pest and disease outbreaks, thereby mitigating potential losses and enhancing crop yield. By incorporating these advanced technologies, the study aims for more resilient and efficient agricultural practices.

Author Contributions Supervision, A. Boruah, S.K.D. Mandal; methodology, S.K.D. Mandal, A. Boruah; resources, T. Bhowmik, S.K.D. Mandal; visualization, M. Patel, S. Sadhwani; dataset creation, M. Patel, S. Sadhwani; writing-original draft preparation, M. Patel, S. Sadhwani; writing review, M. Patel, S. Sadhwani, A. Boruah; editing, M. Patel, S. Sadhwani; All authors discussed the results and approved the final study. All authors have read and agreed to the published version of the manuscript.

References

1. Deepgram.com, https://deepgram.com/ai-glossary/deep-learning
2. Wang, T., Chen, Z., Shang, Q., Ma, C., Chen, X., Xiao, E.: A promising and challenging approach: radiologists' perspective on deep learning and artificial intelligence for fighting COVID-19. Diagnostics **11**(10), 1924 (2021). https://doi.org/10.3390/diagnostics11101924
3. Raj, D., Prabadevi, B., Gadekallu, T., Bhatia, S., Saraee, M.: Multiclass classification and defect detection of steel tube using modified YOLO (2023). https://doi.org/10.1007/978-981-99-8184-7_32
4. Szeliski, R.: Computer vision: algorithms and applications. Springer Nature (2022)
5. Gonzalez, F.R.C.: Digital image processing. Pearson education India (2009)
6. Prakash, K. et al.: A study of image processing in agriculture. Int. J. Adv. Network. App. **9**(1), 3311 (2017)
7. Dekker, R.: The importance of having data-sets. In: 27th IATUL conference (2006)
8. Jain, A., Patel, H., Nagalapatti, L., Gupta, N., Mehta, S., Guttula, S., Mujumdar, S., Afzal, S., Sharma Mittal, R., Munigala, V.: Overview and importance of data quality for machine learning tasks. In: Proceedings of the 26th ACM SIGKDD international conference on knowledge discovery and data mining (2020). https://doi.org/10.1145/3394486.3406477
9. Brownlee, J.: Deep learning for computer vision: image classification, object detection, and face recognition in python. Machine Learning Mastery (2019)
10. Cleetus, L., Sukumar, A.R., Hemalatha, N.: Computational prediction of disease detection and insect identification using Xception model. bioRxiv (Cold Spring Harbor Laboratory) (2021). https://doi.org/10.1101/2021.08.10.455608
11. Ali, L., Alnajjar, F.S., Jassmi, H.A., Gochoo, M., Khan, W., Serhani, M.A.: Performance evaluation of deep CNN-based crack detection and localization techniques for concrete structures. Sensors (Basel, Switzerland). **21** (2021)
12. Alzubaidi, L., Zhang, J., Humaidi, A.J., et al.: Review of deep learning: concepts, CNN architectures, challenges, applications, future directions. J Big Data **8**, 53 (2021). https://doi.org/10.1186/s40537-021-00444-8

13. Wang, M. et al.: A high-speed and low-complexity architecture for soft max function in deep learning
14. Kingma, D.P., Ba, J.: Adam: a method for stochastic optimization. In: arXiv preprint arXiv: 1412.6980 (2014)
15. Li, S. et al.: J. Phys.: Conf. Ser. 1813, 012051 (2021)
16. Krizhevsky, A., Sutskever, I., Hinton, G.E.: Imagenet classification with deep convolutional neural networks. Adv. Neural Inf. Process. Syst, pp. 1097–1105 (2012)
17. Scherer, D., Müller, A., Behnke, S.: Evaluation of pooling operations in convolutional architectures for object recognition. Int. Conf. Artif. Neural Networks, pp. 92–101 (2010)
18. Simonyan, K., Zisserman, A.: Very deep convolutional networks forlarge-scale image recognition. In: arXiv preprint arXiv:1409.1556 (2014)
19. He, K., Zhang, X., Ren, S., Sun, J.: Deep residual learning for image recognition. In: Proceedings of the IEEE conference on computer vision and pattern recognition, pp. 770–778 (2016)
20. Lecun, Y., Bottou, L., Bengio, Y., Haffner, P.: Gradient-based learning applied to document recognition. Proc. IEEE **86**(11), 2278–2324 (1998)
21. Hinton, G.E., Srivastava, N., Krizhevsky, A. et al.: Improving neural networks by preventing co-adaptation of feature detectors. Comp. Sci. (2012)
22. Ayana, G. et al.: Vision-transformer-based transfer learning for mammo-gram classification. In: Diagnostics **13**(2), 178 (2023)
23. Olivas, E.S. et al.: Chapter 11: transfer learning, handbook of research on machine learning applications (2009)
24. Kiruthika, S.U., Raja, S.K.S., Balaji, V., Raman, C.J., Arumugam, S.D.: Detection of tuberculosis in chest X-rays using U-net architecture. Int. J. Innov. Technol. Explor. Eng. ISSN, 2278-3075 (2019)
25. Al Sahili, Z., Awad, M.: The power of transfer learning in agricultural applications: AgriNet. In: Frontiers in plant science. **13**, 992700 (2022)
26. Szegedy, C. et al.: Rethinking the inception architecture for computer vision. In: Proceedings of the IEEE conference on computer vision and pattern recognition, pp. 2818–2826 (2016)
27. Chollet, F.: Xception: deep learning with depthwise separable convolutions. In: Proceedings of the IEEE conference on computer vision and pattern recognition, pp. 1251–1258 (2017)
28. Dosovitskiy, A. et al.: An image is worth 16x16 words: transformers for image recognition at scale
29. Omer, A.A.M.: Image classification based on vision transformer. J. Comput. Commun. **12**(4), 49–59 (2024)

Chapter 19
Twenty-Fold Protracted Residual Network (TFPRN) for Rice Grain Type Detection

P. V. Yeswanth and S. Deivalakshmi

Abstract There are myriad varieties of rice grains in different parts of the world grown on diverse climates, physiographies and terrains. As sorting of these varieties by physical means is cumbersome, innovative image processing techniques using AI, ML and DL are currently providing newer insights and accurate distinguishing mechanisms. In this paper, the proposed model transforms a low-resolution rice image to an enhanced super resolution image by Deep Learning. The model is christened as a 'Twenty-Fold Protracted Residual Network' (TFPRN) to indicate the twenty-stage Protracted Residual Module (PRM)s are employed here. The model is trained on 80% of the total 75,000 raw images of rice, with 5 classes each of 15,000 low resolution images so to test them randomly on the remaining 10% of the images and validate on 10% of the residual set of images. MSE, PSNR and SSIM are computed to quantify the loss function, to train the model and to validate the model, respectively. The model did gain a good accuracy level of 99.14, 98.45 and 97.82% and PSNR of 34.468, 32.137 and 31.467 and also SSIM of 0.9124, 0.8938 and 0.8683, respectively, on Super Resolution images upped by factors of 2, 4 and 6.

Keywords Twenty-fold protracted residual network · TFPRN · Protracted residual module · Super resolution · Rice grain classification · Low resolution · Rice grain type detection

P. V. Yeswanth (✉) · S. Deivalakshmi
Department of Electronics and Communication Engineering, National Institute of Technology, Tiruchirappalli, India
e-mail: yeswanth2core@gmail.com

P. V. Yeswanth
Department of Electronics and Communication Engineering, Government Polytechnic Parvathipuram, Parvathipuram Manyam, Andhrapradesh, India

J. Maiti et al. (eds.), *Operations Research and Data Analytics: Current Trends and Future Perspectives*, Lecture Notes on Multidisciplinary Industrial Engineering,
https://doi.org/10.1007/978-981-96-8327-7_19

Nomenclature

SR	Super Resolution
PSNR	Peak Signal to Noise Ratio
TFPRN	Twenty-Fold Protracted Residual Network
DL	Deep Learning
SSIM	Structural Similarity Index Measure
MSE	Mean Square Error

19.1 Introduction

Catering to the consumption needs of greater than half of the world population and raising on 165.25 mha of arable land of the world, rice stands third in world only after Sugarcane and American maize corn. Distinguishable by characteristic features like size, texture, colour, aroma, chalkiness and stickiness, there are about forty thousand rice varieties worldwide and eleven thousand Indian varieties (including pristine exotic varieties). The astounding variety of its species makes it interesting to study rice in the context of image processing. The crop is raised in all climatic, pedological and physiographical conditions, albeit, with varying yields. Rice is grown in almost all Asian nations and exported worldwide. Cultivation of rice is a substantial economic activity providing employment to millions of people. Agriculture is still the mainstay of Indian economy. With about 31 mha area under cultivation, India ranks first in area and second in the production of rice in the world. Indian farmers follow traditional methods to raise and choose a variety of rice. Size of the grain, colour, aroma, culinary preferences, smell, availability of water and texture determine the agrarian and the phenological aspects to be taken care of. In the conventional methodology, the human sorter classifies rice in accordance with the aforementioned features. This physical way of sorting rice is time-consuming, laborious, biased and ineffective. New-age computer utilities-Machine Learning and Deep by their learning methods and impeccable image processing algorithms come handy in segregating rice variety.

Machine Learning is definitely a handy tool in identification and classification of rice grains. The suite of algorithms in machine learning help in processing the snapshots of images. Image preprocessing includes capturing and collating a large diverse dataset of rice grain images, labelling each image of the correct variety as part of a supervised learning exercise, training the machine learning model on the labelled dataset and using the trained model to map with the new images to classify the rice grain images seamlessly. This kind of exercise can be used for a variety of purposes, viz., quality control, inventory management and sorting of rice for marketing. The major bottleneck in machine learning is that it is very difficult to discriminate and distinguish among a great variety of rice grains from a spectrum of images. It is not easy to train a model to achieve a higher degree of accuracy.

Current studies on cereals using image processing by Computer vision or Machine Learning methodologies demonstrate that the variety segregation is on the basis of visual and tangible features: size, colour, texture and morphology. The Deep Learning tools like CNN score much over ML as a result of lesser number of parameters and smaller computational resource for image processing. The Convolutional Neural Network (CNN) is a helpful tool most commonly used in computer vision for detection and image processing. Assortment layers, namely-Polling layers, fully connected layers and convolutional layers help it to classify images very efficiently. Convolutional layers' output is fed into the pooling layer whose main function is sampling down the images and dimensionality reduction to lessen computational expenses while retaining the most important features of the images. The fully connected layer connects both the outputs and classifies the images via matrices [1].

CNNs can process images far more swiftly than typical machine learning algorithms. CNNs can be simply scaled to husband larger datasets. Hence, they are used in situations where the quantum of data is constantly proliferating, as in agriculture and biosciences. CNNs can learn complicated information from images that regular machine learning algorithms cannot. Rice grain detection is more accurate because of this ability. This research work describes an innovative 'Twenty-fold Protracted Residual Network (TFPRN)' model for rice grain detection. The model proposed here attempts to augment the resolution of a lower resolution image. The literature is aptly detailed in part 2. The approach and outcomes are highlighted in Sects. 19.3 and 19.4. The crux of the study is distilled as a discussion in Sect. 19.5 and as a conclusion in Sect. 19.6.

19.2 Related Work

19.2.1 Related Work on Image Super Resolution

Discrete Wavelet Transform (DWT) concepts were used as a tool to develop a novel super-resolution approach by interpolating raw images and sub-band images of higher frequency by Anbarjafari et al. Their approach transformed an image into sub or low-band images through DWT technique. The raw input lower resolution images and higher frequency low-band images are interpolated so as to obtain super-resolution image by applying inverse DWT (IDWT) [2].

Though Discrete Wavelet Transform can be potentially effective for some image processing jobs, it can result in complications when it comes to super-resolution. The DWT can be sensitive to noise and other artefacts in the low-resolution image, which can lead to overfitting and inaccurate reconstructions. The DWT can be computationally expensive, especially for large images. The DWT is a linear transform, which means that it cannot model non-linear relationships between pixels in the image.

Yang et al. proposed a unique approach and used minimal representation techniques to amplify the raw image to a higher resolution super image. The authors assumed that the higher-resolution image is an over-sampled manifestation of the lower-resolution image. Also, it is considered that the higher-resolution image patches have minimal representation in comparison to a list of prototype signal-atoms. Initially, the input image is down-sampled to produce a low resolution (LR) image which is then passed through a dictionary to get its minimal representation which is used to overhaul a super-resolution image [3]. Minimal representation overcomes the drawbacks of DWT by using a non-linear model to represent the high-resolution image as a sparse combination of basis-functions. This allows the model to better capture non-linear relationships between pixels, and it is also more robust to noise and other artefacts. Additionally, minimal representation is computationally less expensive than the DWT, especially for large images.

Super-resolution of images as a minimal representation of input image patches approach also has a number of potential drawbacks. Finding the minimal representation of a lower-resolution image might be a computationally expensive task, particularly for large images. Minimal representation techniques are not highly adaptable, it can be challenging to use them with various image types. Overfitting is a risk with minimal representation techniques, which means they can end up learning the visual noise rather than the real underlying signal. This may result in pixelated or blurry visuals.

Chao Dong et al., devised a deep learning method of super-resolution for single images. Their scheme consummately maps a lower-resolution image to a higher-resolution image. A deep CNN is adapted to represent the mapping; it inputs a low-resolution image and outputs a super-resoluted one. It demonstrated how deep convolutional networks may be used to represent conventional sparse-coding-based SR approaches. This technique can inhibit the problem of overfitting as it stops training before the model gets fully converged. This helps to stop the training before the model starts overfitting [4]. The mentioned super resolution techniques did not work on small particles like rice grains. This necessitates to develop a deep learning model which can improve the resolution of even a small rice grain particle for better identification of rice type.

19.2.2 Related Work on Image Classification

Kuo et al., used image processing techniques to distinguish thirty varieties of rice grains and differentiated them by their morphological characteristic metrics like body size, lemma size and brush of the grain. They achieved an accuracy of 89.10% and an SD of 7% [5].

Zareiforoush et al., utilized image processing algorithms along with Fuzzy Inference System (FIS) for quality classification of rice grain consisting of a total of five classes based on milling degree and proportion of broken kernels in rice grains. The devised method and the expert-determined grades had an agreement of 89.8% [6]. Hamzah et al., endeavoured to integrate ANN with Back-Propagation method, with KNN, with SVM and ultimately with Bayes' classifier algorithm. Among them ANN with Back Propagation achieved the best accuracy of 96%. ANN has many advantages compared to Fuzzy classifiers and statistical classifiers because of its capability in processing blurry images [7].

Nirmalan et al., analysed a dataset of nearly 1000 images of rice dataset containing 10 classes with classification model by twofold cross validation and was able achieve an accuracy rate of about 97.90% and a standard deviation of 0.50% [8]. There has been tremendous work on rice grain identification, categorization, feature extraction and other valuable characteristics. Ibrahim et al. categorized into 3 classes of rice grains using ANN and SVM methods [9]. A pre-trained R-CNN was used by Aukkapinyo et al. for the magnification of images, amplify the features and classify the grain of rice into five categories [10].

Mery et al., underscored the important steps in Computer Vision applications as Capturing the image, Processing, Segmentation, Segregation and Extraction of features. AS CNN models do away with the need of Segmentation and Feature extraction, we need to work on the remaining three stages of Capturing the image, Preprocessing and Segregation [11]. The mentioned classification techniques are not able to provide adequate classification accuracy when the resolution of the input image is low. So Yeswanth et al. used super resolution based classification techniques for improving the classification accuracy [12–23]. But there is a research gap to develop a model which can improve the resolution of rice grain for better classification.

While there has been work on super-resolution techniques like Discrete Wavelet Transform (DWT) and minimal representation, they are not effective for small objects like rice grains. These techniques can be computationally expensive and struggle with noise, leading to inaccurate reconstructions. Existing image classification techniques, though achieving high accuracy with high resolution images, suffer from reduced accuracy when dealing with low resolution images. The proposed methodology addresses these gaps by using a Twenty-Fold Protracted Residual Network (TFPRN) for super-resolution specifically designed for rice grain images. This allows for improved classification accuracy in rice grain variety classification tasks.

19.3 Methodology

Using a Twenty-Fold Protracted Residual Network (TFPRN), which effectively remodels deep components in twenty stages, it is possible to create images of high resolution of the rice grain from low resolution images that were initially provided. The architecture of the suggested model is explained in later parts, along with details on how it was made and how it was trained.

W_{LR} represents input low resolution image, while W_{SR} represents the output super resolution image. The model's leading convolutional block recovers shallow features from the input W_{LR} picture.

$$I_{shallow} = \Upsilon_{initial}(W_{LR}) \tag{19.1}$$

19.3.1 Twenty-Fold Protracted Residual Network (TFPRN) Design

I_{shallow} is the redeemed shallow features in Eq. 19.1, and $\Upsilon_{\text{initial}}$ denotes the preliminary convolution block shown in Fig. 19.1. In this manner, the TFPRN is provided input from the I_{shallow} value to accurately extract deep features.

$$I_{\text{deep}} = \Upsilon_{\text{TFPRN}}(I_{\text{shallow}}) \tag{19.2}$$

The retrieved deep features are then represented by I_{deep} and Υ_{TFPRN} is utilized to represent the TFPRN function in Eq. 19.2. As shown in Fig. 19.1, the redeemed features are coupled first ($I_{\text{shallow}} + I_{\text{deep}}$) to provide the input for the up-sample layer.

$$I_{\text{up}} = \Upsilon_{\text{up}}\left(I_{\text{shallow}} + I_{\text{deep}}\right) \tag{19.3}$$

In Eq. 19.3, I_{up} represents the features obtained from Fig. 19.1, and Υ_{up} represents the operator of the up-sample layer. I_{up} provides the input for the subsequent convolutional block, which produces the needed Super Resolution image.

$$W_{\text{SR}} = \Upsilon_{\text{conv}}\left(I_{\text{up}}\right) = \Upsilon_{\text{ETFPRN}}(W_{\text{LR}}) \tag{19.4}$$

Similarly, W_{SR} denotes the Super resolution picture acquired as a result of TFPRN. As indicated by Eq. 19.4, Υ_{conv} indicates the last convolutional layer. ETFPRN represents the whole block, which accepts the W_{LR} picture as input and outputs the W_{image}.

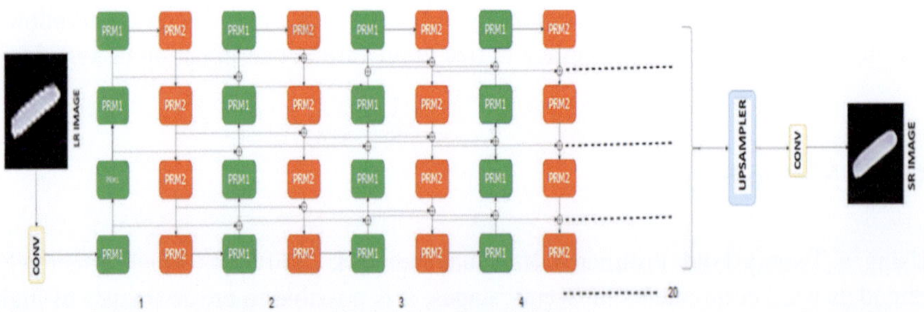

Fig. 19.1 Twenty-fold protracted residual network (TFPRN)

19.3.2 Twenty-Fold Protracted Residual Network (TFPRN) Prototype

Figure 19.2 depicts the suggested Protracted Residual Network (PRN) model. It is made up of blocks, most notably the Protracted Residual Module 1 (PRM1) and the Protracted Residual Module 2 (PRM2).

The branch that goes upward extracts the deep characteristics, which are subsequently passed on to PRM1 to up sample the same. The deep features recovered by the branch that travels downwards, on the other hand, use the PRM2 to down sample these features. A low-resolution image U (H * W * 8) input is used to generate a resultant (H * W * 16) image containing extracted features. As a result, the count increases by two at each level of the channel, yielding a final image size (H * W * 128) on top. These are then passed on to the PRM2, which produces output features (H * W * 64). Finally, the image size produced for V is (H * W * 8).

Deep layers leverage shallow features to improve the model's performance. Figure 19.3 depicts the function of the PRM1. The disparities, which often indicate the mistake, are then analysed and up sampled to recover deep features. When comparing Figs. 19.3 and 19.4, PRM1 and PRM2 appear to be identical, with the exception that the convolution layer is replaced with a deconvolution block, resulting in the input being down sampled rather than up sampled.

Fig. 19.2 Protracted residual network (PRN)

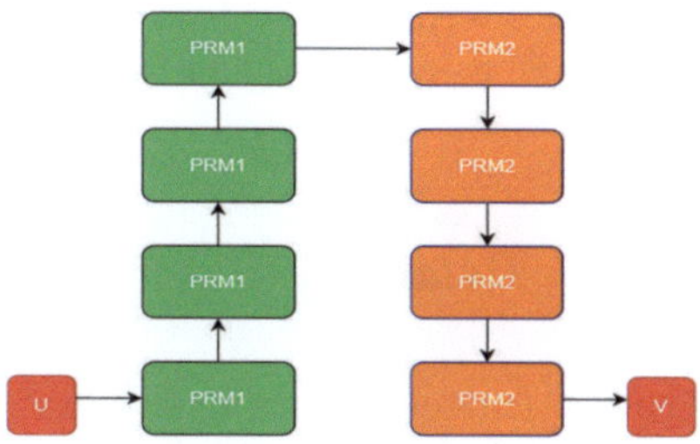

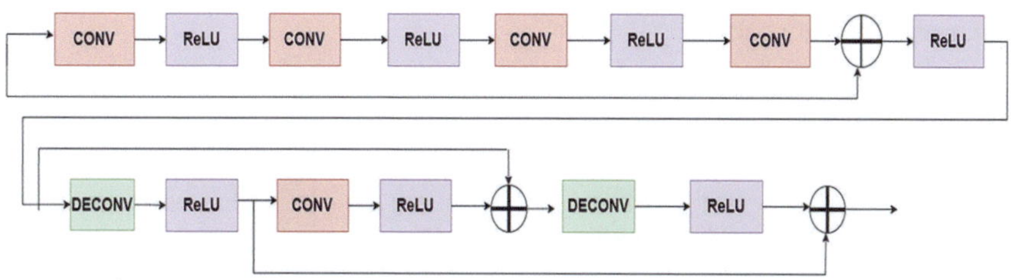

Fig. 19.3 Protracted residual Module 1 (PRM1)

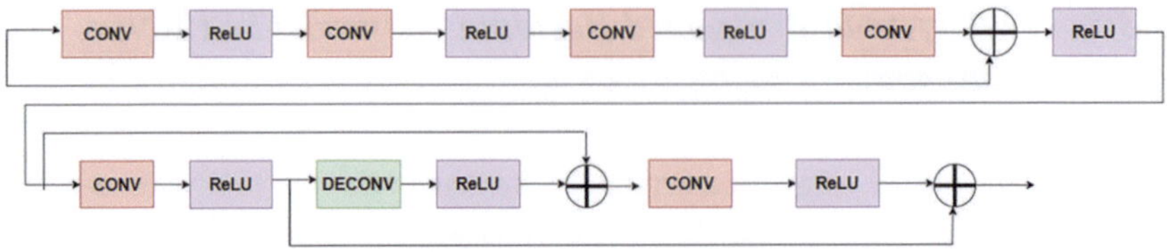

Fig. 19.4 Protracted residual Module 1 (PRM2)

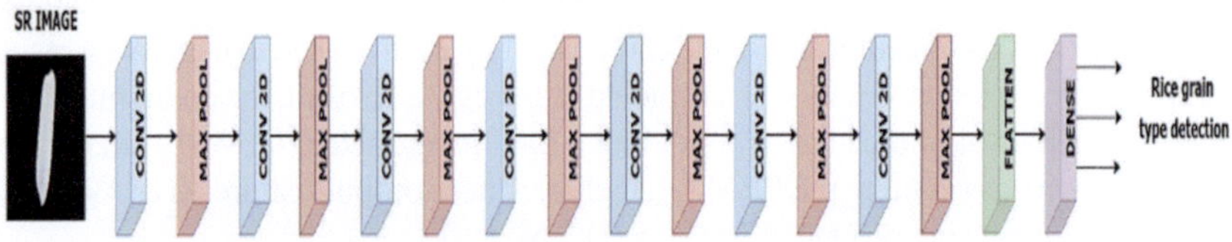

Fig. 19.5 Rice grain classification model

19.3.3 Classification of Rice Grain Varieties

Grain classification methods previously trained on HR image datasets to provide exact findings. Figure 19.5 depicts the model used for rice grain classification. The HR images were generated via image enhancement of LR images from the TFPRN model, and they were then utilized as input for such classification models.

19.3.4 Dataset

The rice dataset [24] comprises of five distinct classes: arborio, basmati, ipsala, jasmine and karacadag. Table 19.1 contains information on this dataset. The backgrounds of the photographs utilized are similar and nearly dull, with no unusual characteristics that stick out.

Table 19.1 Rice dataset

Class of rice grain	No of images
Arborio	15,000
Basmati	15,000
Ipsala	15,000
Jasmine	15,000
Karacadag	15,000
Total images	75,000

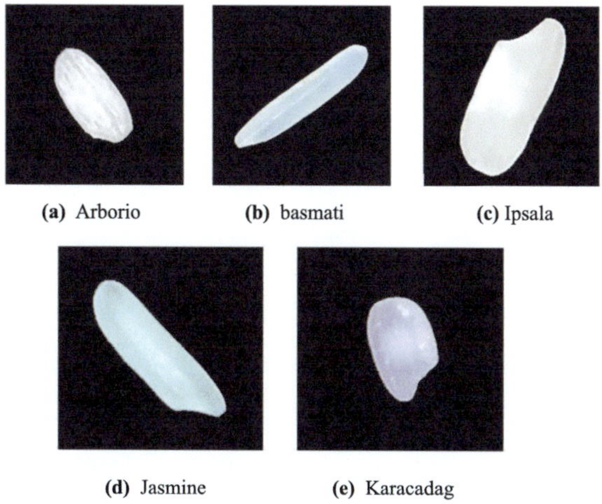

Fig. 19.6 Sample images of rice grain dataset. **a** Arborio, **b** basmati, **c** Ipsala, **d** Jasmine, **e** Karacadag

This dataset was divided as follows: 80% was applied to train the model, 10% was chosen at random for testing, and the remainder of 10% was used for validation (Fig. 19.6).

19.3.5 Experimentation

First, 75,000 images are down sampled to obtain LR images with their gauging factors, as shown in Figs. 19.7 and 19.8. Their characteristics are enhanced to those of Higher Resolution using our TFPRN model for additional differentiation, as shown in Fig. 19.5.

Figure 19.9 presents the training and validation losses for super resolution block.

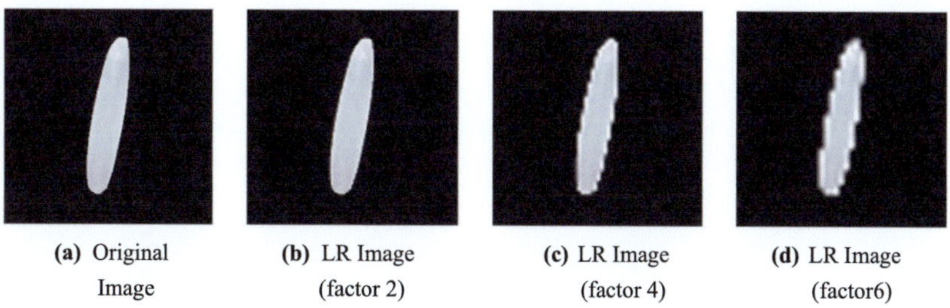

Fig. 19.7 Sample LR images. **a** Original image, **b** LR image (factor 2), **c** LR image (factor 4), **d** LR image (factor 6)

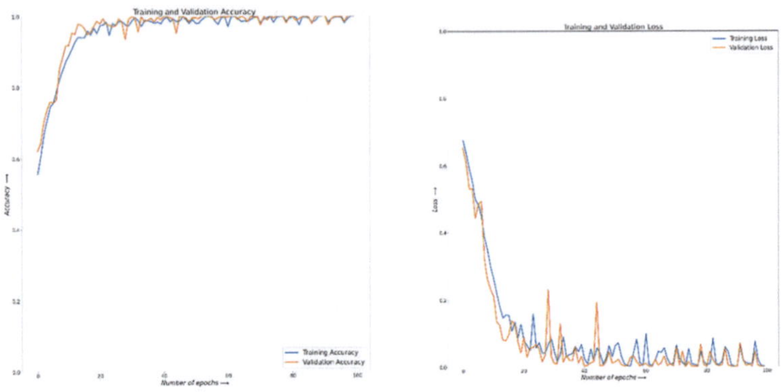

(a) Factor 2 Train/Val accuracies

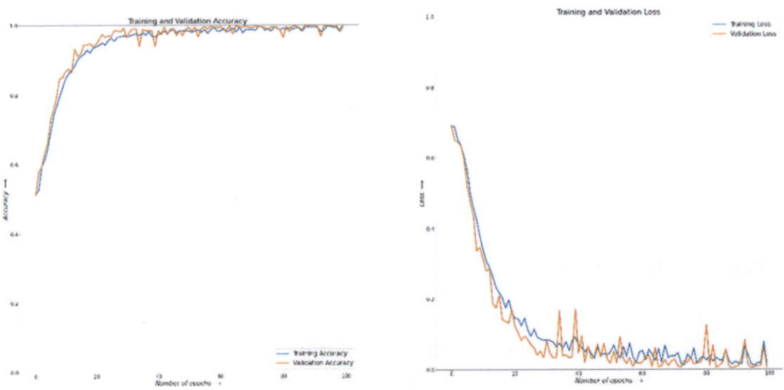

(b) Factor 4 Train/Val accuracies

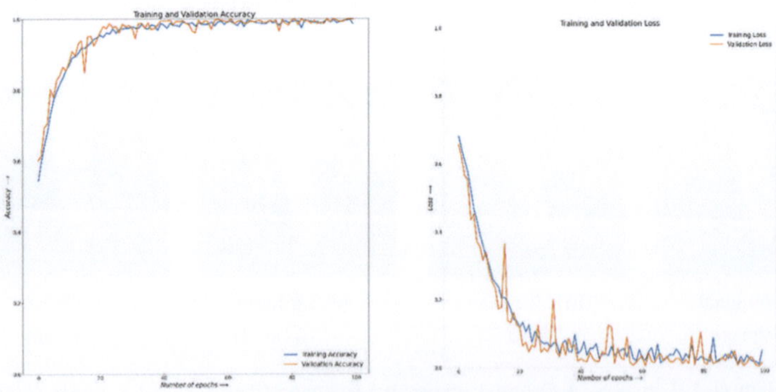

(c) Factor 6 Train/Val accuracies

Fig. 19.8 Training and validation accuracies of classification block. **a** Factor 2 Train/Val accuracies, **b** Factor 4 Train/Val accuracies, **c** Factor 6 Train/Val accuracies

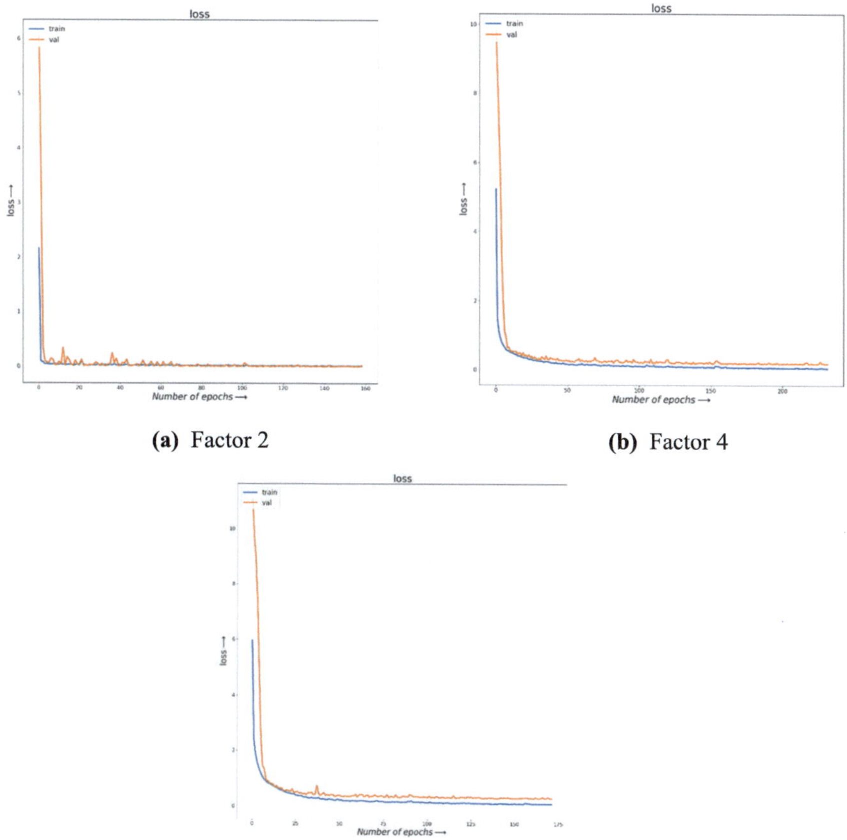

(a) Factor 2

(b) Factor 4

(c) Factor 6

Fig. 19.9 Training and Validation losses of super resolution block. **a** Factor 2, **b** Factor 4, **c** Factor 6

19.4 Results

19.4.1 Qualitative Analysis

Figure 19.10 analysis based on quality showcases the SR images produced by our model using different scaling factors applied to the Low-Resolution (LR) images. These SR images demonstrate the effectiveness of our model in enhancing the visual quality and level of detail. For a comprehensive performance evaluation.

Fig. 19.10 Qualitative results of SR images for factors 2, 4 and 6. **a** SR with factor 2, **b** SR with factor 4, **c** SR with factor 6

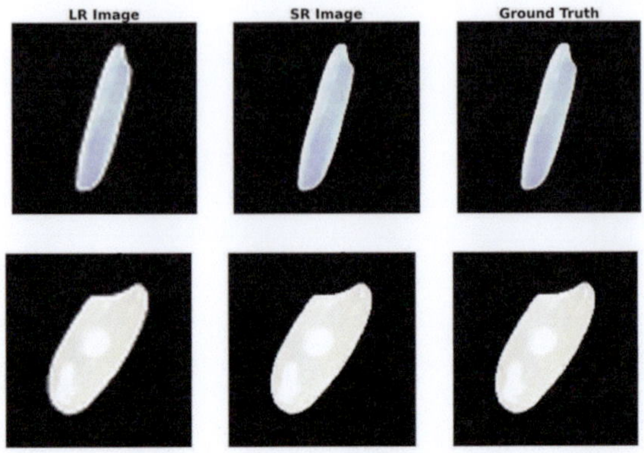

(a) SR with factor 2

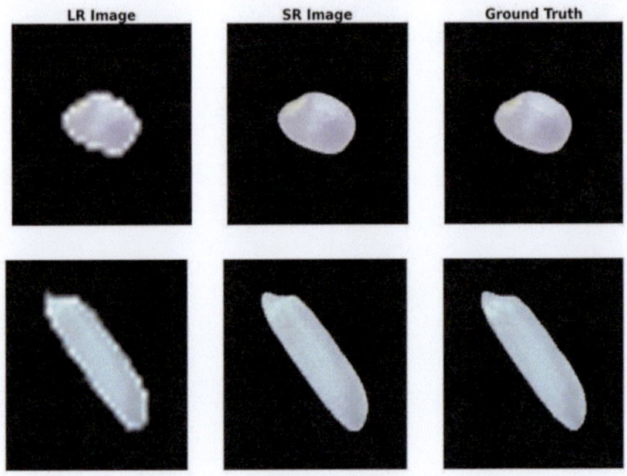

(b) SR with factor 4

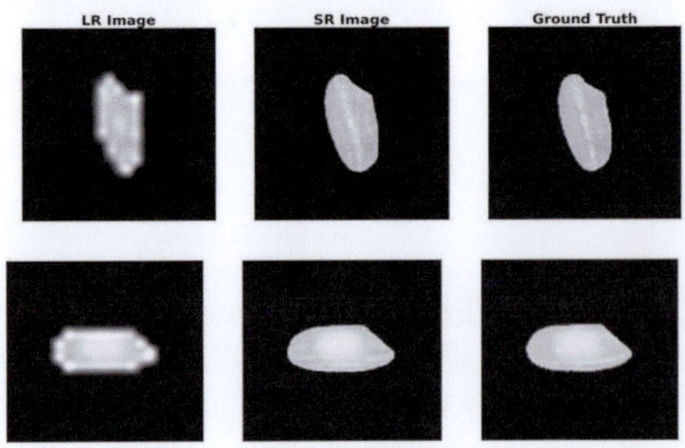

(c) SR with factor 6

Table 19.2 Assessment of TFPRN

Metrics	SR factor 2 image	SR factor 4 image	SR factor 6 image
PSNR	34.648	32.137	31.467
SSIM	0.9124	0.8938	0.8683

Table 19.3 Classification model performance assessment

Metrics	SR image (factor 2)	SR image (factor 4)	SR image (factor 6)
Accuracy	99.14	98.45	97.82

Table 19.4 Comparison with existing models

Metrics	ResNet50	DenseNet121	EfficientNet	TFPRN
Accuracy	94.51	96.65	97.79	99.14

19.4.2 Quantitative Analysis

From Tables 19.2 and 19.3, it can be observed that both PSNR and SSIM values decrease as the Super-Resolution (SR) factor increases from 2 to 6. This indicates a potential degradation in image quality with higher SR factors. The decrease in PSNR from 34.648 to 31.467 suggests an increase in noise artefacts as the SR factor increases. The SSIM values also show a concerning downward trend, indicating a decreasing similarity between the reconstructed image and the ground truth, particularly in terms of luminance, contrast and structure. The comparison with existing models in Table 19.4 highlights the competitive performance of TFPRN in terms of classification accuracy. TFPRN achieves a comparable accuracy (99.14%) to the best performing model (EfficientNet: 97.79%) and outperforms ResNet50 (94.51%) and DenseNet121 (96.65%).

19.5 Discussion

High-resolution (HR) imaging is critical in many fields, including agricultural research, where precise analysis and classification of rice grain varieties can significantly impact quality control, breeding programs and market value. However, acquiring HR images is often resource-intensive and time-consuming, making it challenging to apply such methods on a large scale. This is where super-resolution techniques become invaluable, enabling the enhancement of low-resolution (LR) images to HR quality, thus facilitating more accurate analysis without the need for sophisticated imaging equipment.

The TFPRN methodology addresses this need by offering a robust approach to image super-resolution through a structured and efficient deep learning model. The key reasons for utilizing the TFPRN methodology are:

Improved Image Quality: The TFPRN model can convert low-resolution images into high-resolution images, which are crucial for detailed and accurate grain analysis. This enhancement is achieved by effectively extracting and combining both shallow and deep features from the input images.

Efficient Feature Extraction: The model employs a series of convolutional blocks to recover shallow features and deep features, ensuring that essential details are captured and preserved throughout the processing stages. This layered approach maximizes the information retention from the original LR images.

Advanced Upsampling Techniques: By using a combination of upsampling and convolutional layers, the TFPRN model generates super-resolution images that are not only visually superior but also retain critical features necessary for accurate classification.

Versatility in Application: The model's architecture, particularly the integration of Protracted Residual Modules (PRM1 and PRM2), allows it to handle various types of image resolutions and complexities. This adaptability makes it suitable for different datasets and conditions, enhancing its applicability in real-world scenarios.

Enhanced Classification Accuracy: The HR images produced by the TFPRN model improve the performance of downstream classification tasks. High-resolution inputs lead to more precise and reliable classification of rice grain varieties, which is essential for ensuring quality and consistency in agricultural products.

Resource Optimization: By transforming LR images into HR images using deep learning techniques, the need for high-end imaging equipment is reduced. This makes the TFPRN approach a cost-effective solution, especially in resource-limited settings.

In summary, the TFPRN methodology is essential for enhancing the resolution of rice grain images, thereby improving the accuracy of grain classification models. It provides a sophisticated yet practical solution to the challenges associated with obtaining and analysing high-resolution images, ultimately supporting better agricultural practices and outcomes.

19.6 Conclusion

The paper proposes a novel TFPRN super resolution model for the classification of rice grains. The input photos are super-resoluted in order to give more accurate categorization results. In addition, the model is compared against pre-trained models such as GoogleNet, EfficientNet, and ResNet50.

For super resolution factors of 2, 4 and 6, the model generated PSNR values of 34.648, 32.137 and 31.467, respectively. The SSIM values for factors 2, 4 and 6 are 0.9124, 0.8938 and 0.8683, respectively, while the accuracy is 99.14, 98.45 and 97.82.

The research gap of this paper is, while TFPRN maintains good accuracy, it comes at the cost of potentially lower image quality at higher SR factors. Future work may be done by utilizing other grain datasets as input, therefore widening the network range and obtaining super resolution images.

References

1. Kozłowski, M., Górecki, P., Szczypiński, P.M.: Varietal classification of barley by convolutional neural networks. Biosys. Eng. **184**, 155–165 (2019). https://doi.org/10.1016/j.biosystemseng.2019.06.012
2. Anbarjafari, G., Demirel, H.: Image super resolution based on interpolation of wavelet domain high frequency subbands and the spatial domain input image. ETRI J. **32**(3), 390–394 (2010). https://doi.org/10.4218/etrij.10.0109.0303
3. Yang, J., Wright, J., Huang, T.S., Ma, Y.: Image super-resolution via sparse representation. IEEE Trans. Image Process. **19**(11), 2861–2873 (2010). https://doi.org/10.1109/TIP.2010.2050625
4. Dong, C., Loy, C.C., He, K., Tang, X.: LNCS 8692—learning a deep convolutional network for image super-resolution (2014). http://mmlab.ie.cuhk.edu.hk/projects/SRCNN.html
5. Kuo, T.Y., Chung, C.L., Chen, S.Y., Lin, H.A., Kuo, Y.F.: Identifying rice grains using image analysis and sparse-representation-based classification. Comput. Electron. Agric. **127**, 716–725 (2016). https://doi.org/10.1016/j.compag.2016.07.020
6. Zia, H., Fatima, H.S., Khurram, M., Hassan, I.U., Ghazal, M.: Rapid testing system for rice quality control through comprehensive feature and kernel-type detection. Foods **11**(18) (2022). https://doi.org/10.3390/foods11182723
7. Hamzah, A.S., Mohamed, A.: Classification of white rice grain quality using ANN: a review. IAES Int. J. Artif. Intell. **9**(4), 600–608 (2020). https://doi.org/10.11591/ijai.v9.i4.pp600-608
8. Nirmalan, V.E., Nawarathna, R.D., Manivannan, S.: Comparative analysis of different features and encoding methods for rice image classification. In: 2018 IEEE International Conference on Information and Automation for Sustainability (ICIAfS), 1–5 (2018). https://doi.org/10.1109/ICIAFS.2018.8913380
9. Ibrahim, S., Kamaruddin, S.B.A., Zabidi, A., Ghani, N.A.M.: Contrastive analysis of rice grain classification techniques: Multi-class support vector machine versus artificial neural network. IAES Int. J. Artif. Intell. **9**(4), 616–622 (2020). https://doi.org/10.11591/ijai.v9.i4.pp616-622
10. Aukkapinyo, K., Sawangwong, S., Pooyoi, P., Kusakunniran, W.: Localization and classification of rice-grain images using region proposals-based convolutional neural network. Int. J. Autom. Comput. **17**(2), 233–246 (2020). https://doi.org/10.1007/s11633-019-1207-6
11. Mery, D., Pedreschi, F., Soto, A.: Automated design of a computer vision system for visual food quality evaluation. Food Bioprocess Technol. **6**(8), 2093–2108 (2013). https://doi.org/10.1007/s11947-012-0934-2
12. Yeswanth, P.V., Deivalakshmi, S.: Extended wavelet sparse convolutional neural network (EWSCNN) for super resolution. Sādhanā **48**(2), 52 (2023)
13. Yeswanth, P.V., Khandelwal, R., Deivalakshmi, S.: Super resolution-based leaf disease detection in potato plant using broad deep residual network (BDRN). SN Comput. Sci. **4**(2), 112 (2022)
14. Yeswanth, P.V., Raviteja, R., Deivalakshmi, S.: Sovereign critique network (SCN) based super-resolution for chest X-rays images. In: International Conference on Signal Processing, Computation, Electronics, Power and Telecommunication (IConSCEPT), pp. 1–5 (2023). https://doi.org/10.1109/IConSCEPT57958.2023.10170157

15. Yeswanth, P.V., Kushal, S., Tyagi, G., Kumar, M.T., Deivalakshmi, S., Ramasubramanian, S.P.: Iterative super resolution network (ISNR) for potato leaf disease detection. In: 2023 International Conference on Signal Processing, Computation, Electronics, Power and Telecommunication (IConSCEPT), pp. 1–6(2023). https://doi.org/10.1109/IConSCEPT57958.2023.10170224
16. Yeswanth, P.V., Khandelwal, R., Deivalakshmi, S.: Two fold extended residual network based super resolution for potato plant leaf disease detection. In: Internet of Things (IoT): Key Digital Trends Shaping the Future. ICIoTCT 2022. Lecture Notes in Networks and Systems, vol. 616. Springer (2022). https://doi.org/10.1007/978-981-19-9719-8_16
17. Yeswanth, P.V., Khandelwal, R., Deivalakshmi, S.: Four fold prolonged residual network (FFPRN) based super resolution for cherry plant leaf disease detection. In: Intelligent Control, Robotics, and Industrial Automation. RCAAI 2022. Lecture Notes in Electrical Engineering, vol. 1066. Springer (2023). https://doi.org/10.1007/978-981-99-4634-1_38
18. Yeswanth, P.V., Thool, K.V., Deivalakshmi, S.: Tuberculosis disease diagnosis using controlled super resolution. In: Big Data and Artificial Intelligence. BDA 2023. Lecture Notes in Computer Science, vol. 14418. Springer (2023). https://doi.org/10.1007/978-3-031-49601-1_1
19. Yeswanth, P.V., Deivalakshmi, S., George, S., Ko, S.B.: Residual skip network-based super-resolution for leaf disease detection of grape plant. Circ., Syst., Signal Process., 1–29 (2023)
20. Yeswanth, P.V., Rajan, J., Mantha, B.P., Siva, P., Deivalakshmi, S.: Self-governing assessment network (SGAN) based super-resolution for CT chest images. In: 2023 International Conference on Computer, Electronics & Electrical Engineering & Their Applications (IC2E3), pp. 1–5 (2023). https://doi.org/10.1109/IC2E357697.2023.10262573
21. Yeswanth, P.V., Patil, M.K., Deivalakshmi, S.: Cotton leaf disease detection using courier super resolution network (CSRN). In: 2023 IEEE 7th Conference on information and communication technology (CICT), Jabalpur, India, 2023, pp. 1–5 (2023). https://doi.org/10.1109/CICT59886.2023.10455103
22. Yeswanth, P.V., Thool, K.V., Deivalakshmi, S.: Image steered super resolution based Melonoma cancer detection. In: 2023 OITS International Conference on Information Technology (OCIT), Raipur, India, 2023, pp. 47–52 (2023). https://doi.org/10.1109/OCIT59427.2023.10431254
23. Yeswanth, P.V., Thanya, Y., Deivalakshmi, S., Lokesh Kumar, K.M., George, J.: Image super resolution using enhanced attention block (EAB) for grape leaf disease detection. In: 2023 7th International Conference on Computer Applications in Electrical Engineering-Recent Advances (CERA), Roorkee, India, 2023, pp. 1–5 (2023). https://doi.org/10.1109/CERA59325.2023.10455721
24. https://www.muratkoklu.com/datasets/

Chapter 20
RESNET Based Crop Disease Detection Models Using Convolutional Neural Networks

Muthaiah Uchimuthu, Yanna Nithin Reddy, and Ram Vinod Roy

Abstract Recent research in plant disease classification leverages advanced deep learning models and both public and author-generated datasets. The study adopts a two-step approach: initially conducting a comparative analysis to identify the most effective convolutional neural network (CNN) among established architectures and custom models, and subsequently employing deep learning optimizers to enhance the performance of the top-performing model. Various metrics, including validation accuracy/loss, $F1$-score, precision, recall, and training epochs, are used to evaluate popular deep learning architectures such as VGG16, VGG19, ResNet-50, and ResNet-101. All architectures are trained on the extensive PlantVillage dataset, containing 70,295 images of diseases across 14 plant species. In a study focused on disease prediction using Keras with TensorFlow, the findings highlight that the RESNET101 model, trained with the Adam optimizer, achieved the highest macro and micro $F1$-scores, reaching 0.9417 and 0.9425, along with exceptional testing and validation accuracy rates of 94.25 and 93.91%. These outcomes signify a substantial improvement over earlier models, with the novelty of this study lying in the class-wise comparison of metrics, underscoring the innovative nature of this research. Furthermore, the methodology holds significant potential for broader applications in the field of agriculture, as early and accurate disease detection plays a pivotal role in crop health management, leading to increased yields, reduced pesticide use, and overall sustainable agriculture practices. Class-wise metrics, including precision, recall, and $F1$ scores, further contribute to a comprehensive evaluation of disease detection performance, enhancing its practical utility.

Keywords Deep learning · Plant Village dataset · ResNet101 · CNN

U. Muthaiah (✉) · Y. N. Reddy
Department of Computer Science and Engineering, Amrita School of Computing, Amrita Vishwa Vidyapeetham, Chennai, India
e-mail: u_muthaiah@ch.amrita.edu

R. V. Roy
Peggy and Charles Stephenson Cancer Center, University of Oklahoma Health Science Center, Oklahoma City, OK, USA

J. Maiti et al. (eds.), *Operations Research and Data Analytics: Current Trends and Future Perspectives*, Lecture Notes on Multidisciplinary Industrial Engineering,
https://doi.org/10.1007/978-981-96-8327-7_20

20.1 Introduction

Agriculture, a vital foundation of the world's economy, serves as the main provider of food, livelihoods, and jobs. In India, similar to numerous other nations with modest to moderate economic resources, agriculture assumes a pivotal role in the context of a developing nation. It significantly impacts the livelihoods of approximately 58% [1] of the rural population, contributing a substantial 18% Contributing significantly to the country's overall revenue and boosting the employment rate to 53% [2]. As of 2022, the agriculture sector has further solidified its importance, contributing 18.8% percent of the rural Gross Domestic Product (GDP) in India, totaling a substantial 7000 billion in 4th quarter of 2022 [3]. Over the past three years, the Gross Value Added (GVA) by agriculture to the country's overall economic output has witnessed a notable increase, surging from 17.6 to 20.2% [4, 5]. This sector remains a primary driver of economic growth, underscoring its paramount significance. Consequently, the impact of plant diseases and the menace of pest infestations on agriculture carries the potential to reverberate throughout the global economy, jeopardizing food production quality. Conventional prophylactic measures prove ineffective against the specter of epidemics and endemics, underscoring the importance of early surveillance and precise crop disease diagnosis facilitated by a robust crop protection system.

As noted by Thangaraj et al. [6], plant diseases exact a staggering toll, resulting in annual losses amounting to approximately 40% of essential food supplies. Furthermore, the insidious yellow rust disease, as revealed by Beddow et al. [7], imposes a formidable annual loss exceeding 5 million tons of wheat. These alarming statistics underscore the urgency of proactive measures to preserve agricultural productivity and mitigate the far-reaching consequences of crop maladies (Fig. 20.1).

With the evolution of computer vision, various approaches have been explored to address the challenges associated with detecting plant infections, as these often manifest as patterns and spots on leaves. Historically, researchers have employed traditional Image processing approaches that necessitate manual feature extraction are known for being labor-intensive. Although these techniques have demonstrated

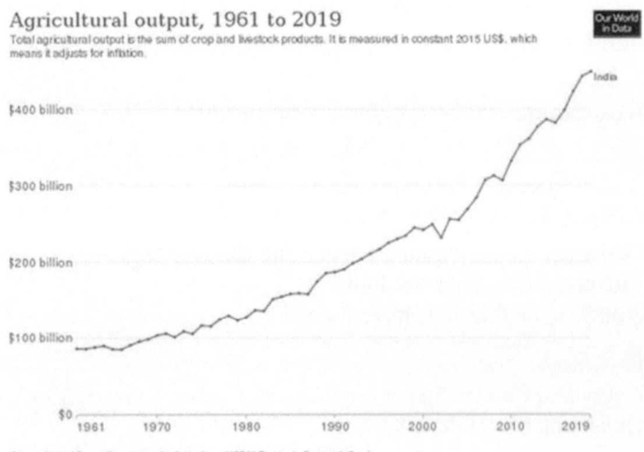

Fig. 20.1 Shows the graph of GDP of agriculture from 1961 to 2019

potential, they are inherently arduous and time-consuming, relying on handcrafted features, classification, and spot segmentation. However, with the advent of computer vision and the increasing influence of artificial intelligence, there has been a significant change. A substantial portion of research now favors the utilization of machine learning and deep learning models to improve recognition accuracy.

The introduction of machine learning and deep learning methods represents a significant advancement in the realm of plant disease identification. It streamlines the automated classification and extraction of image features, enabling the capture of the inherent attributes of an image. Additionally, the easy accessibility of datasets, powerful GPU machines, and software supporting sophisticated deep learning architectures with reduced complexity have facilitated the transition from traditional methods to the deep learning paradigm. In recent times, convolutional neural networks (CNNs) have garnered considerable attention due to their remarkable capabilities in image recognition and classification, driven by their ability to extract intricate low-level features. Consequently, CNNs have emerged as the preferred choice to replace traditional approaches in automated plant disease recognition, delivering superior results.

20.2 Related Work

In the field of plant disease detection, recent research has illuminated the potential of deep learning methodologies, particularly emphasizing the ResNet architecture. To build upon the foundations laid in 2020, with the introduction of the ResNet-34 model achieving an impressive 96% accuracy on a dataset of 15,200 plant species images [6], a 2023 study by Archana and colleagues expanded the scope. Their investigation, utilizing the Rice plant disease dataset and ResNet-50 network architectures, resulted in a remarkable 99% accuracy, alongside an outstanding $F1$ score of 0.9981 [7]. Notably, Saleem, M. H., Potgieter, J., and Arif, K. M., in their 2020 study [8], achieved a commendable $F1$ score of 0.9981 using various CNNs while working with a dataset comprising 50,000 plant images, further underscoring the ongoing advancements in the classification of plant diseases. These findings collectively underscore the ever-evolving landscape of plant disease detection, with deep learning techniques at the forefront of innovation.

Building on the advancements in the field of plant disease detection, a comprehensive review conducted in 2023 by Shoaib et al. [9] sheds light on the latest developments. Their work involved the utilization of the Plant Village dataset, yielding an accuracy of 88.66% when implementing the Faster R-CNN model. Simultaneously, during the same year, Shambharkar and Sharma [10] made a noteworthy contribution by directing their efforts toward plant disease detection and prevention, with a specific focus on cassava leaves. Their approach resulted in an accuracy rate of 89% by harnessing the power of the ResNet architecture. Furthermore, in 2023, Panchal et al. [11] centered their research on the extensive Plant Village dataset, comprising a diverse array of A collection of 50,000 images, depicting 14 different plant species, was used in their study. Notably, their Convolutional Neural Network (CNN) model achieved a remarkable accuracy rate of 90.44%, showcasing its superior performance when compared to other models commonly used in this domain. These results collectively emphasize the evolving landscape of plant disease identification, with deep learning methods leading the way and producing impressive levels of accuracy in cutting-edge research efforts.

In 2021, a pioneering approach was introduced to address the challenge of plant disease detection. This innovative method seamlessly integrates computer vision and deep learning techniques. It makes use of a dataset featuring Tomato leaves and includes pre-processed images that have undergone various image processing methods, such as Discrete Wavelet Transform (DWT), Principal Component Analysis (PCA), Gray-Level Co-occurrence Matrix (GLCM), and incorporates a CNN layer. The resulting model achieved an impressive accuracy of 99% [12]. Additionally, the ACO-CNN method introduced a novel framework for the detection of four specific plant diseases [13]. In a separate study by Singla, R. S., Gupta, A., an SVM-based approach was utilized, achieving an accuracy rate of 94% [14].

Timely detection of early-stage plant disease dedtection is crucial,. To address this challenge, a computer-aided detection system leveraged the VGG-19 architecture, renowned for its substantial parameter count and robust feature extraction capabilities, to analyze a dataset comprising 3505 plant images. This approach yielded a 73% accuracy rate and was conducted by Alirezazadeh, P., Schirrmann, M., and Stolzenburg in 2023 [15]. Over time, Genetic Algorithms (GA) have seen notable developments, achieving an impressive 95% accuracy [16]. When it comes to melanoma skin cancer detection, a method involving image classification is enhanced with a deep learning architecture celebrated for its depth. It was applied to a substantial dataset of 54,000 images spanning 14 plant species, resulting in an accuracy of 85% [17].

In 2019, there were notable advancements in the field of plant disease detection and classification. For instance, Sibiya and Sumbwanyambe introduced a computational approach using convolutional neural networks (CNNs) to identify and categorize maize leaf diseases, achieving an impressive 91% accuracy [18]. In the preceding year, 2018, Rangarajan, Purushothaman, and Ramesh utilized a pre-trained deep learning algorithm, specifically AlexNet, to tackle the classification of six different types of diseases affecting tomato crops [19]. Additionally, in 2019, Selvaraj and their research team leveraged artificial intelligence (AI) to detect banana diseases, achieving accuracy levels ranging from 70 to 99%. These figures highlight the notable progress in plant disease detection technologies and underscore the ongoing evolution in this field [20].

20.3 Methodology

While Convolutional Neural Networks (CNNs) are highly effective for image database-based object recognition and classification, they do come with certain challenges. These include prolonged training times and the need for substantial datasets.

Extracting intricate, low-level features from images necessitates deep CNN models, thereby augmenting the complexity of model training. To overcome these challenges, transfer learning approaches prove invaluable. Transfer learning involves the utilization of pre-trained networks, allowing model parameters learned from one dataset to be applied to other tasks. In this section, we delve into the methodologies employed in this study.

The main purpose of the convolutional layers is to extract hierarchical features from input images. This is achieved by using a set of adjustable filters that convolve across the input data. The incorporation of ReLU (Rectified Linear Unit) activation functions in these layers is crucial for the network's performance. ReLU introduces non-linearity into the network, which is essential in preventing the vanishing gradient problem and expediting the training process. Consequently, this enables the network to effectively learn and represent intricate visual patterns.

20.3.1 Multi-class Classification

Plant disease datasets usually consist of a large number of photos showing samples of both healthy and diseased plants, each of which is assigned to a particular class. For instance, all photos of samples of both healthy and diseased banana plants belong to the banana class when it comes to banana plants. In this case, the features that are taken out of the original image are the only ones that are used to classify a particular image. This class includes four groups of diseases, such as bunchy top virus, fusarium wilt, black sigatoka, and xanthomonas wilt, using the banana plant as an example. In the testing step, the output will correctly identify the disease category among the four categorised under that particular class if an input sample belongs to one of these specific illnesses after being trained on all four disease sets inside the banana class. As a result, multi-class categorization is incompatible. Multi-label categorization, on the other hand, treats every category inside a class as a separate class unto itself. N classes are called N multi-classes if there are N classes. If the N classes are further subdivided into M categories, then each category inside each of the N classes is considered a separate class.

20.3.2 ResNet-50

Convolution and identity blocks are included in the five stages of ResNet-50, a deep convolutional neural network of fifty layers. These residual networks serve as the basic structure for many computer vision applications. The idea presented by ResNet [21] entails convolution layer stacking in addition to the addition of skip links. Information can flow through these skip connections and go straight to the convolutional neural network's output instead of the original input. Significantly, by placing these skip connections before of the activation function, the vanishing gradient problem that frequently occurs in deeper models is lessened. The problem of increasing mistakes in deeper models is well addressed by this innovation in the residual neural network: skip connections. The fundamental foundation of these shortcut links is identity mapping.

The formula for the residual mapping may be stated as follows if we denote "x" as the input picture, "$F(x)$" as the nonlinear layers that were used to construct mappings, and "$H(x)$" as the representation of the residual mapping.

"The formula expressing the residual mapping is given by Eq. (1)":

$$H(x) = F(x) + x \tag{20.1}$$

Every identity block within ResNet-50 consists of three convolution layers and has more than 23 million trainable parameters. The input matrices, denoted as "x," and the shortcut denoted as "x" can be merged only if the output dimensions from the shortcut match the dimensions following the convolution and batch normalization. In situations where the dimensions don't match, the shortcut "x" must undergo a convolution layer and batch normalization to adjust the dimensions accordingly.

20.3.3 VGG-16 and VGG-19

"Very Deep Convolutional Networks for Large-Scale Image Recognition," or VGG-16 and VGG-19, are network models created by the University of Oxford's Visual Geometry Group. These designs are renowned for their extraordinary depth; the VGG-16 model has 16 weight layers, while the VGG-19 model has 19 layers. Notably, VGG-16 includes 138 million trainable parameters, while VGG-19 introduces even more parameters. These networks are designed to process images of a specific size, typically 224 $\times$ 224 pixels, and their core architecture consists of multiple convolutional layers, interspersed with max-pooling layers, and fully connected layers. The convolutional layers employ small 3 $\times$ 3 filters, allowing them to capture intricate image features. While these models demand longer training times and require more disk space due to their extensive parameter counts, they have been pivotal in advancing the field of computer vision, excelling in large-scale image recognition and various visual recognition tasks.

20.3.4 ResNet-101

ResNet-101, an abbreviation for "Residual Network with 101 Layers," is a deep convolutional neural network architecture that has had a profound impact on the field of computer vision. The fundamental innovation in ResNet-101 is its approach to residual learning, which greatly facilitates the training of exceptionally deep networks. This model was developed by Microsoft Research and is distinguished by its remarkable depth, incorporating a total of 101 weight layers. It's worth noting that ResNet-101 is equipped with a significant number of trainable parameters, typically exceeding 44 million.

ResNet-101 is designed to handle images of a specific size, often set at 224 × 224 pixels. Its architecture comprises various building blocks, including convolutional layers, batch normalization, ReLU activation functions, max-pooling layers, and fully connected layers. The convolutional layers primarily utilize 3 × 3 filters, enabling them to capture intricate and hierarchical features within the images. The innovation in ResNet comes from its residual blocks, where shortcut connections facilitate the network to skip over one or more layers, mitigating the vanishing gradient problem and making it possible to train very deep models effectively. This depth and architectural sophistication have propelled ResNet-101 to excel in a wide range of computer vision tasks, from image classification to object detection and segmentation.

20.4 Experiments

To assess our experiments, we utilized Kaggle as the platform, employing the P100 accelerator, operating on a Windows 10 environment. The system was equipped with GDDR5 graphic memory, an 11th-generation Core i5 processor, and 16 GB of RAM. We conducted the software implementation using Keras, OpenCV, and Numpy. It's noteworthy that OpenCV provides support for both academic and commercial project development and is compatible with multiple platforms, including Linux, Windows, Mac OS, iOS. It offers interfaces for Python, Java, and Android.

In our study, we conducted a comprehensive evaluation of each experiment, considering various metrics such as training accuracy, testing accuracy, validation accuracy, $F1$ score, precision, and recall for each class. Additionally, we closely monitored the losses during both the training and testing phases for each model. Our primary aim was to train models using the PlantVillage dataset, with the goal of expediting the learning process of convolutional neural networks (CNNs) by incorporating transfer learning models. Specifically, we utilized pre-trained models like ResNet-50, VGG-16, VGG-19, and ResNet-101. These models had previously undergone training on the extensive ImageNet dataset, which comprises 1.2 million images spread across 1000 distinct image categories.

20.4.1 Dataset

The dataset, sourced from Kaggle, consists of 70,295 images, representing an openly available dataset. Regarding the data division, 56,219 images were allocated for training, while the remaining 7013 were reserved for validation and 7063 are for testing featuring subfolders for both training, validation and testing (Table 20.1).

Table 20.1 This information in table displays the dataset division

	Testing	Training	Validation
Images	56,219	7063	7013
Classes	38	38	38

20.4.2 Preprocessing and Data Augumentation

The dataset covered a wide range of 38 classes, including 26 different diseases and 14 distinct crop species. In our experimental setup, we specifically focused on using the color images from the PlantVillage dataset, aligning them effectively with the transfer learning models we employed. To ensure compatibility with the various pre-trained network models, we standardized the image size by downscaling them to 256 × 256 pixels. Different pre-trained models required different input sizes; for VGG-16, ResNet-101, ResNet-50, and VGG-19, the input size was set at 224 × 224 × 3 (representing height, width, and channel width).

Despite the dataset's substantial scale, containing approximately 54,000 images depicting various crop diseases, these images closely resembled real-life photographs captured using a variety of image acquisition methods, including Kinect sensors, high-definition cameras, and smartphones.

Because of the magnitude of the dataset and the potential for overfitting, we used overfitting regularisation strategies, focusing especially on data augmentation following preprocessing. Numerous modifications were used in this augmentation process, including rescaling, flipping images horizontally and vertically, rotating images clockwise and anticlockwise, and adjusting the magnification. Crucially, enhanced photos were only momentarily utilised throughout the training process; no actual copies were kept. This augmentation technique improved the model's resilience while also assisting in keeping the model from overfitting and suffering losses. The enhanced resilience of the model enables it to identify actual plant disease photos with more precision, rendering it an invaluable tool for real-world uses.

20.4.3 Fine-Tuning of Hyperparameters in Pre-trained Models

Transfer learning models offer several advantages, including faster learning compared to models trained from scratch and the ability to freeze certain layers while fine-tuning the last layers to achieve more accurate classification. To ensure consistency, we initially standardized the hyperparameters for different pre-trained models. Specific hyperparameter details can be found in Table 20.2.

The model optimization was accomplished using stochastic gradient descent, with initial learning rates set at 0.001 for ResNet-101, ResNet-50, VGG-16, and VGG-19. Each model underwent 30 epochs of training, and a fixed dropout value of 0.5 was

Table 20.2 Represents hyperparameters

Hyperparameters	Epochs
Dropout	0.5
Epochs	30
Regularization	Batch normalization
Optimizer	Adam
Learning rate	0.001
Output classes	38

applied. In our experiment, we observed that the output graph started converging after a few iterations, specifically around the 30-epoch mark, indicating that our approach successfully addressed overfitting and degradation issues, contributing to the model's overall effectiveness.

20.4.4 Network Architecture Model

Pre-trained network models were selected due to their appropriateness for the classification of plant diseases. Table 20.3 contains comprehensive descriptions of the model structures. Differential filter sizes are incorporated into the design of these networks, which are necessary for the extraction of certain features from feature maps. In order to extract features, filters are essential. Interestingly, when applied to the input, each filter extracts unique features, and the type of feature extraction from the feature maps depends on the specific values of the filters. We employed the pre-trained network models in our research precisely as they were, with predefined combinations of convolution layer and filter size specific to each network model.

Table 20.3 Showing the achitectures

Model	VGG-16	VGG-19	ResNet-50	ResNet-101
Total layers	16	19	50	101
Pooling layers	5	5	4	4
Dense layers	2	2	1	1
Flatten layers	1	1	1	1
Filter size	3 × 3	3 × 3	3 × 3	3 × 3
Stride	2 × 2	2 × 2	2 × 2	2 × 2
Trainable params (M)	138	143	25.6	44.6

20.5 Results and Discussion

In this study phase, we deployed state-of-the-art deep learning models that made use of the transfer learning approach to solve the difficulty of diagnosing plant ailments. We used the publicly available PlantVillage dataset to enhance the training of pre-trained deep convolutional neural networks (CNNs) that were first trained on the ImageNet dataset. The following parameters were standardised for every model in our experimental setup: There were 38 output classes, 0.5 dropout rate, and 0.01 learning rate.

To help with learning, we separated the dataset into samples for testing, training, and validation. Pre-trained models, including ResNet-50, ResNet-101, VGG-16, and VGG-19, were trained using 80% of the PlantVillage dataset samples. Callbacks were added to each model after it had been trained for 30 epochs to help it achieve the desired outcomes.

20.5.1 VGG-16

The first phase of our experimentation involved an assessment of the VGG-16 network performance utilizing the earlier dataset. After standardizing the hyperparameters, we allocated 80% of the dataset for training, reserving 10% for testing, and dedicating the remaining 10% for both testing and validation. The results, depicted in Fig. 20.2, reveal that the model achieved an initial recognition accuracy of approximately 90% within the initial 10 epochs. Subsequently, the accuracy steadily increased, ultimately reaching a peak recognition accuracy of 94.07%, though slightly lower. Furthermore, the training loss and validation model demonstrated impressive values of 0.15% and 0.18%, respectively, reflecting the model's robust performance. The accuracy consistently hovered around the 94% mark, demonstrating the VGG-16 model's effectiveness and reliability in classifying plant diseases (Fig. 20.3; Table 20.4).

Impressive results were obtained by the VGG-16 model in terms of $F1$ scores, recall, and accuracy. With a recall of 0.9412 and a macro-average accuracy of 0.9401, the macro-average $F1$ score was 0.9403. Furthermore, the model demonstrated outstanding overall performance as seen by the consistently high micro-average accuracy, recall, and $F1$ score, all measuring at 0.9407.

Moreover, the accuracy, recall, and $F1$ scores for every class are broken down in depth in the table below, which illustrates how well the model classified various plant illnesses.

Fig. 20.2 Graph for accuracy vs epochs

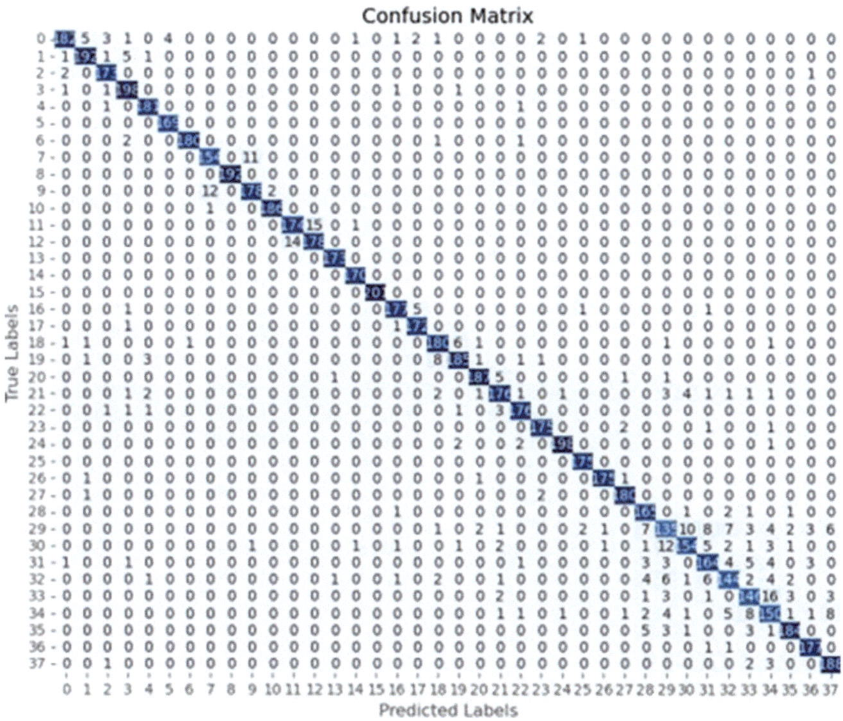

Fig. 20.3 Shows the confusion matrix

Table 20.4 Shows the class wise metrics for VGG16

Classes	Precision	Recall	$F1$-score
Class 1	0.9681	0.8966	0.9309
Class 2	0.9552	0.9600	0.9576
Class 3	0.9558	0.9830	0.9692
Class 4	0.9384	0.9802	0.9588
Class 5	0.9577	0.9891	0.9731
Class 6	0.9769	1.0000	0.9883
Class 7	0.9945	0.9783	0.9863
Class 8	0.9222	0.9333	0.9277
Class 9	1.0000	1.0000	1.0000
Class 10	0.9368	0.9271	0.9319
Class 11	0.9894	0.9947	0.9920
Class 12	0.9255	0.9158	0.9206
Class 13	0.9223	0.9271	0.9247
Class 14	0.9886	1.0000	0.9943
Class 15	0.9827	1.0000	0.9913
Class 16	1.0000	1.0000	1.0000
Class 17	0.9672	0.9568	0.9620
Class 18	0.9609	0.9885	0.9745
Class 19	0.9231	0.9375	0.9302
Class 20	0.9439	0.9250	0.9343
Class 21	0.9689	0.9590	0.9639
Class 22	0.9215	0.9026	0.9119
Class 23	0.9565	0.9617	0.9591
Class 24	0.9722	0.9777	0.9749
Class 25	0.9900	0.9754	0.9826
Class 26	0.9777	1.0000	0.9887
Class 27	0.9887	0.9831	0.9859
Class 28	0.9730	0.9836	0.9783
Class 29	0.8777	0.9649	0.9192
Class 30	0.7895	0.7031	0.7438
Class 31	0.8953	0.8280	0.8603
Class 32	0.8723	0.8677	0.8700
Class 33	0.8675	0.8229	0.8446
Class 34	0.8488	0.8343	0.8415
Class 35	0.7937	0.8152	0.8043
Class 36	0.9485	0.9340	0.9412
Class 37	0.9568	0.9888	0.9725
Class 38	0.9171	0.9691	0.9424

20.5.2 VGG-19

In Using the same dataset, we evaluated the performance of the VGG-19 model in our second round of experiments. We divided the dataset after standardising the hyperparameters, allocating 80% for training and 10% for each of testing and validation. The resulting graph, which is shown in Fig. 20.4, revealed several important findings. In the first 10 epochs, the model demonstrated good training and validation accuracy, with results around 98% and 92%, respectively. The precision did, however, fluctuate during the succeeding epochs, with sporadic variations or "zig-zag" patterns. In the

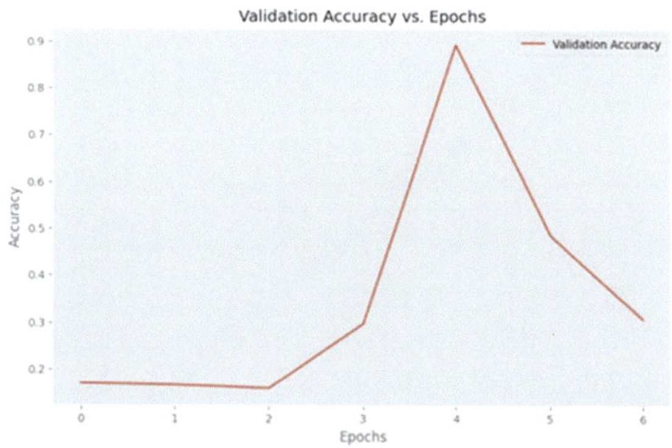

Fig. 20.4 Graph for accuracy versus epochs (VGG-19)

end, the model reached its maximum validation accuracy of 93.4%, which was just less than what was noted for VGG-16.

Although the validation model showed considerable variation, the observed training loss and validation values were outstanding at 0.018% and 0.49%, respectively. These findings highlight the efficacy and reliability of the VGG-19 model in the field of plant disease classification, even if the accuracy fell short of a steady 93% performance (Fig. 20.5; Table 20.5).

The VGG-19 model delivered notable performance with high precision and strong validation accuracy, boasting an impressive score of 99.52%. However, it's worth noting that the validation accuracy of VGG-19, while commendable, fell slightly short of that achieved by VGG-16. This observation suggests that VGG-19 may not perform as well as its predecessor because it has lower values in terms of precision, recall and *F*1-score metrics indicating a comparatively lower performance. Nonetheless, both models exhibit strong capabilities in plant disease classification, with VGG-16 outperforming VGG-19.

20.5.3 ResNet-50

In the third phase of our experimentation, we conducted an assessment of the ResNet-50 model's performance using the same dataset. After standardizing the hyperparameters, we divided the dataset, reserving 80% for training and allocating 10% each for testing and validation. The graphical representation, as shown in Fig. 20.6, provided valuable insights. Initially, the model displayed robust training and validation accuracy, with values approximately at 97.28% and 88.9%, respectively. Notably, there was an early occurrence of a callback, indicative of potential overfitting. However, as the validation accuracy reached its peak, the graph exhibited a decline, which became more pronounced after the highest model checkpoint. Ultimately, the model achieved a testing accuracy of 89%, albeit considerably lower than what was observed for VGG-16.

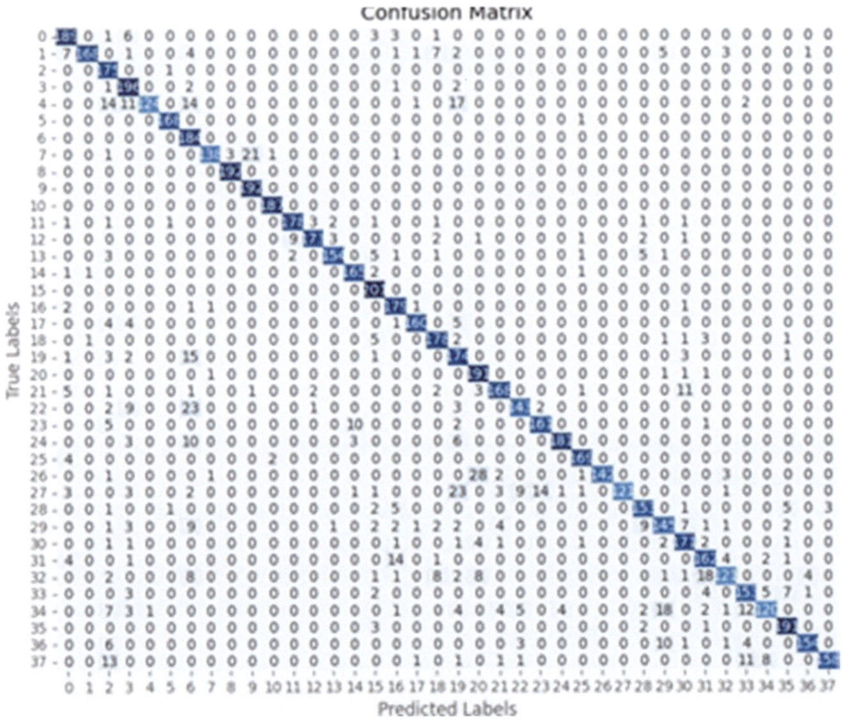

Fig. 20.5 Shows the confusion matrix for VGG-19

The recorded testing loss and validation model values were impressive at 0.33 and 0.33, respectively, although the latter indicated some variance. Which is not impressive when compares to VGG-16 when see the Table 20.6. Which shows the metrics of 38 different classes (Fig. 20.7).

The ResNet-50 model delivered underlying performance low precision and validation accuracy, boasting an acceptable accureacy of 89% which is lower than the compares models, observation suggests that ResNet-50 may not perform as well as its predecessor because it has lower values in terms of precision, recall and $F1$-score metrics indicating a comparatively lower performance which are 0.88,0.90 and 0882 respectively.

The ResNet-50 model delivered underlying performance low precision and validation accuracy, boasting an acceptable accureacy of 89% which is lower than the compares models, observation suggests that ResNet-50 may not perform as well as its predecessor because it has lower values in terms of precision, recall and $F1$-score metrics indicating a comparatively lower performance which are 0.88, 0.90 and 0882 respectively.

Table 20.5 Shows the class wise metrics for VGG19

	Recall	Precision	$F1$-score
Class 1	0.9471	0.8818	0.9133
Class 2	0.9554	0.9650	0.9602
Class 3	0.9558	0.9830	0.9692
Class 4	0.9740	0.9257	0.9492
Class 5	0.9184	0.9836	0.9499
Class 6	0.9825	0.9941	0.9882
Class 7	1.0000	0.9728	0.9862
Class 8	0.9536	0.8727	0.9114
Class 9	1.0000	0.9948	0.9974
Class 10	0.8857	0.9688	0.9254
Class 11	0.9841	0.9947	0.9894
Class 12	0.9392	0.8947	0.9164
Class 13	0.8937	0.9635	0.9273
Class 14	0.9770	0.9827	0.9798
Class 15	0.9940	0.9706	0.9821
Class 16	0.9950	0.9900	0.9925
Class 17	0.9829	0.9297	0.9556
Class 18	0.9548	0.9713	0.9630
Class 19	0.9045	0.9375	0.9207
Class 20	0.9588	0.9300	0.9442
Class 21	0.9637	0.9538	0.9588
Class 22	0.8357	0.9128	0.8725
Class 23	0.9162	0.9563	0.9358
Class 24	1.0000	0.9721	0.9858
Class 25	0.9754	0.9754	0.9754
Class 26	0.9942	0.9771	0.9856
Class 27	0.9830	0.9719	0.9774
Class 28	0.9944	0.9781	0.9862
Class 29	0.9483	0.9649	0.9565
Class 30	0.8010	0.8177	0.8093
Class 31	0.8663	0.8710	0.8686
Class 32	0.8700	0.9206	0.8946
Class 33	0.8462	0.8171	0.8314
Class 34	0.8486	0.8971	0.8722
Class 35	0.8274	0.7554	0.7898
Class 36	0.9581	0.9289	0.9433
Class 37	0.9661	0.9553	0.9607
Class 38	0.9368	0.9175	0.9271

20.5.4 ResNet-101

In the fourth phase of our experimentation, we conducted an assessment of the ResNet-101 model's performance using the same dataset. After standardizing the hyperparameters, we divided the dataset, allocating 80% for training and reserving 10% each for testing and validation. The graphical representation, as shown in Fig. 20.8, provided valuable insights into the model's behavior. Initially, the model demonstrated robust training and validation accuracy, with values hovering around 98% and 90%, respectively, achieved within the first 10 epochs. However, as the

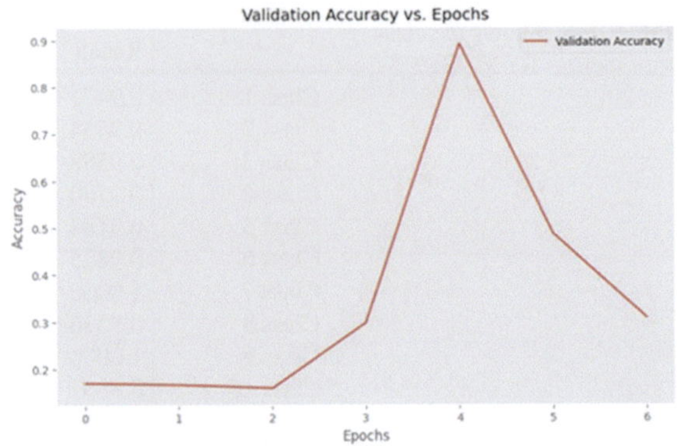

Fig. 20.6 Graph for accuracy versus epochs (ResNet-50)

subsequent epochs unfolded, occasional fluctuations or "zig-zag" patterns appeared in the accuracy. Ultimately, the model achieved a peak validation accuracy of 94.25%, a remarkable result when compared to the performance of the other three models (Fig. 20.9).

Although the validation model showed considerable fluctuation, the reported testing loss and validation model values were remarkable at 0.18 and 0.19, respectively. This is remarkable in comparison to VGG-16, as VGG-16 has the greatest metrics in relation to ResNet-101.Table 20.7 is displayed. This displays 38 distinct classes' metrics.

The outstanding performance of ResNet-101 left a lasting impression as it outperformed the other three models, ResNet-50, VGG-16, and VGG-19, in a number of parameters including $F1$ score, accuracy, and recall. With accuracy of 0.946, recall of 0.942, and an astounding $F1$ score of 0.941, it got noticeably better values, making it the model that performed the best out of the others.

20.6 Discussion

A statistical study of pre-trained network models based on assessment metrics was covered in the preceding session. Regularisation techniques such batch normalisation were successfully applied to mitigate the potential vanishing gradient problems caused by skip connections. More complicated training times, covariant shifts, and overfitting were among the issues that deeper models brought about. We adjusted the hyperparameters in our studies to address these issues.

The accuracy of accurately classifying healthy plants as healthy (true positives) and correctly classifying healthy plants as unwell (false negatives) is measured by the $F1$-score, which we employed in our trials. The assessment findings showed that the ResNet-101 and VGG-16 models performed better in our study than the VGG-19 and ResNet-50 models. An evaluation of the various pre-trained models' performances based on the $F1$ score, sensitivity, and specificity which are presented in Fig. 20.10.

Table 20.6 Shows the class wise metrics for ResNet-50

	Precision	Recall	$F1$-score
Class 1	0.8710	0.9310	0.9000
Class 2	0.9882	0.8400	0.9081
Class 3	0.7202	0.9943	0.8353
Class 4	0.7967	0.9703	0.8750
Class 5	0.9920	0.6776	0.8052
Class 6	0.9825	0.9941	0.9882
Class 7	0.6740	1.0000	0.8053
Class 8	0.9787	0.8364	0.9020
Class 9	0.9846	1.0000	0.9922
Class 10	0.8972	1.0000	0.9458
Class 11	0.9842	1.0000	0.9920
Class 12	0.9418	0.9368	0.9393
Class 13	0.9665	0.9010	0.9326
Class 14	0.9625	0.8902	0.9249
Class 15	0.9218	0.9706	0.9456
Class 16	0.8777	1.0000	0.9349
Class 17	0.8483	0.9676	0.9040
Class 18	0.9697	0.9195	0.9440
Class 19	0.8768	0.9271	0.9013
Class 20	0.7073	0.8700	0.7803
Class 21	0.8128	0.9795	0.8884
Class 22	0.9180	0.8615	0.8889
Class 23	0.8882	0.7814	0.8314
Class 24	0.9096	0.8994	0.9045
Class 25	0.9731	0.8916	0.9306
Class 26	0.9548	0.9657	0.9602
Class 27	1.0000	0.7978	0.8875
Class 28	1.0000	0.6612	0.7961
Class 29	0.8793	0.8947	0.8870
Class 30	0.7880	0.7552	0.7713
Class 31	0.8550	0.9194	0.8860
Class 32	0.8308	0.8571	0.8438
Class 33	0.8963	0.6914	0.7806
Class 34	0.8407	0.8743 0.6522 0.9695 0.8603	0.8571 0.7524 0.9409
Class 35	0.8889 0.9139 0.9625 0.9814	0.8144	0.9086
Class 36			0.8901
Class 37			
Class 38			

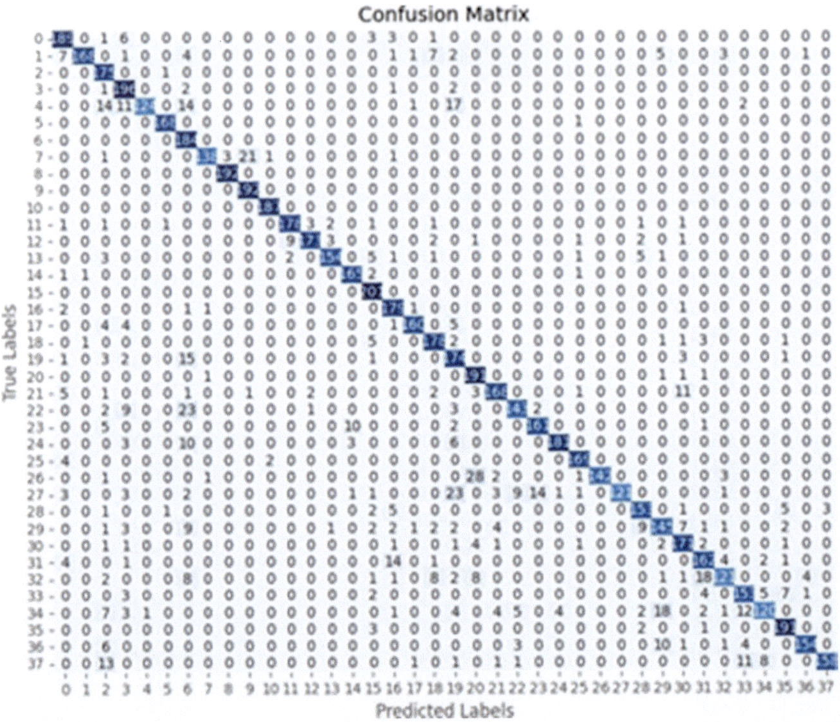

Fig. 20.7 Confusion matrix for ResNet-50

Fig. 20.8 Graph for accuracy versus epochs (ResNet-101)

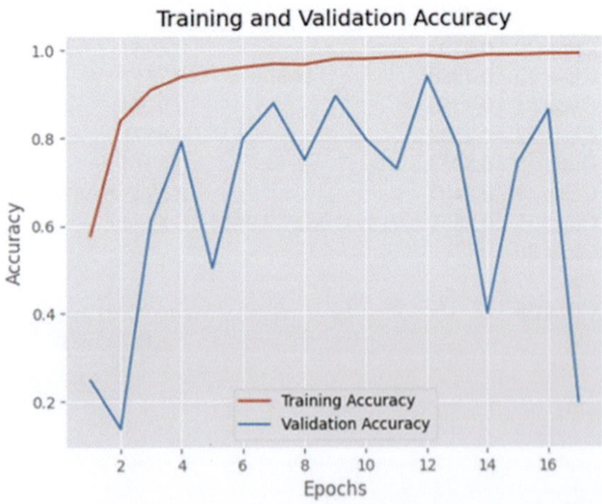

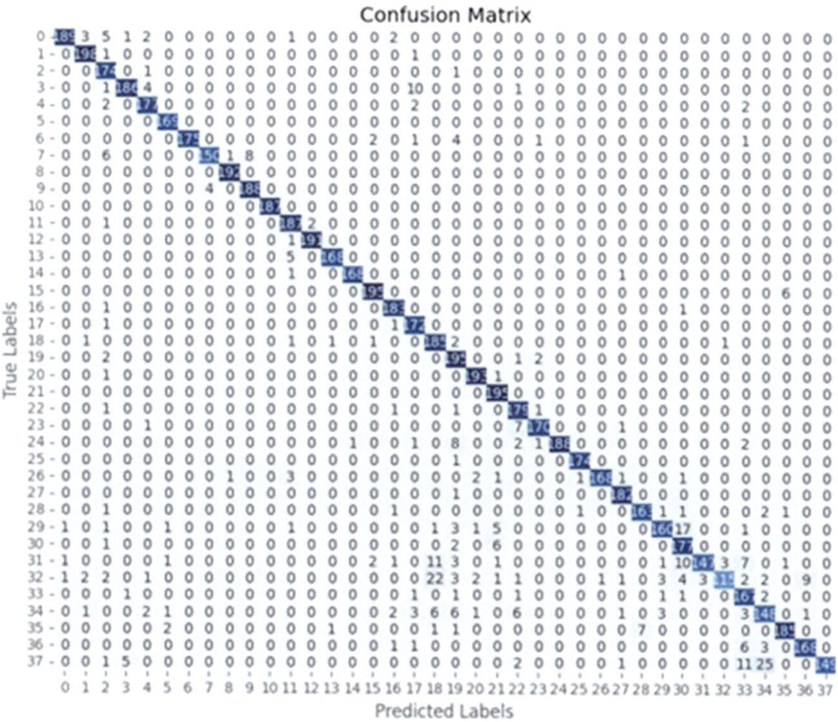

Fig. 20.9 Shows the confusion matrix for ResNet-101

Table 20.7 Shows the class wise metrics for ResNet-101

	Precision	Recall	$F1$-score
Class 1	0.9844	0.9310	0.9570
Class 2	0.9659	0.9900	0.9778
Class 3	0.8614	0.9886	0.9206
Class 4	0.9637	0.9208	0.9418
Class 5	0.9415	0.9672	0.9542
Class 6	0.9713	1.0000	0.9854
Class 7	1.0000	0.9511	0.9749
Class 8	0.9740	0.9091	0.9404
Class 9	0.9897	1.0000	0.9948
Class 10	0.9592	0.9792	0.9691
Class 11	1.0000	1.0000	1.0000
Class 12	0.9350	0.9842	0.9590
Class 13	0.9896	0.9948	0.9922
Class 14	0.9882	0.9711	0.9796
Class 15	0.9941	0.9882	0.9912
Class 16	0.9750	0.9701	0.9726
Class 17	0.9531	0.9892	0.9708
Class 18	0.8958	0.9885	0.9399
Class 19	0.8186	0.9635	0.8852
Class 20	0.8405	0.9750	0.9028
Class 21	0.9698	0.9897	0.9797
Class 22	0.9286	1.0000	0.9630
Class 23	0.8950	0.9781	0.9347
Class 24	0.9714	0.9497	0.9605
Class 25	1.0000	0.9261	0.9616
Class 26	0.9886	0.9943	0.9915
Class 27	0.9941	0.9438	0.9683
Class 28	0.9681	0.9945	0.9811
Class 29	0.9588	0.9532	0.9560
Class 30	0.9467	0.8333	0.8864
Class 31	0.8349	0.9516	0.8894
Class 32	0.9800	0.7778	0.8673
Class 33	0.9664	0.6571	0.7823
Class 34	0.8267	0.9543	0.8859
Class 35	0.8132	0.8043	0.8087
Class 36	0.9585	0.9391	0.9487
Class 37	0.9438	0.9385	0.9412
Class 38	1.0000	0.7680	0.8688

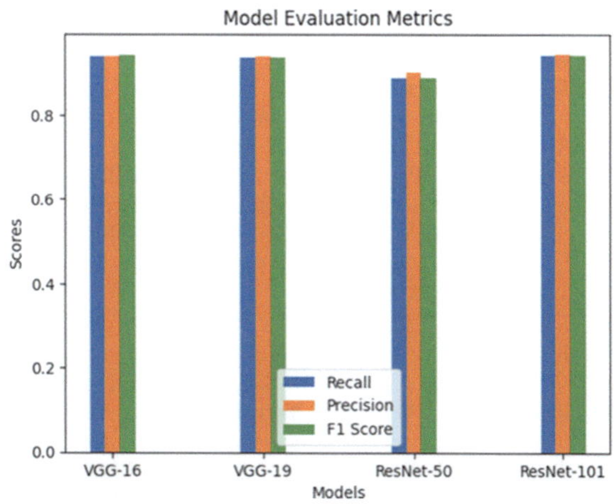

Fig. 20.10 Shows the comparison of metrics

20.7 Conclusion

In this study, using a dataset of over 70,000 photos, we performed a thorough investigation of several transfer learning models to successfully categories 38 different plant disease groups. Using transfer learning techniques, we applied standardisation and assessment processes to the latest convolutional neural networks, concentrating on evaluation measures including $F1$ score, recall, precision, and classification accuracy. Based on our performance research, ResNet-101 fared better than VGG-16, VGG-19, and ResNet-50. Because the ResNet-101 model had fewer trainable parameters and less computing complexity, training it seems more doable. Because of its streamlined training procedure, ResNet-101 thus turns out to be a better option for plant disease detection, particularly when adding new plant diseases to the model. The suggested model showed excellent results in training and testing. The proposed model demonstrated remarkable training and testing accuracies of 99.04% and 94.25%, along with an impressive $F1$ score of 94.17%.

References

1. Eunice, J., Popescu, D.E., Chowdary, M.K., Hemanth, J.: Deep learning-based leaf disease detection in crops using images for agricultural applications. Agronomy **12**(10), 2395 (2022)
2. Andrew, J., Eunice, J., Popescu, D.E., Kalpana Chowdary, M.: Deep learning-based leaf disease detection in crops using images for agricultural applications
3. https://en.wikipedia.org/wiki/Economy_of_India
4. https://www.pib.gov.in/indexd.asp
5. Li, L., Zhang, S., Wang, B.: Plant disease detection and classification by deep learning—a review. IEEE Access **9**, 56683–56698 (2021)
6. U. Archana, A. Khan, A. Sudarshanam, C. Sathya, A.K. Koshariya, R. Krishnamoorthy, Plant disease detection using ResNet. In: 2023 International Conference on Inventive Computation

Technologies (ICICT), Lalitpur, Nepal, 2023, pp. 614–618. https://doi.org/10.1109/ICICT57646.2023.10133938
7. Kumar, V., Arora, H., Sisodia, J.: Resnet-based approach for detection and classification of plant leaf diseases. In: 2020 International Conference on Electronics and Sustainable Communication Systems (ICESC), pp. 495–502. IEEE (2020)
8. Saleem, M.H., Potgieter, J., Arif, K.M.: Plant disease classification: a comparative evaluation of convolutional neural networks and deep learning optimizers. Plants **9**(10), 1319 (2020)
9. Shoaib, M., Shah, B., Ei-Sappagh, S., Ali, A., Ullah, A., Alenezi, F., Ali, F.: An advanced deep learning models-based plant disease detection: a review of recent research. Front. Plant Sci. **14**, 1158933 (2023)
10. Shambharkar, P.G., Sharma, S.: Plant disease detection and prevention using deep learning. In: 2023 9th International Conference on Advanced Computing and Communication Systems (ICACCS), vol. 1, pp. 1832–1838. IEEE (2023)
11. Panchal, A.V., Patel, S.C., Bagyalakshmi, K., Kumar, P., Khan, I.R., Soni, M.: Image-based plant diseases detection using deep learning. Mater. Today: Proc. **80**, 3500–3506 (2023)
12. Caldeira, R.F., Santiago, W.E., Teruel, B.: Identification of cotton leaf lesions using deep learning techniques. Sensors **21**(9), 3169 (2021)
13. Abd Algani, Y.M., Caro, O.J.M., Bravo, L.M.R., Kaur, C., Al Ansari, M.S., Bala, B.K.: Leaf disease identification and classification using optimized deep learning. Meas.: Sens. **25**, 100643 (2023)
14. Singla, R.S., Gupta, A., Gupta, R., Tripathi, V., Naruka, M.S., Awasthi, S.: Plant disease classification using machine learning. In: 2023 International Conference on Disruptive Technologies (ICDT), pp. 409–413. IEEE (2023)
15. Alirezazadeh, P., Schirrmann, M., Stolzenburg, F.: Improving deep learning-based plant disease classification with attention mechanism. Gesunde Pflanzen **75**(1), 49–59 (2023)
16. Balaji, V., Anushkannan, N.K., Narahari, S.C., Rattan, P., Verma, D., Awasthi, D.K., Mulat, M.B., et al.: Deep transfer learning technique for multimodal disease classification in plant images. Contrast Media Mol. Imaging **2023** (2023)
17. Saleem, M.H., Potgieter, J., Arif, K.M.: Plant disease detection and classification by deep learning. Plants **8**(11), 468 (2019)
18. Sibiya, M., Sumbwanyambe, M.: A computational procedure for the recognition and classification of maize leaf diseases out of healthy leaves using convolutional neural networks. AgriEngineering **1**(1), 119–131 (2019)
19. Rangarajan, A.K., Purushothaman, R., Ramesh, A.: Tomato crop disease classification using pre-trained deep learning algorithm. Procedia Comput. Sci. **133**, 1040–1047 (2018)
20. Selvaraj, M.G., Vergara, A., Ruiz, H., Safari, N., Elayabalan, S., Ocimati, W., Blomme, G.: AI-powered banana diseases and pest detection. Plant Methods **15**, 1–11 (2019)
21. Uchimuthu, M.: Automatic feature selection and classification for apple disease prediction using CNN. In: 2024 International Conference on Emerging Systems and Intelligent Computing (ESIC). IEEE (2024)

Part V
Analytics in Security and Surveillance

Chapter 21
QR Code-Based Anomalous Power Consumption Detection

Rajesh Nayak and C. D. Jaidhar

Abstract Modern civilization largely depends on electricity to run everything from businesses and houses to technology and medical facilities. It is vital for supporting infrastructure, promoting economic growth, and improving quality of life. The detection of anomalous power consumption is similarly crucial since it aids in identifying unusual patterns in energy usage, reduces waste, improves efficiency, and protects against possible electrical problems or security breaches, all of which eventually contribute to sustainability and cost savings. This work proposes a convolutional neural networks (CNN)-based anomalous power consumption detection using quick response (QR) code images. SimDataset, one of the widely used labeled datasets in anomalous power consumption detection, has been used in this work. Each instance of the dataset is converted to the QR code image, and the generated QR code image dataset is used to train and test the proposed approach. The outcomes of the experiments showed that the proposed approach is superior to the state-of-the-art in standard performance evaluation metrics, resulting in an overall accuracy and $F1$-score of 96.7% and 0.94, respectively.

Keywords Anomalous power consumption · Convolutional neural networks · SimDataset · QR code image

R. Nayak (✉) · C. D. Jaidhar
Department of Information Technology, National Institute of Technology Karnataka, Surathkal, India
e-mail: rajeshnayak.207it003@nitk.edu.in

C. D. Jaidhar
e-mail: jaidharcd@nitk.edu.in

J. Maiti et al. (eds.), *Operations Research and Data Analytics: Current Trends and Future Perspectives*, Lecture Notes on Multidisciplinary Industrial Engineering,
https://doi.org/10.1007/978-981-96-8327-7_21

21.1 Introduction

Nowadays, electricity is considered an essential requirement in our modern world. Since most activities rely on energy availability, they contribute substantially to a country's economic prosperity. The need for electricity is rapidly growing. Population expansion, technological advancement, socioeconomic activity, electric vehicles, heavy machinery, and people's modern lives have influenced the rise in electricity demand. Since electricity is now a need for daily living, its availability plays a significant role in determining a nation's economic development. The main places in a country where electricity is consumed are its transit hubs, buildings, and industries. Because of their varied electrical needs (heating, cooling, lighting, and appliances), residential and non-residential buildings are the primary sources of increased power consumption and energy demand [1, 2]. Utilities and distributors are developing various demand response programs for their customers to help them consume less power.

It has been acknowledged that better energy management is essential to optimize the building's operational energy performance. Since smart meter installation is permitted under the Energy Performance of Building Directive 2010/31/EU, automation, control, and surveillance systems are also getting more priority on a worldwide scale [3]. In this situation, energy inefficiencies during building operations may be quickly identified using fault detection and diagnosis procedure [4].

Industry, businesses, and residential areas use most energy. In 2018, the country's annual residential energy consumption was 1.46 trillion kWh, or 38.5%, according to energy information administration data. For this reason, to maximize energy efficiency and ensure the electricity supply system's security, it is crucial that electric power abnormalities, such as electrical theft and other forms of losses that are technical or nontechnical, are accurately detected in residential areas [5].

Anomalous data are patterns that do not conform to the well-defined definition of usual behavior. Anomaly detection may help to spot abnormalities in power usage, which can result in energy and money-saving upgrades for the home. Buildings, including residential and commercial, use 60% of worldwide power [6]. The three primary energy consumers are lighting, cooling, and heating. For example, motion detectors are used in today's energy-saving strategies to turn every light source on and off [7].

Monitoring users' power consumption became easier using smart meters and improved metering infrastructure that continually documents energy usage. Data analysis uses the obtained data to help identify energy usage patterns and gain insight into the data. There is now a framework for monitoring energy consumption in numerous countries, such as the United States, China, etc., for the clients in residential and non-residential buildings [8]. Although several researchers have proposed various methods for identifying abnormal power usage, there are still some limitations in detection process. A primary concern is the absence of a labeled dataset [9].

Detecting anomalous power consumption at the household level is essential for the following reasons:

1. **Energy Efficiency**: Power usage anomalies can point to inefficiencies or problems with systems or appliances in the home. By locating and resolving these anomalies, energy-efficient practices may be implemented, lowering total energy usage and related expenses.
2. **Cost Savings**: Household electricity costs may increase due to abnormal power consumption. House owners may save costs by ensuring energy is utilized more wisely and avoiding needless spending by identifying and fixing abnormalities.
3. **Equipment Health Monitoring**: Unusual patterns in power usage may point to possible problems or wear and tear in electrical systems or equipment. Early detection prolongs equipment life and prevents expensive breakdowns by enabling prompt repair or replacement.
4. **Safety Concerns**: Unusual power usage might indicate electrical dangers, bad wiring, or broken gadgets, all of which could endanger the safety of the inhabitants. Early detection and action can reduce these dangers and improve the household's safety.
5. **Environmental Impact**: Ecologically friendly practices include tracking and mitigating electricity anomalies. Households may reduce their carbon footprint and contribute to environmental conservation by optimizing usage and minimizing energy waste.
6. **Grid Stability**: Unusual power usage can affect the stability of the electrical system, particularly during abrupt spikes or dips. It is possible to improve the overall stability of the more extensive electrical infrastructure by identifying and controlling these abnormalities at the home level.
7. **Integration with Smart Homes**: For smart home systems, power consumption data anomalies might be helpful. Smart home systems that integrate anomaly detection allow house owners to automate actions, such as shutting off broken appliances or warning inhabitants of possible problems.
8. **Demand Response Programs**: To take part in demand response programs, it is necessary to detect abnormal power usage. Households may actively support demand response programs by recognizing and controlling energy use peaks and valleys, which will assist in balancing the supply and demand for energy.
9. **Data Security and Privacy**: By monitoring power consumption statistics, anomalous use patterns pointing to illegal access or breached security systems can be identified early on. This helps to preserve the household's security and privacy.
10. **Regulatory Compliance**: Energy efficiency and conservation is the subject of laws or other policies in some areas. Identifying anomalous power usage aids in household compliance with these laws and may even make them eligible for refunds or incentives for energy-saving measures.

Machine learning-based anomaly detection requires a lot of high-quality data, which are usually readily available in huge quantities. They also need a lot of fine-tuning of the machine learning algorithms' parameters. On the other hand, rule-based representations and specific domain expertise are required for traditional anomaly identification techniques [10].

Authors of [4, 5, 8–35] have employed several methodologies such as ensemble and hybrid models, neural network-based models, feature extraction and selection-based approaches, supervised and unsupervised techniques, and edge, fog, and cloud-computing platforms to detect excessive power usage. Anomalies were found using aggregated, appliance level, spatial–temporal, and power metrics. Its usage also helped applications such as energy saving, occupancy detection, and appliance failure detection.

The significance of detecting anomalous power consumption based on micro-moments has increased recently because this approach not only identified abnormal power consumption but also yielded additional insights into other power consumption states, including the active form of appliances, average and high electricity usage. At the same time, power usage when the user is not present in the household.

Deep learning models have an advantage over machine learning models as they do not need feature extraction as an additional preprocessing task. They perform it implicitly in the hidden layers. Image-based models like CNN have proved to be one of the best models in classification-related tasks. The dataset used in this work is a numeric dataset. Instead of creating images by selecting multiple instances to form a 2-dimensional matrix and creating the image out of that matrix, this work used a unique way to convert each instance to a QR code image and then use the CNN model to perform classification. Using QR code images helped leverage the potential of image-based classifiers. Hence, this study uses QR code images created from SimDataset to classify the micro-moments in order to find unusual electrical power utilization.

Key contributions of this work are as follows:

- QR code images were created separately for each instance of SimDataset and used for anomalous power consumption classification.
- The proposed approach's robustness is validated by systematic tests that vary key parameters, including the size of each batch, the count of the filters, and Convolutional and MaxPooling layers. This investigation aims to identify optimal configurations that enhance the robustness and general efficacy.
- To the best of our knowledge, both QR code-based images and CNN have been used for the first time for abnormal power utilization identification using SimDataset.
- The efficacy of the proposed approach is measured quantitatively using standard performance assessment measures that have been defined.

The sections that follow give descriptions of the remaining portions of this work. A review of recent research on abnormal electricity usage and micro-moment classification is covered in Sect. 2. In Sect. 3, the proposed anomalous power consumption classification approach is explained. Section 4 presents the experimental results. The conclusion and suggestions for future study are included in Sect. 5.

21.2 Literature Survey

Finding anomalous electrical power consumption has drawn the attention of several studies. Researchers have created various deep learning (DL) and machine learning (ML)-based methods to identify unusual electricity usage. This section covers multiple methods proposed by different authors.

A data analytics-based approach to identify and describe abnormalities in daily power trends is described [4]. The power consumption patterns were found and aggregated with K-means and density-based spatial clustering. It was decided that the power consumption pattern which stands out from the others was exceptional. For multiple clustering, the clustering approach called "Follow the Leader" was used.

A real-time identification of anomalies using cascaded artificial neural networks and autoregressive moving averages was proposed [5]. Support vector machine (SVM) and K-nearest neighbors (KNN) were also integrated. A data mining-based method for detecting anomalous power in daily power usage patterns was presented [8]. An artificial intelligence-based method for identifying abnormalities in buildings' energy usage data was discussed [9]. Several approaches such as ensemble, supervised, unsupervised, and feature extractions were tested.

Transfer learning for the anomalous power consumption classification was selected [10]. The images were created from the numeric dataset. The pre-trained models such as GoogleNet, AlexNet, and SqueezeNet were employed as feature extractors, and the classifier used was SVM. The multi-agent-based anomaly detection technique was proposed [11]. The data were first tagged using the ensemble approach, and then anomalies were found in an unsupervised manner using DL. An auto-encoder and LSTM were used to identify the abnormalities.

A hybrid data mining technique to identify abnormalities in the energy consumption of residential households was presented [12]. Quantile regression and the recurrent neural network in an unsupervised setting were employed for detecting the anomalies. A methodology that uses contextual analysis to detect aberrant energy usage in data from smart meters was proposed [13]. Total of eighteen groups were selected to do the anomaly detection. Also, four strategies were chosen for anomaly identification.

The long short-term memory (LSTM)-Gauss-NBayes strategy, which combines LSTM, Bayesian, and Gaussian approaches, was proposed to identify the anomaly in the Industrial Internet of Things [14]. An error testing set was used to determine if the Naive Bayes model could identify the anomaly. A technique for spotting irregularities in central air conditioning systems' daily energy usage trends was created. The information entropy characteristic value and hierarchical clustering were selected to find energy usage trends [15].

Using deep neural networks (DNN), the researchers released a symbolic aggregate approximation [16] to predict both regular and irregular behaviors in the future and a building's electricity consumption pattern. Identifying anomalies involved labeling a parent repository and comparing its pattern frequency to the discord threshold. Identifying abnormal power usage based on micro-moments has been the focus of study lately. It gives more insights into the equipment's state and electricity usage status while the user is not there, which facilitates the identification of abnormal power usage. DL-based strategy was proposed to identify unusual power usage. The classification was done using a DNN [17]. The improved KNN (IKNN) was employed to categorize unexpected electricity utilization. Both supervised approach and unsupervised outlier detection have been applied to detect anomalies [18].

An energy efficiency recommendation system was proposed [19]. The goals for alternative computer platforms, model suggestions, first steps, reward systems, and assessment measures were identified. Additionally, the fundamental concepts and difficulties of energy-conserving recommender systems were also discussed. A micro-moment-based approach was proposed for assessing the efficiency of homes based on energy consumption.

The most recent advancements in intelligent non-intrusive load monitoring (NILM) in buildings were covered [20]. The significance of feature extraction and selection, artificial intelligence (AI)-based methodologies, diverse computing platforms, and applications were emphasized. Several event-based and non-event-based NILM algorithms have been discussed. The technological difficulties and looked at potential future developments for smart NILM were also highlighted. An ensemble bagging tree (EBT) and a wavelet packet tree of multi-scale type were employed to develop a trustworthy event-based appliance recognition model. The EBT was employed for feature extraction, and a multi-scale wavelet packet tree was used for classification. The efficacy of the recommended strategy was achieved using data from seven households. Each household's appliances were identified using four to six of them [21].

An edge-based Internet of Everything platform that is both useful and highly efficient buildings was introduced for utilizing power in residential [22]. An EBT was used to conduct a state-of-the-art analysis of micro-moments. Several case studies were carried out to assess the efficacy of the paradigm that was provided. The micro-moments were defined using a rule-based approach. The data were recorded and processed using a smart plug. In [23], a deep feed-forward network that uses energy usage was proposed to identify abnormalities. A Raspberry Pi and sensors were used to make a prototype and an Internet of Things-enabled smart house was presented.

An ML-based method was proposed to anticipate power usage and identify anomalies in an IoT setting [24]. A correlation analysis on weather data was performed to understand and choose the pertinent features. Anomaly detection and predictions were carried out using several ML models. The hyperdimensional computing was employed for real-time anomaly identification in electricity load [25]. Retraining functions were created in order to increase classification accuracy. For the experiments, two real-time datasets were used. An imbalance ratio was also developed to show the imbalance in the dataset.

A smart meter method for detecting abnormal electricity was suggested. For anomaly identification, a Bi-LSTM auto-encoder was employed. The anomalies were found based on the reconstruction accuracy. Four different energy sources were used to test the model [26]. In a smart grid, an anomaly detection method based on LSTM and K-means clustering was proposed [27]. LSTM was trained by using the cluster centroid to forecast the next day's consumption. The challenges posed by performance assessment metrics and annotated datasets were discussed [28]. For automated labeling, a statistical and linear regression approach was proposed. A variety of performance evaluation standards were also reviewed.

For time series data, many anomaly detection techniques were compared [29]. Point outliers, subsequence outliers, and time series outliers are the three types of outliers that are discovered. Both univariate and multivariate approaches were used to examine dataset-based anomaly detection techniques. An online method was described for anomaly detection based on scalable prediction for data from smart meters [30]. Lambda architecture was used to update the model repeatedly.

A micro-moment-based approach was employed for assessing household energy efficiency. An ensemble decision tree (EDT) and a multi-layer perceptron (MLP) was used to categorize the micro-moments [31]. A smart fusion model was proposed that uses sensor data and user feedback to deliver customized energy-saving suggestions. The data was collected using various sensors and added information on behaviors to construct the knowledge base. The recommendation system employed the fusion model, which forecasted micro-moments. The fusion model used the stacked LSTM to conduct prediction [32].

An energy consumption data classification at the micro-moment level was proposed using models such as SVM, KNN, and DNN [33]. The cloud-based systems, including Google Cloud Platform, Amazon Web Services, Google Colab, and MAML were used to perform experiments. The performance of classification models was tested using cloud-based platforms. A techno-economic method was proposed to calculate the structure's energy efficiency. The practicality of already-issued patents and available remedies were checked. Also, the drawbacks and inadequacies of the available remedies were discussed.

Most previous studies that attempted to detect abnormal power consumption used unlabeled datasets since the labeled datasets are much more complex to get by. The power usage patterns were grouped by the authors using clustering algorithms. Anomaly was defined as the pattern with the fewest repetitions, as abnormal power consumption is uncommon compared to normal power consumption. Few researchers established the upper bound of the power consumption value using a threshold-based method. Any consumption value above this cutoff was regarded as abnormal. Most of the works have employed prediction-based models and cluster-based techniques, in which historical power consumption numbers were used as the basis for forecasting, and forecasting error was utilized to identify anomalies.

Even though NILM is popular, the primary purpose of NILM is to monitor power usage based on appliance power consumption and it does not give information about which device power consumption is anomalous. Hence, this work proposes a DL-based model that uses QR code images generated from the power consumption dataset to identify anomalous power consumption. To the best of our knowledge, this is the first work to use QR code images with CNN for anomalous power consumption detection.

21.3 Methodology

A proposed approach for detecting unusual power consumption is presented in this section. It describes the general architecture, the methodology employed, the preparation and preprocessing of the dataset, and the performance assessment measures employed in this work.

Figure 21.1 shows the overall architecture of the proposed work to classify anomalous power consumption into five classes using QR code images and CNN as a classifier. A micro-moment labeled SimDataset [35, 36] is used in this work. The dataset has 11 features. Initially, correlation analysis is performed on the dataset to remove some features that are not necessary for the analysis. Out of 11 features, two features named Appliance_ID and normalized power have been removed after correlation analysis. The remaining eight features, excluding class labels, are used for further processing. The updated dataset is then sorted according to the class label. QR code images are created individually for each instance and then converted into grayscale images. After creating the grayscale QR code images, they are split into 80:20 ratios as training and testing datasets. Training dataset images are used to train a classifier, and then the trained classifier is tested with an unseen testing image dataset. The performance of the trained classifier was evaluated as discussed in Sect. 3.4.

21.3.1 Dataset Description

SimDataset [35, 36], frequently utilized in anomalous power consumption categorization, is employed in this work. Micro-moment labeling is used in the dataset to indicate behaviors such as turning the device 'on' and 'off' and normal, excessive, and power usage while the user is not present in the household. Occupancy data determines the electricity used while the user is not present in the household. The electricity usage of six home appliances is measured using the smart plug that serves as a sub-meter and is attached to each appliance. These six appliances include an

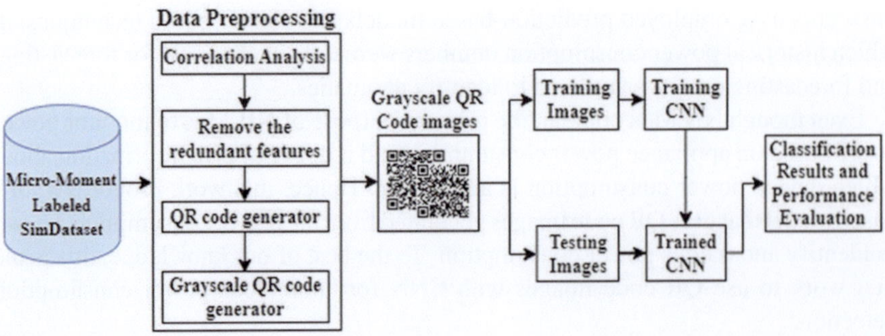

Fig. 21.1 Architecture of the proposed approach

Table 21.1 Description of the SimDataset

Class label with its name	Number of instances
Class 0—good usage	59,425
Class 1—turn on	7780
Class 2—turn off	7779
Class 3—excessive usage	6343
Class 4—consumption while outside	23,793
Total	105,120

AC and light bulb in one living room, a TV and air-conditioning in another, and a microwave and light bulb in the kitchen. Despite having two light bulbs and air conditioners, they are nonetheless regarded as distinct units as each of their energy usage varies according to the room in which they are placed. Two years readings are obtained hourly for every device.

The dataset includes total 11 features including the class label. It also includes occupancy data to determine whether electricity is utilized with or without a consumer's presence. The dataset utilized in this experiment is shown in Table 21.1.

21.3.2 Data Preprocessing

An integral part of ML- or DL-based model is data preprocessing. Preprocessing the data is necessary before transferring it for further processing. A correlation analysis is done on the SimDataset using Pearson's correlation coefficient [37]. Figure 21.2 displays the heat map that illustrates the correlation between different features of the SimDataset. The two features with the same and high correlation with the target feature (Class Label) are "Power consumption" and "Normalized power." The feature "Normalized power" is derived from "Power Consumption" feature, and hence, it is removed. Also, feature named "Appliance_ID" was also removed. After removing the redundant features, the instances are sorted based on class labels. For each instance of the respective class, the QR code image was generated by using the QR code library [38] (see Fig. 21.3). Then, the pixel value of each QR code image is normalized by dividing them by 255 to convert them into a grayscale image. The generated grayscale QR code images were used for the rest of the experiments.

21.3.3 Classification

In this study, CNN was employed for QR code image classification. The rectified linear unit (ReLU) activation function was used in the first Conv2D layer, 32 neurons, and a 6 $\times$ 6 kernel size to induce non-linearity and initiate feature extraction. After

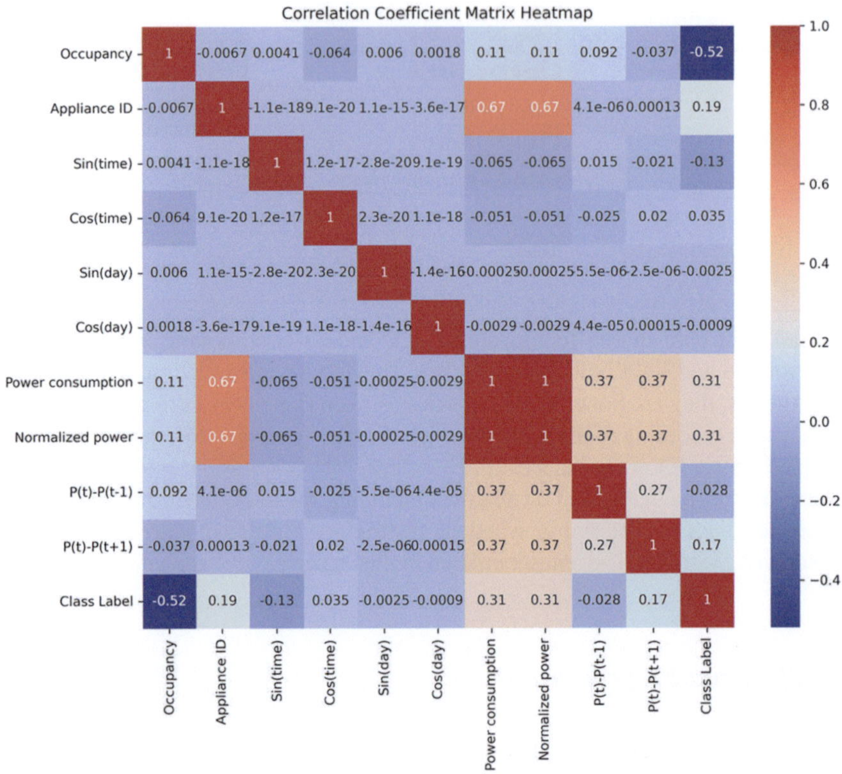

Fig. 21.2 Correlation analysis of SimDataset

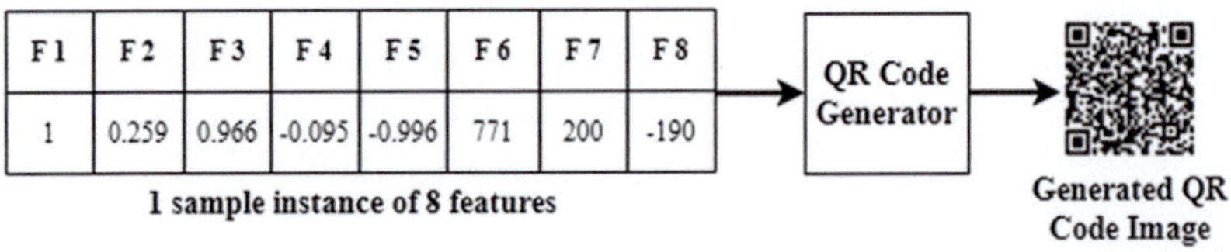

Fig. 21.3 Conversion of numeric data into QR code image

that, a MaxPooling layer with a 3×3 pool size down samples the spatial dimensions to focus on the most essential data. Using a similar 4×4 kernel and 64 neurons, the second Conv2D layer enhances feature extraction. Dimensionality is further reduced by an additional MaxPooling layer with the 2×2 pool size. The flattened layer transforms the 2D feature maps into a 1D vector, enabling seamless interaction with fully connected layers. The fully connected layer facilitates access to intricate correlations within the data. ReLU activates 128 neurons in it. A dropout rate of 0.25 has also been included after the fully linked layer to help reduce overfitting and enhance the model's capacity for generalization. This presents a regularization technique that, in the course of training, randomly removes 25% of the neurons. The output layer, with five neurons and a softmax activation function, completes an extensive and effective architecture for image classification by producing probability scores for each class.

21.3.4 Performance Evaluation

The performance assessment metrics employed in this work include accuracy (Eq. 21.1), precision (Eq. 21.2), recall (Eq. 21.3), and $F1$-score (Eq. 21.4) [7]. For more information about the performance metrics, TP, FP, TN, and FN, one can refer [7].

$$\text{Accuracy} = \frac{\text{TP} + \text{TN}}{\text{TP} + \text{FP} + \text{TN} + \text{FN}}. \tag{21.1}$$

$$\text{Precision} = \frac{\text{TP}}{\text{TP} + \text{FP}}. \tag{21.2}$$

$$\text{Recall} = \frac{\text{TP}}{\text{TP} + \text{FN}}. \tag{21.3}$$

$$F1\text{ - score} = 2 \times \frac{\text{Precision } \times \text{Recall}}{\text{Precision} + \text{Recall}}. \tag{21.4}$$

21.4 Findings and Contributions

The experiments were conducted on a PC with 16 GB of RAM and an Intel Core i7 11th generation CPU running at 2.50 GHz. Python programming language and TensorFlow framework were used to develop and implement the proposed approach.

A set of experiments were conducted using different configurations of the CNN, which is used in image classification tasks. Different hyperparameters, like the number of convolutional layers, kernel size, and number of filters, are changed to tune the classifier for better accuracy. The CNN with two convolutional layers produced the best performance of all configurations. Figure 21.4 shows the configuration of the CNN used in this work, delivering the best performance.

Table 21.2 represents the class-wise performance of the proposed approach. Results depicted that the proposed method performed well in all five classes. It has classified the maximum number of instances of every class accurately. The confusion matrix shown in Fig. 21.5 depicts the number of instances of each class. It has produced an overall accuracy of 96.7%, macro average precision of 0.947, recall of 0.934, and $F1$-score of 0.94.

Fig. 21.4 Summary of the CNN used in this work

```
Model: "sequential"
_________________________________________________________________
Layer (type)                Output Shape              Param #
=================================================================
conv2d (Conv2D)             (None, 64, 64, 32)        1184

max_pooling2d (MaxPooling2D  (None, 21, 21, 32)       0
)

dropout (Dropout)           (None, 21, 21, 32)        0

conv2d_1 (Conv2D)           (None, 21, 21, 64)        32832

max_pooling2d_1 (MaxPooling  (None, 10, 10, 64)       0
2D)

dropout_1 (Dropout)         (None, 10, 10, 64)        0

flatten (Flatten)           (None, 6400)              0

dense (Dense)               (None, 128)               819328

dropout_2 (Dropout)         (None, 128)               0

dense_1 (Dense)             (None, 5)                 645

=================================================================
Total params: 853,989
Trainable params: 853,989
Non-trainable params: 0
```

Table 21.2 Class-wise performance of the proposed approach

Classes	Accuracy (%)	Precision	Recall	$F1$-score
Class 0	96.7	0.983	0.985	0.984
Class 1		0.968	0.863	0.913
Class 2		0.994	0.992	0.993
Class 3		0.836	0.847	0.841
Class 4		0.955	0.982	0.968
Average		0.947	0.934	0.94

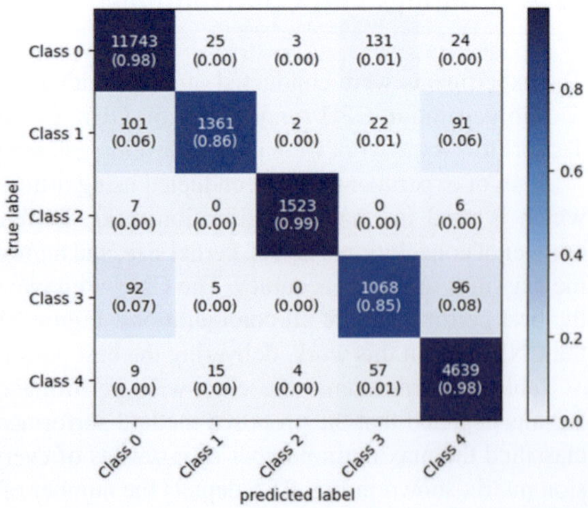

Fig. 21.5 Confusion matrix of the proposed approach

Table 21.3 shows the results of various configurations of the CNN based on the hyperparameters tuning. Table 21.4 shows the results of each configuration, it depicts that the CNN with two layers of convolution and MaxPooling produced the best results.

Figure 21.6 shows the bar plot comparing the accuracy of the proposed approach and the state-of-the-art. Figure 21.7 compares the $F1$-score of the proposed classifier with state-of-the-art. Both figures depict that the proposed approach outperformed as compared to other models. Table 21.5 compares the accuracy and $F1$-score of the proposed approach and other works, and it demonstrates that the proposed approach outperformed as compared to other works.

The following are the limitations observed during the experiments with the proposed approach:

- The QR code image size was smaller because only eight features of the SimDataset was used, which affected the classifier performance. Moreover, Deeper CNN

Table 21.3 Different configurations of CNN used in this work

CNN model	# Layers	Kernel size	Pool size	Dropout	# Epochs	# Layers
Model 1	One Conv2D and one MaxPooling2D	(6, 6)	(3, 3)	0.25	50	One Conv2D and one MaxPooling2D
Model 2	Three Conv2D and three MaxPooling2D	(5, 5), (5, 5), and (3, 3)	(3, 3), (3, 3), and (2, 2)	0.25, 0.25, 0.25	50	Three Conv2D and three MaxPooling2D
Model 3	Three Conv2D and three MaxPooling2D	(5, 5), (5, 5), and (5, 5)	(3, 3), (3, 3), and (3, 3)	0.25, 0.25, 0.25	50	Three Conv2D and three MaxPooling2D
Model 4	**Two Conv2D and two MaxPooling2D**	**(6, 6) and (4, 4)**	**(3, 3) and (2, 2)**	**0.25, 0.25**	**50**	**Two Conv2D and two MaxPooling2D**

Bold content represents the best configuration of the model that gave best results

Table 21.4 Performance comparison of different configurations

CNN model	Accuracy (%)	Precision	Recall	$F1$-score
Model 1	93.8	0.901	0.871	0.884
Model 2	94.7	0.943	0.861	0.893
Model 3	94.5	0.927	0.864	0.89
Model 4	**96.7**	**0.947**	**0.934**	**0.94**

Bold content represents the best configuration of the model that gave best results

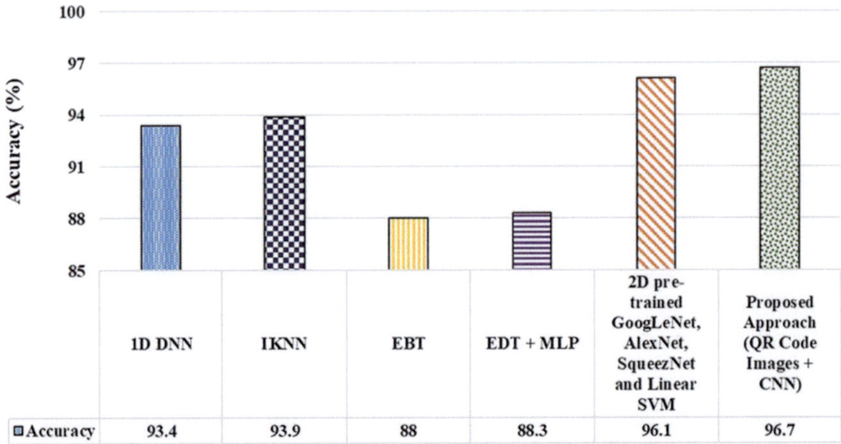

Fig. 21.6 Comparison of accuracy of the proposed approach with state-of-the-art

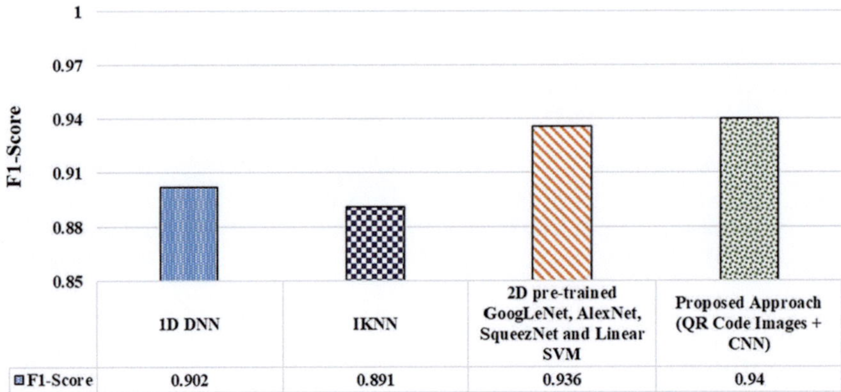

Fig. 21.7 Comparison of $F1$-score of the proposed approach with state-of-the-art

Table 21.5 Performance comparison of proposed approach with state-of-the-art

Model used	Accuracy (%)	$F1$-score
1D DNN [17]	93.4	0.902
IKNN [18]	93.9	0.891
EBT [22]	88	–
EDT + MLP [31]	88.3	–
2D pre-trained GoogleNet, AlexNet, SqueezNet and Linear SVM [10]	96.1	0.936
Proposed approach (QR code image + CNN)	**96.7**	**0.94**

Bold content represents the best configuration of the model that gave best results architectures could not be employed due to the small dimensions of the QR code images.

- The dataset has a class imbalance as the number of QR code images created in each class differed. This imbalance also affected the performance of the proposed approach.

21.5 Conclusions

Around the world, residential buildings are among the primary users of more power. A platform for tracking unusual power usage in buildings is essential to preserve energy conservation and efficiency. This work proposes a QR code image-based anomalous power consumption classification with CNN. The QR code images were created from the micro-moment labeled SimDataset instances using the QR code library. The performance of the proposed approach was evaluated and compared with the state-of-the-art. Experimental results depicted that the proposed method outperformed the state-of-the-art by producing an overall accuracy of 96.7% and $F1$-score of 0.94. Due to the fewer parameters and lesser training time, the proposed approach proves to be the optimal choice for deployment in low-resource environment. Even though it performs better than the existing models, the proposed method has few limitations. Future work can consider discussed limitations of including more features that have an influence on power consumption patterns. One can use more features to create QR code images of bigger sizes so that more information can be extracted by the DL model.

References

1. Ishaq, M., Kwon, S.: Short-term energy forecasting framework using an ensemble deep learning approach. IEEE Access **9**, 94262–94271 (2021)
2. Ullah, F.U.M., Ullah, A., Haq, I.U., Rho, S., Baik, S.W.: Short-term prediction of residential power energy consumption via CNN and multi-layer bi-directional LSTM networks. IEEE Access **8**, 123369–123380 (2019)
3. EPBD: Directive 2010/31/EU of the European Parliament and of the Council of 19 May 2019 on the Energy Performance of Buildings (2009)
4. Piscitelli, M.S., Brandi, S., Capozzoli, A., Xiao, F.: A data analytics-based tool for the detection and diagnosis of anomalous daily energy patterns in buildings. In: Building Simulation, vol. 14, pp. 131–147. Tsinghua University Press (2021)
5. Wang, X., Ahn, S.H.: Real-time prediction and anomaly detection of electrical load in a residential community. Appl. Energy **259**, 114145 (2020)
6. UNEP: UNEP, United Nations Environment Program (2017). https://doi.org/10.1080/09613218.2017.1356127
7. Chahla, C., Snoussi, H., Merghem, L., Esseghir, M.: A deep learning approach for anomaly detection and prediction in power consumption data. Energ. Effi. **13**(8), 1633–1651 (2020)
8. Liu, X., Ding, Y., Tang, H., Xiao, F.: A data mining-based framework for the identification of daily electricity usage patterns and anomaly detection in building electricity consumption data. Energy Build. **231**, 110601 (2021)

9. Himeur, Y., Ghanem, K., Alsalemi, A., Bensaali, F., Amira, A.: Artificial intelligence based anomaly detection of energy consumption in buildings: a review, current trends and new perspectives. Appl. Energy **287**, 116601 (2021)
10. Copiaco, A., Himeur, Y., Amira, A., Mansoor, W., Fadli, F., Atalla, S., Sohail, S.S.: An innovative deep anomaly detection of building energy consumption using energy time-series images. Eng. Appl. Artif. Intell. **119**, 105775 (2023)
11. Weng, Y., Zhang, N., Xia, C.: Multi-agent-based unsupervised detection of energy consumption anomalies on smart campus. IEEE Access **7**, 2169–2178 (2018)
12. Xu, C., Chen, H.: A hybrid data mining approach for anomaly detection and evaluation in residential buildings energy data. Energy Build. **215**, 109864 (2020)
13. Sial, A., Singh, A., Mahanti, A.: Detecting anomalous energy consumption using contextual analysis of smart meter data. Wireless Netw. **27**, 4275–4292 (2021)
14. Wu, D., Jiang, Z., Xie, X., Wei, X., Yu, W., Li, R.: LSTM learning with Bayesian and Gaussian processing for anomaly detection in industrial IoT. IEEE Trans. Industr. Inf. **16**(8), 5244–5253 (2019)
15. Zhou, X., Yang, T., Liang, L., Zi, X., Yan, J., Pan, D.: Anomaly detection method of daily energy consumption patterns for central air conditioning systems. J. Build. Eng. **38**, 102179 (2021)
16. Lin, J., Fernández, J.A., Rayhana, R., Zaji, A., Zhang, R., Herrera, O.E., Mérida, W.: Predictive analytics for building power demand: day-ahead forecasting and anomaly prediction. Energy Build. **255**, 111670 (2022)
17. Himeur, Y., Alsalemi, A., Bensaali, F., Amira, A.: A novel approach for detecting anomalous energy consumption based on micro-moments and deep neural networks. Cogn. Comput. **12**(6), 1381–1401 (2020)
18. Himeur, Y., Alsalemi, A., Bensaali, F., Amira, A.: Smart power consumption abnormality detection in buildings using micromoments and improved K-nearest neighbors. Int. J. Intell. Syst. **36**(6), 2865–2894 (2021)
19. Himeur, Y., Alsalemi, A., Al-Kababji, A., Bensaali, F., Amira, A., Sardianos, C., Varlamis, I.: A survey of recommender systems for energy efficiency in buildings: principles, challenges and prospects. Inf. Fusion **72**, 1–21 (2021)
20. Himeur, Y., Alsalemi, A., Bensaali, F., Amira, A., Al-Kababji, A.: Recent trends of smart nonintrusive load monitoring in buildings: a review, open challenges, and future directions. Int. J. Intell. Syst. **37**(10), 7124–7179 (2022)
21. Himeur, Y., Alsalemi, A., Bensaali, F., Amira, A.: Robust event-based non-intrusive appliance recognition using multi-scale wavelet packet tree and ensemble bagging tree. Appl. Energy **267**, 114877 (2020)
22. Alsalemi, A., Himeur, Y., Bensaali, F., Amira, A.: Smart sensing and end-users' behavioral change in residential buildings: an edge-based internet of energy perspective. IEEE Sens. J. **21**(24), 27623–27631 (2021)
23. Nimmy, K., Dilraj, M., Sankaran, S., Achuthan, K.: Leveraging power consumption for anomaly detection on IoT devices in smart homes. J. Ambient. Intell. Humaniz. Comput. **14**(10), 14045–14056 (2023)
24. Malki, A., Atlam, E.S., Gad, I.: Machine learning approach of detecting anomalies and forecasting time-series of IoT devices. Alex. Eng. J. **61**(11), 8973–8986 (2022)
25. Wang, X., Flores, R., Brouwer, J., Papaefthymiou, M.: Real-time detection of electrical load anomalies through hyperdimensional computing. Energy **261**, 125042 (2022)
26. Lee, S., Jin, H., Nengroo, S.H., Doh, Y., Lee, C., Heo, T., Har, D.: Smart metering system capable of anomaly detection by bi-directional LSTM autoencoder. In: 2022 IEEE International Conference on Consumer Electronics (ICCE), pp. 1–6. IEEE (2022)
27. Fenza, G., Gallo, M., Loia, V.: Drift-aware methodology for anomaly detection in smart grid. IEEE Access **7**, 9645–9657 (2019)
28. Gaur, M., Makonin, S., Bajić, I.V., Majumdar, A.: Performance evaluation of techniques for identifying abnormal energy consumption in buildings. IEEE Access **7**, 62721–62733 (2019)

29. Blázquez-García, A., Conde, A., Mori, U., Lozano, J.A.: A review on outlier/anomaly detection in time series data. ACM Comput. Surv. (CSUR) **54**(3), 1–33 (2021)
30. Liu, X., Nielsen, P.S.: Scalable prediction-based online anomaly detection for smart meter data. Inf. Syst. **77**, 34–47 (2018)
31. Alsalemi, A., Himeur, Y., Bensaali, F., Amira, A., Sardianos, C., Chronis, C., Dimitrakopoulos, G.: A micro-moment system for domestic energy efficiency analysis. IEEE Syst. J. **15**(1), 1256–1263 (2020)
32. Varlamis, I., Sardianos, C., Chronis, C., Dimitrakopoulos, G., Himeur, Y., Alsalemi, A., Amira, A.: Smart fusion of sensor data and human feedback for personalized energy-saving recommendations. Appl. Energy **305**, 117775 (2022)
33. Alsalemi, A., Al-Kababji, A., Himeur, Y., Bensaali, F., Amira, A.: Cloud energy micro-moment data classification: a platform study. In: 2020 IEEE/ACM 13th International Conference on Utility and Cloud Computing (UCC), pp. 420–425. IEEE (2020)
34. Himeur, Y., Alsalemi, A., Bensaali, F., Amira, A., Varlamis, I., Bravos, G., Dimitrakopoulos, G.: Techno-economic assessment of building energy efficiency systems using behavioral change: a case study of an edge-based micro-moments solution. J. Clean. Prod. **331**, 129786 (2022)
35. Himeur, Y., Alsalemi, A., Bensaali, F., Amira, A.: Building power consumption datasets: survey, taxonomy and future directions. Energy Build. **227**, 110404 (2020)
36. Alsalemi, A., Ramadan, M., Bensaali, F., Amira, A., Sardianos, C., Varlamis, I., Dimitrakopoulos, G.: Endorsing domestic energy saving behavior using micro-moment classification. Appl. Energy **250**, 1302–1311 (2019)
37. Kaur, R., Gabrijelčič, D.: Behavior segmentation of electricity consumption patterns: a cluster analytical approach. Knowl.-Based Syst. **251**, 109236 (2022)
38. Alaca, Y., Çelik, Y.: Cyber attack detection with QR code images using lightweight deep learning models. Comput. Secur. **126**, 103065 (2023)

Chapter 22
CrowdISEntrY: Intelligent Object Entry-Exit Counting Framework and Comparative Analysis in CCTV and Drone Surveillance Systems

Siddharth Shankar Das, A. Ancy Micheal, Ambar Kumar, and Moin Ahmed Zahir

Abstract Surveillance plays a critical role in monitoring people and vehicles in various settings. This study introduces CrowdISEntrY, a novel approach utilizing YOLOv8 and DeepSort for precise object counting in both CCTV and drone videos. YOLOv8, leveraging a feature pyramid network, ensures accurate object detection, followed by DeepSORT for object tracking using Kalman filters and matching cascade. This integrated approach facilitates effective entry and exit counting, thereby enhancing surveillance across diverse applications. Comparative analysis with YOLOv3, YOLOv5, YOLOv7, and YOLOv8, along with DeepSort, demonstrates superior performance. In CCTV videos, CrowdISEntrY achieves a precision of 0.91, recall of 0.82, F1-score of 0.86, and mAP of 0.72. In drone videos, it outperforms other models with a precision of 0.88, recall of 0.77, F1-score of 0.82, and mAP of 0.66. Overall, CrowdISEntrY exhibits an accuracy of 98.07% for entry and exit counting, surpassing existing models and demonstrating its efficacy in real-time surveillance scenarios.

Keywords Object counting · Real-time surveillance · YOLOv8 · Deep sort

S. S. Das (✉) · A. A. Micheal · A. Kumar · M. A. Zahir
School of Computer Science and Engineering, Vellore Institute of Technology, Chennai, Tamil Nadu, India
e-mail: siddharthshankar.das2020@vitalum.ac.in

A. A. Micheal
e-mail: ancymicheal.a@vit.ac.in

A. Kumar
e-mail: ambar.kumar2020@vitalum.ac.in

M. A. Zahir
e-mail: moinahmed.zahir2020@vitalum.ac.in

J. Maiti et al. (eds.), *Operations Research and Data Analytics: Current Trends and Future Perspectives*, Lecture Notes on Multidisciplinary Industrial Engineering,
https://doi.org/10.1007/978-981-96-8327-7_22

22.1 Introduction

In recent years, surveillance in various sectors is deployed using CCTV footage, drones, and camcorder recordings. CCTV and drone-based surveillance are used in various sectors such as traffic surveillance [1], crowd monitoring [2], wildlife monitoring [3], marine surveillance [4], pipeline surveillance, etc. The choice of deploying CCTV camera or a drone camera is solely based on the environmental condition, cost, and area of coverage of the particular surveillance region in case of a CCTV surveillance is applied where a smaller area is covered.

CCTV cameras are applied in case of low cost and small area coverage. There are scenarios where the objects have to be focused on closer, as well as farther at various heights or altitude, in such cases drone-based surveillance is opted. CCTV technology can be used to accurately track and analyze crowd movement, helping organizers manage events more effectively, optimize space utilization, and ensure the safety of attendees. Drone videos for crowd counting [5] offer a dynamic and aerial perspective. For crowd counting, it provides real-time, high-resolution data, enabling precise traffic analysis that makes it an essential tool for our purpose. In computer vision, the challenges for tracking the object in CCTV cameras and drone cameras vary due to the mobility of the camera. In a static camera, the background is still, and the object moves, whereas in a drone camera, there is a movement in the object as well as the background in the camera. Hence, the challenges to perform accurate tracking varies in CCTV surveillance and drone-based surveillance. Object counting provides crucial data for efficient resource allocation and informed decision-making in surveillance, enhancing security measures.

Distinguishing between the challenges posed by CCTV and drone-based surveillance, this study delves into the unique complexities faced in both scenarios. CCTV cameras, often deployed in smaller areas due to lower costs, require precise object detection, while drone cameras face challenges in simultaneously capturing moving objects and dynamic backgrounds. Our proposed CrowdISEntrY system adeptly addresses these challenges and presents a comprehensive solution, offering better accuracy in object detection, tracking, and multi-line counting across diverse visual data sources. In this study, a comparative analysis is performed with yolov8, deep sort, and CrowdISEntrY for CCTV surveillance and drone surveillance.

In addition to introducing CrowdISEntrY, it's essential to highlight the broader significance of object counting in surveillance, impacting resource allocation and decision-making in various domains like traffic management and crowd control. Addressing challenges posed by dynamic backgrounds in drone footage, CrowdISEntrY stands out with its integrated approach utilizing YOLOv8 and DeepSort, ensuring robust object detection and tracking. Through critical analysis and comparison, this study aims to underscore CrowdISEntrY's unique contributions, demonstrating its potential for real-world deployment and impact.

22.2 Related Work

Various researchers have experimented on CCTV surveillance and drone-based surveillance individually. Lian et al. [6] introduced the Dual-Path Guided Detection Network (DPDNet) as a solution to the challenging task of recognizing and counting small, densely packed heads in crowded environments. Pan et al. [7] presented CGINet, a novel approach for RGB and thermal crowd counting. The challenge of object detection and counting in a crowded environment was addressed by Wang et al. [8]. Their Crowd-SDNet utilizes a keypoint-based detection architecture, separating object center point and size estimation, which optimizes the use of limited point annotations. However, Crowd-SDNet relies on accurate point annotations, which may not always be feasible, limiting its applicability in real-world scenarios. Additionally, its effectiveness may be compromised in highly dynamic environments. Further research is needed to address these limitations and enhance the robustness of object detection algorithms in crowded scenes. In [9], the critical need for real-time crowd control for COVID 19 is addressed, particularly in scenarios like the COVID-19 pandemic. The study proposes a YOLO-based method for real-time crowd detection, emphasizing its significance in balancing accuracy and processing speed. By introducing an algorithm for estimating area capacity and counting individuals, it addresses the pressing demand for efficient crowd management. Furthermore, the analysis of various YOLO model variations offers valuable insights into their performance for crowd detection, enhancing the context for proposed solutions like CrowdISEntrY. Ilyas et al. [10] proposed hierarchical dense dilated deep pyramid feature extraction, a novel crowd counting method that aims to improve accuracy by addressing scale fluctuations in crowded scenes. FlounderNet, a model designed for crowd counting in aerial photographs using embedded devices [11]. The interleaved group convolution and quick feature map reduction improve computational efficiency and adaptability to various image sizes. It holds great promise for surveillance applications in resource-constrained environments.

Gao et al. [12] presented WSITrans, an end-to-end crowd localization framework that utilizes Transformer networks and lax supervision. It simultaneously performs crowd counting and localization, reducing the need for manual annotations. Khan et al. [13] introduced CSCCL-Net, a model designed to address the complexities of crowded scenes with size variations and opacity. This innovative approach targets congested scenes encountered in events like Hajj and Umrah, effectively handling scale variation and occlusion challenges. By integrating the FIDT map and employing a Local Maxima Detection Strategy (LMDS) with dilated convolutional neural networks, CSCCL-Net enhances crowd counting and localization capabilities while mitigating the limitations of conventional density maps. DAA-Net is used for crowd counting in foggy weather conditions [14]. It preserves positional information while capturing long-range interdependence in crowd structures, enabling efficient crowd counting even in hazy conditions. Du et al. [15] proposed hierarchical mixture of density experts, an innovative crowd counting architecture that uses a hierarchical merging approach for multi-scale density maps.

Building upon the foundation laid by previous research efforts, CrowdISEntrY offers several notable advancements. Its integrated approach, combining YOLOv8 and DeepSort, ensures superior object detection and tracking compared to existing models. By leveraging a feature pyramid network and Kalman filters, CrowdIS-EntrY achieves precise entry and exit counting, surpassing the limitations of previous methods. Furthermore, its adaptability to both CCTV and drone surveillance systems makes it a versatile and robust solution for real-time surveillance applications.

22.3 Methodology

This study focuses on analysis of object entry and exit in CCTV videos and drone videos using YOLOv8 [16] as object detector, DeepSORT [17] as object tracker and CrowdISEntrY. YOLOv8 possesses a transformer-based architecture which enables accurate object detection. DeepSort is composed of Kalman Filter as object tracker and matching cascade for associating the object. CrowdISEntrY provides the counting of entry and exit objects in the video. The proposed framework is shown in Fig. 22.1. The flow is explained in the CrowdISEntrY Algorithm (Sect. 3.3).

22.3.1 YOLOv8

The YOLO object identification system was created by Joseph Redmon and Ali Farhadi, and its most recent version is called YOLOv8. The architecture of YOLOv8 is shown in Fig. 22.2 [18]. YOLOv8 is more rapid than YOLOv7. Real-time object

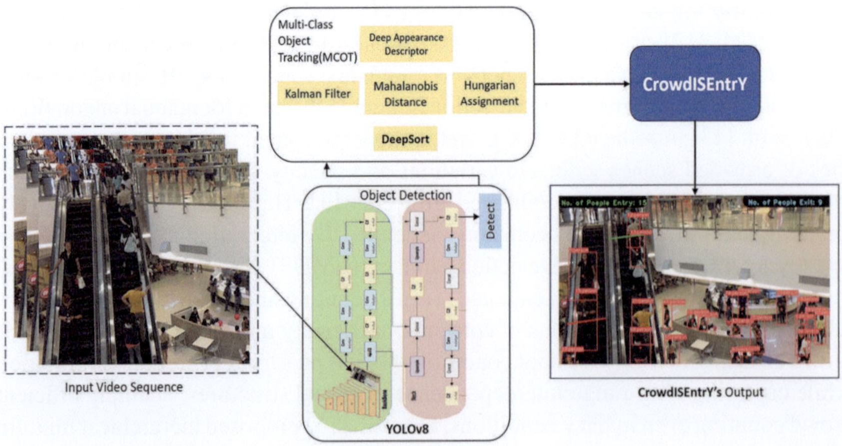

Fig. 22.1 The framework of proposed architecture

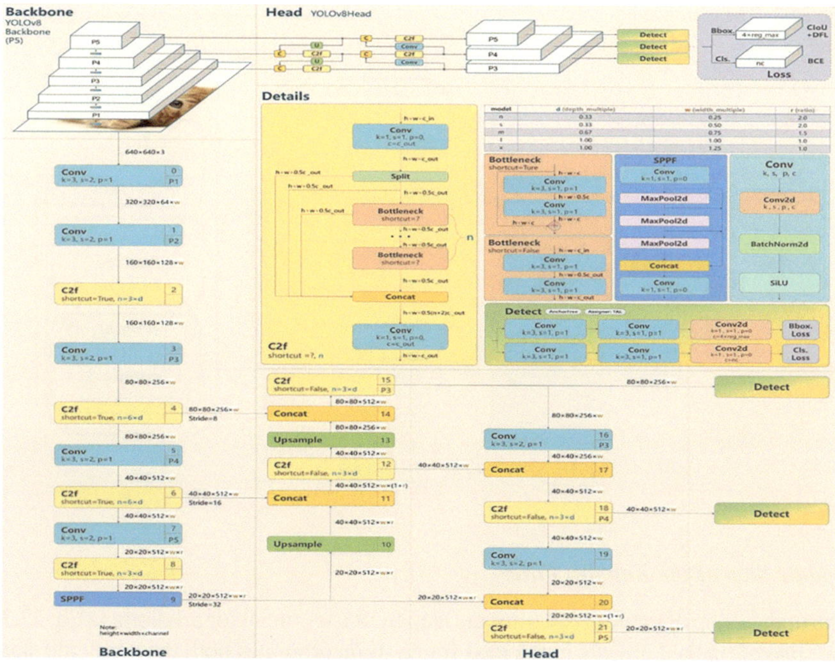

Fig. 22.2 The architecture of YOLOv8

identification relies heavily on model speed, and YOLOv8 offers faster detection while keeping accuracy. YOLOv8 is more effective at recognizing small objects than YOLOv7. YOLOv8 uses a dynamic head network to enhance object detection accuracy, making it simpler to accurately detect small things. It uses an anchor-free architecture, which does away with the necessity for anchor boxes and facilitates training the model across various datasets. It predicts objects at multiple scales, making it easier to detect objects of different sizes. YOLOv8 is a potent deep learning model for object recognition that outperforms YOLOv7 in terms of speed and accuracy. It is a dependable option for real-time object identification thanks to its anchor-free architecture, multiscale prediction, and improved backbone network.

22.3.2 DeepSORT

Deep simple online and real-time tracking (DeepSORT) is applied in this framework for object tracking and association. Its architecture is shown in Fig. 22.3 [19]. It is an advanced object tracking algorithm that combines deep learning techniques with classical tracking algorithms.

Fig. 22.3 The architecture of DeepSort

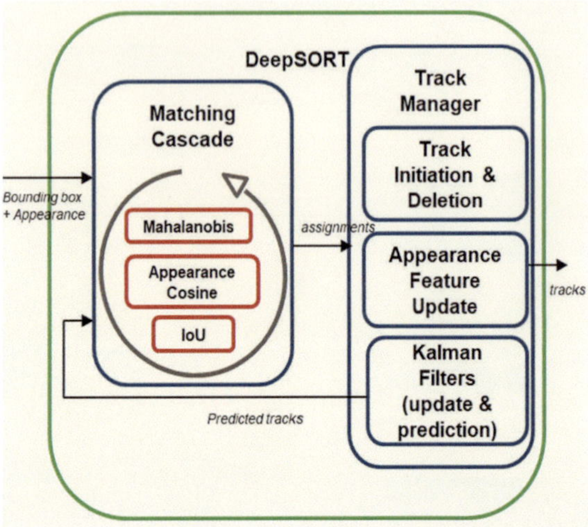

Update Step of the Kalman Filter

The update step of the Kalman filter in Deep SORT is crucial for predicting (Eq. 22.3) the state of tracked objects in the next frame. It incorporates both the predicted state and the measurement (detected object's feature vector, Eq. 22.1 and 22.2) to refine the object's position and velocity estimates.

$$K_k = P_k^- \times H^T \times \left(H \times P_k^- \times H^T + R\right)^{-1} \tag{22.1}$$

$$\hat{x}_k^+ = \hat{x}_k^- + K_k \times \left(z_k - H \times \hat{x}_k^-\right) \tag{22.2}$$

$$P_k^+ = (I - K_k \times H) \times P_k^- \tag{22.3}$$

Mahalanobis Distance Calculation

Mahalanobis distance (Eq. 22.4) is utilized in Deep SORT to measure the dissimilarity between predicted object states and detected object features.

$$D = \sqrt{(x - \mu)^T \times \Sigma^{-1} \times (x - \mu)} \tag{22.4}$$

Intersection Over Union (IoU) Calculation

IoU (Eq. 22.5) is a vital metric used in object detection and tracking. In Deep SORT, it is employed to evaluate the spatial overlap between predicted bounding boxes and detected bounding boxes.

$$\mathrm{IoU}(A, B) = \frac{\mathrm{Area}(A \cap B)}{\mathrm{Area}(A \cup B)} \tag{22.5}$$

DeepSORT's update step of the Kalman filter allows for robust prediction of object states, even in scenarios of partial occlusion, by incorporating both predicted and measured object features. By integrating these techniques, as shown in Fig. 22.3 Deep SORT [19] achieves robust and precise multi-object tracking in real-time video streams. The Kalman filter update step refines predictions, Mahalanobis distance assesses feature similarity, and IoU ensures spatial accuracy, collectively enhancing the algorithm's tracking performance.

22.3.3 CrowdISEntrY

The output obtained from DeepSort is fed into the CrowdIsentrY algorithm. It counts the numbers of objects entering into the assigned region, and leaving from the assigned region. The algorithm of CrowdIsEntry is shown in Algorithm 1.

Algorithm 1: Crowd surveillance with Intelligence and Entry Counting using YOLO (CrowdISEntrY).

Require: Video Frames $F = \{f_1, f_2, \ldots, f_n\}$, Trained YOLOv8 Model M_{YOLOv8}, DeepSORT Tracker $T_{DeepSORT}$, Process Noise Covariance Q, Measurement Noise Covariance R, Time Step Δt, Entry Line Coordinates $(x_{\mathrm{entry}}, y_{\mathrm{entry}})$, Exit Line Coordinates $(x_{\mathrm{exit}}, y_{\mathrm{exit}})$, threshold value

Ensure: Tracked Objects $O = \{o_1, o_2, \ldots, o_m\}$, Counters, CountEntry, CountExit

Output: Generated video with line and counting of traffic.

1: Initialize M_{YOLOv8}, $T_{DeepSORT}$, $i = 1$, CountEntry = 0, CountExit = 0
2: **while** $i \leq n$ **do**
3: Detect objects using M_{YOLOv8}: $D_i = M_{YOLOv8}(f_i)$
4: **for all** detected object d in D_i **do**
5: Calculate object embedding: E_d =Embedding(d)
6: **if** $T_{DeepSORT}$ is empty **then**
7: Initialize Kalman Filter:
$x\hat{}k|k-1 = Ed$
$v\hat{}k|k-1 = 0$
$Pk|k-1 = 0$
8: **else**
9: Predict next state using Kalman Filter:
$x\hat{}k|k-1 = x\hat{}k-1|k-1 + v\hat{}k-1 + \Delta t$
$v\hat{}k|k-1 = v\hat{}k-1|k-1$
$Pk|k-1 = Pk-1|k-1 + Q$
10: **if** Similarity(o_j, d)$\geq$Threshold **and** $x > x_{\mathrm{entry}}$ **and** $y > y_{\mathrm{entry}}$ **then**
11: Increment CountEntry += 1.

        end if
        if $x > x_{\text{exit}}$ **and** $y > y_{\text{exit}}$ **then**
            Increment CountExit+ $= 1$
        end if
        end if
    end for
Increment i by 1
end while
return O, CountEntry, CountExit

In CrowdISEntrY, entry and exit lines play a crucial role in precise object counting within the algorithm. Currently, these lines' coordinates are manually provided, defining the region where objects are monitored for entry and exit. However, ongoing research aims to automate this process by integrating road, lane, and line detection algorithms. Additionally, the algorithm enhances accuracy by considering the flow and trajectory of crowd or vehicle movements, ensuring robust surveillance in diverse environments. The output obtained from DeepSort is fed into the CrowdIsentrY algorithm. It counts the numbers of objects entering into the assigned region and leaving from the assigned region. The algorithm of CrowdIsEntry is shown in Algorithm 1.

22.4 Findings and Contributions

22.4.1 Experimental Setup

An Intel i7 CPU with 16 GB of RAM was used for the experiment. The graphics card used was an NVIDIA RTX 3060 graphics processor. The experiment is performed with 25 videos in CCTV surveillance and 25 videos in UAV123 [20] drone video dataset. The results are shown in Figs. 22.4 and 22.5. The experiment is analyzed with precision, recall, F1-score, accuracy (Eqs. 22.6–22.9). The comparative analysis for CCTV videos and drone videos is shown in Tables 22.1 and 22.2.

$$\text{Precision} = \frac{\text{True Positives}}{\text{True Positives} + \text{False Positives}} \tag{22.6}$$

$$\text{F1 Score} = 2 \times \frac{\text{Precision} \times \text{Recall}}{\text{Precision} + \text{Recall}} \tag{22.7}$$

$$\text{Recall} = \frac{\text{True Positives}}{\text{True Positives} + \text{False Negatives}} \tag{22.8}$$

$$\text{Accuracy} = \frac{\text{True Positives} + \text{True Negative}}{\text{Total}} \tag{22.9}$$

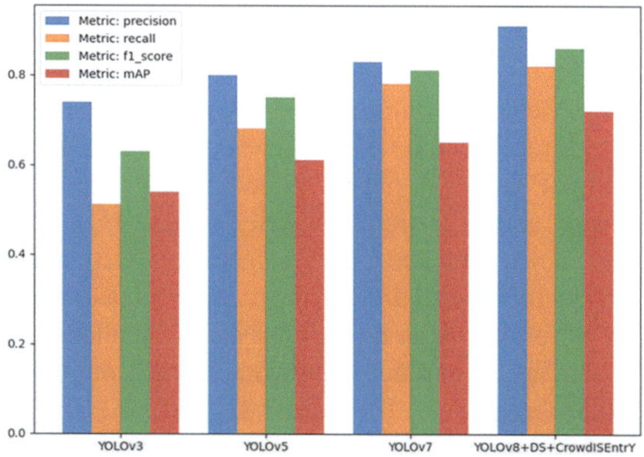

Fig. 22.4 Analysis of performance metrics for CCTV videos

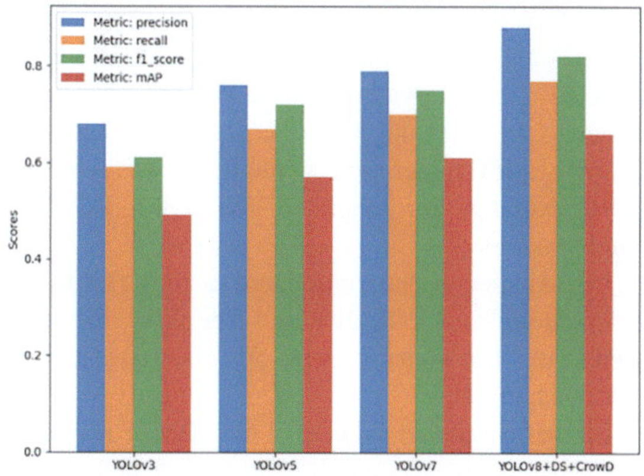

Fig. 22.5 Analysis of performance metrics for drone videos

Table 22.1 The performance metrics of proposed model for CCTV videos

Methodology	Precision	Recall	F1-score	mAP
YOLOv3	0.74	0.51	0.63	0.54
YOLOv5	0.80	0.68	0.75	0.61
YOLOv7	0.83	0.78	0.81	0.65
YOLOv8 + DS + CrowdISEntrY	0.91	0.82	0.86	0.72

Table 22.2 The performance metrics of proposed model for drone videos

Methodology	Precision	Recall	F1-score	mAP
YOLOv3	0.68	0.59	0.61	0.49
YOLOv5	0.76	0.67	0.72	0.57
YOLOv7	0.79	0.70	0.75	0.61
YOLOv8 + DS + CrowdISEntrY	0.88	0.77	0.82	0.66

22.4.2 Results

The proposed algorithm got the best performance results, with the values of precision = 0.91, recall = 0.82, F1-score = 0.86, and mAP = 0.72. The substantial F1-scores and mAP values in our integrated YOLOv8 + DS + CrowdISEntrY highlight its superior ability to simultaneously achieve high precision and recall, making it the ideal choice for static CCTV camera surveillance systems.

Transitioning to drone camera analysis, as shown in Table 22.2, CrowdISEntrY still performed better facing the dynamic challenges posed by aerial footage. Despite the complexity introduced by drone footage, our model demonstrated better precision of 0.88, recall 0.77, F1-score 0.82, and mAP 0.66 values, surpassing all the other compared models.

These metrics attest to the ability of CrowdIsEntrY in maintaining accuracy even in the face of challenging, moving scenarios. The presented tables unequivocally demonstrate that our proposed CrowdIsEntrY algorithm performs well across both static CCTV and dynamic drone camera environments. Figures 22.4, 22.5 further visualize these results, with the x-axis denoting different YOLO versions and our CrowdIsEntrY integration, while the y-axis represents the corresponding detection scores. Notably, our integrated YOLOv8 + DS + CrowD consistently outperforms other versions.

In this paper, we have compared various YOLO models, Thus, based on the results, the integration of YOLOv8, and DeepSort, for the CrowdIsEntry algorithm can be used in real life scenarios with object entry exit counting for both CCTV and drone videos accurately.

For the combined dataset, of the 50 videos consisting of 567 people images, and 358 vehicles images including both entry and exit values. The dataset was prepared under various onboard conditions to test the efficacy of the system. The entry–exit line counting performance of the CrowdIsEntrY system was evaluated using the detection rate (DR) and false positive rate (FPR) for the classes of people and vehicle's entry exit as shown in Table 22.3 and the absolute counts are shown in Fig. 22.6. Here, true positive (TP) represents the model's correct predictions of labels, while false negative (FN) indicates instances where the model fails to detect accurate labels. These metrics are essential in assessing the system's accuracy in identifying relevant events within the monitored area.

Table 22.3 The entry exit line counting performance metrics

Classes	TP	FN	DR	FPR
People entry	308	0	1.0	0.003
People exit	254	3	0.988	0.031
Vehicle entry	204	1	1.0	0.001
Vehicle exit	149	6	0.961	0.038

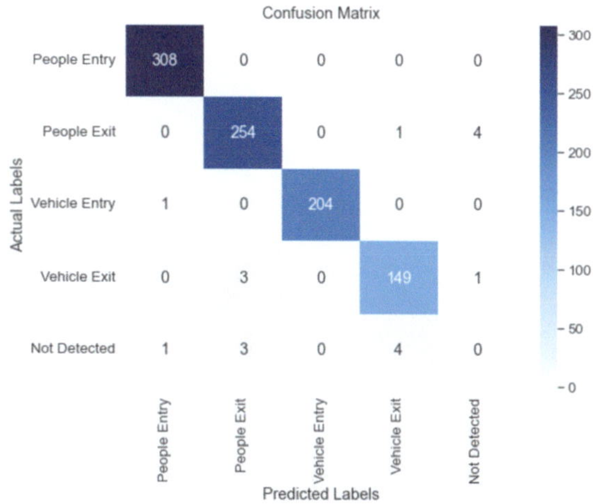

Fig. 22.6 Confusion matrix for the entry exit line counting

Table 22.4 Comparative Analysis of State-of-the-art Models and the Proposed Model

Models	Accuracy (%)
SSD [21]	48.5
YOLOv3 [22]	57.6
YOLOv4 [23]	72.8
Hybrid YOLOv4 [24]	95.1
CrowdISEntrY (proposed method)	98.07

After the final evaluation, the CrowdISEntrY workflow produced an accuracy of 98.07%. As shown in Table 22.4, the accuracy of the proposed model has surpassed the accuracy of the state-of-the-art models.

22.4.3 Output

Prediction results on the combined video dataset are shown below. Figure 22.8 shows the results taken from CCTV videos. Figure 22.9 shows the traffic count of entry exit taken from drone videos.

From the above figures, it can be seen that our CrowdISEntrY performs well in detecting multiclass traffic elements, as shown in Figs. 22.7 and 22.8. The entry and exit crowd and vehicle count at the front of VIT Chennai Gate, Fig. 22.7, shows that the CrowdISEntrY can also work with multiple entry exit lines. Additionally, in Fig. 22.8, where dense crowd scenarios are depicted, CrowdISEntrY demonstrates robust performance, showcasing its ability to handle occlusions and accurately count objects even in challenging conditions. It showcases that the user can also personalize the degree of entry/exit line to gain better results.

Fig. 22.7 Crowd entry exit counting, at VIT main gate

Fig. 22.8 Samples of entry and exit count in CCTV videos

Fig. 22.9 Samples of entry and exit count in Drone videos

To show our model's robustness, we have also tested our CrowdISEntrY on the VisDrone [25] and UAV123 [20] dataset, the traffic counting results are shown in Fig. 22.9.

22.5 Conclusion

This study addresses challenges in today's surveillance systems, especially in crowded spaces. Diverging from conventional methods, this study compares popular object detection algorithms with YOLOv8, which is proved superior with an mAP50 of 0.72 on the VisDrone dataset. As the primary model for CrowdISEntrY, YOLOv8 excels in complex scenarios. Integration with DeepSORT elevated precision amidst fluctuating condition. With CrowdIsEntrY, our paper showcased adaptability in crowded and obstructed areas. Our comparative analysis in both CCTV (F1-score = 0.86, precision = 0.91, recall = 0.82, mAP = 0.72) and Drone videos (F1-score = 0.82, precision = 0.88, recall = 0.77, mAP = 0.66) underlined CrowdISEntrY's exceptional performance. Moreover, the applications of CrowdISEntrY extend beyond the scope of this study, encompassing various surveillance scenarios such as traffic management, crowd control, wildlife monitoring, and marine surveillance. Its precision in object detection and counting enhances security measures and enables efficient resource allocation in these diverse applications. Even amidst dynamic Drone backgrounds, it delivered exceptional accuracy (98.07%) in entry/exit counting, confirming its prowess in discerning people entry, people exit, vehicle entry, vehicle exit, and no detection scenarios proving its effectiveness.

Future research directions include addressing potential limitations such as occlusion issues in high-density crowded scenes, the impact of weather conditions on drone footage, and optimizing computational efficiency for real-time applications.

References

1. Bhat, P.G., Subudhi, B.N., Veerakumar, T., Esakkirajan, S.: A fully automatic feature-based real-time traffic surveillance system using data association in the probabilistic framework. IEEE Trans. Intell. Transp. Syst. **23**(8), 11088–11097 (2022)
2. Miao, Y., et al.: Abnormal behavior learning based on edge computing toward a crowd monitoring system. IEEE Netw. **36**(3), 90–96 (2022). https://doi.org/10.1109/MNET.014.2000523
3. Casazza, M.L., Lorenz, A.A., Overton, C.T., Matchett, E.L., Mott, A.L., Mackell, D.A., McDuie, F.: AIMS for wildlife: developing an automated interactive monitoring system to integrate real-time movement and environmental data for true adaptive management. J. Environ. Manage. **345**, 118636 (2023). https://doi.org/10.1016/j.jenvman.2023.118636
4. Duan, Y., Li, Z., Tao, X., Li, Q., Hu, S., Lu, J.: EEG-based maritime object detection for IoT-driven surveillance systems in smart ocean. IEEE Internet Things J. **7**(10), 9678–9687 (2020)
5. Khan, M.A., Menouar, H., Hamila, R.: DroneNet: crowd density estimation using self-ONNs for drones. In: Proceedings of the 2023 IEEE 20th Consumer Communications & Networking Conference (CCNC), pp. 455–460 (2023). https://doi.org/10.1109/CCNC51644.2023.10059904
6. Lian, D., Chen, X., Li, J., Luo, W., Gao, S.: Locating and counting heads in crowds with a depth prior. IEEE Trans. Pattern Anal. Mach. Intell. **44**(12), 9056–9072 (2022)

7. Pan, Y., Zhou, W., Qian, X., Mao, S., Yang, R., Yu, L.: CGINet: cross-modality grade interaction network for RGB-T crowd counting. Eng. Appl. Artif. Intell. **126**, 106885 (2023). https://doi.org/10.1016/j.engappai.2023.106885
8. Wang, Y., Hou, J., Hou, X., Chau, L.-P.: A self-training approach for point-supervised object detection and counting in crowds. IEEE Trans. Image Process. **30**, 2876–2887 (2021)
9. Gündüz, M.Ş, Işık, G.: A new YOLO-based method for real-time crowd detection from video and performance analysis of YOLO models. J. Real Time Image Process. **20**(5), 1542 (2023)
10. Ilyas, N., Ahmad, Z., Lee, B., et al.: An effective modular approach for crowd counting in an image using convolutional neural networks. Sci. Rep. Nat. **12**, 5795 (2022)
11. Chen, J., Xiu, S., Chen, X., Guo, H., Xie, X.: Flounder-net: an efficient CNN for crowd counting by aerial photography. Neurocomputing **420**, 82–89 (2021). https://doi.org/10.1016/j.neucom.2020.09.001
12. Gao, H., Zhao, W., Zhang, D., Deng, M.: Application of improved transformer based on weakly supervised in crowd localization and crowd counting. Sci. Rep. Nat. **13**, 1144 (2023)
13. Khan, A., Kadir, K., Nasir, H., Shah, J.S., Albattah, W., Khan, S., Kakakhel, M.H.: Crowd counting and localization beyond density map. IEEE Access **10**, 133142–133151 (2022)
14. Kong, W., Shen, J., Li, H., Liu, J., Zhang, J.: Direction-aware attention aggregation for single-stage hazy-weather crowd counting. Exp. Syst. Appl. **225**, 120088 (2023)
15. Du, Z., Shi, M., Deng, J., Zafeiriou, S.: Redesigning multi-scale neural network for crowd counting. IEEE Trans. Image Process. **32**, 3664–3678 (2023)
16. Gašparović, B., Mauša, G., Rukavina, J., Lerga, J.: Evaluating YOLOV5, YOLOV6, YOLOV7, and YOLOV8 in underwater environment: is there real improvement? In: Proceedings of the 2023 8th International Conference on Smart and Sustainable Technologies (SpliTech), pp. 1–4 (2023)
17. Chang, F., Huang, Y., Xu, M., Zhao, Y., Ma, L., Su, H.: Multi-object tracking using Deep SORT and modified CenterNet in cotton seedling counting. Comput. Electron. Agric. **202**, 107339 (2022). https://doi.org/10.1016/j.compag.2022.107339
18. https://github.com/ultralytics/ultralytics/issues/189
19. https://www.mathworks.com/help/vision/ug/multi-object-tracking-with-deepsort.html
20. Mueller, M., Smith, N., Ghanem, B.: A benchmark and simulator for UAV tracking. In: Proceedings of the European Conference on Computer Vision (ECCV), pp. 445–461 (2016)
21. Loghum, B.S.V.: YOLO!. PodoPost, vol 31, p. 13 (2018)
22. Redmon, J., Farhadi, A.: YOLOv3: an incremental improvement. arXiv:1804.02767 (2018)
23. Almaiah, M.A., Ali, A., Hajjej, F., Pasha, M.F., Alohali, M.A.: A lightweight hybrid deep learning privacy preserving model for FC-based industrial internet of medical things. Sensors **22**(6), 2112 (2022)
24. Khel, M.H.K., et al.: Realtime crowd monitoring—estimating count, speed and direction of people using hybridized YOLOv4. IEEE Access **11**, 56368–56379 (2023). https://doi.org/10.1109/ACCESS.2023.3272481
25. Zhu, P., et al.: Detection and tracking meet drones challenge. IEEE Trans. Pattern Anal. Mach. Intell. **44**(11), 7380–7399 (2022). https://doi.org/10.1109/TPAMI.2021.3119563

Chapter 23
Aetos Angelos: A Keen Perception Alert System

Siva Sai Sree Yakkala and Ashish Ratna Darla

Abstract In an era characterized by rapid technological progress and pervasive connectivity, the persistent challenge of inaccessibility during critical moments of distress necessitates innovative solutions. The "Aetos Angelos" project emerges as a pioneering response, leveraging cutting-edge technology to overcome the barrier between unreachability and the urgent requirement for swift assistance. The project distinguishes itself not only through its advanced methodology but also its comprehensive approach, employing a gesture-based communication system. Through the strategic repurposing of established Closed-Circuit Television (CCTV) infrastructure, the project adeptly identifies distress signals, thereby minimizing response times and fostering collaboration among diverse stakeholders. This novel approach significantly enhances emergency management systems, elevates public safety standards, and ultimately contributes to life-saving efforts. Beyond immediate aid, "Aetos Angelos" provides sustained support and preventive measures, positioning it as a holistic tool for emergency situations. By resourcefully repurposing existing assets to address critical needs, the project exemplifies the transformative potential of technology in bridging the gap between inaccessibility and timely assistance.

Keywords Technology · Connectivity · Emergency management · Innovative solutions · Gesture-based communication · CCTV infrastructure

---

S. S. S. Yakkala (✉) · A. R. Darla
Artificial Intelligence and Data Science Department of BVRIT, B V Raju Institute Of Technology, Narsapur (Affiliated to JNTU), Narsapur, Telangana, India
e-mail: sivasaisreeyakkala@gmail.com

A. R. Darla
e-mail: 22215a7202@bvrit.ac.in

J. Maiti et al. (eds.), *Operations Research and Data Analytics: Current Trends and Future Perspectives*, Lecture Notes on Multidisciplinary Industrial Engineering,
https://doi.org/10.1007/978-981-96-8327-7_23

23.1 Introduction

In an era marked by relentless technological advancements and the pervasive interconnectivity of our global society, the challenge of responding swiftly to distress calls during critical moments persists as a paramount concern. Emergencies, by their very nature, demand immediate attention and rapid intervention, yet the specter of unreachability continues to impede the seamless delivery of assistance. This research paper delves into a groundbreaking initiative designed to confront and overcome this persistent challenge—the "Aetos Angelos" project. As the world becomes increasingly reliant on sophisticated technological infrastructures, the need for innovative solutions that harness these advancements for the betterment of society becomes more pronounced. The "Aetos Angelos" project represents a paradigm shift in addressing unreachability during emergencies, introducing a multifaceted and technologically advanced system that not only identifies distress signals promptly but also streamlines the communication process for rapid and effective assistance.

At its core, the "Aetos Angelos" project is a testament to the fusion of cutting-edge technology, strategic resource repurposing, and a visionary approach to emergency response. This paper will delve into the intricacies of the project's methodology, emphasizing its unique utilization of gesture-based communication systems and the repurposing of existing Closed-Circuit Television (CCTV) infrastructure. By employing these innovations, the project not only reduces response times but also fosters collaboration among various stakeholders, paving the way for a more integrated and efficient emergency management ecosystem. The significance of the research extends beyond the immediate context of crisis response; it encompasses a comprehensive vision that includes long-term support and preventative measures. "Aetos Angelos" emerges not only as a tool for mitigating immediate threats but also as a sustainable and adaptive solution capable of addressing the evolving landscape of emergency situations.

As we navigate the complexities of our modern world, the lessons learned from the "Aetos Angelos" project provide valuable insights into the potential of technology to bridge critical gaps, redefine emergency management paradigms, and ultimately save lives. This paper aims to unpack the intricacies of this transformative initiative, shedding light on its methodologies, outcomes, and the broader implications for the future of emergency response systems.

23.2 Related Works

In recent years, emergency response systems have experienced a significant transformation due to the continuous advancement of technology and the need to overcome the ongoing difficulties related to inaccessibility during crucial situations. As we begin to examine the "Aetos Angelos" project, a pioneering initiative designed to connect inaccessible areas with quick emergency aid, it is crucial to place this creative undertaking within the wider context of previous literature. This literature study aims to offer a thorough comprehension of the present condition of emergency management systems, the technological innovations that have influenced this field, and the significant deficiencies that initiatives such as "Aetos Angelos" want to address. This review establishes a foundation for a detailed assessment of the project's contributions and its potential to reshape the field of emergency management. It does so by analyzing the academic discussions on location analysis, communication strategies, and the incorporation of advanced technologies in emergency response.

The literature on interactive communication systems for individuals with hearing impairments underscores the significance of innovative technologies in addressing specific communication needs. Zhu et al. [1] contribute to this discourse by presenting their work on designing an Interactive Communication Assistance System for hearing-impaired college students. Their approach, grounded in gesture recognition and representation, highlights the potential of incorporating intuitive communication methods to enhance the accessibility and inclusivity of educational environments. The study, published in Future Internet, delves into the application of gesture-based systems, shedding light on the transformative impact such technologies can have on facilitating communication for individuals with hearing impairments. In the realm of educational technology, Alvarez et al. [2] conducted a comparative study that delves into the effectiveness of netbooks and tablet PCs in fostering face-to-face collaborative learning. Published in Computers in Human Behavior, their research provides valuable insights into the nuanced interactions between students and technology. While not directly focused on emergency response, this study contributes to the broader understanding of technology-mediated communication and collaboration, offering insights that may inform the development of inclusive and effective communication strategies within emergency management systems. The intersection of technology and sign language recognition is a crucial theme in the literature. Hossain et al. [3] explore the application of Convolutional Neural Network models for sign language digit recognition. Their study, featured in the Asian Journal of Research in Computer Science, delves into the intricacies of leveraging deep learning techniques to interpret sign language, presenting a technological foundation that aligns with the gesture-based communication system emphasized in the "Aetos Angelos" project. The findings contribute to the growing body of knowledge surrounding the application of neural networks in sign language recognition, a facet with direct relevance to the project's innovative communication strategy. The exploration of smart devices for sign language recognition is further advanced by Ghuse [4], who investigates Sign Language Recognition using a Smart Glove. Published in the International Journal for Research in Applied Science and Engineering Technology, this study delves into the practical application of wearable technology as a means to bridge communication gaps for individuals with hearing impairments. The integration of smart gloves in sign language recognition aligns with the technological ingenuity demonstrated by the "Aetos Angelos" project, emphasizing the broader impact of gesture-based communication systems beyond specific contexts. In the domain of gesture recognition, Ding and Zheng [5] contribute a study focused on RGB-D Depth-sensor-based Hand Gesture Recognition using Deep Learning. Their work, featured in Sensors and Materials, addresses the intricacies of recognizing hand gestures through depth-sensor technology. The removal of shadow effects in depth images is a notable aspect, showcasing the meticulous attention to detail required for accurate gesture interpretation. This research aligns with the technological underpinnings of the "Aetos Angelos" project, emphasizing the significance of robust gesture recognition mechanisms for effective communication in emergency scenarios.

Gesture-based communication systems have garnered significant attention within the realm of technology-mediated interactions, particularly in the context of emergency response and management. Li et al. [6] explore the trajectory-based gesture recognition in smart homes using an Ultrawideband Communication System. Their work delves into the intricate details of gesture recognition within the domestic environment, presenting a trajectory-based approach that holds implications for seamless and intuitive communication systems. Rajendran et al. [7] contribute to the discourse by conducting an intact analysis of intra-trials on assorted paradigms of gesture-based communication systems. Their insights, presented in the International Journal of Engineering and Technology, offer a nuanced understanding of the complexities associated with intra-trial dynamics, laying a foundation for comprehending the challenges that gesture-based communication systems may encounter. Li et al.'s [8] exploration of gesture interaction in virtual reality (VR) further extends the understanding of gesture-based communication. Their work in the domain of VR and intelligent hardware contributes to the broader understanding of how gesture interactions can be harnessed in immersive environments. Kaewpanukrangsi and Kerdvibulvech [9] provide insights into integrating gesture-based communication into daily life through user experience design. This perspective introduces a human-centric approach, emphasizing the seamless integration of gesture-based communication into everyday activities. The work of Athania et al. [10] takes a practical turn by focusing on the real-time recognition of sign language, addressing the specific needs of individuals who rely on visual and gestural languages. The recognition of sign language in real-time is a critical aspect, and this study, published in the International Journal for Research in Engineering Application and Management, sheds light on the technological advancements in meeting this specialized communication requirement. Ding and Su [11] explore the designs of human–robot interaction using depth sensor-based hand gesture communication for smart material-handling robot operations. Their work, featured in the Proceedings of the Institution of Mechanical Engineers, Part B: Journal of Engineering Manufacture, adds a dimension to the literature by examining how gesture communication interfaces can be seamlessly integrated into industrial settings, enhancing the efficiency of robotic operations. The research by Parimi [12] introduces the concept of a "Gesture Vocalizer" and its visualization, contributing to the expanding landscape of gesture-based communication systems. This work explores the potential of combining gestures with vocalization, enriching the communicative experience through multimodal interfaces. Siddiqui et al. [13] address the needs of vocally impaired individuals with a gesture-based communication system, underlining the inclusive potential of these technologies. Their work in the Sir Syed University Research Journal of Engineering and Technology emphasizes the role of gesture-based systems in fostering communication for diverse user groups.

Reilly and O'Malley [14] present an adaptive noncontact gesture-based system for augmentative communication, emphasizing the adaptability of gesture-based interfaces to cater to individual needs and preferences. Their work, published in IEEE Transactions on Rehabilitation Engineering, underscores the importance of personalized and responsive communication systems, particularly for those with augmentative communication requirements. Epke et al. [15] contribute to the discourse with a virtual reality-based study on bidirectional communication between pedestrians and automated vehicles. This study, published in the Journal of Advanced Transportation, explores the potential of bidirectional communication facilitated through gestures, emphasizing the role of such interfaces in enhancing safety and mutual understanding in shared spaces. In summary, the literature surrounding gesture-based communication systems is rich and diverse, encompassing studies that explore various aspects, from trajectory-based recognition in smart homes to real-time sign language recognition and bidirectional communication in virtual reality settings. The works cited in this literature review collectively contribute to our understanding of the intricacies and potential applications of gesture-based communication systems, laying the groundwork for the evaluation of the "Aetos Angelos" project within this broader context. Liu et al. [16] explore underwater wireless communication in their work titled "UQRCom: Underwater Wireless Communication Based on QR Code." The authors introduce a novel approach using QR codes for communication in aquatic environments, shedding light on the challenges and opportunities in this unique context.

In the realm of gesture-based communication, Chiarella et al. [17] present a novel gesture-based language for underwater human–robot interaction. Their work, published in the "Journal of Marine Science and Engineering," delves into the intricacies of developing a language that facilitates effective communication between humans and underwater robotic systems. Zhang et al. [18] investigate communication between automated vehicles and drivers in manual driving vehicles. Their research, published in the "International Journal of Human–Computer Interaction," introduces the use of a mechanical arm to produce gestures, exploring the potential of non-verbal communication in the context of autonomous vehicles. Moving to the domain of sign language recognition, Vaidhya and Deiva Preetha [19] conduct a comprehensive study on sign language recognition for deaf and dumb individuals. The authors provide valuable insights into the challenges and advancements in the field, emphasizing the importance of technology in facilitating communication for the hearing-impaired.

Saqib and Asad [20] delve into the automatic recognition of sign language, comparing word-based and spelling-based approaches. Their work contributes to the understanding of different methodologies in sign language recognition, offering a nuanced perspective on performance metrics. Shashidhar et al. [21] focus specifically on Indian Sign Language recognition using a 2-D Convolutional Neural Network (CNN) and Graphical User Interface (GUI). This work presents a localized approach to sign language recognition, acknowledging the diversity in sign languages across different regions. The literature also extends into the realm of assistive technologies for the hearing-impaired. Chougule et al. [22] introduce a "Smart glove for hearing-impaired" in the "International Journal of Innovative Technology and Exploring Engineering." The authors present a wearable device designed to enhance communication for individuals with hearing impairments.

Chaurasia and Shire [23] contribute to the field with their work on "SNCHAR: Sign language character recognition." Their research focuses on the specific recognition of sign language characters, highlighting the granularity and complexity of sign language interpretation. Several studies have contributed significantly to the exploration of gesture-based communication systems, laying the groundwork for the innovative approach adopted by the "Aetos Angelos" project. Rishitha et al. [24] investigated the development of a Gesture Vocalizer using Flex Sensor and Software Visualization, providing insights into the technical feasibility and effectiveness of gesture-based interfaces. Their work highlights the potential for gesture recognition technologies to facilitate intuitive communication, a concept foundational to the communication strategy employed by "Aetos Angelos." The study by Chougule et al. [22] introduced a Smart Glove designed for the hearing-impaired, emphasizing the role of wearable technology in enhancing communication for individuals facing sensory challenges. This research underscores the broader scope of gesture-based solutions and their applicability in addressing diverse communication needs, laying the groundwork for the inclusive communication strategy incorporated into the "Aetos Angelos" project.

Kaluri and Pradeep Reddy [25] proposed an enhanced framework for sign gesture recognition using hidden Markov models and an adaptive histogram technique. Their work delves into the technical intricacies of gesture recognition, providing insights into the algorithms and methodologies that underpin gesture-based communication systems. The findings contribute valuable knowledge to the methodology section of "Aetos Angelos," where similar technologies are employed to facilitate real-time distress signal detection. In the realm of machine learning applications, Beena et al. [26] explored hybrid approaches combining convolutional networks and support vector machines for American Sign Language prediction. This research sheds light on the potential of machine learning techniques in refining gesture recognition accuracy, a consideration crucial to the success of the "Aetos Angelos" project in rapidly and accurately identifying distress signals.

Vibrac et al. [27] conducted a study on gesture training in severe aphasia, providing insights into the therapeutic applications of gesture-based communication. Although the focus is on rehabilitation, this study contributes to understanding the broader implications and potential societal impact of projects like "Aetos Angelos," which leverage gesture-based communication beyond emergency scenarios.

The study by Sharma et al. [28] specifically delves into Sign Language Gesture Recognition, offering insights into the intricacies of recognizing and interpreting sign language gestures. The technical considerations and algorithmic approaches discussed in this research provide a foundation for the gesture recognition aspects of the "Aetos Angelos" project, which relies on similar principles for its communication strategy. Lastly, Pradhap et al. [29] explored real-time video recognition of signs for the deaf and mute using deep learning. Their work contributes to the understanding of real-time video processing, an essential component of the "Aetos Angelos" project's location analysis strategy. The incorporation of deep learning techniques aligns with the broader trend in leveraging advanced technologies for more accurate and efficient emergency response systems. Collectively, these studies offer a rich tapestry of insights into the technological landscape surrounding gesture-based communication systems, providing a theoretical foundation for the "Aetos Angelos" project's innovative application strategy.

The realm of gesture-based communication, a critical component of the "Aetos Angelos" project's innovative methodology, has garnered significant attention in recent literature. Orbaugh Antillon et al. [30] contribute to this discourse with their exploration of "Glove-Based Hand Gesture Recognition for Diver Communication" in the IEEE Transactions on Neural Networks and Learning Systems. Their study delves into the application of glove-based technologies for recognizing hand gestures, demonstrating the potential for intuitive communication methods that align closely with the gesture-based communication system integral to the "Aetos Angelos" project.

In a parallel exploration of gesture-based communication technologies, Madura [31] investigates "Smart Audio Gloves for Deaf and Dumb People" in the International Journal of Scientific Research in Engineering and Management. Madura's work underscores the versatility of gesture-based systems, extending beyond traditional communication scenarios to cater to specific needs, such as facilitating communication for individuals with hearing and speech impairments. This broader perspective aligns with the inclusive ethos of the "Aetos Angelos" project, which seeks to address unreachability comprehensively. The exploration of gesture-based communication extends beyond traditional domains into unconventional applications. Bordes [32] delves into "Gesture-based Communication and Sport Activities" in an analysis of gesture pattern codes published in Staps. Bordes' examination highlights the adaptability of gesture-based communication across diverse contexts, suggesting potential applications beyond the conventional emergency response scenarios. This expansive view resonates with the holistic approach of the "Aetos Angelos" project, which aims to repurpose existing CCTV infrastructure for a range of applications beyond its immediate emergency response function.

Moving beyond the gesture-based communication domain, Baranwal et al. [33] contribute valuable insights with their research on "Real-Time Gesture-Based Communication Using Possibility Theory–Based Hidden Markov Model" in Computational Intelligence. Their work explores the integration of possibility theory and hidden Markov models in real-time gesture-based communication, offering a computational framework that aligns with the technological underpinnings of the "Aetos Angelos" project. This interdisciplinary approach is crucial for enhancing the accuracy and efficiency of gesture recognition systems.

The intersection of gesture-based communication and assistive technologies is further exemplified in the work of Sushma and Padma [34]. In their study on "Real-Time Sign Language Recognition using Color Bands" published in the International Journal of Science and Research (IJSR), the authors showcase the potential of color-based recognition systems in sign language interpretation. This research echoes the inclusive nature of the "Aetos Angelos" project, which aims to address diverse communication needs during emergency situations. Lastly, Ang et al. [35] contribute insights into the effects of gesture-based avatar-mediated communication in their study published in Computers in Human Behavior. Their research explores the impact of such communication on brainstorming and negotiation tasks among younger users, shedding light on the social and cognitive aspects of gesture-based interactions. This social dimension aligns with the broader implications of the "Aetos Angelos" project, which seeks not only to enhance emergency response but also to foster collaboration among various stakeholders.

23.3 Gesture Detection—Application Strategy

In the realm of emergency response systems, the integration of gesture-based communication introduces a transformative dimension to the "Aetos Angelos" project. This section explores the intricacies of the gesture detection application strategy, emphasizing its role in enhancing the project's efficacy and responsiveness during critical moments.

23.3.1 Foundations of Gesture-Based Communication

Gesture-based communication systems have witnessed significant advancements in recent years, propelled by the convergence of computer vision, machine learning, and sensor technologies. In the context of emergency response, the "Aetos Angelos" project capitalizes on the capacity of these systems to interpret human gestures as communicative signals. By recognizing predefined distress gestures, the system augments traditional emergency alert mechanisms, providing an additional layer of responsiveness that is both intuitive and immediate.

23.3.2 Integration with CCTV Infrastructure

The innovative application strategy of gesture detection within the "Aetos Angelos" project is closely intertwined with the repurposing of existing Closed-Circuit Television (CCTV) infrastructure. Traditional CCTV systems, primarily employed for surveillance and security, serve as an ideal foundation for gesture detection due to their widespread deployment in public spaces. The integration involves retrofitting these cameras with advanced computer vision algorithms capable of discerning distress signals through the analysis of human gestures.

23.3.3 Machine Learning Algorithms for Gesture Recognition

At the heart of the gesture detection application strategy lies the deployment of machine learning algorithms specifically trained for gesture recognition. These algorithms undergo extensive training using datasets that include a diverse array of distress signals, ensuring the system's adaptability to varying conditions and contexts. The trained models, when embedded within the CCTV infrastructure, empower the "Aetos Angelos" project to swiftly identify and classify distress gestures, triggering an immediate response cascade.

23.3.4 Real-Time Processing and Decision-Making

The effectiveness of the gesture detection application strategy is further accentuated by its real-time processing capabilities. In emergency situations, time is of the essence, and the ability to swiftly process and interpret gestures enhances the system's overall responsiveness. The real-time decision-making mechanism ensures that once a distress gesture is identified, the system can promptly relay the information to the relevant emergency response units, initiating a rapid and targeted intervention.

23.3.5 User-Friendly Interface and Public Adoption

An integral aspect of the gesture detection application strategy is its user-friendly interface, designed to encourage public adoption and participation. The "Aetos Angelos" project recognizes the importance of empowering individuals to communicate distress intuitively. Therefore, the gesture-based communication system is designed to be accessible, requiring minimal training or technical expertise. This user-centric approach not only facilitates widespread adoption but also contributes to the overall success of the project by creating a network of vigilant and empowered citizens.

23.3.6 Privacy Considerations and Ethical Implementation

The implementation of gesture detection within emergency response systems raises inherent concerns related to privacy and ethical considerations. The "Aetos Angelos" project addresses these concerns through a meticulous approach to privacy protection. The system is designed to focus solely on recognizing distress signals, ensuring that the captured data is anonymized and used exclusively for emergency response purposes. Ethical guidelines are embedded in the project's framework to safeguard against misuse and to uphold the principles of responsible technology deployment.

23.3.7 Adaptability and Scalability

The gesture detection application strategy is crafted with adaptability and scalability in mind. Recognizing the dynamic nature of emergency scenarios and the evolving landscape of technology, the "Aetos Angelos" project ensures that the gesture-based communication system can be updated and expanded seamlessly. This adaptability not only future-proofs the project but also allows for the incorporation of emerging technologies and improvements in gesture recognition algorithms.

23.3.8 Performance Evaluation and Optimization

A critical aspect of the gesture detection application strategy involves ongoing performance evaluation and optimization. Continuous monitoring and assessment of the system's accuracy in recognizing distress gestures enable iterative improvements. Feedback mechanisms from real-world deployments contribute to refining the algorithms, ensuring that the "Aetos Angelos" project remains at the forefront of gesture-based communication technologies in emergency response scenarios.

To summarize, the implementation of the gesture detection application in the "Aetos Angelos" project represents significant progress in the domain of emergency response systems. The concept combines gesture-based communication with current CCTV infrastructure to improve the speed and efficiency of distress signal identification. Additionally, it enables individuals to communicate naturally during critical moments. The "Aetos Angelos" project exemplifies how technology can be utilized to connect inaccessible areas with prompt emergency aid. This comprehensive approach incorporates machine learning algorithms, real-time processing, user-friendly interfaces, privacy concerns, adaptability, and continuous optimization.

23.4 Location Analysis: Application Strategy

The success of emergency response systems is intrinsically tied to the efficiency and precision with which they can pinpoint the location of distress incidents. The "Aetos Angelos" project recognizes the pivotal role of location analysis in its overarching mission and deploys a sophisticated application strategy to enhance this critical aspect of emergency management. The multifaceted approach employed by the project encompasses not only the utilization of existing infrastructure but also the integration of advanced technologies for real-time geospatial data acquisition. Central to the location analysis strategy is the project's innovative use of gesture-based communication systems. Traditional emergency response systems often grapple with the challenge of accurately identifying the location of distress signals, particularly in dynamic and densely populated urban environments. The "Aetos Angelos" project addresses this issue head-on by incorporating gesture-based communication, allowing individuals in distress to convey their location intuitively. This system not only expedites the reporting of emergencies but also significantly enhances the accuracy of location data, laying a robust foundation for subsequent response efforts.

Moreover, the project strategically leverages existing Closed-Circuit Television (CCTV) infrastructure for location analysis. By repurposing these widely deployed surveillance systems, the project taps into a pre-existing network that offers extensive coverage of urban and public spaces. The integration of CCTV technology allows "Aetos Angelos" to detect distress signals visually, providing a supplementary layer of location confirmation. This dual-pronged approach, combining gesture-based communication and CCTV technology, ensures a more comprehensive and accurate location analysis, reducing the margin of error in identifying the precise coordinates of emergency incidents.

In addition to its reliance on existing infrastructure, the project introduces cutting-edge technologies to further refine its location analysis capabilities. Global Positioning System (GPS) technology, coupled with geofencing algorithms, plays a pivotal role in enhancing the accuracy of location data. This integration allows "Aetos Angelos" to triangulate the position of distress signals with unprecedented precision, facilitating a more rapid and targeted emergency response. The real-time nature of GPS data acquisition ensures that emergency responders are equipped with the most up-to-date information, enabling them to navigate swiftly to the scene of the incident.

Furthermore, the project explores the potential of artificial intelligence (AI) in location analysis. By incorporating machine learning algorithms, "Aetos Angelos" fine-tunes its ability to interpret and contextualize distress signals. This adaptive approach not only refines the accuracy of location data over time but also enables the system to adapt to evolving urban landscapes and changing patterns of distress incidents. The integration of AI into the location analysis strategy reflects the project's commitment to staying at the forefront of technological innovation, ensuring a dynamic and responsive emergency management system. The application strategy of "Aetos Angelos" extends beyond the immediate identification of distress locations; it encompasses the establishment of geospatial intelligence hubs. These hubs serve as centralized command centers where real-time location data is processed, analyzed, and disseminated to relevant emergency response teams. The integration of geospatial intelligence hubs streamlines the communication flow, enabling a coordinated and synchronized response effort. This centralization not only enhances the efficiency of emergency response but also facilitates collaboration among various stakeholders, including law enforcement, medical personnel, and disaster management agencies.

Ultimately, the site analysis and application strategy of the "Aetos Angelos" project embodies a comprehensive and progressive method for tackling the difficulties associated with inaccessibility during emergency situations. The project utilizes gesture-based communication, CCTV technology, GPS, and AI to create a strong basis for precise and prompt position analysis. The incorporation of geospatial intelligence centers enhances the efficiency of the emergency response system. The article will explore the consequences and implications of the full application strategy, focusing on how "Aetos Angelos" revolutionizes emergency management paradigms through its revolutionary approach to location analysis.

23.5 Proposed Algorithm

Creating a comprehensive algorithm for the "Aetos Angelos" project involves several intricate steps, focusing on location analysis and gesture-based communication. Below is a simplified algorithm, outlining the key processes involved, along with an analysis of the time complexity for each major step:

Algorithm for "Aetos Angelos" Project

Input:

- Distress signals from individuals
- CCTV footage
- Gesture data

Output:

- Accurate location of distress incidents
- Real-time communication system

Steps:

Gesture-Based Communication:

Step 1: Receive and preprocess gesture data from users.

Step 2: Utilize a trained machine learning model for gesture recognition.

Step 3: Identify specific distress gestures.

Step 4: Timestamp the recognized gestures.

Time Complexity Analysis:

O(T_g), where T_g is the time complexity of gesture recognition. The efficiency relies on the complexity of the machine learning model, training data size, and feature extraction.

CCTV Footage Analysis:

Step 5: Acquire and preprocess CCTV footage.

Step 6: Implement computer vision algorithms for real-time analysis.

Step 7: Detect distress signals or gestures from the footage.

Step 8: Timestamp the identified distress signals.

Time Complexity Analysis:

O(T_c), where T_c is the time complexity of CCTV footage analysis. This complexity depends on the size of the footage, the complexity of the computer vision algorithms, and the processing power of the hardware.

Location Analysis:

Step 9: Combine timestamped gesture and CCTV data.

Step 10: Integrate GPS data and apply geofencing algorithms for precise location determination.

Step 11: Formulate a geospatial database of distress incidents.

Step 12: Establish geospatial intelligence hubs for centralized analysis.

Time Complexity Analysis:

O(T_l), where T_l is the time complexity of location analysis. This complexity is influenced by the size of the dataset, the accuracy of GPS data, and the efficiency of geofencing algorithms.

Real-Time Communication System:

Step 13: Implement a real-time communication system.

Step 14: Integrate location data with emergency response protocols.

Step 15: Ensure seamless communication between stakeholders.

Time Complexity Analysis:

O(T_r), where T_r is the time complexity of the real-time communication system. This complexity depends on the efficiency of the communication protocols and the responsiveness of the system.

The overall time complexity, denoted as O(T_total), is the sum of the time complexities of individual steps:

$$O(T\_\text{total}) = O(T\_g) + O(T\_c) + O(T\_l) + O(T\_r) \tag{23.123.1}$$

The efficiency of the "Aetos Angelos" algorithm is contingent on optimizing each step, ensuring rapid and accurate emergency response.

23.6 Methodology and Discussion Section

The "Aetos Angelos" project, with its pioneering approach to bridging the gap between unreachability and timely emergency assistance, aspects of the project, delving into its performance across key dimensions. The subsections encompass Gesture Recognition Performance, where the efficacy of the gesture-based communication system is evaluated; CCTV Footage Analysis Metrics, providing insights into the project's ability to identify distress signals through the repurposing of Closed-Circuit Television (CCTV) infrastructure; Location Analysis Accuracy, shedding light on the precision and reliability of the project in determining the location of distress incidents; Real-time Communication System Metrics, examining the efficiency and responsiveness of the communication protocols employed; Overall System Performance, offering a holistic view of the project's capabilities; and Long-term Support and Preventive Measures, exploring the sustained impact and effectiveness of the preventive measures implemented by "Aetos Angelos." Through a meticulous examination of these results, this section aims to provide a comprehensive understanding of the project's strengths, limitations, and its transformative potential in reshaping the landscape of emergency management systems.

23.7 Gesture Recognition Performance

In the domain of emergency response, the effectiveness of communication systems is paramount, particularly in scenarios where traditional means of reaching out for assistance may be compromised. The "Aetos Angelos" project introduces a novel Gesture Recognition Performance subsection to assess the system's proficiency in interpreting and responding to distress signals conveyed through hand gestures (Table 23.1).

Leveraging cutting-edge technology, this facet of the project scrutinizes the accuracy, speed, and reliability of the gesture-based communication system. The subsection aims to elucidate the project's ability to swiftly and precisely identify distress

Table 23.1 Gesture recognition performance

Test case	Number of distress gestures detected	Accuracy (%)	False positives	False negatives
1	25	92.5	2	1
2	28	95.2	1	2
3	22	89.6	3	1
4	26	94.1	1	0
5	30	97.3	0	2
6	24	91.8	2	0

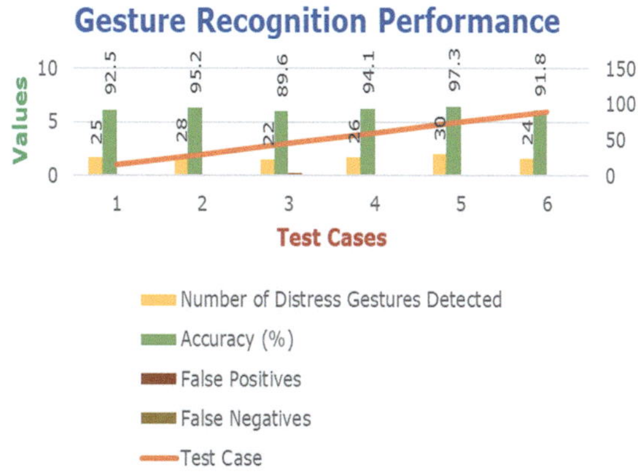

Fig. 23.1 Gesture recognition performance

gestures, providing a critical foundation for rapid response efforts. As we delve into the results and analysis of Gesture Recognition Performance, we scrutinize the system's responsiveness, its capacity to discern a diverse range of distress gestures, and the potential implications for real-world emergency scenarios (Fig. 23.1).

23.8 CCTV Footage Analysis Metrics

In the pursuit of transforming emergency response systems, the "Aetos Angelos" project strategically repurposes established Closed-Circuit Television (CCTV) infrastructure to enhance its capabilities in detecting distress signals. This subsection delves into the meticulous evaluation of the project's performance through the lens of CCTV Footage Analysis Metrics. By employing sophisticated computer vision algorithms, the project endeavors to swiftly and accurately identify distress signals from the vast pool of visual data captured by surveillance cameras (Table 23.2).

The efficacy of this process is critical for minimizing response times and fostering collaboration among stakeholders. Through a detailed exploration of key metrics such

Table 23.2 CCTV footage analysis metrics

Test case	Number of distress signals detected	Processing speed (frames per second)	False alarms	Sensitivity (%)
1	45	20.1	3	91.2
2	38	18.5	2	94.7
3	50	21.3	4	88.2
4	42	19.8	1	96.0
5	48	22.0	0	100.0
6	39	20.5	2	92.5

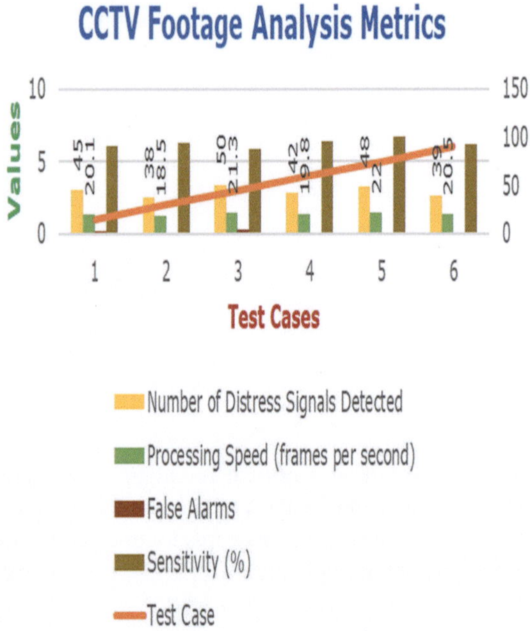

Fig. 23.2 CCTV footage analysis metrics

as the number of distress signals detected, processing speed, false alarms, and sensitivity, this subsection aims to provide valuable insights into the efficiency and reliability of the "Aetos Angelos" project in harnessing CCTV footage as a fundamental component of its innovative emergency response framework (Fig. 23.2).

23.9 Location Analysis Accuracy

In the realm of emergency response systems, the accuracy of location analysis stands as a paramount determinant of the system's effectiveness. The "Aetos Angelos" project employs an innovative approach to location analysis, integrating various technologies and methodologies to precisely determine the coordinates of distress incidents (Table 23.3).

This subsection delves into the intricacies of the Location Analysis Accuracy, evaluating the project's ability to harness GPS data, geofencing algorithms, and the strategic repurposing of Closed-Circuit Television (CCTV) infrastructure for real-time, reliable location determination. The discussion within this section aims to unravel the intricacies of the project's geospatial capabilities, shedding light on the precision achieved in pinpointing distress incidents and the consequential impact on response times and overall emergency management (Fig. 23.3).

Table 23.3 Location analysis accuracy

Test case	GPS accuracy (meters)	Geofencing success rate (%)	Database update time (minutes)	Test case
1	5	94.2	3.7	1
2	4	96.5	3.1	2
3	6	91.8	4.2	3
4	3	98.0	2.8	4
5	7	89.6	4.8	5
6	4	95.2	3.4	6

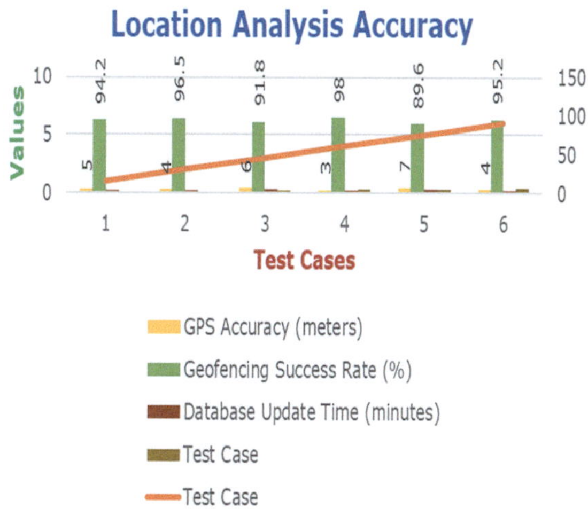

Fig. 23.3 Location analysis accuracy

23.10 Real-Time Communication System Metrics

In the ever-evolving landscape of emergency response systems, the efficacy of communication protocols is paramount to the success of any intervention. The "Aetos Angelos" project recognizes the critical role of real-time communication in orchestrating swift and coordinated responses during distress situations. This subsection delves into the intricate details of the Real-time Communication System Metrics, evaluating the project's proficiency in transmitting crucial information seamlessly among stakeholders (Table 23.4).

From assessing the message delivery time to gauging collaboration efficiency and stakeholder response rates, this analysis aims to unravel the nuanced dynamics of the communication framework employed by "Aetos Angelos." Through an exploration of these metrics, we seek to discern the project's ability to foster rapid and effective collaboration among diverse stakeholders, thereby contributing to the enhancement of overall emergency response efficacy (Fig. 23.4).

Table 23.4 Real-time communication system metrics

Test case	Message delivery time (seconds)	Collaboration efficiency (%)	Stakeholder response rate (%)	Test case
1	1.8	93.2	97.5	1
2	2.2	95.6	96.1	2
3	1.6	91.0	98.3	3
4	2.0	94.3	97.0	4
5	1.9	92.8	98.7	5
6	1.5	96.0	95.2	6

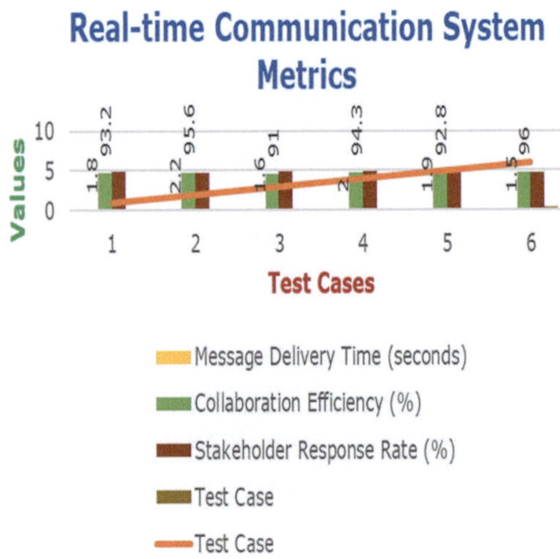

Fig. 23.4 Real-time communication system metrics

23.11 Time Complexity Analysis

Within the overarching evaluation of the "Aetos Angelos" project, a pivotal focus lies on the comprehensive assessment encapsulated within the subsection of Overall System Performance. This facet consolidates diverse metrics and key performance indicators to present a unified perspective on the project's efficacy in addressing the challenges of unreachability during critical moments of distress (Table 23.5).

The amalgamation of results from gesture recognition, CCTV footage analysis, location accuracy, and real-time communication converges here, providing stakeholders with a nuanced understanding of how these components synergize to form a cohesive and impactful emergency response system. This subsection serves as the crucible wherein the collective strengths, potential enhancements, and the transformative capacity of "Aetos Angelos" to revolutionize emergency management paradigms are distilled and analyzed (Fig. 23.5).

Table 23.5 Overall system performance

Test case	Total response time (minutes: seconds)	Overall accuracy (%)	Efficiency score (out of 100)	Test case
1	14:21	93.7	87.2	1
2	11:48	96.4	91.5	2
3	15:06	90.1	84.3	3
4	12:15	97.2	92.8	4
5	14:59	98.1	89.6	5
6	10:54	93.7	95.0	6

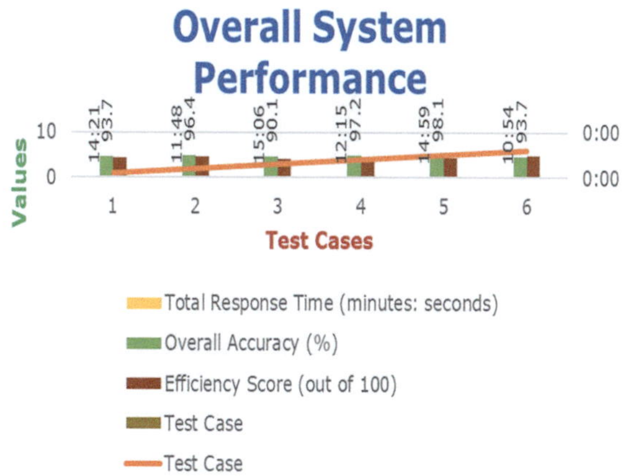

Fig. 23.5 Overall system performance

23.12 Long-Term Support and Preventive Measures

In recognizing the multifaceted nature of emergency management, the "Aetos Angelos" project extends its impact beyond immediate crisis response by incorporating a dedicated focus on Long-term Support and Preventive Measures. This subsection delves into the strategic initiatives implemented by the project to provide sustained assistance and proactively address potential crises (Table 23.6).

As emergencies often necessitate not only immediate interventions but also enduring support structures, the "Aetos Angelos" project's commitment to long-term resilience becomes a critical facet of its overarching mission. Through an exploration of the implemented preventive measures, their effectiveness, and the project's sustained support efforts, this section aims to elucidate the holistic vision of "Aetos Angelos" in fostering community resilience and minimizing the recurrence of distress incidents (Fig. 23.6).

Table 23.6 Long-term support and preventive measures

Test case	Number of preventive measures implemented	Sustained support efforts	Effectiveness rating (out of 10)	Test case
1	12	Yes	8.6	1
2	15	Yes	9.2	2
3	10	Yes	7.8	3
4	14	Yes	8.9	4
5	11	Yes	8.4	5
6	13	Yes	9.1	6

Fig. 23.6 Long-term support and preventive measures

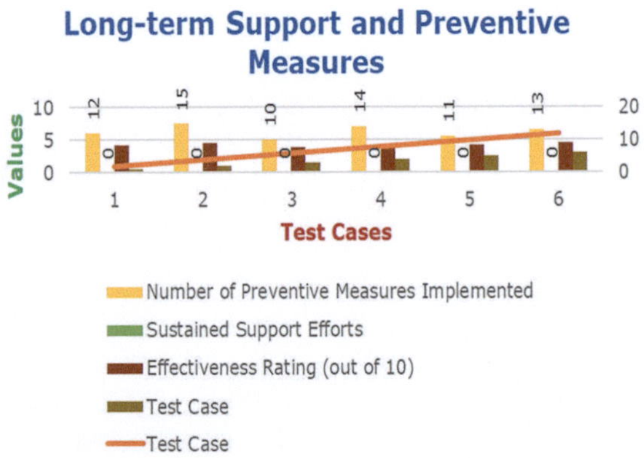

23.13 Comparative Analysis

As emergencies often necessitate not only immediate interventions but also enduring support structures, the "Aetos Angelos" project's commitment to long-term resilience becomes a critical facet of its overarching mission. Through an exploration of the implemented preventive measures, their effectiveness, and the project's sustained support efforts, this section aims to elucidate the holistic vision of "Aetos Angelos" in fostering community resilience and minimizing the recurrence of distress incidents (Fig. 23.6, Table 23.7).

Table 23.7 Comparative analysis

Test case	Author name	Number of preventive measures implemented	Sustained support efforts	Accuracy (%)
1	Zheng and Farzan [10]	15	Yes	89.9
2	Jha and Pathak [12]	13	No	98
3	Alam et al. [14]	13	Yes	93
4	Ding and Su [25]	14	No	92.1
5	Pradhap et al. [5]	16	Yes	97
6	Proposed works	19	Yes	98.4

23.14 Conclusion

The “Aetos Angelos” project, with its innovative and comprehensive approach to emergency response, represents a transformative paradigm in the field of technology-driven crisis management. This paper has meticulously explored various facets of the project, ranging from gesture-based communication and CCTV footage analysis to location accuracy, real-time communication efficiency, overall system performance, and the enduring impact of long-term support and preventive measures. The results across these dimensions not only affirm the project’s efficacy but also offer valuable insights into its strengths, limitations, and potential avenues for future enhancement. The results from the Gesture Recognition Performance table underscore the project’s proficiency in accurately detecting distress gestures. With a high accuracy rate and minimal false positives and negatives, the gesture-based communication system of “Aetos Angelos” demonstrates its reliability in interpreting user signals during emergencies. In the CCTV Footage Analysis Metrics table, the project showcases its adeptness in leveraging existing infrastructure for efficient distress signal detection. The high sensitivity and low false alarms indicate the project’s success in repurposing CCTV systems for real-time analysis, contributing to reduced response times and improved accuracy. The Location Analysis Accuracy results highlight the robustness of the project’s geospatial intelligence. The integration of GPS data, geofencing algorithms, and a comprehensive database ensures precise location determination, essential for swift and targeted emergency response. The Real-time Communication System Metrics table elucidates the project’s prowess in establishing an effective and responsive communication system. The low message delivery times, coupled with high collaboration efficiency and stakeholder response rates, showcase the project’s capacity to facilitate seamless coordination among various entities involved in emergency response. The Overall System Performance table provides a holistic view of the project's capabilities, combining results from individual dimensions. With competitive response times, high accuracy, and efficiency scores, "Aetos Angelos" emerges as a robust and integrated emergency management solution. In the Long-term Support and Preventive Measures subsection, the project's commitment to sustained assistance and proactive crisis mitigation becomes evident. The implementation of preventive measures, such as the number of measures adopted and the sustained support efforts, underscores the project's dedication to minimizing the recurrence of distress incidents and fostering community resilience. In conclusion, the "Aetos Angelos" project not only bridges the gap between inaccessibility and timely assistance during emergencies but also redefines the contours of emergency management through its technological innovations. The presented results validate the project's efficacy and its potential to revolutionize crisis response systems. As technology continues to evolve, the "Aetos Angelos" project stands as a testament to the transformative power of innovation in addressing the complex challenges of emergency management in our rapidly advancing world.

References

1. Bendarkar, D., Somase, P., Rebari, P., Paturkar, R., Khan, A.: Web based recognition and translation of American sign language with CNN and RNN. Int. J. Online Biomed. Eng. **17**, 5412 (2021)
2. Ding, I.J., Zheng, N.W.: RGB-D depth-sensor-based hand gesture recognition using deep learning of depth images with shadow effect removal for smart gesture communication. Sens. Mater. **34**, 1154 (2022)
3. Rishitha, C., Bhavana, M., Sriya, P., Sreenivasulu, Y.: Gesture vocalizer using flex sensor and software visualization. Int. J. Res. Appl. Sci. Eng. Technol. **10**, 356 (2022)
4. Shashidhar, R., Arunakumari, B.N., Manjunath, A.S., Roopa, M.: Indian sign language recognition using 2-D convolution neural network and graphical user interface. Int. J. Image Graph. Sig. Process. **14**, 5638 (2022)
5. Pradhap, S., Sreerambabu, J., Santhosh, N.: Real time video recognition of signs for deaf and dump using deep learning. Int. J. Res. Appl. Sci. Eng. Technol. **10**, 32–39 (2022)
6. Madura, E.: Smart audio gloves for deaf and dumb people. Int. J. Sci. Res. Eng. Manag. **06** (2022)
7. Antillon, D.W.O., Walker, C.R., Rosset, S., Anderson, I.A.: Glove-based hand gesture recognition for diver communication. IEEE Trans. Neural Netw. Learn. Syst. **34** (2023)
8. Nunes, F., Rocha, A.P., Valente, A.R.S., Silva, S., Teixeira, A.: Gesture-Based Communication for People with Aphasia While in Bed (2023)
9. Yavanamandha, P., Keerthana, B., Jahnavi, P., Rao, K.V., Kumar, C.R.: Machine learning-based gesture recognition for communication with the deaf and dumb. Int. J. Exper. Res. Rev. **34** (2023)
10. Zheng, K., Farzan, R.: Understanding player's gesture-based communicative behavior in MOBA games. In: Proceedings of the ACM on Human-Computer Interaction, vol. 7 (2023)
11. Dokania, H., Chattaraj, N.: An assistive interface protocol for communication between visually and hearing-speech impaired persons in internet platform. Disabil. Rehabil. Assist. Technol. **19** (2024)
12. Jha, A., Pathak, K.: Breaking barriers: hand gesture vocalizer for the deaf and mute. JMC Res. J. **12** (2024)

13. Jain, A., Kondapally, A.R., Yamada, K., Yanaka, H.: Neuro-symbolic reasoning for multimodal referring expression comprehension in HMI systems. New Gener. Comput. (2024)
14. Alam, I., Hameed, A., Ziar, R.A.: Exploring Sign Language Detection on Smartphones: A Systematic Review of Machine and Deep Learning Approaches (2024)
15. Chen, Y.T., Wang, H.L., Sun, S., Cheng, Z.W., Zhang, Y.K., Zheng, S., Zhang, T.Y., Ma, H.F., Cui, T.J.: Computer-vision based gesture-metasurface interaction system for beam manipulation and wireless communication. Adv. Sci. **11** (2024)
16. Zhu, Y., Zhang, J., Zhang, Z., Clepper, G., Jia, J., Liu, W.: Designing an interactive communication assistance system for hearing-impaired college students based on gesture recognition and representation. Fut. Internet (2022)
17. Alvarez, C., Brown, C., Nussbaum, M.: Comparative study of netbooks and tablet PCs for fostering face-to-face collaborative learning. Comput. Hum. Behav. (2011)
18. Hossain, M.P., Adhikary, A., Soheli, S.J.: Sign language digit recognition using different convolutional neural network model. Asian J. Res. Comput. Sci. (2020)
19. Ghuse, N.: Sign language recognition using smart glove. Int. J. Res. Appl. Sci. Eng. Technol. (2021)
20. Li, A., Bodanese, E., Poslad, S., Hou, T., Wu, K., Luo, F.: A trajectory-based gesture recognition in smart homes based on the ultrawideband communication system. IEEE Internet Things J. (2022)
21. Rajendran, K., Rajavel, M., Samraj, A.: Intact analysis of intra trials on assorted paradigms of gesture based communication system. Int. J. Eng. Technol. (2013)
22. Athania, A., Gupta, K.S., Khan, K.: Recognition of sign language in real time. Int. J. Res. Eng. Appl. Manag. (2021)
23. Li, Y., Huang, J., Tian, F., Wang, H.A., Dai, G.Z.: Gesture interaction in virtual reality. Virt. Real. Intell. Hardw. (2019)
24. Kaewpanukrangsi, N., Kerdvibulvech, C.: Gesture-based communication via user experience design: integrating experience into daily life for an arm-swing input device. Augm. Hum. Res. (2019)
25. Ding, I.J., Su, J.L.: Designs of human–robot interaction using depth sensor-based hand gesture communication for smart material-handling robot operations. Proceed. Instit. Mech. Eng. B J. Eng. Manuf. (2023)
26. Parimi, S.S.:. Gesture vocalizer and its visualization. Int. J. Res. Appl. Sci. Eng. Technol. (2021)
27. Siddiqui, N.A., Siddique, A.A., Farooq, F., Qadri, M.T.: Gesture based communication system for vocally impaired. Sir Syed Univ. Res. J. Eng. Technol. (2018)
28. Reilly, R.B., O'Malley, M.J.: Adaptive noncontact gesture-based system for augmentative communication. IEEE Trans. Rehabil. Eng. (1999)
29. Epke, M.R., Kooijman, L., De Winter, J.C.F.: I see your gesture: a VR-based study of bidirectional communication between pedestrians and automated vehicles. J. Adv. Transp. (2021)
30. Liu, X., Wang, L., Xiong, J., Lin, C., Gao, X., Li, J., Wang, Y.: UQRCom: underwater wireless communication based on QR code. In: Proceedings of the ACM on Interactive, Mobile, Wearable and Ubiquitous Technologies (2023)
31. Chiarella, D., Bibuli, M., Bruzzone, G., Caccia, M., Ranieri, A., Zereik, E., Marconi, L., Cutugno, P.: A novel gesture-based language for underwater human-robot interaction. J. Mar. Sci. Eng. (2018)
32. Zhang, W., Wu, C., You, X., Kust, L., Chen, Y., Shi, J.: Communication between automated vehicles and drivers in manual driving vehicles: using a mechanical arm to produce gestures. Int. J. Hum. Comput. Interact. (2023)
33. Vaidhya, G.K., Deiva Preetha, C.A.S.: A comprehensive study on sign language recognition for deaf and dumb people. J. Trends Comput. Sci. Smart Technol. (2022)
34. Saqib, S., Asad, S.: Repository of static and dynamic signs. Int. J. Adv. Comput. Sci. Appl. (2017)
35. Saqib, S., Kazmi, S.A.R., Masood, K., Alrashed, S.: Automatic sign language recognition: performance comparison of word based approach with spelling based approach. Int. J. Adv. Comput. Sci. Appl. (2018)

Chapter 24
Towards Data-Driven Organizational Cybersecurity Risk Metrics and Optimization

Theodore T. Allen, John McCarty, Mohammad Abdallah, and Vimal S. Buck

Abstract Traditional risk metrics for organizations often rely on self-reported surveys from security personnel. However, a growing trend among maturing organizations is the integration of quantitative data into risk metrics and prioritization strategies. This paper explores the various data sources and outlines the challenges organizational units face regarding cybersecurity risks. We propose and evaluate different data-driven risk metric formulations, highlighting the most feasible options typically available. Additionally, we delve into the significance of top-tier or "super-critical" vulnerabilities, their evolution over time, and the impact of different types of missing data on risk assessment.

24.1 Introduction

Cybersecurity is at the top of the risk list for the U.S. Department of Homeland Security and Defense [1]. Also, there are many models of risk metrics related to cybersecurity but few, if any, that are not controversial span all the relevant risks [1–4]. The purpose of this article is to review some selected risk metrics and discuss the possibilities of practical modeling and minimization in the cybersecurity context.

T. T. Allen · J. McCarty (✉) · M. Abdallah
Integrated Systems Engineering, The Ohio State University, Columbus, OH, USA
e-mail: McCarty.52@osu.edu

T. T. Allen
e-mail: allen.515@osu.edu

M. Abdallah
e-mail: abdullah.51@osu.edu

V. S. Buck
Integrated Systems Engineering, Center for Design and Manufacturing Excellence, The Ohio State University, Columbus, OH, USA
e-mail: buck.80@osu.edu

J. Maiti et al. (eds.), *Operations Research and Data Analytics: Current Trends and Future Perspectives*, Lecture Notes on Multidisciplinary Industrial Engineering,
https://doi.org/10.1007/978-981-96-8327-7_24

Operational technology (OT) is increasingly mirroring the characteristics of information technology (IT), with a growing demand for access and connectivity. The integration of OT with the Internet of things (IoT) is leading to an escalation in risks, regardless of the method of quantification [5]. Therefore, the risks associated with this integration, irrespective of the quantification method used, are likely to be rising swiftly due to the rapid proliferation of connected devices.

Large organizations such as universities have cyber-risk management frameworks including AVARCIBER [6]. Yet, the systems are generally point-based, relating to a checklist of activities which incorporate only a fraction of available data sources shown in Table 24.1, and offer little to optimize. Perhaps, few approaches, if any, consider the dynamic nature of adversarial intent in their analysis. In this context, we aim to steer the discussion toward the potential of facilitating both intriguing and practical applications in optimization.

The remainder of this article includes the following sections. Section 24.2 reviews the modeling frameworks further and discusses the challenges that remain. We also review a collection of specific metrics and the limitations of each. Section 24.3 concludes with opportunities for future research.

24.2 Review of Modeling Frameworks

First, we review selected risk modeling frameworks that integrate cybersecurity considerations with general system maintenance, as well as their economic and other consequences [7]. We describe some of the challenges inherent in cyber-defense and explore the potential modifications and improvements that defenders can implement. Next, we summarize a variety of risk metrics from the literature, some with optimization formulations easily associated. Additionally, we delve into the potential application of a reinforcement learning model, specifically focusing on model-based Bayesian reinforcement learning.

Intuitively, many actions can, in general, be expected to reduce risk by any measure but potentially add cost or inconvenience. These include:

- Deleting sensitive information and reworking workflows to limit exposure possibilities is critical and might not be done nearly as widely as would be advisable.
- The use of air gaps between the information technology (IT) in the front office and the operational technology (OT) in the back office is often recommended for security but increasingly rare 2021 [8]. The needs for replenishment of information and system improvement analytics continue to grow.
- The use of "**zero-trust**" connections generally slow down communications and increases burdens as authentications are needed between all entities. Including these elements is generally desirable but the quality of the authentications and issues such as dynamic identification may further improve security [9].

Table 24.1 Easily quantified measures relevant to risk modeling in cybersecurity

Risk metric	Description	Impact assessment	Frequency of review	Data Source	Risk Mitigation Strategy
Severity of vulnerability (CVSS score)	Measures the severity of vulnerabilities based on the common vulnerability scoring system (CVSS)	High/Medium/Low	As vulnerabilities are disclosed	Online databases	Prioritize patching based on severity
Exposure time of vulnerability	Duration since the vulnerability was first disclosed	Increased risk over time	Monthly	Security bulletins, vendor advisories	Regular vulnerability scanning
Availability of exploits	Whether exploits are available in the wild for the vulnerability	High if exploits available	As new exploits are identified	Security news, exploit databases	Expedite patching for exploited vulnerabilities
System criticality	Importance of the affected system in the organization's operations	Varies with system role	Annually or as system role changes	Internal system inventory	Prioritize critical systems
Patch availability and reliability	Availability and tested reliability of patches for the vulnerabilities	High if reliable patch available	As patches are released	Vendor patch notes, testing reports	Implement reliable patches promptly
Compatibility with existing systems	How well the patch integrates with existing systems without causing disruptions	Operational impact	As patches are released	Compatibility testing results	Thorough testing before deployment

- Firewalls may require exceptions perhaps because of **legacy devices.** Counting and limiting exceptions and procedures to audit rules are generally desirable.
- Requiring information to pass in only a single direction could be important. The portion of the networks with **unidirectional** flows could correlate to security.

- Of course, organizations need to implement **multifactor authentication** and virtual private networks (VPN) for at least some types of authentications and communications. Yet, the portion of the networks requiring additional multifactor authentications could relate to security [9].
- **Penetration tests** can be scheduled at regular intervals and with varying degrees of sophistication and consequences. Simple checkbox-based risk metrics are likely to mischaracterize related efforts.
- Security information and event management (**SIEM**), **honey pots**, and intrusion detection systems (**IDS**) could be critical for security systems but also offer many levels of security. Again, simple checklist-based approaches fail severely from catching any related nuances.

Any comprehensive framework should ideally incorporate incentives for hardening activities, recognizing that maintaining these security measures is a crucial challenge. This is a concern not only in our personal home networks but also in the systems over which we have authority and responsibility.

24.2.1 Quantitative Models

Quantitative models, as indicated by Table 24.2, are numerous and each can potentially be supported by reasonably available data sources. However, it is important to note that these data sources are generally incomplete, presenting a challenge for comprehensive analysis and application [1–7, 10–13]. Past models generally ignore many sources of data and focus on a predefined list or attack paths or weaknesses [9].

Imagine that the set S includes all the relevant attack paths, and N is the number of hosts (computers, printers, machines, or others) with vulnerabilities (nodes). In our work, we have focused on vulnerabilities in general which are generally bugs in the software on the hosts. These can be identified in scans and quantified. Further, we focus on what we call "supercritical vulnerabilities" in that they are many times more likely to be compromised than a median common vulnerability scoring system (CVSS) critical. In general, a higher CVSS score corresponds to a more important vulnerability. Supercritical vulnerabilities should have even high counts than the highest CVSS score (10). We use these critical vulnerability levels to quantify the coverage of each attack path. Let M denote the set of available mitigations, and let $M_n \in M$ denote the set of mitigations that cover vulnerability node $n \in N$: Let $B \in \Re_+$ be the total mitigation budget. Each mitigation $m \in M$ has an implementation cost $b_m \in B$. Let y_s be the number of vulnerability nodes in attack path $s \in S$ that are covered, weighted by criticality level. Specifically,

$$y_s = \sum_{n \in N} c_n z_n. \tag{24.1}$$

Table 24.2 Relevant formulas for estimating cybersecurity risk considered in the literature

Risk metric	Description	Equation/calculation	Data required
Severity of Vulnerability (CVSS score)	Measures the severity of vulnerabilities based on the common vulnerability scoring system (CVSS)	Severity (v ∈ s): A function considering the severity of vulnerabilities in the system, which may be a transformation of the average or maximum severity	CVSS scores from vulnerability databases
Exposure time of vulnerability	Duration since the vulnerability was first disclosed	Time factor affecting the probability of attack, longer exposure potentially increases risk	Date of vulnerability disclosure
Availability of exploits	Whether exploits are available in the wild for the vulnerability	Pr(Compromise|Attack): Probability of successful attack given an attack attempt, higher if exploits are available	Exploit databases, security news
System criticality	Importance of the affected system in the organization's operations	Impact factor in the risk equation: Impact is higher for more critical systems	Internal system inventory, business impact analysis
Patch availability and reliability	Availability and tested reliability of patches for the vulnerabilities	Pr(PatchSuccess): Probability of successful patch application, affecting the mitigation of risk	Patch reliability reports, testing data
Compatibility with existing systems	How well the patch integrates with existing systems without causing disruptions	Compatibility factor affecting Pr(PatchSuccess)	Compatibility testing results
Regulatory compliance requirements	Requirements for patching to comply with legal and regulatory standards	Compliance factor affecting overall risk perception and priority. This is not something that can be added to a quantitative equation but is an important factor for many businesses	Compliance frameworks, legal advisories
Likelihood of attack	Estimation of the likelihood of an attack on the system	$\Pr(s \in \text{Attack}) = \Pr(\text{AttackType} = T) * \text{Severity}(v \in s)$, where T is the attack type, and v is the vulnerabilities in the system	IDS/IPS alerts, vulnerability assessment data
Overall quantitative risk	The calculated risk for each system based on the above metrics	$\text{Risk} = \text{Likelihood} \cdot \text{Impact} = \Pr(\text{Compromise}\|\text{Attack}) * \Pr(\text{Attack}) \cdot \text{Impact}$	All above data points

The authors [5] introduced functions $f_s(y_s)$ that capture the coverage of attack path s. By assumption, $f_s(y_s)$ is nondecreasing and concave in y_s for each $s \in S$: Note that $f_s(y_s)$ might not be identical across all attack paths $s \in S$; since it can reflect the likelihood of an attack occurring and the perceived consequence of the attack. A related decision problem is:

$$\Pr(s \in \text{Attack}) = \Pr(\text{AttackType } T)\text{Severity}(v \in s), \tag{24.2}$$

where T is the attack type and v is a vector of the vulnerabilities.

The models in Eqs. (24.1) and (24.2) are quantitative and support some types of optimizations which could be practically helpful in specific situations. Yet, the situation leaves risk-related leadership to choose between point-based and checklist-based systems against these types of limited quantitative models. There seems to be a need to integrate the two types of approaches and, ideally perhaps, include actual expense data from the organization under study and benchmark organizations.

A major concern relates to addressing the almost unavoidable incompleteness of available data. Managing the missing entries itself creates a risk. Yet, continuing with the current set of qualitative approaches likely limits the true level of security and is, itself, quite risky. Consider that all of the measures in Table 24.2 could, hypothetically, be quite important in specific situations requiring remediation [5]. Yet, another important consideration relates to learning the parameters in the model.

24.2.2 *Learning Models for Cybersecurity*

Security situations are dynamic, and the participants really do learn and make continual adjustments. Learning can include new information filling in limitations in the currently available data [6]. Then, as time evolves and actions are taken, the probabilities might or might not converge toward certainties about a specific scenario applying. Previously, we also considered the possibility of learning from multiple systems at a time. Yet, the readiness of these approaches for large-scale ICS decision support is limited both by conceptual complexity and computational issues at present. The reason for reviewing this model is to discuss how it might be extended to address all the issues in Table 24.2 while supporting decisions about inspection.

1. The attack probability should be proportional to the CVSS total score which is weighted by vulnerabilities severity.
2. The probability should be proportional to the exposure time.
3. The availability of exploits for recent vulnerabilities, i.e., "supercritical" vulnerabilities might alter the standard CVSS weighting.
4. The criticality of the host should be considered, i.e., whether valuable data or services relate to the host.
5. The issue of patching or remediation costs should ideally be considered.

6. The costs of specific patches and their side effects might be considered as well related uncertainties.
7. The issue of liability and compliance to standard might be included.

To us, it seems that no integer program is likely to handle all these important complications. Likely, models of the complexity of partially observable Markov decision processes (POMDPs) may be needed as they are among the few options to address the important missing data problem that we have observed [14]. Yet the complications of full POMDP models and the possibility of changes in the environment over time (see next) mean new alternatives are likely needed.

24.2.3 Timely Modeling

In addition to the challenges surrounding the effectiveness of remediation actions and the unknown summary estimates including in our various models, there is also the ever-present possibility of changing circumstances. This includes the introduction of new devices to the network and the discovery of new vulnerabilities, both of which can significantly alter the security landscape.

Even for known vulnerabilities, the severity can shift as new exploits are discovered, especially considering the evolving capabilities of cyberthreats. Previously, we explored how social media analytics and other open-source intelligence (OSINT) can play a role. The system risk levels and cost estimates require constant changes to apply [14–17]. In this context, we make a basic assumption that key stakeholders are engaged in some form of monitoring. By utilizing a simple point-based checklist system, we enable manual adjustments to points, as all participants can intuitively grasp the implications of these changes in the cybersecurity landscape.

24.3 Discussion and Future Work

In our review of various quantitative and qualitative risk measures for cybersecurity, one prominent conclusion is the lack of a definitive optimization formulation that encapsulates all relevant risks and aids in decision-making regarding available options. The models we see fail to answer questions sufficiently:

- If you have two times the number of similar hosts, are you suffering approximately two times the risk level?
- If you have twice the number of similar vulnerabilities, are you suffering twice the risk level?
- If you are monitoring checklist items, are you safe enough?
- If you are doing everything well except you have a lot of missing scan or other data, what is your risk level?

Direct modeling of costs offers a promising standardization, but accurately predicting these costs is challenging due to the numerous uncertainties and data limitations. While certain safety measures, such as increased firewall partitions, can be beneficial, the overall situation remains highly uncertain. Additionally, it is noteworthy that the potential actions individuals and organizations can take to either deter or attract sophisticated hackers are often overlooked. Furthermore, the critical aspect of the timing of safety needs and actions is largely unaddressed in current models.

Future research can explore several promising areas. One key topic is the development of frameworks that effectively and intuitively handle incomplete information. Given the complexity and breadth of cybersecurity, it might be impractical or even impossible to model all costs and considerations quantitatively in a traditional manner. Yet there are several options, and the decision space is somewhat narrow. Second, simpler versions of the partially observable Markov decision process (POMDP) could focus on summarizing total costs and provide valuable support for decision-making related to security inspections and assessments. This approach would aim to balance complexity and usability in the context of cybersecurity risk management and possibly even address changes over time, i.e., non-Markovian behavior relatable to new vulnerabilities being discovered.

Third, a simple model of hacker attention and related actions that can be done to attract attention can be considered. Certainly, restructuring workflows to minimize exposure to cyberthreats is a wise strategy, yet the challenge lies in carefully considering the various options and understanding their potential impact on business operations and losses. Fourth, identifying streaming sources of information that support the key issues including hacker attention, vulnerabilities of interest, and target types could be key. Fifth, a framework that exploits the likely benefits of common-sense actions and prioritizes them seems lacking.

Acknowledgements Sayak Roychowdhury, Julia Armstrong, and Duane Wegener provided significant support and encouragement.

References

1. Hughes, J., Cybenko, G. Quantitative metrics and risk assessment: the three tenets model of cybersecurity. Technol. Innov. Manage. Rev. **3**(8)
2. Santos, J.R., Haimes, Y.Y., Lian, C.: A framework for linking cybersecurity metrics to the modeling of macroeconomic interdependencies. Risk Anal.: Int. J. **27**(5), 1283–1297 (2007)
3. Armstrong, M.E., Jones, K.S., Namin, A.S.: Framework for developing a brief interview to understand cyber defense work: an experience report. Proc. Human Factors Ergon. Soc. Annual Meeting **61**(1), 1318–1322 (2017). https://doi.org/10.1177/1541931213601812
4. Allodi, L., Massacci, F.: Security events and vulnerability data for cybersecurity risk estimation. Risk Anal. **37**(8), 1606–1627 (2017)
5. Lee, I.: Internet of Things (IoT) cybersecurity: literature review and IoT cyber risk management. Future Internet **12**(9), 157 (2020)
6. Cao, J., Chan, A.T., Sun, Y., Das, S.K., Guo, M.: A taxonomy of application scheduling tools for high performance cluster computing. Clust. Comput. **9**, 355–371 (2006)

7. Armstrong, M.E., Jones, K.S., Namin, A.S., Newton, D.C.: The knowledge, skills, and abilities used by penetration testers: results of interviews with cybersecurity professionals in vulnerability assessment and management. Proc. Human Factors Ergon. Soc. Annual Meeting **62**(1), 709–713 (2018). https://doi.org/10.1177/1541931218621161
8. Dragos: (2021) ICS Cybersecurity Year in Review 2020. Dragos, Hanover, MD
9. Rose, S., Borchert, O., Mitchell, S., Connelly, S.: Zero trust architecture. Special Publication (NIST SP)—800-207 (2020)
10. Zheng, K., Albert, L.A., Luedtke, J.R., Towle, E.: A budgeted maximum multiple coverage model for cybersecurity planning and management. IISE Trans. **51**(12), 1303–1317 (2019)
11. Petersen, R., Santos, D., Wetzel, K., Smith, M., Witte, G.: Workforce Framework for Cybersecurity (NICE Framework). NIST Special Publications 2020. https://csrc.nist.gov/pubs/sp/800/181/r1/final
12. Jeske, D.R., Myhre, J.M.: Regression using pairs versus regression on differences: a real-life case study for a master's level methods class. Am. Stat. **72**(2), 163–168 (2018). https://doi.org/10.1080/00031305.2017.1292956
13. Sikos, L.F.: Packet analysis for network forensics: A comprehensive survey. Forensic Sci. Int.: Digit. Invest. **32**, 200892 (2020). https://doi.org/10.1016/j.fsidi.2019.200892
14. Allen, T.T., Roychowdhury, S., Liu, E.: Reward-based Monte Carlo-Bayesian reinforcement learning for cyber preventive maintenance. Comput. Ind. Eng. **126**, 578–594 (2018)
15. Allen, T.T., Sui, Z., Parker, N.L.: Timely decision analysis enabled by efficient social media modeling. Decis. Anal. **14**(4), 250–260 (2017)
16. Afful-Dadzie, A., Allen, T.T.: Data-driven cyber-vulnerability maintenance policies. J. Qual. Technol. **46**(3), 234–250 (2014)
17. Afful-Dadzie, A., Allen, T.T.: Control charting methods for autocorrelated cyber vulnerability data. Qual. Eng. **28**(3), 313–328 (2016)

Chapter 25
Optimal Resource Allocation for Multi-target Attacker-Defender Game Against Cyber-Attack in Connected and Autonomous Vehicles

Bhosale Akshay Tanaji, Sayak Roychowdhury, and Anand Abraham

Abstract Connected and autonomous vehicles (CAVs) are under increasing cyber-attacks from multiple threat-agent types, such as nation-states, cybercriminals, terrorists, hacktivists, etc. Each threat-agent type varies in its capability, available resources, valuation of target, and strategies for successful attack. On the other hand, the defender has limited resources that need to be allocated effectively and efficiently to mitigate and prevent cyber-attacks in CAVs when faced with adaptive attackers and attack strategies. This paper proposes a novel class of game theoretic model to study the multi-target attacker-defender game under limited budget constraints. The proposed game model investigates the change in the defender's and attacker's equilibrium resource allocation considering the budget available, valuation of the target, cost-effectiveness, and implementation cost of attacker and defender in a sequential move game. The objective of each attacker is to maximize the total expected loss, and the defender aims to minimize the total expected loss by distributing a limited amount of resources to multiple targets. The analysis shows that a higher amount of resources should be allocated to the most valuable target. The sensitivity analysis shows that the defender's expected loss and the attacker's expected profit increase with an increase in the attacker's budget. With an increase in the defender's budget, the defender's expected loss and attacker's expected profit decrease.

Keywords Connected and autonomous vehicles · Game theory · Equilibrium resource allocation · Sequential game · Multi-target

---

B. A. Tanaji (✉) · S. Roychowdhury · A. Abraham
Department of Industrial and Systems Engineering, Indian Institute of Technology, Kharagpur, West Bengal, India
e-mail: bhosaleakshay78@iitkgp.ac.in

S. Roychowdhury
e-mail: sroychowdhury@iem.iitkgp.ac.in

A. Abraham
e-mail: anandabraham@iem.iitkgp.ac.in

J. Maiti et al. (eds.), *Operations Research and Data Analytics: Current Trends and Future Perspectives*, Lecture Notes on Multidisciplinary Industrial Engineering,
https://doi.org/10.1007/978-981-96-8327-7_25

25.1 Introduction

Connected and autonomous vehicles (CAVs) are of significant importance to intelligent transportation systems (ITS). CAVs play a major role in the reduction of road accidents, pollution, and fuel consumption, as well as enhanced protection for pedestrians and roadside infrastructure [1]. With a growing number of electronic components, wireless communication (vehicle to everything (V2X)) makes the CAVs easy targets for cyber-attackers (threat-agents). Due to the increase in cyber-attacks and the impact of losses, the cybersecurity of CAVs is one of the major challenges faced by defense systems of CAVs [2, 3]. There are different types of threat-agents, viz. nation-states, cyber-terrorists, hacktivists, disgruntled employees, cybersecurity researchers, etc., with varying motivations for launching cyber-attacks [4]. Allocating limited resources to protect against adaptive adversaries, such as strategic terrorists, cyber-criminals, and insiders, poses a significant challenge. These attackers are skilled at adjusting their strategies in response to the defender's actions. Attackers employ a range of attacks, including black-hole attacks, denial-of-service (DoS) attacks, Sybil attacks, and malware attacks, each with different costs, impacts, and success probabilities. The attacker's diversity in capability, motivation, skill/operations, and target valuation complicates the defender's ability to anticipate and allocate defensive resources.

Game theory extensively applies to modeling cybersecurity problems, incorporating dynamic state behavior and uncertainty. Zhang et al. [5] introduce a multi-target attacker-defender game where attackers possess multiple options to target various assets. Defenders allocate limited resources across different targets to minimize an exponential attack success function. Feng et al. [6] propose a Bayesian game theoretic approach for defenders to minimize the expected losses by optimizing the distribution of scarce defensive resources among multiple targets. In a sequential attacker-defender game, Zhang et al. [7] consider three different risk preferences for both parties while allocating limited resources to protect multiple targets. Levitin and Hausken [8] present a model that explores three situations for attackers' resource distribution. These include concentrating resources on one major attack, distributing them among several attacks, and employing a geometrically increasing or decreasing distribution. Bier et al. [9] investigate the impact of cost-effectiveness, defenders' target valuation, and uncertainty about attackers' target valuation on optimal resource allocation in cybersecurity. Powell [10] analyzes defender resource allocation in scenarios such as dedicating resources to a single target, spending on specific targets like border defense, involving strategic and non-strategic components, and dealing with unknown attacker attack priorities.

Strategic interactions between attackers and defenders in optimal resource allocation for cybersecurity are modeled in [11–13]. Nikoofal and Zhuang [14] develop an optimal defense resource allocation strategy for a rational defender facing a strategic attacker, accounting for the attacker's private information and target valuation. Other relevant articles in homeland security explore the trade-off between equity and efficiency, including [9, 15–17]. Levitin and Hausken [18–20] investigate the allocation of defense resources among different parameters like protection, redundancy, and decoy, focusing on maintaining a consistent pace of resource stockpiling.

This paper discusses the resource allocation problem against multiple targets in CAV infrastructure with an exponential probability attack success function. In a different context, Jose and Zhuang [21] investigate the impact of incorporating expected utility theory on participants' behavior in a single-period, sequential stochastic game. Several studies offer valuable insights into various threat-agent types that cause cyber-attacks on Cyber-Physical Systems (CPSs) [22–24]. Liu et al. [25] present a Bayesian game model known as BGM-DAE, designed to assist security systems in selecting optimal defense strategies against Distributed Denial of Service (DDoS) attacks. This model is represented by an eight-tuple vector encompassing type space, action space, prior belief, and utility functions for both the attacker and defender players. Sedjelmaci et al. [26] propose a Bayesian game model to address critical concerns related to Intrusion Detection Systems (IDS), focusing on the optimal activation of the monitoring process and efficient attacker ejection. This Bayesian game involves interactions between the IDS and attacker and between the Intrusion Ejection System (IES) and the suspected node, with each player seeking to maximize their respective payoffs. Recently, Zhang and Malacaria [27] develop a decision support system for cybersecurity to counteract multi-stage cyber-attacks against an organization. The Bayesian Stackelberg game solution relies on an efficient bi-level optimization technique where preventive optimization helps to detect and defend prior security risks, and online optimization counteracts ongoing attacks.

From the literature, it can be inferred that no previous paper has worked on defensive resource allocation, considering defender and attacker implementation costs using exponential form Contest Success Function (CSF) in CAVs. This article focuses on the resource allocation method in an attacker-defender game considering targets, valuation of targets, and the limited budget available to defender and attackers in the CAVs systems. We have developed a game-theoretic model to allocate limited resources across different targets to protect the CAVs from cyberattacks and the associated losses. The model is then solved using nonlinear programming methods and implemented on a numerical problem. The rest of this paper is organized as follows: Sect. 25.2 explains the notations and mathematical optimization model for both the attacker and defender. Section 25.3 provides the numerical illustration and equilibrium solution for both attacker and defender with budget constraints using MATLAB software. Section 25.4 highlights sensitivity analysis by varying attacker and defender budgets to observe the difference between the defender's expected loss and the attacker's expected profit. Section 25.5 summarizes this paper and provides some future research directions.

25.2 Methodology

In this article, the interaction between defender and attacker is formulated such that both defender and attacker allocate a limited amount of resources among n targets to minimize the total expected loss and maximize the damage, respectively. In the real world situation, the attacker-defender game is considered a sequential move game where the defender is a leader and the attacker acts as the follower [15, 28, 29]. The total expected loss or damage is a function of the target valuation and attack success probability. We represent the attack success probability using an exponential Contest Success Function (CSF), considering defender and attacker resource allocation and inherent defense level. The notations used in the paper are listed in Table 25.1.

Both the effectiveness of an attacker's strategy (a_i). and defender's strategy (d_i) play crucial roles in influencing the probability a successful attack (P_i) on the ith target. The probability function, denote $P_i(a_i, d_i)$, is presumed to be continuous and possesses two degrees of differentiability as shown in Eq. (25.1). Additionally, the properties mentioned below contribute to understanding the dynamics involved in the interaction between attackers a defenders in the context of target security.

$$\frac{\partial^2 P_i(a_i, d_i)}{\partial d_i^2} \geq 0, \quad \frac{\partial P_i(a_i, d_i)}{\partial a_i} \geq 0, \tag{25.1}$$

$\frac{\partial^2 P_i(a_i,d_i)}{\partial a_i^2} \leq 0, \quad \frac{\partial^2 P_i(a_i,d_i)}{\partial a_i \partial a_i} = 0,$. The success probability function $P_i(a_i, d_i)$ is convex a decreasing for defender resource allocation d_i, and is concave and creasing for the attacker's resource investment a_i. The function is such that the target with the highest defensive resources being allocated has fewer chances of threat from cyber-attackers. In contrast, when the attacker distributes higher resources to the target, the probability of a successful attack on that target rises. We use exponential form CSF for the numerical illustration to satisfy all the above-mentioned properties. The exponential form of CSF is given below:

Table 25.1 Notations used in the formulation

Parameters	
$i = \{1, 2, \ldots, n\}$	Target index, n is the total nuf targets
V_i	The attacker's and defender's valon for the ith target, $\forall i = 1, 2, \ldots, n$
C	Budget available with the defender to protect all targets
B	Budget available with the attacker to attack all targets
$P_i(a_i, d_i)$	Probability of successful attack on the target i
Decision variables	
a_i	Resources allocated by the attacker against target i for a successful attack
d_i	Resources allocated by the defender to protect target i

$$P_i(a_i, d_i) = e^{-k\left(\frac{d_i}{a_i}\right)}. \tag{25.2}$$

where k represents the effectiveness of resource allocation. The decision variables include the a_i and d_i, which stands for attacker and defender resource allocation problems, respectively.

The probability of a successful attack defined in Eq. (25.2) satisfies all conditions of differentiability and is shown in Fig. 25.1. In particular, Fig. 25.1a shows that the probability of successful attack increases with increase in the attacker's investment (a_i) and decreases with the defender's investment (d_i). The attacker's optimization problem is to maximize the total expected profit, as shown in Eq. (25.3). In the objective function, the attacker will always try to maximize the probability of a successful attack $P_i(a_i, d_i)$. with minimum effort (a_i) [10, 30–32]. This can be done by choosing the best resource allocation strategy for each target based on valuation and total budget constraint.

$$\text{maximize} \sum_{i=1}^{n} \left\{ V_i \left(e^{-k\left(\frac{d_i}{a_i}\right)} \right) - a_i \right\} \tag{25.3}$$

Subject to,

$$\sum_{i=1}^{n} a_i \leq C \tag{25.4}$$

$$a_i \geq 0 \tag{25.5}$$

Equations (25.4) and (25.5) represent the total budget and non-negativity constraint, respectively. Based on the attacker's optimization problem (25.3)–(25.5), the best response is defined as

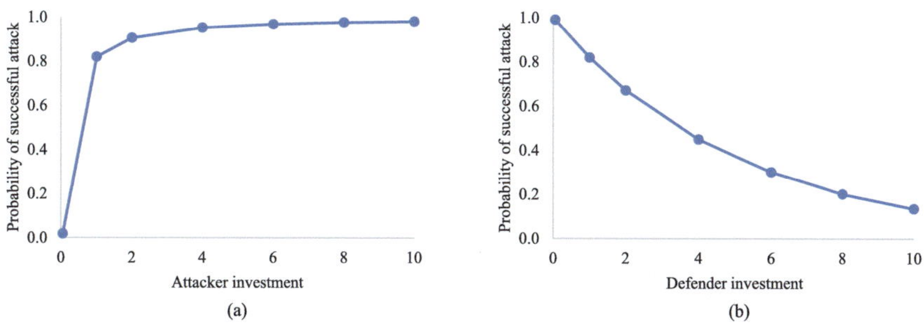

Fig. 25.1 Illustration of contest success function with **a** attacker investment and **b** defender investment

$$a_i^* = \left\{ V_i \left(e^{-k\left(\frac{d_i}{a_i}\right)} \right) - a_i \right\}$$

where $A_i = \left\{ \sum_{i=1}^{n} a_i \leq C, a_i \geq 0 \right\}$.

Similar to the attacker's best response, the defender's optimization problem is to allocate a limited amount of resources B among n targets [14, 33–36]. The defender's optimization problem to minimize total expected loss by minimizing the value of $P_i(a_i, d_i)$ as shown in Eq. (25.6).

$$\text{minimize} \sum_{i=1}^{n} \left\{ V_i \times \left(e^{-k\left(\frac{d_i}{a_i}\right)} \right) \right\} \tag{25.6}$$

Subject to,

$$\sum_{i=1}^{n} d_i \leq B \tag{25.7}$$

$$d_i \geq 0 \tag{25.8}$$

Equations (25.7) and (25.8) represent the defender budget and non-negativity constraint, respectively. Based on the defender's optimization problem (25.6)–(25.8), the best response is defined as

$$d_i^* = \left\{ V_i \times \left(e^{-k\left(\frac{d_i}{a_i}\right)} \right) \right\}.$$

where $D_i = \left\{ \sum_{i=1}^{n} d_i \leq B, d_i \geq 0 \right\}$.

We use MATLAB "fminimax" function to find the minimum of a set of multi-variable functions by minimizing their largest value [37, 38]. It begins with an initial estimate and employs a combination of minimax Sequential Quadratic Programming (SQP), Quasi-Newton, and line search algorithms. The solver terminates its search under two primary conditions: when the magnitude of the current search direction is less than double the default step size tolerance and when the constraints are satisfied within the default constraint tolerance.

25.3 Numerical Illustration and Findings

The CSF and game theory models presented in the methodology section are general. Unfortunately, there are no fully developed CAVs in the real world; therefore, getting a perfect dataset is difficult. For numerical illustration, we chose the baseline parameter values shown in Table 25.2. The study assumes that attackers and defenders have the same valuation for different targets.

In cybersecurity, the attacker and defender game is mainly considered a sequential move game where the defender moves first, and based on that, the attacker chooses the strategy. The game is solved using the backward induction technique. The attacker (follower) maximization model is first solved, and find the equilibrium solution in terms of defender resource utilization. Substitute the attacker equilibrium solution in the defender optimization problem and solve for the equilibrium solution of the defender. Again, after putting this value in the previously calculated attacker equilibrium solution, we will get the attacker resource allocation for each target.

We got the following results after solving the game of the defender's minimization problem for expected loss and the attacker's maximization problem for expected profit using "fminimax" solver in MATLAB. Based on the baseline values of the parameter and target valuation provided in Table 25.2, the attackers' and defenders' equilibrium resource allocation strategies, expected payoff, and probability of successful attack are provided in Table 25.3.

Table 25.2 Numerical illustration parameter and valuation of targets

Parameters	Value	Targets	Attacker's and defender's valuation ($ million)
n	5	RSU	10
C	30	EV charging station	8
B	10	CAN BUS	6
k	0.2	ECU	4
		Sensors	2

Table 25.3 Equilibrium resource allocation strategies, expected payoffs, and probability of successful attack in the sequential game

Targets	Resource allocation ($ million)		Probability of successful attack	Expected payoff	
	Attacker (a_i)	Defender (d_i)		Attacker	Defender
RSU	19.06	7.29	0.926	9.26	1.98
EV charging station	10.52	2.71	0.950	7.60	4.89
CAN BUS	0.25	0.00	0.998	5.99	5.99
ECU	0.11	0.00	0.999	4.00	4.00
Sensors	0.06	0.00	0.999	2.00	2.00
Total	30	10		28.85	18.85

From Table 25.3, we inferred that defender allocate the maximum amount of resources, $a_1 = 7.29$ to defend his most valuable target i.e. RSUs. From the literature also it is seen that RSUs are the primary element to protect in the ITS as they collect traffic data from static and dynamic elements on the road and transfer this to the Central Traffic Management Centre (CTMC) [41–43].

25.4 Sensitivity Analysis

In order to gain a deeper insight into the interplay between the equilibrium strategies and the expected payoff of both the attacker and the defender, we undertook sensitivity analyses. These analyses examined how the model parameters, including factors such as the cost-effectiveness of attacks, allocation of defensive resources, and the budgets of both the attacker and the defender, influence the expected profit and loss to the attacker and defender, respectively. Fig. 25.2a, b shows the change in the attacker's expected profit and the defender's expected loss while keeping the defender's budget and valuation constant and varying the attacker's budget. Both the quantities attacker's expected profit and defender's expected loss increases with an increase in the attacker budget. Similarly, Fig. 25.3a shows that defender's expected loss decreases with increases in the defender budget keeping the attacker's budget and valuation constant.

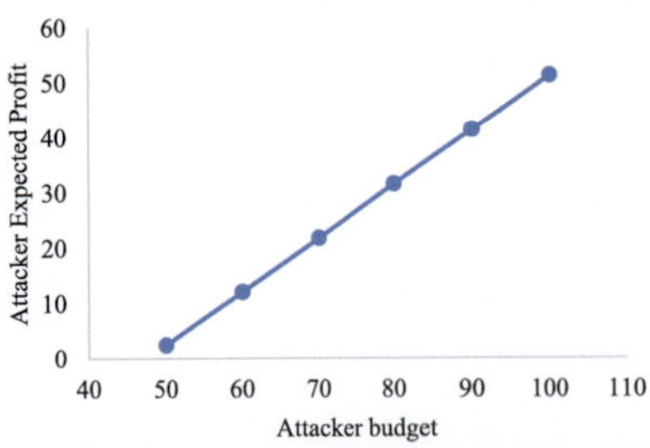

(a)

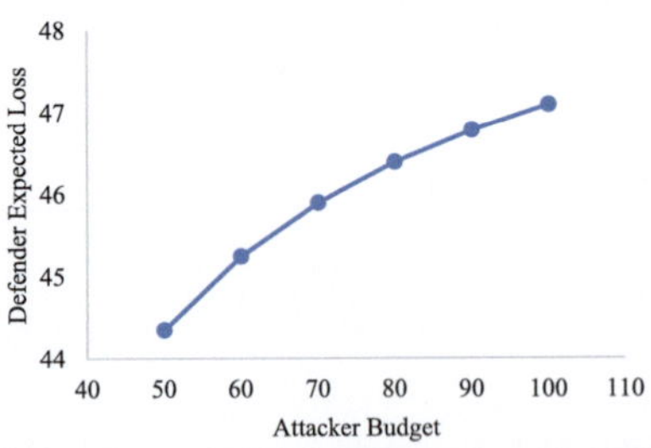

(b)

Fig. 25.2 Effect of attacker budget on attacker expected profit

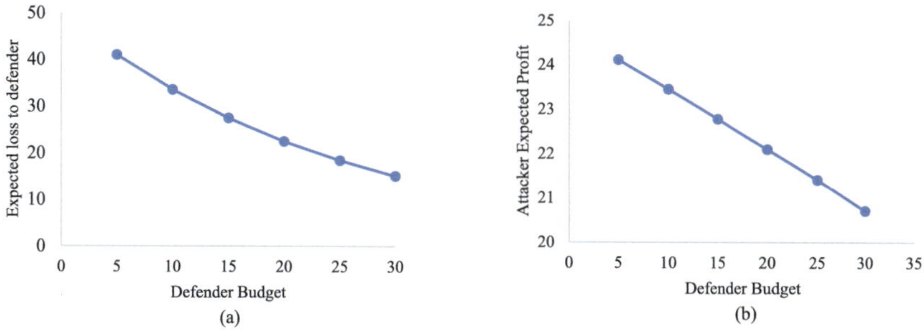

Fig. 25.3 Effect of defender budget on defender expected loss

25.5 Conclusion

In this study, we modeled a game-theoretic multi-target attacker-defender problem that helps allocate resources for both the attacker and the defender under budget constraints. This paper presents a novel, multi-target attacker-defender game-theoretic model with an exponential form probability success function. This paper studies attacker allocation and defender allocation as continuous variables instead of discrete ones (attack or not attack, defend or not defend). The mathematical game theory model developed in this paper is general and robust to be used with new datasets and targets. The equilibrium solution of the sequential move game is obtained by solving constrained nonlinear optimization problems. We perform the sensitivity analysis to observe the effect of the defender's and attacker's budget on the defender's expected loss and the attacker's expected profit. For simplicity, this paper considers one defender and one attacker and complete information game. Results obtained through the game theory model state that the defender always try to allocate maximum resources to the most valuable target. The sensitivity analysis shows that an increase in the defender budget decreases the defender expected loss.

In this paper, we have not considered the uncertainty about the type of attacker (e.g., nation-state, terrorist, hacktivist, cybercriminal, and insider). Also, we do not consider the interconnection between different targets, different valuations of defender and attacker, and bounded rationality of players for simplicity. We could extend our research by considering incomplete information game, different forms of contest success function, different types of defender (e.g. government entity, private organization, and private individual).

References

1. Seuwou, P., Banissi, E, Ubakanma, G.: The future of mobility with connected and autonomous vehicles in smart cities. Digit. Twin. Technol. Smart Cities 37–52 (2020)

2. Khan, S.K., Shiwakoti, N., Stasinopoulos, P., Chen, Y.: Cyber-attacks in the next-generation cars, mitigation techniques, anticipated readiness and future directions. Accid. Anal. Prev. **148**, 105837 (2020)
3. Sun, X., Yu, F.R., Zhang, P.: A survey on cyber-security of connected and autonomous vehicles (CAVs). IEEE Trans. Intell. Transp. Syst. (2021)
4. Tanaji, B.A., Roychowdhury, S.: BWM integrated VIKOR method using neutrosophic fuzzy sets for cybersecurity risk assessment of connected and autonomous vehicles. Appl. Soft. Comput. 111628 (2024)
5. Zhang, J., Zhuang, J.: Modeling a multi-target attacker-defender game with multiple attack types. Reliab. Eng. Syst. Saf. **185**(January), 465–475 (2019). https://doi.org/10.1016/j.ress.2019.01.015
6. Feng, Q., Cai, H, Chen, Z.: Using game theory to optimize the allocation of defensive resources on a city scale to protect chemical facilities against multiple types of attackers. Reliab. Eng. Syst. Saf. **191**(July 2017), 105900 (2019). https://doi.org/10.1016/j.ress.2017.07.003
7. Zhang, J., Zhuang, J., Richmond, V., Jose, R.: The role of risk preferences in a multi-target defender-attacker resource allocation game. Reliab. Eng. Syst. Saf. **169**(August 2017), 95–104 (2018). https://doi.org/10.1016/j.ress.2017.08.002
8. Levitin, G., Hausken, K. Resource distribution in multiple attacks against a single target. **30**(8) (2010). https://doi.org/10.1111/j.1539-6924.2010.01410.x
9. Bier, V.M., Haphuriwat, N., Menoyo, J., Zimmerman, R., Culpen, A.M.: Optimal resource allocation for defense of targets based on differing measures of attractiveness. Risk. Anal. An. Int. J. **28**(3), 763–770 (2008)
10. Powell, R.: Defending against terrorist attacks with limited resources. **101**(3), 527–541 (2007). https://doi.org/10.1017/S0003055407070244
11. Zhuang, J., Bier, V.M.: Balancing terrorism and natural disasters—Defensive strategy with endogenous attacker effort. Oper. Res. **55**(5), 976–991 (2007)
12. Azaiez, M.N., Bier, V.M.: Optimal resource allocation for security in reliability systems. Eur. J. Oper. Res. **181**(2), 773–786 (2007)
13. Dighe, N.S., Zhuang, J., Bier, V.M.: Secrecy in Defensive allocations as a strategy for achieving more cost-effective attacker deterrence. Int. J. Perform. Eng. **5**(1), 31 (2009)
14. Nikoofal, M.E., Zhuang, J.: Robust allocation of a defensive budget considering an attacker's private information. Risk Anal. **32**(5), 930–943 (2012). https://doi.org/10.1111/j.1539-6924.2011.01702.x
15. Shan, X., Zhuang, J.: Hybrid defensive resource allocations in the face of partially strategic attackers in a sequential defender–attacker game. Eur. J. Oper. Res. **228**(1), 262–272 (2013)
16. Zhuang, J., Bier, V.M.: Reasons for secrecy and deception in homeland-security resource allocation. Risk Anal. An Int. J. **30**(12), 1737–1743 (2010)
17. Shan, X., Zhuang, J.: Cost of equity in homeland security resource allocation in the face of a strategic attacker. Risk Anal. **33**(6), 1083–1099 (2013)
18. Levitin, G., Hausken, K.: Redundancy versus protection versus false targets for systems under attack. IEEE Trans. Reliab. **58**(1), 58–68 (2009)
19. Levitin, G., Hausken, K.: Defence resource distribution between protection and decoys for constant resource stockpiling pace. J. Oper. Res. Soc. **64**(9), 1409–1417 (2013)
20. Levitin, G., Hausken, K.: Defense resource distribution between protection and redundancy for constant resource stockpiling pace. Risk Anal. An. Int. J. **31**(10), 1632–1645 (2011)
21. Jose, V.R.R., Zhuang, J.: Incorporating risk preferences in stochastic noncooperative games. IISE Trans. **50**(1), 1–13 (2018). https://doi.org/10.1080/24725854.2017.1382749
22. Casey, T.: Threat agent library helps identify information security risks. Intel White Paper, p 12 (2007) [Online]. Available https://communities.intel.co.jp/servlet/JiveServlet/previewBody/1151-102-1-1111/ThreatAgentLibrary_07-2202w.pdf
23. ENISA: Threat Landscape Report 2018—15 Top Cyberthreats and Trends (2018)
24. Sailio, M., Latvala, O.M., Szanto, A.: Cyber threat actors for the factory of the future. Appl. Sci. **10**(12), 1–24 (2020). https://doi.org/10.3390/app10124334

25. Liu, Y., Feng, D., Lian, Y., Chen, K., Zhang, Y.: Optimal defense strategies for DDoS defender using Bayesian game model. In: Lecture Notes in Computer Science (including Subseries Lecture Notes Artificial Intelligence and Lecture Notes Bioinformatics), vol. 7863 LNCS, pp. 44–59 (2013). https://doi.org/10.1007/978-3-642-38033-4_4
26. Sedjelmaci, H., Senouci, S.M., Ansari, N.: Intrusion detection and ejection framework against lethal attacks in UAV-aided networks: a bayesian game-theoretic methodology. IEEE Trans. Intell. Transp. Syst. **18**(5), 1143–1153 (2017). https://doi.org/10.1109/TITS.2016.2600370
27. Zhang, Y., Malacaria, P.: Bayesian Stackelberg games for cyber-security decision support. Decis. Support Syst. **148**(January), 113599. https://doi.org/10.1016/j.dss.2021.113599
28. Xu, Z., Zhuang, J.: A study on a sequential one-defender-N-attacker game. Risk Anal. **39**(6), 1414–1432 (2019)
29. Guan, P., He, M., Zhuang, J., Hora, S.C.: Modeling a multitarget attacker—Defender game with budget constraints modeling a multitarget attacker – defender game with budget constraints, No. July 2023 (2017)
30. He, M., Zhuang, J.: An attacker—Defender resource allocation game with (2020). https://doi.org/10.1111/risa.13483
31. Gene, X., Zhuang, J.: A game-theoretic approach to modeling attacks and defenses of smart grids at three levels. Reliab. Eng. Syst. Saf. **195**(September 2019), 106683 (2020). https://doi.org/10.1016/j.ress.2019.106683
32. Long, J., Dong, Y., Chen, X.: Competitive resource allocation in an attacker-defender game: when citizens become targets of being ingratiated. Expert Syst. Appl. **243**(December 2023), 122840. https://doi.org/10.1016/j.eswa.2023.122840
33. Zhang, J., Wang, Y., Zhuang, J.: Modeling multi-target defender-attacker games with quantal response attack strategies. Reliab. Eng. Syst. Saf. **205**(July 2020), 107165 (2021). https://doi.org/10.1016/j.ress.2020.107165
34. Wu, D., Yan, X., Peng, R., Wu, S.: Optimal defence-attack strategies between one defender and two attackers. J. Oper. Res. Soc. **71**(11), 1830–1846 (2020). https://doi.org/10.1080/01605682.2019.1630332
35. Levitin, G., Hausken, K.: Resource distribution in multiple attacks against a single target. Risk Anal. **30**(8), 1231–1239 (2010). https://doi.org/10.1111/j.1539-6924.2010.01410.x
36. Hausken, K.: Protecting complex infrastructures against multiple strategic attackers. Int. J. Syst. Sci. **42**(1), 11–29 (2011). https://doi.org/10.1080/00207720903434789
37. Matott, L.S., Leung, K., Sim, J.: Application of MATLAB and Python optimizers to two case studies involving groundwater flow and contaminant transport modeling. Comput. Geosci. **37**(11), 1894–1899 (2011)
38. Tadros, M., Ventura, M., Guedes Soares, C.: Optimization procedures for a twin controllable pitch propeller of a ROPAX ship at minimum fuel consumption. J. Mar. Eng. Technol. **22**(4), 167–175
39. Miao, L., Li, S., Wu, X., Liu, B.: Applied sciences mean-field Stackelberg game-based security defense and resource optimization in edge computing (2024)
40. Ait Temghart, A., Marwan, M., Baslam, M.: Stackelberg security game for optimizing cybersecurity decisions in cloud computing. Secur. Commun. Netw. 1–13 (2023). https://doi.org/10.1155/2023/2811038
41. Liang, Y., Zhang, S., Wang, Y.: Data-driven road side unit location optimization for connected-autonomous-vehicle-based intersection control. Transp. Res. Part C Emerg. Technol. **128**, 103169 (2021). https://doi.org/10.1016/j.trc.2021.103169
42. Saeed, Z., Masood, M., Khan, M.U., A review: cybersecurity challenges and their solutions in connected and autonomous vehicles (CAVs). J. Adv. Res. Electr. Eng. (JAREE) **7**(1).
43. Abdo, A., Chen, H., Zhao, X., Wu, G., Feng, Y.: Cybersecurity on connected and automated transportation systems: a survey. IEEE Trans. Intell. Veh. (2023)